Developments in Petroleum Science, 2

ABNORMAL FORMATION PRESSURES

Implications to Exploration, Drilling, and Production
of Oil and Gas Resources

FURTHER TITLES IN THIS SERIES

1 A. GENE COLLINS
GEOCHEMISTRY OF OILFIELD WATERS

2 W.H. FERTL
ABNORMAL FORMATION PRESSURES

3 A.P. SZILAS
PRODUCTION AND TRANSPORT OF OIL AND GAS

4 C.E.B. CONYBEARE
GEOMORPHOLOGY OF OIL AND GAS FIELDS
IN SANDSTONE BODIES

5 T.F. YEN and G.V. CHILINGARIAN (Editors)
OIL SHALE

6 D.W. PEACEMAN
FUNDAMENTALS OF NUMERICAL RESERVOIR SIMULATION

7 G.V. CHILINGARIAN and T.F. YEN (Editors)
BITUMENS, ASPHALTS AND TAR SANDS

8 L.P. DAKE
FUNDAMENTALS OF RESERVOIR ENGINEERING

9 K. MAGARA
COMPACTION AND FLUID MIGRATION

10 M.T. SILVIA and E.A. ROBINSON
DECONVOLUTION OF GEOPHYSCIAL TIME SERIES IN THE
EXPLORATION FOR OIL AND NATURAL GAS

11 G.V. CHILINGARIAN and P. VORABUTR
DRILLING AND DRILLING FLUIDS

Developments in Petroleum Science, 2

ABNORMAL FORMATION PRESSURES

Implications to Exploration, Drilling, and Production of Oil and Gas Resources

WALTER H. FERTL

Director of Interpretation and Field Development,
Dresser Atlas Division,
Dresser Industries, Inc., Houston, Texas, U.S.A.

with a contribution by

GEORGE V. CHILINGARIAN

Professor of Petroleum Engineering, University of Southern California,
Los Angeles, Calif., U.S.A.

and

HERMAN H. RIEKE, III

Professor, School of Mines, West Virginia University,
Morgantown, W. Va., U.S.A.

ELSEVIER SCIENTIFIC PUBLISHING COMPANY
Amsterdam — Oxford — New York 1976

first edition 1976
second impression 1981

ELSEVIER SCIENTIFIC PUBLISHING COMPANY,
1 Molenwerf,
P.O. Box 211, 1000 AE Amsterdam, The Netherlands

ISBN 0-444-41328-6

With 233 illustrations and 36 tables

Copyright © 1976 by Elsevier Scientific Publishing Company, Amsterdam

All rights reserved. No part of this publication may be reproduced, stored in a retrieval system, or transmitted, in any form or by any means, electronic, mechanical, photocopying, or otherwise without the prior written permission of the publisher, Elsevier Scientific Publishing Company, Jan van Galenstraat 335, Amsterdam

Printed in The Netherlands

PREFACE

The present book by Dr. Walter Fertl represents the first modern collection and comprehensive review of laboratory studies, combined with extensive field observations, dealing with the vast subject area related to the origin, detection, and quantitative evaluation of abnormal formation pressures and their profound technical and managerial decision-making impact on exploration, drilling, production, and reservoir engineering concepts. Location, detection, and magnitude of abnormal formation pressures is critical to the oil industry's quest to probe new and remote areas and to test the potential of ever-deeper targets in today's enhanced search for oil and gas resources.

The author, who ranks among the world's leading experts in this subject field, has synthesized both published and unpublished data, and at the same time has incorporated a wealth of his own personal experiences, which are based on both worldwide field observations and laboratory data.

The manuscript is well organized and documented and provides an excellent, in-depth treatment of obvious value to both the newcomer and the expert. As such, the volume should serve as a valuable text and source book for both industrial and academic circles, including geologists, engineers, teachers, and students alike. It also provides a basis for stimulation of thought and it points out numerous opportunities for future research in a variety of fields. The information presented in this splendid text gives an excellent, clear, and up-to-date discussion of this rapidly developing subject matter, which has been previously unorganized and scattered throughout the literature in earth sciences and/or petroleum engineering journals.

The initial chapter presents a well-organized introduction to the large variety of possible modes and causes of abnormal formation pressure environments in the subsurface, and simultaneously establishes the basic parameters of these in-situ pressure conditions with precise and intelligent descriptions and definitions. At the same time Chapter 1 also provides a brief but rather comprehensive historical review of the subject.

This is followed in Chapter 2 by a condensed look at the compaction of argillaceous sediments. Pertinent information is presented in a simple manner, and incorporates mathematical developments of compaction models, a treatise of stresses in sediments, and basic concepts of fluid flow through porous media. Furthermore, the importance and inherent complexity of the chemistry of interstitial fluids and the behavior of these solutions during compaction is highlighted and thoroughly treated on the basis of theoretical concepts, experimental data, and actual field observations.

The next three chapters focus on qualitative detection and quantitative evaluation of abnormal formation pressure environments, which are frequently encountered by the drill bit in the subsurface. This section of the book pragmatically reflects the author's outstanding expertise at its best. Chapter 3 concentrates on the application of geophysical data, particularly from seismic methods, to pinpoint formation tops, and, under specific circumstances, also predict the presence, depth, and magnitude or abnormal formation pressures prior to spudding a well. Besides an introduction to the basic concepts, the author has presented comparative data on predicted and actually encountered overpressures for several deep wells in the United States of America and other countries.

In Chapter 4, the author discusses a large variety of drilling indicators, which are currently being utilized by the oil industry to evaluate downhole pressure conditions *while drilling* a well. These techniques, many of which are used daily in every corner of the world, include several drilling parameters (including logging-while-drilling concepts) and a large number of drilling mud and shale cutting parameters. Besides the multitude of conventional overpressure evaluation techniques, several interesting novel detection methods, which were developed by the author, are candidly discussed and illustrated in the light of field and laboratory data.

Chapter 5 takes an authoritative look at some of the best qualitative and quantitative pressure detection and evaluation techniques, which are based on various wireline-logging methods. Here, the author reviews basic tool responses, theoretical considerations and shows the recommended mathematical analyses of specific formation parameters using selected field cases. An important part of this chapter, and surely a highlight to the operational personnel in the oil industry, deals with the limitations and possible pitfalls in applying these techniques. Application of these methods is not always simple and straightforward, and complications may often arise due to unrecognized logging sonde problems, severe borehole conditions, and unusual formation characteristics. The author summarizes this splendid discussion with the conclusion that even though these methods are *after-the-fact* techniques, i.e. the wellbore has to be drilled prior to logging, interpretation of well logs represents one of the best methods to detect and realistically evaluate abnormal formation pressures.

Having thus established means to detect and evaluate overpressures, the author moves on to Chapter 6 where he discusses the impact of such hostile downhole environments on drilling and completion concepts. This chapter provides the right blend of fundamental introduction and a "how-to-do-it" review of basic concepts, such as the pore pressure and fracture pressure relationship, and the prediction methods available for fracture pressure gradients in conventional and deep-water offshore drilling operations. The author tops off this discussion with a look at typical casing programs in several problem areas of known overpressures, and also comments on cementing operations and other specific completion and production considerations. Very appropriately this chapter closes with some pertinent thoughts and information on contingency planning for drilling and producing high-pressure sour-gas-bearing formations and an interesting treatise on relief-well planning.

Present-day reservoir engineering concepts in overpressured reservoir rocks are profusely illustrated in Chapter 7. This includes a look at shale water influx, rock compressibility and failure, peripheral water influx, and predictions in overpressured depletion-type gas reservoirs.

Geologists and management will find Chapter 8 a bonanza for stimulating thoughts, supported by a multitude of field data, which relate to appraisal of hydrocarbon distribution in the subsurface and related economic factors in both normal- and overpressured environments. Starting with a look at hydrocarbon accumulation and distribution as a function of depth and in-situ temperature and pressure conditions, the author leads up to the shale resistivity ratio method. Today, the importance of the latter method to managerial decision-making, both in exploration and drilling activities, is well established. The concept is used by the petroleum industry in many areas as a supplementary completion guide and is a decisive factor in "dry-hole" money negotiations between companies. The author eloquently supports the shale resistivity ratio concept with selected field case studies from Texas, Louisiana, and California. The impact of this concept is highlighted by the discussion of its value in combination with the analysis of conventional data obtained from various sources, such as mud logs, digital well-log analysis, and subsequent well tests in a specific wildcat prospect.

In the final Chapter 9, we are brought up-to-date on the worldwide occurrence and evaluation of abnormal formation pressures, as they have been so far encountered by the oil industry while drilling deep wells and wildcatting in remote areas spanning the globe.

All chapters are well written, extensively referenced, and supported by a multitude of excellent illustrations and useful nomographs. The book provides fascinating perspectives on the interrelation of many, equally important technical areas and concepts, and their superimposed impact on today's enhanced quest to find and exploit additional reserves of the earth's valuable hydrocarbon resources.

The author has done a marvellous job in selecting the highlights from a vast amount of literature and blending them with his own research findings and field observations. The splendid result is this first book, and without any doubt the future classic reference, written on the subject. As such, this book is essential and highly recommended to exploration and engineering personnel in the oil industry and petroleum-related governmental agencies, and industrial scientists, teachers, and students alike.

GEORGE V. CHILINGARIAN
Professor of Petroleum Engineering, University
of Southern California, Los Angeles, California,
U.S.A.

FOREWORD AND ACKNOWLEDGEMENTS

Detection and quantitative evaluation of overpressured formations is critical to exploration, drilling, and production operations for hydrocarbon resources. Worldwide experience indicates a significant correlation between the presence and magnitude of formation pressures, and the shale/sand ratio of sedimentary sections. Distribution of oil and gas is related to regional and local subsurface pressure and temperature environments. Knowledge of expected pore pressure and fracture gradients is the basis for efficiently drilling wells with correct mud weights, engineered casing programs, and for proper completions which must be effective, safe and allow for killing of the well without excessive formation damage. In reservoir engineering, formation pressures influence compressibility and the failure of reservoir rocks and may be responsible for water influx from adjacent overpressured shale sections as an additional driving mechanism in hydrocarbon production.

Even with a contribution by acknowledged experts in their specific field*, it is no simple task for an author to compile a text on abnormal-pressured formations with the principal objective of bringing together geological, chemical, and engineering sciences and to illustrate concepts and practical applications in the petroleum industry that have resulted; therefore, in some instances the reader may refer to the number of excellent articles primarily devoted to his specific subject area.

Yet, to the professional, teacher, and student interested in a *"how-to-do-it"* discussion and a condensed review of subject matter, which has previously been unorganized and scattered throughout the literature, this book will illustrate the interrelation of the *academic* and *practical, technological* values of several disciplines and their profound impact on managerial decision-making in the petroleum industry. Concepts presented were selected from a vast amount of literature, orally discussed ideas, and our own experiences in both the field and laboratory.

Inasmuch as *observation is the first step in research*, the present discussion focuses on basic principles and their practical field application. After some deliberation, the author has chosen not to use a highly mathematical approach (its details being available in the references cited), which one may

---

*Dr. G. V. Chilingarian, Petroleum Engineering Department, University of Southern California, Los Angeles, California, and Dr. H. H. Rieke, III, Petroleum Engineering Department, West Virginia University, Morgantown, West Virginia.

find acceptable or perhaps even commendable. Wherever feasible, data are presented in both metric and English units.

The author acknowledges permission by the management of Continental Oil Company to publish this manuscript and wishes to thank various technical organizations and several trade journals for allowing reproduction of illustrations and tables. Special gratitude goes to Bulletin of the American Association of Petroleum Geologists, Journal of Petroleum Technology, Oil and Gas Journal, Petroleum Engineer, and World Oil.

As this book is not only the labor of professional but also of personal interest, it had to find the moments needed for its final execution in night hours and holidays over the last few years. To my wife Irma, I express my deepest appreciation for her patience, understanding, and encouragement, during this time.

Anno 1975
Ponca City, Oklahoma   WALTER H. FERTL

CONTENTS

Preface IX

Foreword and acknowledgements XIII

Chapter 1. Abnormal formation pressure environments
Introduction 1
Pressure concepts 1
    Hydrostatic pressure 1
    Overburden pressure 2
    Formation pressure 3
Subnormal formation pressures (subpressures) 5
Abnormally high formation pressures (surpressures) 14
    Pressure seals 14
    Concepts for abnormal formation pressure origin 16
Acknowledgements 43
References 43

Chapter 2. Compaction of argillaceous sediments
(G. V. Chilingarian and H. H. Rieke, III)
Introduction 49
Compaction model development 52
    The Katz and Ibrahim compaction model 53
Mathematical description of compaction 55
Flow of fluids through porous media 63
Overburden potential 66
Effect of compaction on the chemistry of interstitial solutions ... 67
    Chemistry of interstitial fluids 67
    Composition of interstitial solutions related to seawater 69
    Changes in the chemistry of solutions squeezed out at different overburden pressures 73
    Salinity distribution in sandstones and associated shales 80
Stresses in sediments 83
    Hydrostatic stress 83
    Resolution of the total stress field 84
Compressibilities of sand and clayey sediments 89
    Effect of rock compressibility on the estimation of petroleum reserves 95
References 96

Chapter 3. Geophysical methods detect and evaluate abnormal formation pressures
Introduction 101
Seismic data analysis 102
Gravity data analysis 114
Other geophysical methods 115
References 115

Chapter 4. Drilling data detect and evaluate abnormal formation pressures

Introduction 117
Drilling parameters 120
    Drilling rate (penetration rate) 120
    d-Exponent (normalized rate of penetration) 122
    Modified d-exponent 125
    Drilling rate equations 128
    Drilling porosity and formation pressure logs 130
    Novel logging-while-drilling concepts 136
    Torque 136
    Drag 136
Drilling mud parameters 137
    Mud-gas cutting 137
    Flow-line mud weight 141
    Pressure kicks 141
    Flow-line temperature 144
    Resistivity, chloride ion, and other novel concepts 147
    Pit level and total pit volume 147
    Hole fill-up 148
    Mud flow rate 149
Shale cuttings parameters 150
    Shale bulk density 150
    Shale factor 156
    Volume, shape, and size of shale cuttings 158
    Novel, miscellaneous detection methods 158
References 172

Chapter 5. Geophysical well-logging techniques detect and evaluate abnormal formation pressures

Introduction 177
Electrical surveys 178
    The short normal curve 178
    The conductivity curve 184
    The spontaneous potential (SP) curve 189
    The shale formation factor method 196
Acoustic (sonic) surveys 198
Formation density measurements 204
Neutron measurements 204
Pulsed neutron measurements 205
Nuclear magnetic resonance techniques 207
Downhole gravity data measurements 209
Well logs recommended for conventional pressure evaluation 210
Limitations and pitfalls of wireline methods 212
    Possible pitfalls 212
    Considerations in plotting shale parameters 220
Conclusions 226
References 227

Chapter 6. Drilling and completion concepts in overpressured formations

Introduction 231
Basic concepts of the pore pressure—fracture pressure relationship 231
    Casing point selection 233

Drill pipe sticking	235
Lost circulation..	237
Prediction methods for fracture pressure gradients.............................	238
The Hubbert and Willis approach.......................................	238
Empirical data correlations..	240
Method of Matthews and Kelly...	240
Method of Eaton..	243
Field data study by Taylor and Smith...................................	247
Method of MacPherson and Berry.......................................	250
Method of Christman..	251
Method of Anderson, Ingram, and Zanier...............................	253
Typical casing programs in overpressured areas..............................	256
General comments..	256
Delaware Basin, West Texas..	261
U.S. Gulf Coast area..	261
Anadarko Basin, U.S. Mid-Continent area..............................	262
Rocky Mountain area...	262
Comments on cementing operations...	262
Specific completion and production considerations.............................	264
Abnormally high permafrost-freezeback pressures around Arctic wells.....	264
Hydrates...	265
Precipitation of salts, sulfur, etc.......................................	265
Packer fluid requirements...	266
General comments on casing and tubing requirements...................	266
Contingency planning for drilling and producing high-pressure sour-gas wells....	268
Relief-well planning..	270
References..	272

Chapter 7. Reservoir engineering concepts in abnormal formation pressure environments

Introduction..	275
Reservoir mechanisms...	276
Shale water influx..	276
Rock compressibility and rock failure..................................	278
Peripheral water influx..	281
Hammerlindl's prediction methods in overpressured depletion-type gas reservoirs	281
Initial well potential versus formation water salinities........................	286
Pressure depletion and abandonment pressures...............................	286
References..	289

Chapter 8. Appraisal of hydrocarbon distribution and related economic factors in normal and abnormal formation pressure environments

Introduction..	291
Depth versus hydrocarbon accumulation.....................................	291
Formation temperature versus hydrocarbon distribution.......................	294
Formation pressure gradient versus hydrocarbon accumulations................	300
Formation temperature and pressure versus hydrocarbon accumulation.........	304
Shale resistivity ratio method — a valuable concept for economic drilling decisions	309
Completion guidelines..	311
Field case, offshore Texas...	313
Field case, offshore Louisiana...	314
Field case, northern California..	314

Use of mud log, digital well-log analysis, and shale resistivity ratio on a wildcat, U.S. Gulf Coast area	316
Economics in overpressure environments	319
References	321

Chapter 9. Global occurrence and evaluation of abnormal formation pressures

Introduction	325
Abnormal formation pressure environments in Europe	327
Onshore areas	327
Offshore Europe	331
Abnormal formation pressure environments in the Soviet Union	334
Abnormal formation pressure environments on the North American continent	336
Abnormal formation pressure environments in South America	339
Abnormal formation pressure environments in Africa	341
Abnormal formation pressure environments in the Far East	342
Abnormal formation pressure environments on the Indian subcontinent	346
Abnormal formation pressure environments in the Middle East	347
Acknowledgements	348
References	348

Appendix

Conversion factors between metric, A.P.I., and U.S. measures	351
Oil, water, and mud specific gravity — weight and pressure table	355
Temperature conversion formulas	356
References	356
References index	357
Subject index	365

Chapter 1. ABNORMAL FORMATION PRESSURE ENVIRONMENTS

Introduction

Many factors can cause abnormal formation pressures, that is, pressures other than hydrostatic. In some areas, a combination of these factors prevails. To place the possible causes of abnormal formation pressures in proper perspective, it is necessary to understand the importance of petrophysical and geochemical parameters and their relationship to the stratigraphic, structural, and tectonic history of a given area or basin.

Because conditions can vary widely, special care should be taken not to assume that the cause of abnormal formation pressures established from experience in a well-known area is necessarily the cause of a similar condition in a nearby basin, which may not yet have been adequately tested by drilling.

Pressure concepts

Hydrostatic pressure

Hydrostatic pressure is caused by the unit weight and vertical height of a fluid column. The size and shape of this fluid column have no effect on the magnitude of this pressure. Hydrostatic pressure, P_{Hy}, equals the mathematical product of the average fluid density and its vertical height, such as:

$$P = \rho \cdot g \cdot D$$

where P = pressure, ρ = average density, g = gravity value, and D = height of the column. In terms of drilling operations, we then can write:

$$P_{Hy}\,(\text{psi}) = C \cdot MW \cdot D \tag{1.1a}$$

where D = vertical height of fluid column in feet, MW = fluid density or mud weight in lb/gal (#/gal, ppg) or lb/ft³ and C = conversion constant (C = 0.052 if MW in lb/gal and C = 0.00695 if MW in lb/ft³).

In the metric system, eq. 1.1a becomes:

$$P_{Hy} = 0.098 \cdot MW \cdot D \tag{1.1b}$$

where D, the vertical fluid column, is in meters and MW, the mud weight, is in kg/dm³.

The *hydrostatic pressure gradient* is affected by the concentration of dissolved solids (i.e. salts) and gases in the fluid column and different or varying temperature gradients. In other words, an increase in dissolved solids (i.e. higher salt concentration) tends to increase the normal pressure gradient, whereas increasing amounts of gases in solution and higher temperatures would decrease the normal, hydrostatic pressure gradient. For example, a pressure gradient of 0.465 psi/ft [0.1074 kg cm^{-2} m^{-1}] assumes a water salinity of 80,000 parts per million (ppm) NaCl at a temperature of 77°F [25°C].

Generally, the typical average hydrostatic gradients which may be encountered during drilling for oil and gas are shown in Table 1.I.

TABLE 1.I

Typical average hydrostatic gradients

Geologic basin	Hydrostatic pressure gradient		Areas in U.S.A.
	(psi/ft)	(kg cm^{-2} m^{-1})	
Fresh and brackish water	0.433	0.10	Rocky Mountains, Mid-Continent
Salt water	0.465	0.1074	Gulf Coast

In general then, the hydrostatic pressure gradient p (in psi/ft) can be defined as:

$$p = 0.433 \cdot SG \tag{1.2}$$

where SG is the specific gravity of a representative column of water.

Overburden pressure

This pressure originates from the combined weight of the formation matrix (rock) and the fluids (water, oil, gas) in the pore space overlying the formation of interest. Mathematically, the overburden pressure (P_o) can be expressed as:

$$P_o = \frac{\text{weight (rock matrix + fluid)}}{\text{area}}$$

$$= D[(1-\phi)\rho_{ma} + \phi \rho_{fl}] \tag{1.3}$$

where D = vertical height of geologic column in feet or meters, ϕ = porosity of formation expressed as a fraction, ρ_{ma} = density of rock matrix in lb/ft^3 or kg/dm^3 and ρ_{fl} = density of fluid in lb/ft^3 or kg/dm^3.

Generally, it is assumed that overburden pressure increases uniformly with depth. For example, average Tertiary deposits on the U.S. Gulf Coast, and

PRESSURE CONCEPTS

elsewhere, exert an *overburden pressure gradient* of 1.0 psi/ft of depth [0.231 kg cm^{-2} m^{-1}]. This corresponds to a force exerted by a formation with an average bulk density of 2.31 g/cm^3. Experience also indicates that the probable maximum overburden gradient in clastic rocks may be as high as 1.35 psi/ft [0.312 kg cm^{-2} m^{-1}].

Worldwide observations over the last few years have resulted in the concept of a *varying* overburden gradient for fracture pressure gradient predictions used in drilling and completion operations. Based upon sedimentation and compaction considerations, one may expect overburden pressure gradients at shallow depth to be substantially smaller than 1.0 psi/ft [0.231 kg cm^{-2} m^{-1}] (Fig.1.1). The latter concept and its practical implications for the oil industry will be discussed in more detail in Chapter 6.

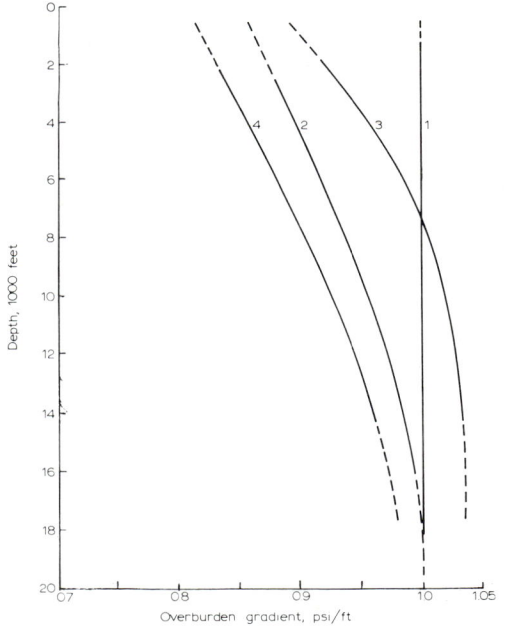

Fig.1.1. Composite overburden load for normally compacted formations (shale/sand sequence). *1* = constant gradient 1.0 psi/ft [0.231 kg cm^{-2} m^{-1}] ; *2* = Texas and Louisiana, U.S.A., Gulf Coast area; *3* = California, U.S.A., Santa Barbara Channel; *4* = North Sea area.

Formation pressure

Formation pressure (P_f) is the pressure acting upon the fluids (formation water, oil, gas) in the pore space of the formation. Normal formation pressures in any geologic setting will equal the hydrostatic head (i.e. hydrostatic pressure) of water from the surface to the subsurface formation. Abnormal formation pressures, by definition, are then characterized by any departure from the normal trend line.

Formation pressures exceeding hydrostatic pressures ($P_f > P_{Hy}$) in a specific geologic environment are defined as abnormally high formation pressures (*surpressures*), whereas formation pressures less than hydrostatic are called subnormal (*subpressures*). Global exploration for hydrocarbon resources, both on- and offshore, has shown that abnormal formation pressures occur worldwide, with surpressures occurring more frequently than subpressures (see Chapter 9).

Fig.1.2 and eq. 1.4 both illustrate how these subsurface pressures and stress concepts are related:

$$P_o = P_f + \delta \qquad (1.4)$$

where P_o = overburden pressure (total vertical stress, lithostatic pressure), P_f = formation pressure (pore fluid pressure, pore pressure), and δ = grain-to-grain pressure (matrix stress, effective stress, vertical rock-frame stress).

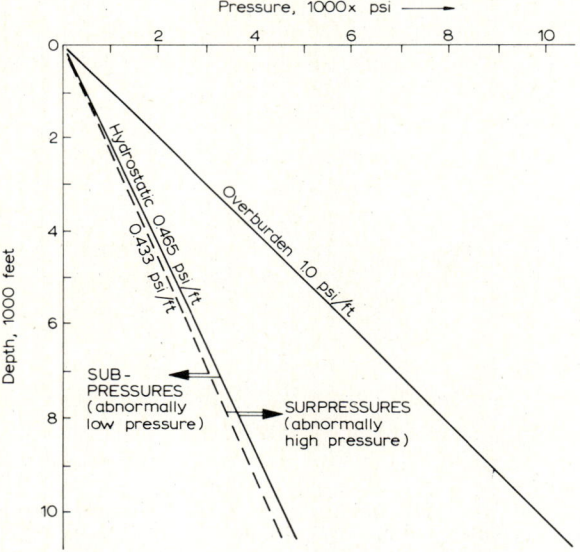

Fig.1.2. Subsurface pressure concepts.

In normal pressure environments ($P_f = P_{Hy}$), the matrix stress supports the overburden load due to grain-to-grain contacts. Any reduction in this direct grain-to-grain stress ($\delta \to 0$) will cause the pore fluid to support part of the overburden, the result being abnormal formation pressures ($P_f > P_{Hy}$). In other words, the overburden may effectively be "floated" by high formation pressures.

There are numerous factors that can cause abnormal formation pressures, such as surpressures and subpressures. Frequently, a combination of several superimposed causes prevails in a given basin and as such is related to the stratigraphic, tectonic, and geochemical history of the area.

Subnormal formation pressures (subpressures)

Worldwide experience indicates that subnormal formation pressures occur less frequently than abnormally high ones. Nevertheless, subpressures have been encountered in many areas while drilling for oil and gas, including the Texas and Oklahoma Panhandle area, parts of the Colorado Plateau, the Uinta Basin in Utah, areas in Arkansas, the Appalachian region, etc., all in the United States; the Lower Cretaceous Viking Formation in Central Alberta, Canada; Middle Miocene formations in the Chokrak and Karagan areas in the U.S.S.R.; arid areas in Iran; and elsewhere.

Subnormal formation pressures may also occur artificially by producing oil, gas, and/or water from permeable subsurface formations (reservoirs). Production of large amounts of reservoir fluids can drastically reduce formation pressure (Krynine and Judd, 1957; Marsden and Davis, 1967). Basically, fluid withdrawal, such as production, causes a decline in pore fluid pressure if no strong water drive tends to compensate for it; frequently, the producing layers compact as a result of this.

Subnormal reservoir pressures occur in many depleted oil reservoirs (Brighenti, 1967). Occasionally, surface indication of such in-situ pressure reduction is manifested by subsidence, which has been observed at Long Beach Harbor area near Los Angeles, California; the Goose Creek area near Galveston Bay, Texas; along the Bolivar Coast on Lake Maracaibo, Venezuela; and elsewhere (Gilluly and Grant, 1949; Van der Knaap and Van der Vlies, 1967; Mayuga and Allen, 1969; Allen, 1972).

Today, special subsidence and compaction monitoring devices are available, including (1) precision casing-joint-length measuring tools for detection of possible deep compaction, and (2) compaction-monitoring wells that measure shallow compaction by means of a cable extensometer. In oil fields where subsidence is suspected, these data are monitored and reviewed continually, and operating measures, such as special waterfloods, are taken when necessary to ensure that no subsidence occurs.

As in the production of hydrocarbons, withdrawal of groundwater may also result in subpressures and subsidence. For example, a comprehensive study was made of the phenomenon of subsidence occurring in the region of the Po Delta, Italy (Caloi, 1967): it indicated the necessity of shutting in the thousands of wells for the extraction of methaniferous waters, to which nearly all the enormous ground subsidences observed in the delta were attributed.

Effects of groundwater withdrawal are also well known in the U.S. Gulf

Coast area (Gabrysch, 1967). In freshwater aquifers that have been tapped by thousands of wells in the near-shore belt of the Gulf Coast Plain, the head has been lowered hundreds of feet over large areas. The largest area is the Houston district in Texas, where the total area of appreciable head decline approximates 5000 square miles [12,950 km^2]. About 22% of water pumped in the Katy—Houston—Pasadena—Baytown area between 1954 and 1959 was derived from compaction of sediments of the heavily pumped interval (Winslow and Wood, 1959). It is believed that the water in clay beds between the sand-bed aquifers drained to the aquifers when the head was lowered by pumping.

In areas of the Texas Panhandle, subnormal formation pressure gradients range from 0.36 psi/ft [0.083 kg cm^{-2} m^{-1}] to 0.39 psi/ft [0.09 kg cm^{-2} m^{-1}]. A specific field case has been reported in the Amarillo—Texas Panhandle field, northwestern Texas, where pressure gradients equal to about half the hydrostatic gradient have been measured (Levorsen, 1967). A possible cause is that a granite-wash producing formation crops out to the east in the Wichita Mountains of Oklahoma at an elevation of about 1000 ft [304.8 m] above sea level. This would approximately correspond to the potentiometric surface of the producing formation of the Amarillo field.

Similar subpressure conditions have been encountered in drilling operations in semi-arid and arid areas, where the water table is very low. Exploratory drilling in the Middle East, for example, has encountered the water table as deep as several hundred or even several thousand feet below the surface, a situation which calls for proper well planning. Under such conditions the normal, hydrostatic pressure gradient starts at the water table only.

Investigations by the Alabama Geological Survey and the U.S. Geological Survey in Alabama indicate that many areas underlain by carbonate rocks are prone to subsidence. Sinkhole collapses are related to natural phenomena such as heavy rainfall, seasonal fluctuations in the water table, earthquakes, or other changes in the hydrogeologic regime affecting residuum stability, and man-imposed effects such as artificial drainage, dewatering, seismic shocks, and breaks in water or sewage pipes.

The formation of sinkholes often results from the collapse of cavities in residual clay, caused by spalling or downward migration of clay through openings in underlying carbonate rocks. The spalling and formation of cavities are caused by (or may be accelerated by) a lowering of the water table, resulting in a loss of support to the clay overlying openings in the bedrock, fluctuation of the level of the water table with respect to the base of residual clay, downward movement of surface water through openings in the clay, or an increase in water velocity in cones of depression to the points of discharge. Collapses have occurred where spalling and the resulting enlargement of cavities have progressed upward until the overlying clay could not support itself and where sufficient vibration, shock, or loading over cavities caused the clay to be jarred loose or forced down.

Another interesting study discussed the predominantly lenticular pre-Pennsylvanian reservoirs of the Appalachian region (Russel, 1972). Fluid pressures in many gas fields in these reservoirs are subnormal, but some are normal and a few are above normal. Each reservoir, across large areas, has a characteristic departure of the mean values of fluid pressures from the normal. Abnormally low fluid pressures tend to occur in lenticular reservoirs closely associated with shales in areas which have undergone erosion. A possible explanation is that erosion causes a reduction in the fluid pressure in the pore space of shales and that this reduction is transmitted to the closely associated reservoir rocks. The pressure reduction in shales may be due to the increase in pore volume and adsorption of water in clay minerals as the overburden pressure decreases and to the adsorption of water during mineral transformations that occur because of the decrease in temperature.

Several formation pressure investigations in Oklahoma have dealt with subpressures, a significant feature of several Oklahoma gas fields in Permian and pre-Permian reservoir rocks. For example, Flores (1967) studied subpressure conditions in the Keyes gas field, located in Cimarron County, Oklahoma, successfully applying electric log parameters for detection and evaluation of these abnormal pressure conditions. Methods used include variations in formation water resistivity derived from the spontaneous potential (SP) curve and the formation factor method as proposed by Foster and Whalen (1966) for quantitative formation pressure evaluation (Fig.1.3). Principles of these methods are discussed in Chapter 5. The pressure gradient of 0.188 psi/ft [0.043 kg cm^{-2} m^{-1}] cited for the Keyes gas field has been one of the lowest values we have encountered.

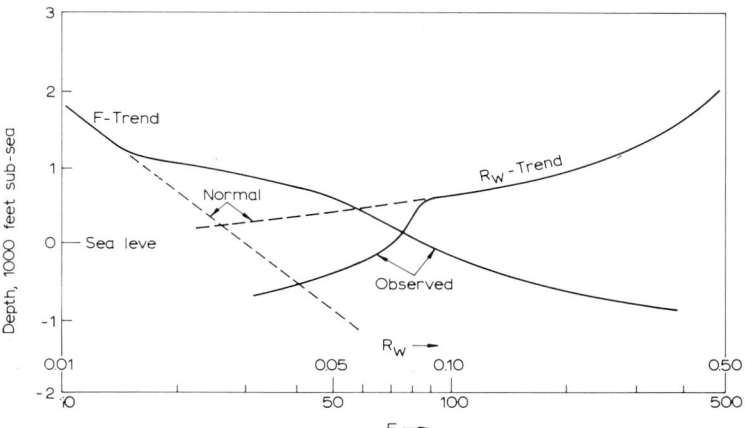

Fig.1.3. Relationship of formation water resistivity (R_w) to formation factor (F) with depth in subpressured Keyes Field, Cimarron County, Oklahoma, U.S.A. (after Flores, 1967).

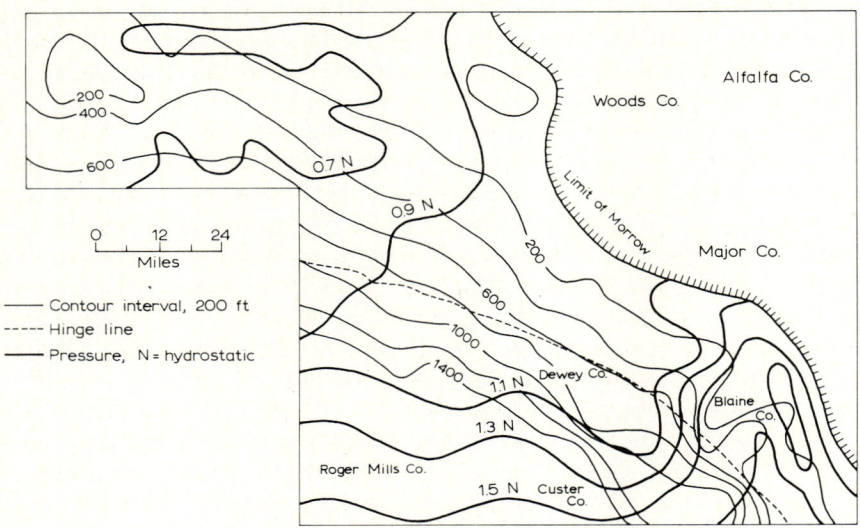

Fig.1.4. Isopach map of the Morrowan series and pressure distribution, northwestern Oklahoma, U.S.A. (modified after Breeze, 1970).

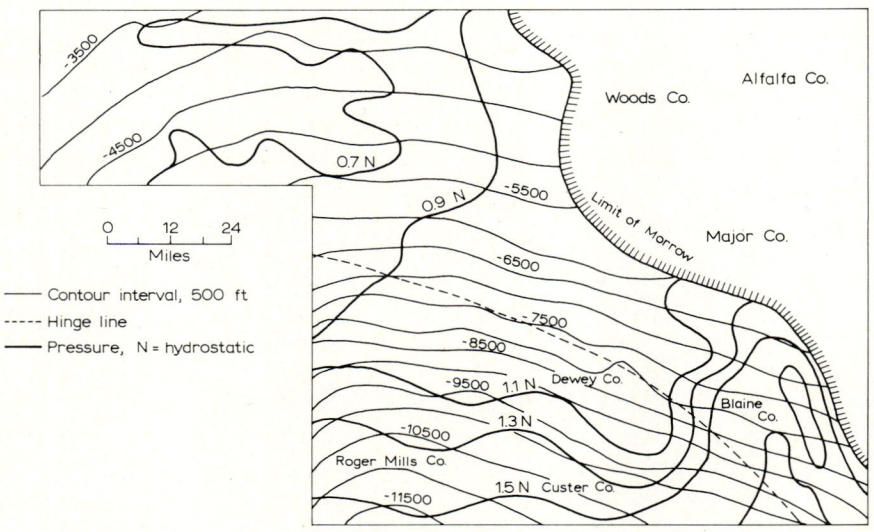

Fig.1.5. Subsurface structure on top of the Morrowan series and pressure distribution, northwestern Oklahoma, U.S.A. (modified after Breeze, 1970).

More recently, Breeze (1970) also investigated the apparent pressure relationships in these Morrow sands of northwestern Oklahoma on a regional basis. In this area the Morrow Formation is defined as the geologic section between the Atoka Formation and the Mississippian unconformity; the Morrow sands, which are the permeable facies within the Morrowan interval, are transgressive, lenticular lenses, mainly producing condensate and gas. Breeze's interesting study focuses attention on a regional transition from abnormally low to abnormally high formation pressures (subpressures → hydrostatic → surpressures). In Figs.1.4 and 1.5 he has shown the results as a generalized, comparative presentation of formation pressure variations superimposed upon the isopach map of the Morrowan series and the subsurface structure of the top of this formation. Breeze suggested that, "because the salinity contrast and the lenticular sand development are very favorable to osmotic action, there is a strong possibility that osmosis is contributing to the pressure distribution. Such a process could be responsible for the uniform, gradual transition from abnormally high to subnormal conditions."

Since this explanation of the observed pressure variations is open to question, another study of the same area (Ferran, 1973) considered possible effects due to uplift and erosion ("decompressional expansion") and associated temperature reduction caused by the depth change of the sediments. The combined effect of these processes may cause a pressure drop which can be expressed by a differential equation, such as:

$$dp = \left(\frac{\partial p}{\partial S_w}\right)_{t,V} dS_w + \left(\frac{\partial p}{\partial t}\right)_{V,S_w} dt + \left(\frac{\partial p}{\partial V}\right)_{t,S_w} dV \qquad (1.5)$$

where dp = total reservoir pressure change in the reservoir rock, and $(\partial p/\partial S_w)_{t,V} dS_w$ = partial change of the pore pressure in the sand reservoir with respect to any change in water saturation. Such a desaturation, i.e. movement of water from the sand into adjacent shales, is due to the decompressional expansion of these shales as caused by stress removal with uplift and subsequent partial removal of overburden by erosion (Fig.1.6). The magnitude of such a "dilation" effect (expansion of rock due to unloading) will depend on the rock type, differential changes in overburden weight and subsurface geothermal regime, and geologic time elapsed. $(\partial p/\partial V)_{t,S_w} dV$ = partial change of pore pressure with respect to pore volume change due to dilation effect in sand reservoirs. Ferran (1973) considered this term negligible. $(\partial p/\partial t)_{V,S_w} dt$ = partial change of pore pressure with respect to temperature reduction at isovolumetric and isosaturation conditions. Pore space and fluid saturation are assumed to remain constant.

In transition zones from normal- to abnormal-pressure environments, a temperature increase exceeding the normal geothermal gradient is a frequently observed phenomenon. Similar conditions have also been

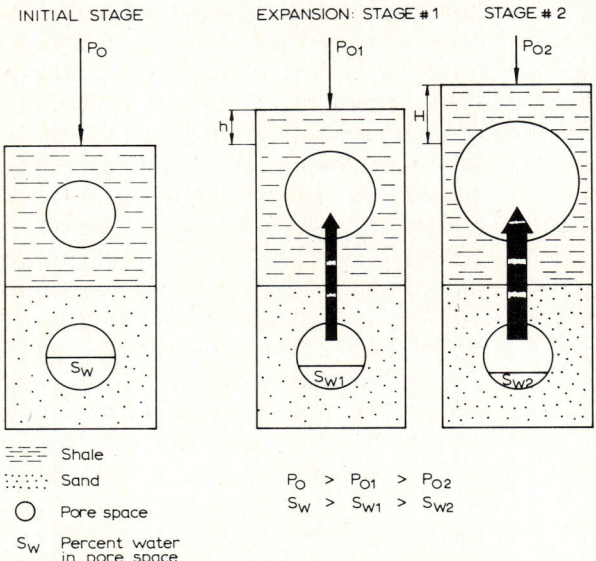

Fig.1.6. Model representing a shale/sand system subject to overburden pressure reduction (after Ferran, 1973).

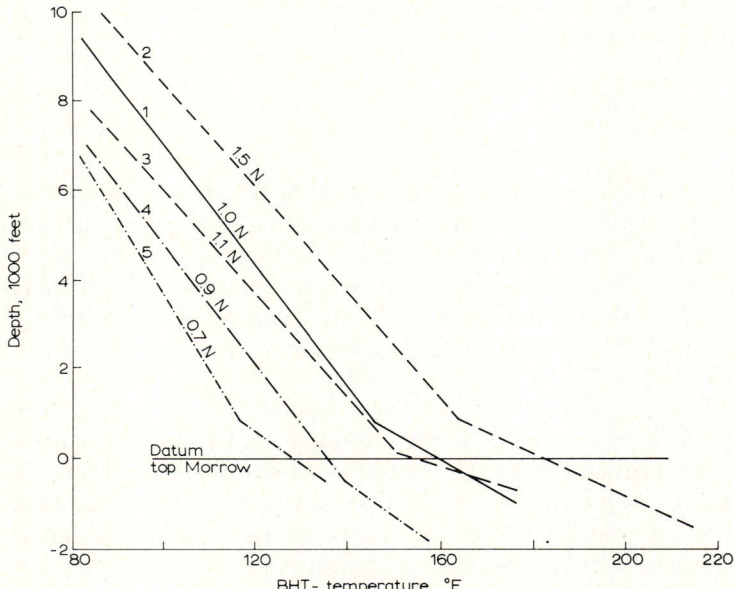

Fig.1.7. Geothermal gradients in sub- and surpressured wells in the Morrowan sands, northwestern Oklahoma, U.S.A. (after Breeze, 1970). N = normal, hydrostatic pressure environment. 1 = Paul No. 1, $1.0N$; 2 = McNeely—Fahl No. 1, $1.5N$; 3 = Cheyenne—Arapaho No. 1, $1.1N$; 4 = Ferguson No. 1, $0.9N$; 5 = Hill No. 1, $0.7N$.

observed in producing wells drilled into subnormal and surpressure environments in the Morrow sands (Fig.1.7). Note the generally lower temperature increases in subnormal-pressured wells. If one assumes the Morrow Formation to be overheated and undercompacted at initial conditions, any formation cooling due to uplift and erosion will then reduce the pore pressure.

Such a concept may be supported and illustrated by the pressure-temperature-density diagram for water (Barker, 1972) shown in Fig.1.8.

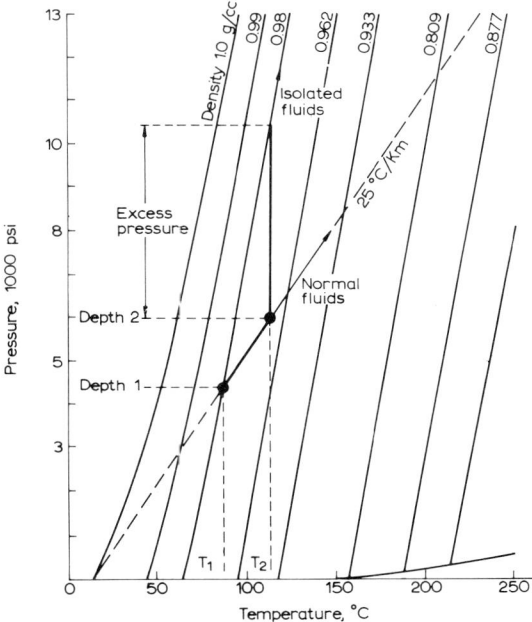

Fig.1.8. Pressure-temperature-density diagram for water (after Barker, 1972). Excess pressure is the higher pressure due to the temperature increase from T_1 to T_2.

Note that during cooling of an isolated fluid system, such as may be caused by uplift and/or erosion, the pressure in the isolated volume will decrease below the normal, hydrostatic value. In a similar fashion, such a temperature decrease may cause a pressure drop in isolated gas zones, with the magnitude of reduction depending on the value of pressure change and the composition of the gas.

Since extensive drilling operations in the Anadarko Basin, Oklahoma, have encountered subpressures in formations of Permian, Pennsylvanian, Mississippian, and even Devonian age, we have plotted some of the data given by Moore (1971) in Figs.1.9 and 1.10.

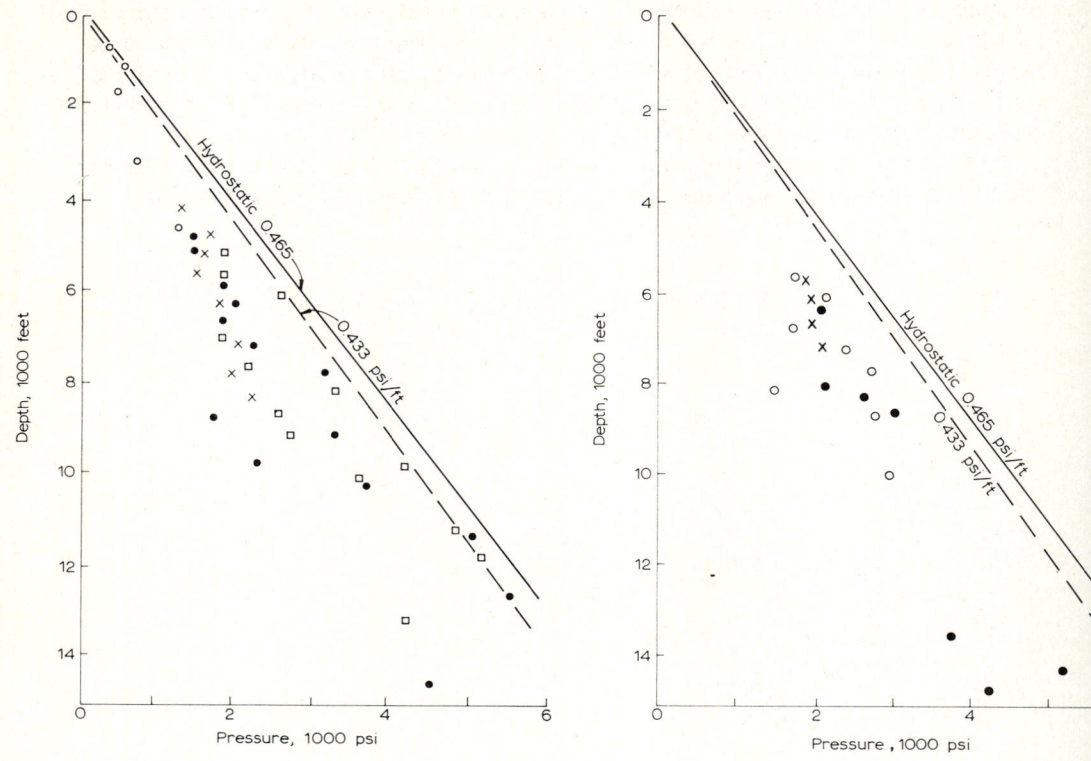

Fig.1.9. Subnormal formation pressures, Anadarko Basin, Oklahoma, U.S.A. ○ = Permian, × = Upper Pennsylvanian, ● = Middle Pennsylvanian, □ = Lower Pennsylvanian.

Fig.1.10. Subnormal formation pressures, Anadarko Basin, Oklahoma, U.S.A. ○ = Upper Mississippian, × = Middle Mississippian, ● = Hunton carbonates.

Another quite interesting field observation of subpressures and an artificial earthquake caused by fluid injection has been reported for the Rocky Mountain Arsenal well (Evans, 1966). This chemical waste-water disposal well was drilled northeast of Denver in the Denver—Julesburg Basin to a total depth of 12,045 ft [3671 m], bottoming in Precambrian gneiss. Well tests made at a depth of 11,002 ft [3353 m] determined the pore pressure to be 4128 psi [290 kg/m^2], indicating subnormal conditions. Evans also suggested that "the highly fractured rocks of the reservoir are at rest on steep slopes under a condition of subhydrostatic fluid pressure. As the fluid pressure is raised within the reservoir, frictional resistance along fracture planes is reduced and, eventually, movement takes place. The elastic wave energy released is recorded as an earthquake."

Subnormal formation pressures have also been encountered in a number of Russian oil and gas reservoirs of Tertiary age. Subpressures are often found in reservoir rocks of Middle Miocene age, especially in the Chokrak and Karagan stages. Occurrence of these subpressures, however, declines with increasing depth of these hydrocarbon reservoirs (Fig.1.11). Soviet investigators have attempted to explain this condition as being a result of the differences in density of the minerals making up the geologic sequence.

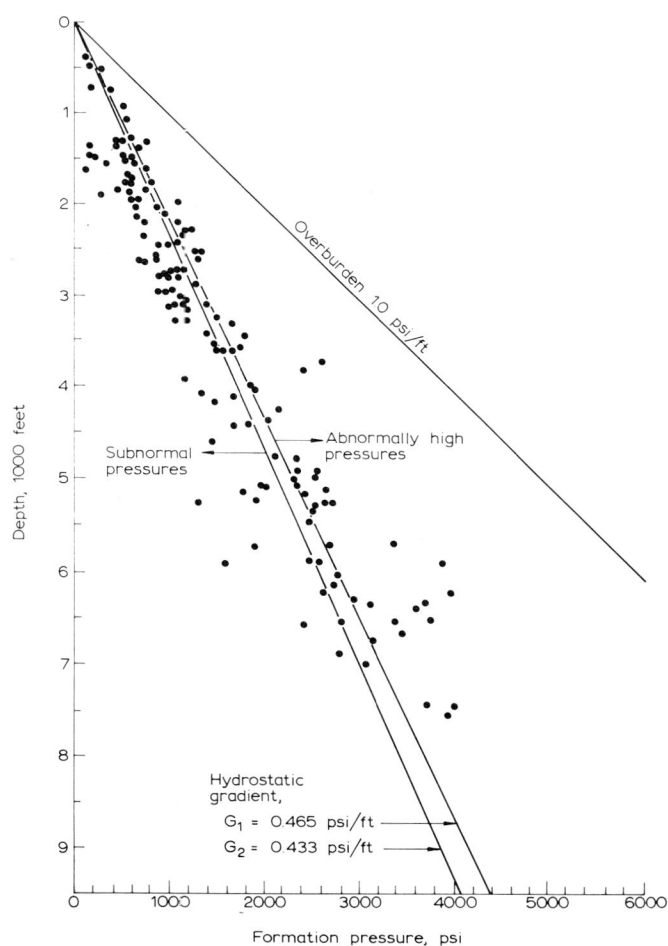

Fig.1.11. Abnormal formation pressure environments in Tertiary formations, U.S.S.R. (after Fertl, 1971). (Courtesy of the Society of Petroleum Engineers of A.I.M.E.)

Abnormally high formation pressures (surpressures)

Abnormally high pore fluid pressures are encountered worldwide in formations ranging in age from the Cenozoic era (Pleistocene age) to as old as the Paleozoic era (Cambrian age).

A normal, hydrostatic pressured geologic environment can be visualized as a hydraulically "open" system; i.e. permeable, fluid communicating formations allow establishment and/or reestablishment of hydrostatic conditions. Conversely, abnormally high formation pressure systems are essentially "closed", preventing, or at least greatly restricting, fluid communication. Here, overburden pressure, in part, is supported by formation fluids in the pore space.

Pressure seals

Normal and abnormal formation pressures coexist only if separated by a permeability barrier which simultaneously acts as a pressure barrier. Theoretically, such a seal could be any material or combination of materials in the earth's crust, restrictive or preventive to movement and passage of substantial volumes of fluids. The origin of a pressure seal is physical, chemical, or may be a result of the combination of the two (Louden, 1972). Since modes and origins of such seals can be manifold, the possible types are listed in Table 1.II (Fertl and Timko, 1972).

From this compilation, it can be seen that problems of pressure seals resolve themselves into three basic questions: "What, When, and How?"

"What" involves a description of the sealing material present, "When"

TABLE 1.II

Suggested types of formation pressure seals (after Fertl and Timko, 1972)

Type of seal	Nature of trap	Examples
Vertical	massive shales and siltstones massive salts anhydrite gypsum limestone, marl, chalk dolomite	Gulf Coast, U.S.A., Zechstein in North Germany, North Sea, Middle East, U.S.A., U.S.S.R.
Transverse	faults salt and shale diapirs	worldwide
Combination of vertical and transverse		worldwide

concerns the stages in the rock's diagenesis during which sealing was accomplished, and "How" relates to the cap-rock chemical composition, the type of aqueous formation fluids from which minerals are precipitated, the prevailing pressure and temperature environments, etc.

The existence of any mineral is a function of the stability field indigenous to the mineral species. In turn, the stability field depends on temperature, pressure, concentration of solution, composition of solution, etc. This stability field also determines whether a mineral will remain in equilibrium with the environment, and whether it will grow or dissolve.

Generally, a pressure-seal formation will depend on many factors, including the highly complex phenomena of deformational response of clay minerals to applied physical loads (such as overburden); behavior of clays and shales as semipermeable membranes; the type and quantities of clay minerals present; extremely low shale permeabilities (10^{-4} to 10^{-6} millidarcy); non-Newtonian behavior of water in finer interstices of clayey sediments; interaction with escaping gases and organic acids; the type, concentration, and self-diffusion of cations; precipitation of dissolved solids (e.g. calcite, silica, feldspar, pyrite, siderite, etc.); composition and possible mixing of brines; and sharp subsurface pressure and temperature changes.

All of these factors, or any combination of them, plus consideration of the geologic time factor, may change the physicochemical environment, causing slow lithological changes in the formation of these cap rocks as pressure seals.

Then, inasmuch as fluids and gases can no longer move through this seal in large quantities, further precipitation and cementation by calcite, silica, etc., will be slowed until essentially halted.

Although such shales are hardened as a result of lithification, they should retain some flexibility (plasticity), permitting the pressure seal to "give" during minor movements in the earth's crust. But larger tectonic activities will break the seals by fracturing and fissuring. This allows the dissipation of abnormal formation pressure, or an in-situ redistribution, both decreasing or increasing in magnitude, as reflected in sequential cap rocks.

The upper limit of such surpressures is usually expected to equal the total weight of the overburden, which is generally assumed to be equivalent to a gradient of 1.0 psi/ft [0.231 kg cm^{-2} m^{-1}]. On a *regional* basis, overpressures may approach the overburden load, which for reason of stability they cannot surpass. However, experience in several areas, such as Pakistan, Iran, Papua, and the U.S.S.R., indicates that *locally* confined, even higher pressure gradients may be encountered. Pore fluid pressure gradients exceeding the overburden gradient by up to 40% are known. Such *superpressures* can be visualized for localized conditions in the presence of a so-called *pressure bridge* (Fig.1.12). The internal strength of rock overlying the surpressured zones assists the overburden load to contain locally these tremendous upward forces.

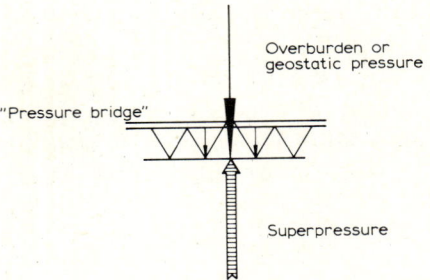

Fig.1.12. Schematic model of pressure bridge (after Fertl, 1971). (Courtesy of the Society of Petroleum Engineers of A.I.M.E.)

Concepts for abnormal formation pressure origin

Several, and often a multitude, of superimposed factors may be the cause of abnormal pore fluid pressures. The phenomena are related to geologic, physical, geochemical, and mechanical processes. Frequently, it is difficult or even impossible to speculate on the importance of each individual cause for a given overpressure environment in a specific geologic area.

For example, abnormal formation pressures in the U.S. Gulf Coast Basin have been discussed by Harkins and Baugher (1969) as the direct result of the structural and depositional history. Genesis of overpressures during geologic time is controlled by the environment of deposition on the paleo-continental shelf and slope, the geometry and lithology of the sediments, regional and local faulting, basin hinge lines, burial and compaction, and subsequent structural deformation.

According to Jones (1969), the following combination of factors may be responsible for overpressures along the U.S. Gulf Coast area: "The Neogene deltaic and neritic marine deposits in the northern Gulf of Mexico Basin form regional aquifer systems in which pore pressures, salinities and temperatures do not compare with those in older sedimentary basins. Rapidly buried sand and clay sequences, sealed off by growth faults, remain undercompacted and contain abnormally high fluid pressures (up to 0.95 times the overburden pressure). Such high pressures force saline formation water toward normally pressured aquifers. The intervening clay beds act like membrane filters to freshen the water, which subsequently dilutes the receiving aquifers. High pressures and trapped radiogenic heat cause diagenesis of montmorillonite to illite. The released fresh water of crystallization either remains in the transformed clay under high pressure because the adjacent sand beds are already geopressured, or flows to and dilutes normally pressured aquifers. Osmotic pressures, which are generated by the membrane-like clays and which may be responsible for halting or even reversing these flows, are considered the dominant factor in the preservation of abnormal high pressures."

An excellent investigation of the high formation pressures in California coast ranges and their tectonic significance has been summarized by Berry (1973) as follows: "Anomalous high fluid potentials exist within the miogeosynclinal Great Valley and eugeosynclinal Franciscan sequences of Jurassic/Cretaceous age within the Coast Ranges and at depth on the west side of the Central Valley, California. These rocks are dominantly mudstones with low fluid transmissibilities.

Certain problems exist as to the probable regional distribution of these high fluid potentials. Low fluid potential areas such as The Geysers geothermal district are present in the Franciscan of northern California within a region generally characterized by high fluid potentials. The low-potential areas are attributed to fracture zones with a channel-type flow whose transmissive characteristics exceed those of intergranular flow. It is concluded that the Franciscan of northern California probably is characterized regionally by near-lithostatic fluid pressures at depth, but fracture zones with both low (i.e. near-hydrostatic) and high (i.e. near-lithostatic) fluid potentials probably exist at various depths from the surface. The Geysers dry-steam occurrence is envisioned as a fracture zone with low fluid potentials by virtue of a decrease in transmissive characteristics of a fracture system with depth, in a local region of high heat flow, possibly caused by the existence at depth of a magma chamber.

An abundance of direct fluid pressure measurements within the Great Valley section of the Sacramento Valley demonstrates the existence of high fluid potentials. The only direct fluid pressure measurement that has been made within the Great Valley section in the central or southern San Joaquin Valley indicates high fluid potentials. The regional chemistry of the lower Tertiary waters of the San Joaquin Valley (membrane effluent type) suggests that these waters have been extruded from a widely distributed series of mudstones and other rocks that are undergoing compaction. The presumed source for this widespread compacting sequence is the underlying Great Valley sediments with their postulated high fluid potentials.

It is concluded that the anomalous high fluid potentials of Tertiary rocks within folds on the west side of the San Joaquin Valley reflect indirectly the presence at depth of high fluid potentials in the underlying Great Valley section. The origin of the folds is attributed to dynamic tectonic compression caused by current deep-seated linear diapirism of Great Valley mudstones and related rocks that possess near-perfect plastic properties by virtue of their near-lithostatic fluid pressures. The closed gravity minimum over the south end of South Dome—Lost Hills anticline is postulated as being the result of a diapir of serpentine or similar material.

It is postulated that a fault zone, named herein the "West Side" fault, probably exists at depth along the west side of the Central Valley. This buried fault is envisioned as having an intermittent near-surface expression in the form of faults such as the Midland fault, or long linear folds such as the Kettleman folds. Diapirism along this fault is presumed to be responsible for these folds.

Subsidence along the West Side fault is postulated as having occurred contemporaneously with deposition of the Great Valley sequence, and thus provided a local trough in which the thick (maximum 60,000 ft) [18,288 m] Great Valley section was deposited. The depositional barrier between the Franciscan and Great Valley sequences is postulated as a zone of serpentinite-ultrabasic rocks that intruded intermittently to form a sediment trap on the continental slope throughout Jurassic/Cretaceous geosynclinal deposition.

The final conclusion reached is that an extensive geographic zone is present in which the pore fluid pressures of the thick Franciscan and Great Valley geosynclinal sediments reach near-lithostatic values. This zone is 400—500 miles [643.74—804.67 km] long and 25—80 miles [40.23—128.75 km] wide; it is bounded on the west by the San Andreas fault and the granitic Salinas block, on the east by the buried West Side fault and the granitic Sierran—Klamath block, on the south by the granitic San Emigdio—Sierran block; the northern boundary is interpreted as being the northern termination of the San Andreas fault in the Cape Mendocino region. Structural deformation of this zone by diapirism and thrusting is facilitated by the lithic plasticity caused by high fluid pressures. Known diapirism and thrusting and possible diapiric folding suggest a late Cenozoic age for the development of the high fluid potentials.

The origin of the anomalous fluid pressures adjacent to the San Andreas fault is attributed to compression between the granitic Sierran—Klamath and Salinas blocks resulting from late Cenozoic extension of the central Great Basin in Nevada and Utah. The San Andreas is a transform fault which separates the independent stress field of the Pacific plate (Salinas block) that is moving northwestward relative to the North American plate (Sierran—Klamath block and the Great Valley—Franciscan sediments). The Sierran—Klamath block also is moving westward or southwestward by continued late Cenozoic central Great Basin extension; this westerly motion is terminated by compression of the rocks on both sides of the San Andreas. This compression has the greatest effect within the Franciscan and Great Valley shale mass just east of the fault; the effect is greatly reduced within the granitic basement and overlying sediments of the Salinas block west of the fault but has been responsible for folding of the sedimentary veneer. The high fluid potentials are caused by the squeezing of this belt of highly compressible shales east of the San Andreas in a vise whose jaws are formed of relatively incompressible granite; these anomalous fluid potentials are envisioned as being late Cenozoic phenomena dynamically active today.

Diapirism and diapiric folding instead of thrusting have been the preferred modes of late Cenozoic structural deformation within this high fluid potential belt. The dominance of diapirism is attributed to the limited crustal shortening related to the development of this compressive field, as opposed to the dominance of the shearing stresses related to plate movements on both sides of the San Andreas fault. Diapirism and more limited thrust faulting related to the current generation of high fluid potentials may develop in the future.

Among the possible consequences of the existence of this postulated extensive zone of near-lithostatic fluid pressures are the shallow-focus earthquakes and extensive aftershocks along the San Andreas fault. The near-continuous fault creep along the San Andreas and related Calveras and Hayward faults also may be a result of these postulated high pore fluid pressures adjacent to these faults.

An important implication of this paper is the demonstration that fluid pressures within rocks can serve as extremely sensitive and unique strain gauges for the detection of local or regional structural movements."

Causes of abnormal formation pressures will also differ for sand/shale sequences as found in the U.S. Gulf Coast area, massive carbonate sections in the Middle East, or conditions found in the North Sea area and the South China Sea. In other words, abnormally high formation pressures may have many origins, frequently a combination of superimposed factors, including the following.

Piezometric fluid level

The effect of a regional potentiometric surface which is abnormally high can cause surpressures. A classic example is an artesian water system. Generally, artesian pressures are present under the following conditions: (1) porous and permeable aquifers are sandwiched between impermeable beds, such as shales, and (2) these aquifers are deformed in such a way as to exhibit a high intake area to produce the necessary hydraulic head.

In such systems, resulting overpressures may be small to moderate, such as in the Artesian Basin in Florida (Leve and Goolsby, 1967) or in the Great Artesian Basin in Queensland, Australia, which covers 600,000 square miles [1554×10^6 km^2] (Slatyer and Mabutt, 1964). However, artesian pressures sometimes become considerable, such as in the North Dakota Basin (Meinzer, 1928).

Structure of reservoir

In sealed reservoir rocks, such as lenticular reservoirs, dipping formations, and anticlines, formation pressures — normal for the deepest part of the zone — will be transmitted to the shallower end, where they will cause abnormal-pressure conditions (Figs.1.13 and 1.14). In the presence of hydrocarbons in anticlines, abnormal formation pressures are encountered in the

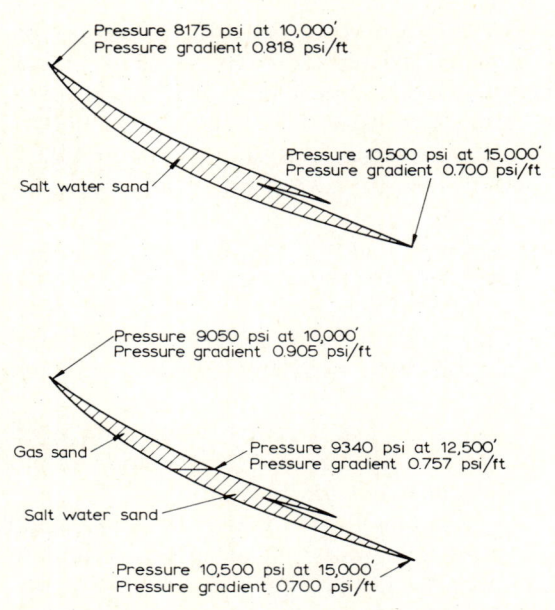

Fig.1.13. Effect of structure on pressure gradients in sands containing fluids under abnormal pressure (after Dickinson, 1953). (Courtesy of the American Association of Petroleum Geologists.)

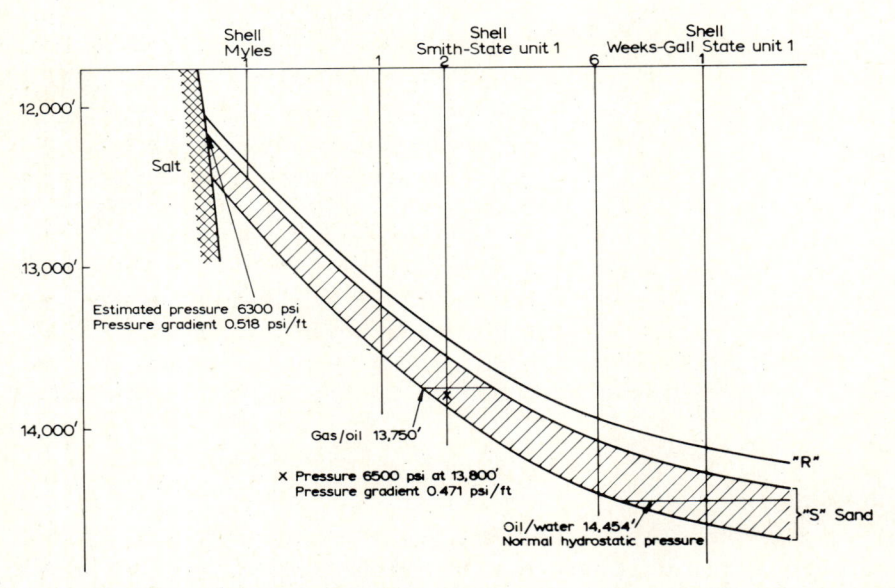

Fig.1.14. Schematic section illustrating abnormal pressure caused by structural elevation in Weeks Island, Iberia Parish, Louisiana, U.S.A. (after Dickinson, 1953). (Courtesy of the American Association of Petroleum Geologists.)

potential pay section, whereas normal hydrostatic pressure conditions still may exist at and below the oil/water contact.

In very large structures, such as are found in the Middle East and elsewhere, overpressures resulting from density differences in oil/water systems and, particularly, in gas/water systems are known to approach the geostatic pressure of the overburden. According to Hubbert and Rubey (1959), control in one Iranian field required 17.3 lb/gal [2.08 kg/dm^3] mud weight at a depth of 640 ft [195 m], which corresponds to 0.9 psi/ft [0.208 kg cm^{-2} m^{-1}] pressure gradient.

Repressuring of reservoir rock

Normal- or low-pressured reservoir rocks, particularly at shallow depth, containing formation water and/or hydrocarbons may sometimes be pressured up and/or repressured due to hydraulic communication with deeper, higher-pressured formations. Such fluid communication is possible (1) behind casing in old wells or boreholes with faulty cement jobs, (2) along "leaky" fault zones, (3) as a result of casing leaks in old wells, and (4) while drilling a sequence of permeable formations exhibiting drastically different pore fluid pressures (causing recharge saltwater flows).

Drilling through shallow recharged zones sometimes requires mud weights of up to 19 lb/gal [2.275 kg/dm^3].

Rate of sedimentation and depositional environment

Jones (1969) suggested that "the hypothesis presented by Dickinson (1953), that abnormally high fluid pressures are caused by the weight of the overburden and fluid entrapment, must be extended to include several other factors, the most important of which is probably the *rate of sedimentation*." This had previously been suggested by Hubbert and Rubey (1959), when they listed conditions favorable for development of overpressures, such as (1) large total thickness of sediments, (2) presence of clay rocks, (3) interbedded sandstones, (4) *rapid loading*, and (5) geosynclinal conditions in general. Fertl and Timko (1972) also stated that rapid deposition of predominantly clays and some sand (i.e. high shale/sand ratio), exceeding the structural downwrap, can cause abnormal pressures.

The data in Table 1.III suggest that the sedimentation rates in the Texas and Louisiana Gulf Coast area have been very rapid since the beginning of the Pleistocene age (Jones, 1969).

According to Rochon (1967) overpressured zones and reservoirs are associated with *periods* of *fast deposition* when hydraulic equilibrium accompanying normal compaction is disturbed by a restriction (a seal consisting of a highly mineralized zone resulting from cementation of shales and sands by salt, calcite, anhydrite, etc.) to the orderly expulsion of water. Similar concepts have been proposed by Bredehoeft and Hanshaw (1968), Magara (1971), and Smith (1973) who state that the creation and mainte-

TABLE 1.III

Depositional rates in areas of thickest sediments, Texas and Louisiana Gulf Coast (after Jones, 1969)

Age	Duration (m.y.)	Thickness (m)		Rate (cm/1000 yr)	
		Louisiana	Texas	Louisiana	Texas
Holocene and Pleistocene	2	2400	600	120	30
Pliocene	12	2400	900	20	7.5
Miocene	12	7500	3000	62.5	25
Anahuac, Frio and Vicksburg	11	5300	4800	48.5	43.5

nance of anomalous pressures depend, to a large degree, upon the *hydraulic conductivity* of the geologic system.

Harkins and Baugher (1969) while studying the geologic significance of abnormal formation pressures, suggested that: "Abnormally high formation pressures are common in the post-Cretaceous sediments of the Gulf of Mexico Basin, particularly so in southern and off-shore Louisiana. Wherever thick clays are deposited rapidly, as in the Gulf Basin, interstitial water is likely to be trapped and isolated from communicating with the surface. In this situation the sediment cannot compact and the contained water is subjected not only to hydrostatic forces, but also to the weight of newly deposited sediment. This results in a formation with abnormal fluid pressure.

A factor that probably further affects development of abnormal pressure in specific water depths is *rate of deposition*. For instance, a very slow rate of deposition in a slope environment might not develop abnormal pressures, even though the sediments being deposited had a very low sand content."

The depositional setting for abnormally high pressures has been discussed by several investigators (Dickinson, 1953; Dickey et al., 1968; Harkins and Baugher, 1969; Martin, 1972). Martin (1972) observed an apparent relationship of the environment of deposition to the occurrence of abnormal high fluid pressures in a sand/shale sequence of the subsurface South Texas Oligocene. The abnormal high pressures are generally encountered in the "down-dip" facies, which consists of deltaic sediments of alternating inner neritic (includes brackish and near-shore marine) and middle neritic (shallow marine) depositional environments.

"There is an abrupt change in the environment of deposition that seems to coincide with the top of the abnormal high-pressure zone" (Martin, 1972).

Fig.1.15 shows the paleoecologic log, induction-electric log, and the formation pressure gradient (FPG) curve in an overpressured well in Petronilla field, Nueces County, Texas.

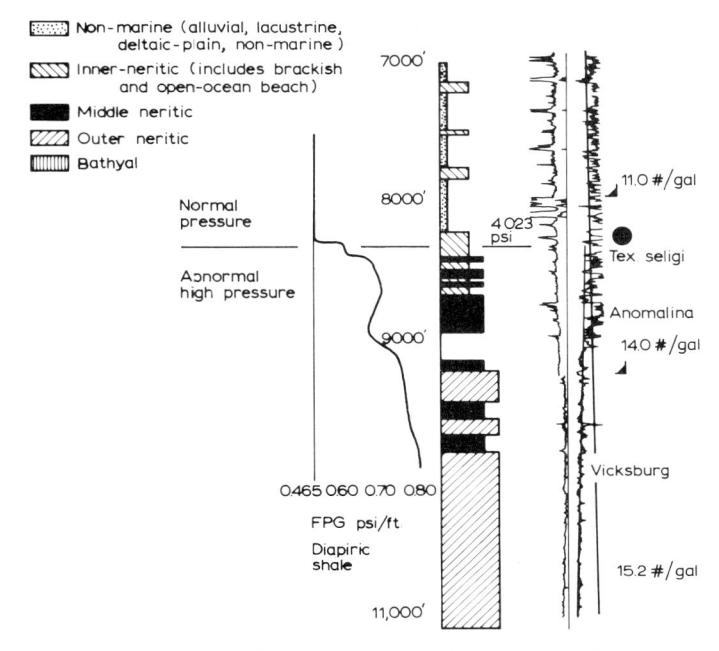

Fig.1.15. Depositional environment and presence of overpressures (after Martin, 1972). Large solid circle indicates the pay zone; the induction-electric log is shown on the right-hand side, the paleoecologic log in the center. (Courtesy of the Society of Petroleum Engineers of A.I.M.E.)

Paleopressures

Such abnormal formation pressures can only exist in older rocks which have been completely enclosed by massive, dense, and essentially impermeable rocks, or in completely sealed formations uplifted to a shallower depth.

Tectonic activities

Abnormally high pore fluid pressures may result from local and regional faulting, folding, lateral sliding and slipping, squeezing caused by down-dropping of fault blocks, diapiric salt and/or shale movements, earthquakes, etc.

Faults. Much has been written about the importance of various types of faults in relationship to abnormal-pressure environments (Dickinson, 1953; Murray, 1961; Carver, 1968; Classen, 1968; Dickey et al., 1968; Harkins and Baugher, 1969; Meyers, 1968; Jones, 1969; Fowler, 1970). Overpressures

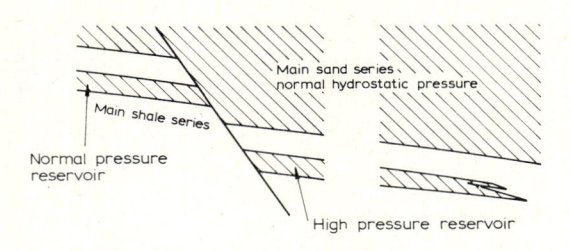

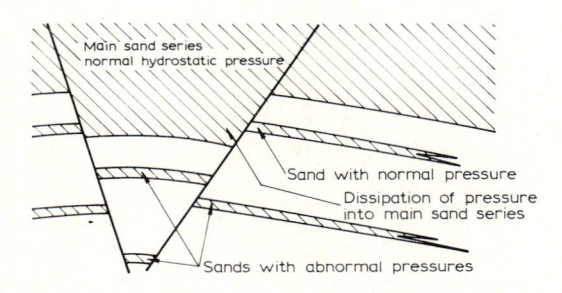

Fig.1.16. Types of fault seals necessary to preserve abnormal-pressure environments (after Dickinson, 1953). Sand to sand contact across a fault results in normal pressures, whereas isolated sands (i.e. sand to shale contact across a fault) are overpressured. (Courtesy of the American Association of Petroleum Geologists.)

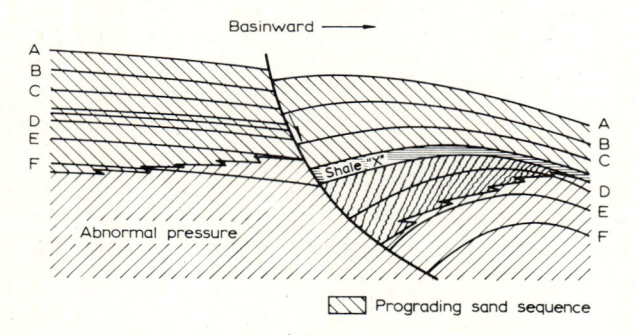

Fig.1.17. Schematic section showing stratigraphic rise of overpressures in relation to prograding sedimentation modified by growth faults (after Harkins and Baugher, 1969). (Courtesy of the Society of Petroleum Engineers of A.I.M.E.)

related to faults are illustrated in Fig.1.16, and a schematic section shows the stratigraphic rise of abnormal pressure related to prograding sedimentation modified by growth fault (Fig.1.17).

According to Dickey et al. (1968), the pattern of the high-pressure zones appears to be related to the peculiar patterns of *faulting contemporaneous with sedimentation* which are characteristic of the U.S. Gulf Coast, with the complex down-to-the-basin *growth* faults (Ocamb, 1961; Thorsen, 1963) of southern Lousiana apparently being caused by slumping along the edge of the continental shelf during sedimentation.

Fault movements occurring during sedimentation are termed *contemporaneous faults* (Hardin and Hardin, 1961). Carver (1968) suggested several causes of regional contemporaneous faults, the most significant of which are basement tectonics, deep salt or shale movement, slump across flexures, slump at the shelf edge, differential compaction, response to crustal loading, or a combination of these factors. Recently, Bishop (1973) studied such faults in North Louisiana and South Arkansas and concluded that "Jurassic contemporaneous faults generally parallel regional structural and depositional strike and are slightly younger basinward (south). Most are downthrown toward the basin, but locally, and regionally in one place, compensating down-to-north faults are present. Beds on downthrown sides are greatly thickened, correlation of individual units is usually impossible, and throws increase with depth. Because the downthrown blocks are tilted, sediments are thickest adjacent to the fault. Fault planes are believed to be curved, and although they are of high angle (60—70°) near the top, they may flatten with depth. As a result of this flattening, together with flowage of underlying Jurassic salt away from downthrown blocks, faults are not known to extend below the salt. They do not cut beds younger than Jurassic and die out upward in a conformable section."

The mechanism for development of regional contemporaneous faults corresponding to overpressured shales and related sediment deformation has been summarized by Bruce (1973) as follows: "Regional contemporaneous faults of the Texas coastal area are formed on the seaward flanks of deeply buried linear *shale masses* characterized by low bulk density and *high fluid pressure*. From seismic data, these masses, commonly tens of miles in length, have been observed to range in size up to 25 miles [40.23 km] in width and 10,000 ft [3048 m] vertically. These features, aligned subparallel with the coast, represent residual masses of undercompacted sediment between sandstone/shale depoaxes in which greater compaction has occurred. Most regional contemporaneous fault systems in the Texas coastal area consist of comparatively simple down-to-basin faults that formed during times of shoreline regression, when periods of fault development were relatively short. In cross-sectional view, faults in these systems flatten and converge at depth to planes *related to fluid pressure* and form the seaward flanks of underlying shale masses. Data indicate that faults formed during regressive phases of deposition were developed primarily as the result of differential compaction of adjacent sedimentary masses. These faults die out at depth near the depoaxes of the sandstone/shale sections.

Where subsidence exceeded the rate of deposition, gravitational faults developed where basinward sea-floor inclination was established in the area of deposition. Some of these faults became bedding-plane type when the inclination of basinward-dipping beds equaled the critical slope angle for gravitational slide. Fault patterns developed in this manner are comparatively complex and consist of one or more gravitational faults with numerous antithetic faults and related rotational blocks.

Post-depositional faults are common on the landward flanks of deeply buried linear shale masses. Many of these faults dip seaward and intersect the underlying low-density shale at relatively steep angles.

Conclusions derived from these observations support the concept of regional contemporaneous fault development through sedimentary processes where thick masses of shale are present and where deep-seated tectonic effects are minimal."

A schematic dip section through the Rio Grande Embayment, illustrating strata thickening across growth faults (Murray, 1961), is shown in Fig.1.18. Three basic types of such regional contemporaneous faults (Bruce, 1973) are

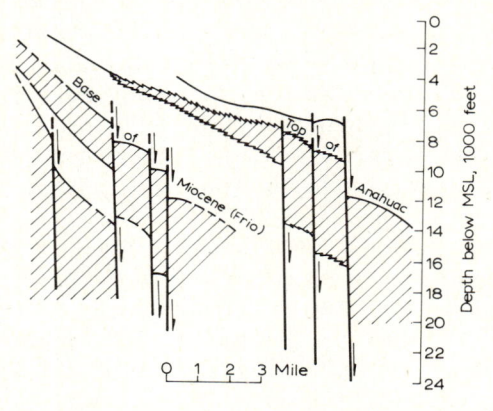

Fig.1.18. Strata thickening across growth faults, Rio Grande Embayment, Texas, U.S.A. (after Murray, 1961).

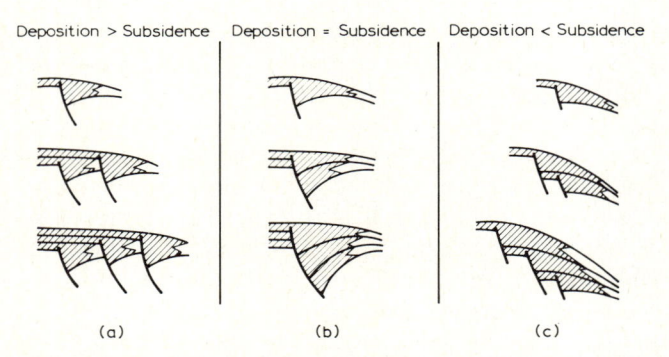

Fig.1.19. Development of three types of contemporaneous fault systems (after Bruce, 1973). (Courtesy of the American Association of Petroleum Geologists.)

presented in Fig.1.19, with differentiation based on rates of deposition of sandy sediments upon unconsolidated clay surfaces. According to Bruce (1973): "Two of these types are considered to be associated with sea floors which were relatively flat at time of deposition, and the third appears to be formed in areas of slope environments where sea-floor subsidence exceeded the rate of deposition.

The first example (Fig.1.19a) represents faults formed during a *regressive sequence* of deposition (progradation locally), when the amount of sediment available for deposition was greater than the space available for accumulation. Under these conditions each successive depoaxis was formed seaward from that of the adjacent underlying unit. Antiregional dip, developed adjacent to the downthrown sides of these faults, varies in relation to the amount of sediment deposited. In areas where *still-stand depositional conditions* prevailed, the rate of faulting was sufficient to accommodate all incoming sediment (Fig.1.19b). In these areas, a strong antiregional dip developed that increased with depth and time. Contemporaneous faults, formed during still-stand and regressive phases of deposition, are common in southern Texas and are considered to have developed primarily through differential compaction associated with relatively flat sea floors.

Faults formed during *transgressive phases* of deposition are present in southern Texas; however, they are less common than the other two forms. Where subsidence exceeded the rate of deposition (Fig.1.19c), the sea floor is considered to have been inclined basinward at an angle related to the rate of subsidence. The primary cause of sea-floor subsidence and tilting was not dependent on differential compaction and differential loading, as described for faults formed during regressive and still-stand phases of deposition, but was controlled by forces below or outside the area of deposition. These forces may have been related to either salt movement or basement tectonics. Other manifestations of contemporaneous faulting can be explained when sea-floor inclination and basinward formational dips are considered with rates of deposition. The most significant of these are gravity-slide faults, many of which become bedding-plane types at depth."

Recently, Hosper (1971) suggested that the Niger Delta area in Nigeria, Africa, has a clay/shale base of considerable thickness. It is well known that overpressures are encountered in the delta area. It is also known that the detailed subsurface structure of the delta is characterized by typical *growth faults* with associated rollover structures, which are interpreted as being caused by gravity.

Based upon the pore fluid pressure hypothesis of Hubbert and Rubey (1959), several tectonic concepts have been discussed and certain conclusions drawn (Roberts, 1972). According to the author, *overthrusts* cannot develop unless the thrust sheet is underlain by a weaker layer or unless abnormal pore fluid pressures are restricted to this layer. Conditions of failure implied by the Hubbert-Rubey hypothesis indicate that both requirements are met if the ready ingress or egress of pore fluid is prevented during impending shear failure. Under these circumstances, sediments capable of further compaction undergo an increase in formation pressure so that the effective value of λ (i.e. fluid pressure/overburden pressure ratio) at failure is unity. Such behavior is typical of shale horizons, which act as the locus of overthrust faults, whereas dilation hardening affects the intervening sandstone or limestone horizons, which in turn form the overthrust sheets. Once shear failure is initiated, movement is essentially frictionless as long as excess pore pressures (overpressure) are maintained.

Shale diapirism (mud lumps, mud volcanoes). Conditions necessary for diapirism are a density inversion including a material of low shear strength, a situation which may be produced when a low-permeability formation is rapidly loaded and depocenters are rapidly shifted (Gretener, 1969). Such conditions are found in delta areas of major rivers, such as the Mississippi, Niger, Nile, Danube, Amazon, etc. High-porosity (high water content) shales of low shear strength are rapidly loaded by sands, causing formation of small shale diapirs called *mud lumps* (Morgan, 1952; Murray, 1961).

Thus diapirism on a small scale produces mud lumps, whereas on a large scale the results are mud volcanoes.

Mud volcanoes are overpressure phenomena caused by an intrusion from depth of mud or a mixture of mud and more solid rock (Fig.1.20). One problem is to distinguish material that has been extruded over the ground surface from material that has been intruded diapirically and is now exposed as a result of erosion of the older enveloping rock; both are what Suter (1960) has called "diapiric rocks" and both are included in the term "sedimentary volcanism" used by, for example, Kugler (1933, 1938) to emphasize the nonigneous nature of the phenomenon. Another problem is to distinguish between (1) accretionary cones of gently extruded mud accompanied by gas and water, and (2) gas seepages accompanied by more or less muddy water but lacking mud cones (Ridd, 1970).

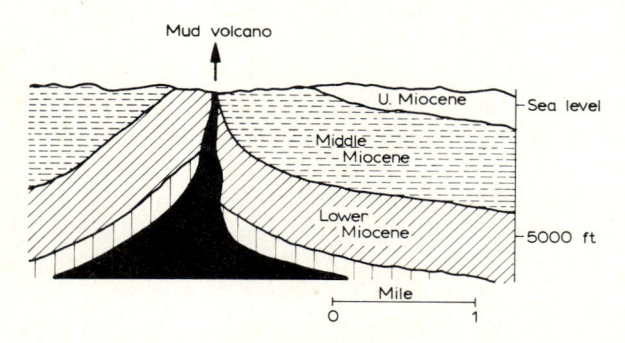

Fig.1.20. Schematic concept of a mud volcano (modified after Ridd, 1970).

Throughout the world, wherever mud volcanoes are found, there has been rapid Tertiary and/or Late Cretaceous sedimentation; and where evidence is available, pore fluid pressures are abnormally high. For example, pressure gradients of 0.9 psi/ft [0.208 kg cm^{-2} m^{-1}] have been measured around mud volcanoes on Apsheron Peninsula in Azerbaidzhan, U.S.S.R. Here one has to recall that a rock with a pore fluid pressure gradient equal to or exceeding 0.80 psi/ft [0.185 kg cm^{-2} m^{-1}] may become dilatant during structural deformation (Handin et al., 1963).

It is unusual in geology to find a phenomenon associated only with rocks of one particular age; however, geologic time is a factor in the escape of abnormally high pore fluid pressures (Ridd, 1970). Nevertheless, mud volcanoes may have erupted in the geologic past, in association with older sediments, their remains being "fossilized". For example, Shelton (1967) discussed diapirism in the Mississippian/Pennsylvanian Springer Group in Oklahoma, formed when the shale was in an overpressured and undercompacted state.

Quiescent mud volcano activity is also believed to be due to abnormal fluid pressure, except that instead of gradual pressure buildup until eruption, mud gently escapes to the surface through fissures. If this quiescent bleeding-off of pressure is insufficient, paroxysmal eruption can occur (Ridd, 1970).

Subsurface gas under pressure has been suggested by many investigators as the driving mechanism responsible for mud volcanoes (Kugler, 1933, 1938; Wilson and Birchwood, 1965; Gorkun and Siryk, 1967); however, exceptions are known (Richard, 1945). In a worldwide study Gansser (1960) listed several criteria which mud volcanoes have in common. These include:

(1) They are generally associated with Tertiary (and Upper Cretaceous) sedimentary strata.

(2) The sedimentary strata are usually of marine origin.

(3) Plastic, pelitic beds predominate.

(4) Gas and connate salt water are invariably present.

(5) The plastic beds are overlain by more competent deposits.

(6) Broad synclines are separated by sharp anticlines in which the deeper plastic sediments push upward.

(7) Increasing stress mobilizes the plastic clay in the core with salt water, gas, and, in many places, oil. The resulting mud is pressed upward in magma-like fashion and, if the equilibrium of the surface is disturbed, it erupts and forms a mud volcano.

(8) Most eruptive centers consist of several volcanic cones.

(9) Shallow and steep-sided cones can be present together.

(10) The eruptions are in part periodic, but commonly irregular. Many large eruptions have occurred after long periods of quiescence.

(11) With the mud, small and large rock fragments are commonly present, usually originating from older strata.

(12) The life of an individual eruptive center is usually short.

(13) Diapiric zones with mud volcanoes generally coincide with areas of negative gravity anomaly.

The magnitude of forces sometimes associated with erupting mud volcanoes is illustrated by a recent field observation in Azerbaidzhan, U.S.S.R. (Yakubov et al., 1973): "After 13 years of no activity, the largest mud volcano of Loktaban erupted in Azerbaidzhan for almost 6 hours. During the eruption, the volcano discharged 125,000 m^3 of breccia in shape of 2 huge tongues, each about 200 m long and 60 m wide. During the eruption, gas flames (1000—1200°C) reached 500 m into the sky."

In several areas, such as the Caucasus and Central Asia, the study of mud volcanoes has resulted in important information concerning the composition and amounts of hydrocarbons occurring, the conditions of their occurrence, and their migration (Kalinko, 1967). As a rule, no major gas deposits were present at the source of the volcanoes. Instead, there were small gas accumulations characterized by rock pressures that were much higher than normal and a constant gas inflow from the deeper parts of the basins. In such cases, potential gas reserves should also include the migrating gas. As suggested by Kalinko (1967), it is economical to exploit such gas pools using one or two wells.

Salt diapirism. Salt is known to behave plastically. During short-time tests at elevated temperatures, halite single crystals (Handin and Hager, 1958) flowed without further stress increase when the yield stress was reached. From field observations it is known that salt will behave plastically over a long period of time at lower temperatures and stresses than required during short periods in the laboratory (Muehlberger, 1959; Trusheim, 1960).

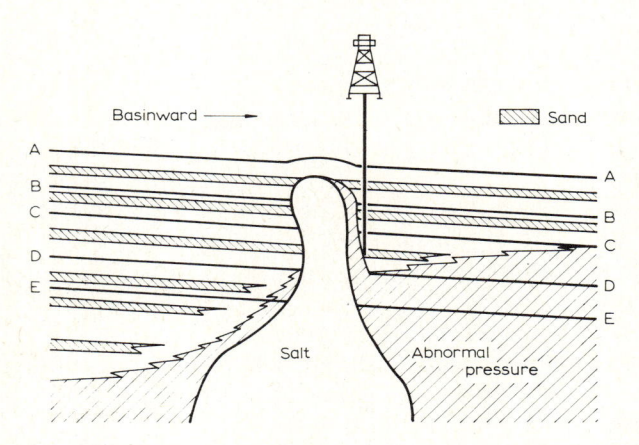

Fig.1.21. Piercement salt dome showing modification of abnormal pressure surface (after Harkins and Baugher, 1969). (Courtesy of the Society of Petroleum Engineers of A.I.M.E.)

Formation pressure anomalies commonly associated with salt domes are shown in Fig.1.21. Note the rapid rise of the pressure boundary reflecting earlier topography — and onlapping sands — associated with the original uplift in this figure (Harkins and Baugher, 1969).

Diapiric movement of salt is often accompanied or preceded by diapiric shale movements (Atwater and Forman, 1959). Salt movement resulted in contemporaneous faulting over intermediate- and high-relief salt anticlines (Hughes, 1968). The importance of overburden in the generation and continuation of such faulting with local structures in the U.S. Gulf Coast area has been emphasized by Hardin and Hardin (1961). According to Halbouty (1967), 409 such salt domes in the U.S. Gulf of Mexico Basin have been encountered while drilling for oil and gas.

Recently, Johnson and Bredeson (1971) studied the structural development of several salt domes in Louisiana's Miocene productive belt. Fig.1.22 shows the geologic section for the Weeks Island salt dome, which is typical for the six fields studied. The authors summarized their findings as follows: "Six shallow salt domes in the Miocene productive belt of coastal Louisiana were studied to attain an understanding of the structural history during the time of deposition of the sediments penetrated by drilling. The domes were

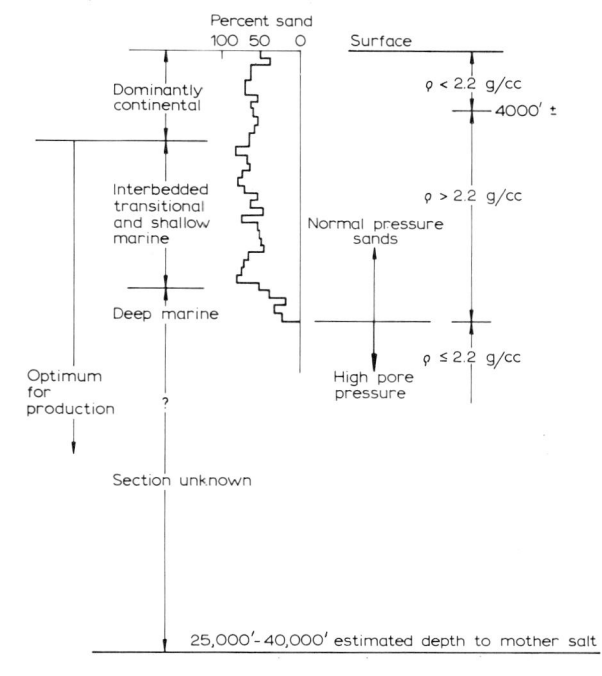

Fig.1.22. Schematic section of Weeks Island salt dome (after Johnson and Bredeson, 1971). (Courtesy of the American Association of Petroleum Geologists.)

Clovelly, Cote Blanche, Lake Washington, Napoleonville, Weeks Island, and White Castle.

Structural growth of the salt domes was contemporaneous with the deposition of sediments. The known stratigraphic section at the domes consists of about 15,000 ft [4572 m] of interbedded sandstone and shale underlain by a predominantly shale section with abnormally high pore pressures. The physical properties and differences in physical properties of the sandstone and shale sequence, predominantly shale section, and salt of these salt-dome structures are partly responsible for the geometry of the structures. Structural details and features which were studied in an effort to understand the history of the salt structures were the 'sheath' configuration, indications of uplift and differential subsidence, development of rim synclines, and faulting and fault patterns.

Most of the flanking sedimentary rocks were either deposited against the core of the structure or 'pierced' by the core shortly after deposition. *The core of the structures consists of salt or a combination of salt with shale from the high pore pressure section underlying the shallower sandstone and shale sequence.* The shale of the core moved and was emplaced in its present position as a 'sheath' by folding or by a combination of folding and faulting. The shape of the core exerted a control on the radius and rate of divergence of sedimentary intervals away from the core. If the side of the core dips at an angle of up to about 60°, the contours on the major faults in the flanking sedimentary rocks curve in tangentially to the core."

Salt domes and associated overpressures have been encountered in drilling operations on a worldwide basis, both on- and offshore.

Sandstone dikes. Sand dikes, similar to shale diapirism, are an overpressure phenomenon. Localized loading of sand in relatively mobile conditions (Laubscher, 1961) causes high pore fluid pressures which are equilibrated by formation of sand dikes. Oomkens (1966) discussed this phenomenon caused by the loading characteristics of migrating dunes that move over impermeable clayey surfaces, thereby locally increasing pore fluid pressures in underlying water sands. The results are small sand dikes intruding from below into the clayey formation.

By the same reasoning, one could predict that sedimentary beds under lava flows might show extreme internal deformation, since lava flow represents an almost classic case of rapid and progressive loading (Gretener, 1969). This is confirmed by field observations (Walton and O'Sullivan, 1950) of sandstone dikes intruding a sill. As the driving mechanism, the authors refer to abnormally high pore fluid pressures caused by the heating of formation water.

Earthquakes. Earthquakes may cause compression in subsurface formations such as aquifers and hydrocarbon reservoirs. Earthquakes have probably caused such sudden changes in pore fluid pressures throughout present and past geologic times.

An apparent effect is the sudden rise in the groundwater level, sometimes causing springs to start flowing again. Such forces generally operate in shallow aquifers but, as observed in petroleum reservoirs, are sometimes also found at greater depth. For example, the 1953 earthquake in California caused production in the nearby Mountain View oil field to double over a period of several weeks following the disturbance.

Earthquakes have been recorded from the vicinity of Rangely, Colorado, U.S.A., since the construction of the Uinta Basin Seismological Observatory in 1962. Waterflooding by injection into the Rangely oil field began in 1958. From subnormal pore pressures in 1958, the reservoir fluid pressure in some peripheral parts of the field has been raised to 0.6 psi/ft [0.139 kg cm^{-2} m^{-1}]. The earthquakes have been precisely located since September, 1969, by a network of 14 stations. The principal seismic activity occurs over a depth interval of 7—4 km along a vertical fault which cuts across the field. The earthquake epicenters are located on the south margin and die out toward the interior of the field, disappearing where the pore pressure is at or below the normal hydrostatic gradient. Focal-plane solutions indicate right-lateral strike-slip faulting. Numerical modeling of the reservoir pressure history indicates that pumping from four wells in the hypocentral region will draw the pressures down to a value approaching normal hydrostatic pressure in one year in an area larger than the error in determining the hypocentral location (Raleigh et al., 1970).

Osmotic phenomena

As early as 1748, Nollet described an experiment in which an animal bladder acted as a semipermeable membrane, in the presence of a pressure differential. The first detailed quantitative study of osmotic pressure is found in a series of investigations by Pfeffer in 1887.

Basically, an osmotic pressure can arise when two solutions of different concentration, or a pure solvent and a solution, are separated by a semipermeable membrane. Irrespective of the mechanism by which the semipermeable membrane operates, the final result is the same. Osmotic flow continues until the chemical potential of the diffusion component is the same on both sides of the barrier.

Of importance to abnormally pressured geologic environments is the consideration that if flow is directed into a closed volume, the pressure inside necessarily increases until equilibrium conditions are obtained. Any disturbance, physical (tectonic activities) and/or chemical, would cause a resumption of flow until equilibrium conditions are reestablished.

Glasstone (1946) described osmosis as the spontaneous flow of water into a solution, or as the flow from a more dilute to a more concentrated solution, when the two are separated from each other by a suitable mem-

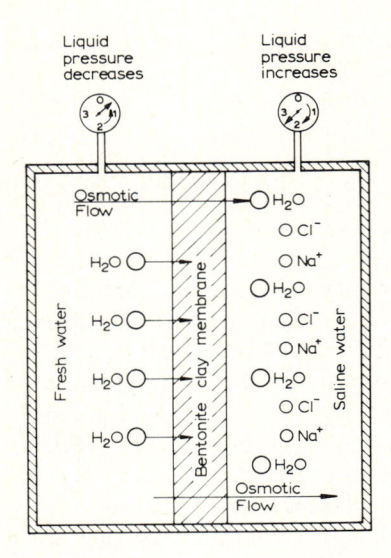

Fig.1.23. Diagram of osmotic flow — fresh water toward saline water (after Jones, 1969).

brane. A schematic representation of this process is shown in Fig.1.23 (Jones, 1969).

Both laboratory and field evidence suggest that shales serve as semipermeable membranes. The relative passage rate of various chemical species through such a membrane will be determined by (1) the availability of species in solution, (2) the relative dissociation of the species, (3) nonionic adsorption, (4) the relative adsorption of divalent versus monovalent cations at cation exchange sites, (5) the relative filtration of different associated species with a given chemical group, and (6) the relative influence of mass effects upon transport rates (Berry, 1969).

For a given solution at a given temperature, the osmotic pressure is almost indirectly proportional to the concentration differential; and for a given concentration differential the osmotic pressure increases with the absolute temperature (Glasstone, 1946). Theoretical osmotic pressure values to be expected across clay membranes (shales) are shown graphically in Fig.1.24 (Jones, 1969). Furthermore, Young and Low (1965) observed that fine-grained natural sediments, such as the clayey siltstone of the Lower Cretaceous Viking Formation of Central Alberta, Canada, act as effective, even if imperfect, semipermeable membranes. Jones (1969) suggests that osmotic flow is the dominant factor in the hydrology of regional aquifer systems of the northern U.S. Gulf Coast Basin, for both hydrostatic and abnormally high-pressured aquifers.

Similar conclusions are drawn by Breeze (1970) for the Morrow sands of northwestern Oklahoma.

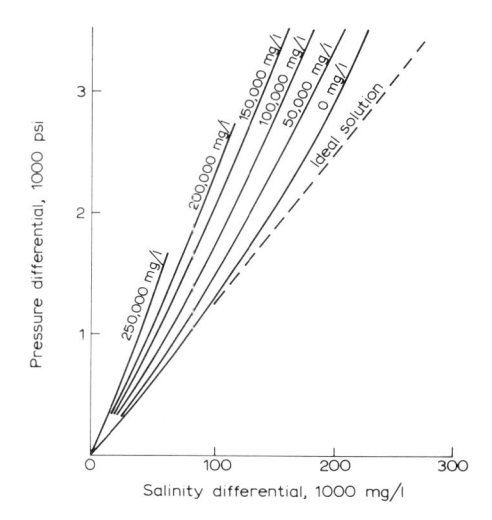

Fig.1.24. Theoretical values of osmotic pressure across a clay membrane (after Jones, 1969).

Jones (1968) states that "the pressure difference across a simple clay bed could, under natural conditions, exceed 3500 psi [246 kg/cm^2]. In known geopressured reservoirs, stepwise increments of osmotic pressure with depth through a series of bedded sands and clays could, as by a multistage pump, produce any of the reservoir pressures observed to date in the northern Gulf of Mexico Basin. Conceivably, osmotic-derived fluid pressure could equal or exceed that due to the weight of the overburden causing reservoir rupture and diapirism, especially where heating had reduced the load-bearing strength of clay beds."

Another practical application of some of these concepts (osmotic phenomena) has been proposed by Pirson (1967) in the use of well logs to locate osmotic entrapment of hydrocarbons.

Zen and Hanshaw (1964) proposed osmosis as an important cause of the high pore fluid pressure responsible for flotation of overthrust fault sheets and referenced overpressures up to 400 atmospheres above hydrostatic where shales separate formation waters of different salinities.

Several other regional studies of osmotic and salt-filtering phenomena of shales and compacted clays have been successfully related to formation pressure and chemical anomalies. Such pressure anomalies cover much of the San Juan Basin, New Mexico (Berry, 1959, 1960), occur in the Western Sedimentary Basin of Canada (Berry and Hanshaw, 1960; Hill et al., 1961; Van Everdingen, 1968), and are present in certain portions of the San Joaquin Valley, California (Bailey et al., 1961), at depth in the Illinois Basin (Bredehoeft et al., 1963), in the Permian Basin (McNeal, 1965), in the

Paradox Basin region of Utah, Colorado, New Mexico, and in many other regional aquifer systems in the world.

Diagenesis phenomena

Basically, diagenesis is the post-depositional alteration of sediment and its constituent minerals. Processes of diagenesis include (1) formation of new minerals, (2) redistribution and recrystallization of the substances in sediments, and (3) lithification.

Diagenesis of clayey sediments. In several oil-bearing regions, a marked temperature-dependent alteration of clay minerals with depth goes from montmorillonite (swelling lattice) through an intermediate mixed-layered phase to illite (nonswelling lattice). Published data for such clay-mineral changes have shown similar results in the U.S. Gulf Coast area (Burst, 1959; Powers, 1959; Weaver, 1960, 1961; Kerr and Barrington, 1961; Powers, 1967; Burst, 1969; Perry and Hower, 1970; Weaver and Beck, 1971), in Upper Cretaceous shales in the African Cameroun (Dunoyer de Segonzac, 1965), Jurassic and Cretaceous shales in the Papuan Basin, southeastern New Guinea (Van Moort, 1971), Jurassic shales of eastern Ciscaucasia (Teodorovich and Konyukhov, 1970), and elsewhere.

It appears that water release from deeply buried clays may be caused partly by diagenesis. Recently, the two-stage water escape model (subsurface dehydration) as proposed previously (Powers, 1959) has been modified by Burst (1969) to a three-stage system (Fig.1.25). According to Burst (1969):

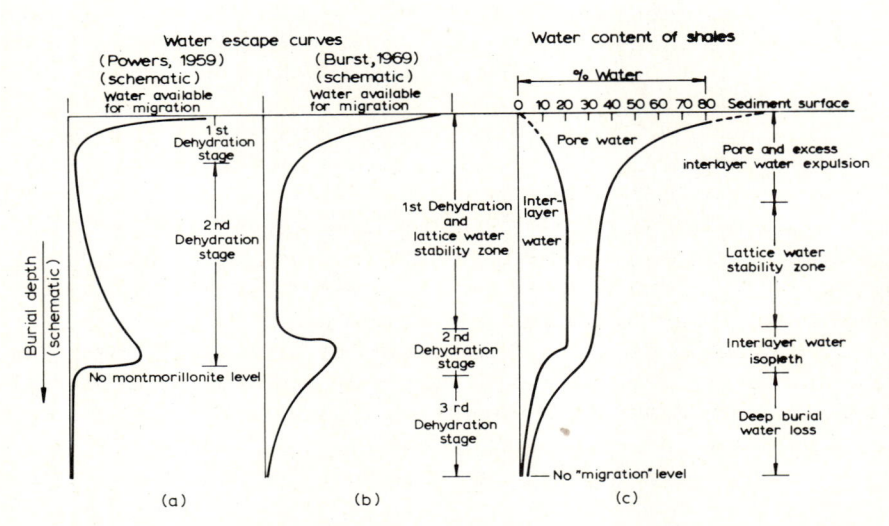

Fig.1.25. Subsurface water distribution (after Powers, 1959; Burst, 1969). (Courtesy of the American Association of Petroleum Geologists.)

"Water, the principal fluid component of the sedimentary section, is thought to migrate in three separate stages. Initially, pore water and excessive (more than two) clay-water interlayers are removed by the action of overburden pressure. This initial water flow (which is essentially completed after the first few thousand feet of burial) reduces the water content of the sediment to about 30% most of which is in the semisolid interlayer form. A second stage of dehydration is thought to occur when the heat absorbed by the buried sediment becomes sufficiently great to mobilize the next-to-last water interlayer in an $M(H_2O)_x - \Delta H_r = I + xH_2O$ fashion:

$$M(H_2O)_x = I + xH_2O + \Delta H_r \tag{1.6}$$

where M = hydrated clay lattice, $(H_2O)_x$ = water held within the hydrated lattice, I = dehydrated lattice, xH_2O = capillary (pore) water, and ΔH_r = heat of reaction (H is negative, consequently the reaction is endothermic).

The first and last dehydration stages are probably unimportant in Gulf Coast oil migration, inasmuch as they occur, respectively, at levels too shallow and too deep to intersect the interval of maximum liquid petroleum availability. The amount of water in movement during the second stage, at a level which does intersect this interval, is 10—15% of the compacted bulk volume and represents a significant fluid displacement capable of redistributing any mobile subsurface component. A measure of the degree to which the second-stage interlayer water has been discharged into the system can be noted on X-ray diffractograms. The movement appears to occur in a relatively-restricted, depth-dependent temperature zone in which the average dehydration temperature of the points measured is 221°F [105°C]."

Water movement during the second stage is capable of redistributing any mobile subsurface components, such as water and hydrocarbons. This large-scale water release will produce fluids at a constant rate under geologic conditions where the reaction can occur. As such, it should influence any overpressure environment, either due to its mere presence in such a closed system or changes imposed on osmotic reactions.

Diagenesis of sulfates*. Diagenesis in sulfate sections may contribute to overpressures due to creation of permeability barriers, a fluid source, and/or a rock-volume change.

For example, Hanshaw and Bredehoeft (1968) suggested the phase change of gypsum ($CaSO_4 \cdot 2H_2O$) to anhydrite ($CaSO_4$) plus water (H_2O) as an example of a fluid source at depth. If a hydraulic conductivity barrier is present, pore fluid pressures will build up. However, if such horizons of low-permeability material are absent or infrequent, it is doubtful if anomalous pore fluid pressures can be maintained for more than a geologic instant.

---

*Sulfates are commonly associated with carbonate reservoirs.

A similar concept was considered by Laubscher (1961b) in the analysis of the origin of the Jura Mountains. The author proposed the possibility of a phase change from gypsum to anhydrite plus water to produce the necessary high pore fluid pressures.

Louden (1972) proposed the rehydration of anhydrite and water to gypsum as a possible source for overpressures. In the semihydrate state ($CaSO_4 \cdot \frac{1}{2} H_2 O$), the rock volume would increase by 15—25% while in the completely hydrated state (i.e. conversion into gypsum), the volume increase would be about 40%. If such rehydration were to take place at depth in a closed system, tremendous pressures could originate. Louden cites the Mississippi region as a typical field example of such rehydration; here the onset of overpressures encountered in the Buckner Formation is flagged by a gradation of anhydrite into gypsum.

Diagenesis of volcanic ash. An unusual abnormal formation pressure phenomenon occurs along the northwest coast of the United States, in areas of the South China Sea region (Java, etc.), and elsewhere. Original sediments may be fine-grained redbed or sandstone, plus large amounts of volcanic ash. Over geologic time, diagenesis of volcanic ash results in three basic main products — clay minerals, methane, and carbon dioxide. As a result, drilling mud is heavily gas cut when penetrating such formations. However, the amount of gas cut is usually not directly related to the magnitude of overpressures (Fertl and Timko, 1972).

Secondary precipitation of cementation materials. Existence of any mineral is a function of the stability field indigenous to the mineral species. This stability field is a function of a combination of variables, such as temperature, pressure, concentration of solution, and composition of solution. This stability field also determines whether a specific mineral will remain in equilibrium with the environment or whether it will grow or dissolve (Garrels and Christ, 1965).

Based on extensive water-analyses data generally available for the U.S. Gulf Coast area, a multitude of dissolved chemical components found to be present may include sodium chloride, sulfate, and bicarbonate; calcium chloride, sulfate and bicarbonate; magnesium chloride and sulfate; trace elements; etc.; and mixtures of all these components. Data by Voroshilov et al. (1968) also suggest the great importance of the influx and mixing of the seawater with formation brines in several Russian oil fields (Surakhan and Karachukhar oil reservoirs), which causes precipitation of salts, chiefly magnesium and calcium carbonates. Previously, Back and Hanshaw (1965) suggested that precipitation of calcite and other minerals may result from the mixing of two waters, neither of which was capable of forming a precipitate before mixing.

In a related study Marenko and Postnikov (1967) attempted to explain the anomalously high reservoir pressure in the deposits of the Markovo oil field, U.S.S.R., by the peculiarities of the growth of rock-salt crystals in the empty spaces of the lithified rock skeleton. The salt crystals grew at the base rather than at the upper end, gradually filling the voids in the lower part of the deposits, displacing oil. Due to the negligible compressibility of oil, the rock pressure in the closed deposits gradually increased as the volume of voids decreased, reaching abnormal values. The mechanism of this process is verified by simple calculations of the possible reduction of pore volume under the action of penetrating salt. The Markovo oil deposits in the Osinskii zone are tightly sealed by rock salt, each deposit representing an independent, closed hydrodynamic system. It is believed that high rock pressures in the salt-bearing carbonate and sulfate-carbonate deposits in other oil fields can be explained by the same processes.

Similar increase in pore fluid pressure may take place due to calcite or silica precipitation.

Important in diagenesis is the occurrence of quartz grains with secondary rims (quartz overgrowth) which has been related to the maximum depth of burial prior to trapping of hydrocarbons in the reservoir rock (Füchtbauer, 1961). This supports conclusions drawn previously by Lowry (1956) that in hydrocarbon-bearing rocks the connate water is not sufficient to continue diagenesis. Füchtbauer also showed the retarding effect of the clay and carbonate content and various types of interstitial solutions on quartz diagenesis. Since at the same time diagenesis continues in water-bearing zones, such a "diagenetic gap" can be the basis to draw conclusions about oil migration and the structural history of petroleum prospects. Several field cases have been discussed in detail by Füchtbauer (1961) and Philipp (1963).

The source of this secondary silica is somewhat speculative, but it probably comes from pressure solution of silica from sand grains at points of contact and/or postdepositional transformations of expandable clays to illite.

Finally, one should recall the thermodynamic model for grain-boundary diffusion induced by a pressure gradient in deep-seated rock as presented by Gresens (1966). "Two assumptions are made: (1) the dispersed phase may be treated as a separate phase for purposes of phase equilibria, and (2) the dispersed phase can support a pressure gradient. The model is developed in terms of 'chemical potential'. When the compositions of the solid phases are held constant, diffusion takes place in the dispersed phase in response to gradients of chemical potential within the dispersed phase. A gradient of concentration is stabilized parallel to the pressure gradient as the system adjusts to a lower energy state. Disequilibrium between the solid and dispersed phases can be increased, even though the free energy of the system is lowered by migration in the dispersed phase. When the solid phases are allowed to change composition (for example, by stress-induced recrystallization), there may be damping effect on diffusion, but there also may be an enhancement of diffusion. *Eventually a compositional gradient will parallel the pressure gradient in both the solid and dispersed phases.* Consideration of the effect of the pressure gradient on the chemical potential in a closed system lends insight into the consequences of extending the discussion to an open system. Components in the dispersed phase that develop steep gradients of chemical potential and have a relatively large tendency to lower the free energy of the system by migration should have a tendency to migrate first. However, the activation energy required for migration and the rate of dissolution of the components from the solid phases into the dispersed phase must also be considered. The theoretical model leads to two possibilities, the effects of which may be observable to the geologist: (1) the concentration and nucleation of certain components in preferred structural sites, and (2) systematic compositional variations of a given mineral that are related to the structure."

Massive areal rock-salt deposition

Massive rock-salt deposits are found over large areas. Salt is totally impermeable (and not semipermeable as shales) to fluids and, unlike other rocks, transforms under pseudoplastic movement (recrystallization effect), thereby exerting pressure equal to the overburden load in all directions. Underlying formations have no fluid escape possibilities, thus remaining unconsolidated and becoming overpressured. In many areas, abnormal pressures are frequently found in shales or silty zones directly below the salt. For example, gas-bearing Permian reservoir rocks below large masses of rock salt are found in North Germany. Associated overpressured shales, despite their Permian age, have retained their fresh, highly porous, and plastic conditions, causing severe shale heaving when penetrated by the drill bit (Thomeer and Bottema, 1961). Exploratory drilling for hydrocarbon resources has encountered similar conditions in the United States, the U.S.S.R., Middle East, North Africa, and elsewhere.

Permafrost environment

Observations in the Arctic indicate that in permafrost regions abnormally high formation pressures do occur as local phenomena. In permafrost regions, unfrozen areas (talik) exist in many places, such as under deep lakes. Drastic changes in climate and/or surface conditions cause permafrost encroachment, thereby trapping an unfrozen area in an essentially closed system. As freezing proceeds, a buildup of abnormally high formation pressures occurs in unfrozen pockets. Due to the proximity to the surface, such pressures may be released violently (Muller, 1947), thereby forming giant frost heaves at ground level. These structures are called pingos and are often found in northern latitudes with heights sometimes exceeding 100 ft [30.48 m] (Mueller, 1959; Mackay, 1963).

A schematic presentation of pingo formation is shown in Fig.1.26

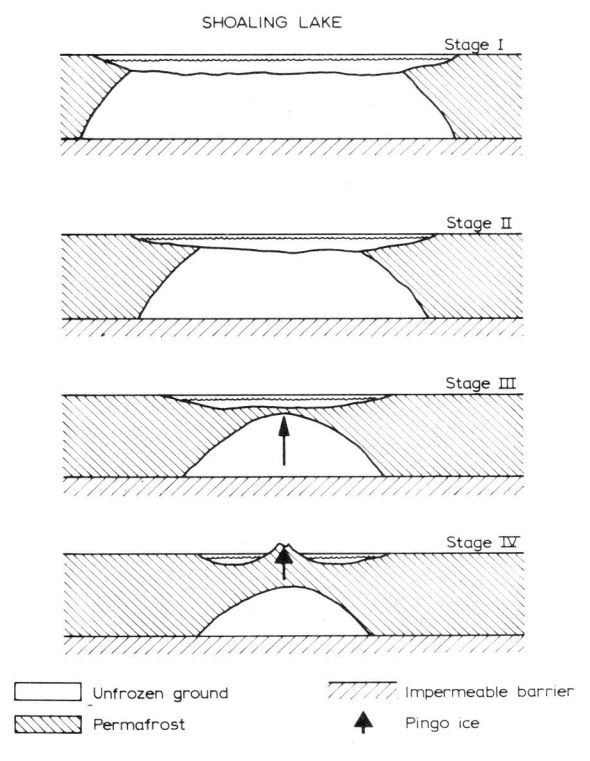

Fig.1.26. Abnormal pressure conditions in permafrost areas — formation of pingos (after Gretener, 1969). (Courtesy of *Bulletin of Canadian Petroleum Geology*.)

(Gretener, 1969). In the first stage the lake is deep enough not to freeze to the bottom in winter; i.e. an unfrozen area exists under the lake. As the lake shoals, it will eventually freeze to the bottom and the permafrost will encroach onto the unfrozen pocket. As freezing proceeds from the surface downward, water is trapped in the unfrozen segment and the water pressure rises (Stage III). The thin frozen roof is bent upwards, and it develops tension cracks which become filled with water; upon the release of the pressure, the water immediately freezes, forming an ice cone and pushing up the uppermost sediments and the vegetative mat, as seen in Stage IV. Mechanical properties of rocks at low temperatures and frozen soil have recently been studied in detail (Chamberlain, 1973; McGinnis et al., 1973; Mellor, 1973; Nakano and Froula, 1973).

Furthermore, arctic oil-well drilling and production may cause extensive thawing of the permafrost, which, if allowed to refreeze, may result in collapse loading which has to be considered in casing design. Such casing damage may result from two mechanisms (Perkins et al., 1973): first, if fluids are confined within pipes and allowed to freeze, the resulting pressure increase may cause casing damage (Bleakley, 1971); second, the mechanism involves refreezing of thawed permafrost or fluids external to the casing which may result in pressures sufficiently high to damage the casing (Goodman and Wood, 1973).

Thermodynamic and biochemical causes

These processes may be considered an integral part of other causes discussed previously, since their effects are present in any geologic system. If the system is essentially closed, these factors are an incremental addition to the buildup of abnormally high formation pressures.

Changes in formation temperature will also change the fluid pressure, with changes being either of local or regional significance. The pressure-temperature-density diagram for water (Fig.1.8) suggests that pressure in an isolated volume increases with increasing temperature more rapidly than in the surrounding fluids. For higher gradients, smaller increases in depth are required to cause the same increase in pressure (Barker, 1972). This mechanism for producing excess pressures may operate in addition to most of the other processes which have been suggested but the exact influence in any given area will depend on how well the system remains isolated.

Breakdown of hydrocarbon molecules into simpler compounds increases their volume. Research indicates such volume changes to be due to catalytic reactions, radioactive decay, bacterial reaction, and/or temperature changes. Thermal cracking of hydrocarbons will lead to an increased volume of hydrocarbons in the order of two to three times the original volume.

On the other hand, decomposition of organic matter through bacterial actions forms pockets of methane under excessive pressure. Such conditions may be observed in drilling shallow formations as well as in the well-known gas seepages in swamps or along beaches.

Other effects

Tidal disturbances of shelf regions in oceans, tsunamis (seismic waves) caused by submarine earthquakes, effects of hurricanes, etc., may also cause minor and probably innumerable temporary changes in formation pressures of underlying sediments.

Meinzer (1928) reported rapid loading and resulting overpressures in a water well near a railroad. The water level in the well rose as a train passed by. For slow, heavy freight trains, the rise was more pronounced than for the faster and lighter passenger trains, while a single locomotive did not produce any noticeable effect.

As early as 1753, travellers noted subsurface effects of varying atmospheric pressure in the oil-seepage region of the Caspian Sea in Russia: "when the weather is thick and heavy (low pressure) the springs boil up higher and the naphtha often takes fire at the surface . . . in clearer weather (high atmospheric pressure) the springs do not boil up over 2—3 ft [0.61—0.91 m]" (Hanway, 1753).

Although these effects on pore fluid pressures are often minor, they are nevertheless present and cause abnormally high formation pressures, either short or long lived.

Acknowledgements

Concepts presented in this discussion were selected from a vast amount of literature on the subject, orally discussed ideas, and our own field experience. Unfortunately, it is virtually impossible to reference all those whose contributions have been substantial.

References

Allen, D.R., 1972. Environmental aspects of oil-producing operations — Long Beach, California. *J. Pet. Technol.*, 24: 125—131.

Atwater, G.I. and Forman, M.J., 1959. Nature of growth of southern Louisiana salt domes and its effect on petroleum accumulation. *Bull. Am. Assoc. Pet. Geol.*, 43: 2592—2622.

Back, W. and Hanshaw, B.B., 1965. Chemical geohydrology. In: V.T. Chow (Editor), *Advances in Hydroscience*, Vol. 2. Academic Press, New York, N.Y., pp.49—109.

Bailey, E.H., Snavely, P.D. and White, D.E., 1961. Chemical analysis of brines and crude oil, Cymric Field, Kern County, California. *U.S. Geol. Surv. Prof. Papers*, 398D: 306—309.

Barker, C., 1972. Aquathermal pressuring — role of temperature in development of abnormal pressure zones. *Bull. Am. Assoc. Pet. Geol.*, 56: 2068—2071.

Berry, F.A.F., 1959. *Hydrodynamics and geochemistry of the Jurassic and Cretaceous System in the San Juan Basin, northwestern New Mexico and southwestern Colorado*. Ph.D. Thesis, Stanford Univ., Stanford, Calif., 213 pp.

Berry, F.A.F., 1960. Geologic field evidence suggesting membrane properties of shales. *Bull. Am. Assoc. Pet. Geol.*, 44: 953—954.

Berry, F.A.F., 1969. Relative factors influencing membrane filtration effects in geologic environments. *Chem. Geol.*, 4: 295—301.

Berry, F.A.F., 1973. High fluid potential in California Coast Ranges and their tectonic significance. *Bull. Am. Assoc. Pet. Geol.*, 57: 1219—1249.

Berry, F.A.F. and Hanshaw, B.B., 1960. Geologic evidence suggesting membrane properties of shales. *21st Int. Geol. Congr.*, Copenhagen, Rep. Sess., Norden, p.209.

Bishop, W.F., 1973. Late Jurassic contemporaneous faults in North Louisiana and South Arkansas. *Bull. Am. Assoc. Pet. Geol.*, 57: 858—877.

Bleakley, W.B., 1971. North Slope operators tackle production problems. *Oil Gas J.*, 69: 89—92, October 25.

Bredehoeft, J.B. and Hanshaw, B.B., 1968. On the maintenance of anomalous fluid pressures, I. Thick sedimentary sequences. *Geol. Soc. Am. Bull.*, 79: 1097—1106.

Bredehoeft, J.D., Blyth, C.R., White, W.A. and Maxey, G.B., 1963. Possible mechanism for concentration of brines in subsurface formations. *Bull. Am. Assoc. Pet. Geol.*, 47: 257—269.

Breeze, A.F., 1970. *Abnormal—subnormal relationships in the Morrow sands of northwestern Oklahoma*. M.S. Thesis, Univ. of Oklahoma, Tulsa, Okla., 122 pp.

Brighenti, G., 1967. Influence of pore pressure decline on the behavior of petroleum reservoir rocks. *Proc. 7th World Pet. Congr.*, Mexico City, 3: 97—107.

Bruce, C.H., 1973. Pressured shale and related sediment deformation — mechanism for development of regional contemporaneous faults. *Bull. Am. Assoc. Pet. Geol.*, 57: 878—886.

Burst, J.F., 1959. Post-diagenetic clay mineral environmental relationships in the Gulf Coast Eocene. *Clays Clay Miner.*, 6: 327—341.

Burst, J.F., 1969. Diagenesis of Gulf Coast clayey sediments and its possible relation to petroleum migration. *Bull. Am. Assoc. Pet. Geol.*, 53: 73—93.

Caloi, P., 1967. The phenomenon of abnormal subsidence with particular reference to the Po delta. *Ann. Geofis.*, 20: 331—400.

Carver, R.E., 1968. Differential compaction as a cause of regional contemporaneous faults. *Bull. Am. Assoc. Pet. Geol.*, 52: 414—419.

Chamberlain, E., 1973. Mechanical properties of frozen ground under high pressure. *Proc. 2nd Int. Can. Organ. Comm. U.S. Planning Comm. Permafrost Conf.*, Yakutsk, U.S.S.R. U.S. Academy of National Sciences, Washington, D.C., pp.295—305.

Classen, J.S., 1968. Formation pressure—production relationship, Lake Mongoulois Field. *SPE 2206, 43rd AIME Fall Meet.*, Houston, Texas, September.

Dickey, P.A., Shriram, C.R. and Paine, W.R., 1968. Abnormal pressures in deep wells of southwestern Louisiana. *Science*, 160: 608—615.

Dickinson, G., 1953. Geological aspects of abnormal reservoir pressures in Gulf Coast Louisiana. *Bull. Am. Assoc. Pet. Geol.*, 37: 410—432.

Dunoyer de Segonzac, G., 1965. Les argiles du Cretace supérieur dans le bassin de Douala (Cameroun): problèmes de diagenèse. *Bull. Serv. Carte Geol. Alsace Lorraine*, 17: 287—310.

Evans, D.M., 1966. The Denver area earthquakes and the Rocky Mountain Arsenal well. *Mt. Geol.*, 3: 23—36.

Ferran, L.H., 1973. *Evaluation of abnormally high and low pressured Morrow sands in northwestern Oklahoma using well logs and water sample data*. M.S. Thesis, Univ. of Tulsa, Tulsa, Okla., 110 pp.

Fertl, W.H., 1971. A look at abnormally pressured formations in the U.S.S.R. *SPE 3613, 46th AIME Fall Meet.*, New Orleans, La., October.

Fertl, W.H. and Timko, D.J., 1972. How downhole temperatures, pressures affect drilling, 1. Origin of abnormal formation pressures. *World Oil*, 174(7): 67—71, June.

Flores, J.G., 1967. *Study of subnormal formation pressures based on geological and electrical log data, Keyes Field, Cimarron County, Oklahoma*. M.S. Thesis, Univ. of Oklahoma, Tulsa, Okla., 89 pp.

Foster, J.B. and Whalen, H.E., 1966. Estimation of formation pressures from electrical surveys — offshore Louisiana. *J. Pet. Technol.*, 18: 165—171.

Fowler, Jr., W.A., 1970. Pressure, hydrocarbon accumulation and salinities — Chocolate Bayou field, Brazoria County, Texas. *J. Pet. Technol.*, 22: 411—432.

Füchtbauer, H., 1961. Zur Quarzneubildung in Erdoellagerstaetten. *Erdoel Kohle*, 14 169—173.

Gabrysch, R.K., 1967. Development of ground water in the Houston District, Texas, 1961—1965. *Texas Water Dev. Board, Rep.*, 63, 79 pp.

Gansser, A., 1960. Ueber Schlammvulkane und Salzdome. *Naturforsch. Ges. Zuerich Vierteljahrssch.*, 105: 1—46.

Garrels, R.M. and Christ, C.K., 1965. *Solutions, Minerals, and Equilibria*. Harper and Row, New York, N.Y., 450 pp.

Gilluly, J. and Grant, U.S., 1949. Subsidence in the Long Beach harbor area, California. *Bull. Am. Assoc. Pet. Geol.*, 60: 461—529.

Glasstone, S., 1946. *Textbook of Physical Chemistry*. Van Nostrand, Princeton, N.J., 557 pp.

REFERENCES

Goodman, M.A. and Wood, D.B., 1973. A mechanical model for permafrost freezeback pressure behavior. *SPE 4589, 48th AIME Fall Meet.*, Las Vegas, Nev., October.

Gorkun, V.N. and Siryk, I.M., 1967. Calculation of depth and volume of released gas during the eruption of mud volcanoes in southern Sakhalin. *Geol. Geofiz.*, 2: 30—42.

Gresens, R.L., 1966. The effect of structurally produced pressure gradients on diffusion in rocks. *J. Geol.*, 74: 307—321.

Gretener, P.E., 1969. Fluid pressure in porous media — its importance in geology: a review. *Bull. Can. Pet. Geol.*, 17: 255—295.

Halbouty, M.T., 1967. *Salt Domes — Gulf Region, United States and Mexico.* Gulf Publishing Co., Houston, Texas, 425 pp.

Handin, J. and Hager, R.V., 1958. Experimental deformation of sedimentary rocks under confining pressure: tests at high temperature. *Bull. Am. Assoc. Pet. Geol.*, 42: 2892—2934.

Handin, J., Hager, R.V., Friedman, M. and Feather, J.N., 1963. Experimental deformation of sedimentary rocks under confining pressure — pore pressure tests. *Bull. Am. Assoc. Pet. Geol.*, 47: 717—755.

Hanshaw, B.B. and Bredehoeft, J.D., 1968. On the persistence of anomalous fluid pressures from a source layer. *49th Annu. AGU Meet.*, Washington, D.C., Abstr. No. T19.

Hanway, J., 1753. *An Historical Account of the British Trade Over the Caspian Sea*, Vol. 7. London, 121 pp.

Hardin, F.R. and Hardin, G.C., 1961. Contemporaneous normal faults of Gulf Coast and their relation to flexures. *Bull. Am. Assoc. Pet. Geol.*, 45: 239—248.

Harkins, K.L. and Baugher, J.W., 1969. Geological significance of abnormal formation pressures. *J. Pet. Technol.*, 21: 961—966.

Hill, G.A., Colburn, W.A. and Knight, J.W., 1961. Reducing oil-finding costs by use of hydrodynamic evaluation. In: *Economics of Petroleum Exploration, Development, and Property Evaluation.* Prentice-Hall, Englewood, Calif., 380 pp.

Hospers, J., 1971. The geology of the Niger delta area. *Inst. Geol. Sci., London, Rep.*, 70: 121—142.

Hubbert, M.K. and Rubey, W.W., 1959. Role of fluid pressure in mechanics of overthrust faulting, 1. Mechanics of fluid filled porous solids and its application to overthrust faulting. *Bull. Geol. Soc. Am.*, 70: 115—166.

Hughes, D.J., 1968. Salt tectonics as related to several Smackover fields along the northeast rim of the Gulf of Mexico Basin. *Trans. Gulf Coast Assoc. Geol. Soc.*, 18: 320—330.

Johnson, H.A. and Bredeson, D.H., 1971. Structural development of some shallow salt domes in Louisiana Miocene productive belt. *Bull. Am. Assoc. Pet. Geol.*, 55: 204—226.

Jones, P.H., 1968a. Hydrodynamics of geopressures in the northern Gulf of Mexico Basin. *SPE 2207, 43rd AIME Fall Meet.*, Houston, Texas, September; also *J. Pet. Technol.*, 21: 802—810, 1969.

Jones, P.H., 1968b. Geochemical hydrodynamics, a possible key to the hydrology of certain aquifer systems in the northern part of the Gulf of Mexico Basin. *Proc. 23rd Int. Geol. Congr.*, Prague, 2: 113—125.

Jones, P.H., 1969. Hydrology of Neogene deposits in the northern Gulf of Mexico Basin. *La. Water Resour. Res. Inst., Bull., GT-2*, Louisiana State Univ., Baton Rouge, La., 105 pp.

Kalinko, M.K., 1967. Mud volcanoes as a source of information on the composition of hydrocarbons, their quantity, and conditions of occurrence. *Sov. Geol.*, 7: 86—96.

Kerr, P.F. and Barrington, J., 1961. Clays of deep shale zone, Caillou Island, Louisiana. *Bull. Am. Assoc. Pet. Geol.*, 45: 1697—1712.

Krynine, D.P. and Judd, W.R., 1957. *Principles of Engineering Geology and Geotechnics.* McGraw-Hill, New York, N.Y., 730 pp.

Kugler, H.G., 1933. Contribution to the knowledge of sedimentary volcanism in Trinidad. *Inst. Pet. J.*, 19: 743—760.

Kugler, H.G., 1938. Nature and significance of sedimentary volcanism. In: *The Science of Petroleum*, Vol. 1. Oxford Univ. Press, London, pp.297—299.

Laubscher, H.P., 1961a. Die Mobilisierung Klastischer Massen, 1. Teil und 2. Teil. *Eclogae Geol. Helv.*, 54: 283—334.

Laubscher, H.P., 1961b. Die Fernschubhypothese der Jurafaltung. *Eclogae Geol. Helv.*, 54 : 221—282.

Leve, G.W. and Goolsby, D.A., 1967. Test hole in aquifer with many water-bearing zones at Jacksonville, Florida. *Ground Water*, 5: 18—22.

Levorsen, A.I., 1967. *Geology of Petroleum*. W.H. Freeman and Co., San Francisco, Calif., 2nd ed., 724 pp.

Louden, L.R., 1972. Origin and maintenance of abnormal pressures. *SPE 3843, 3rd Symp. on Abnormal Subsurface Pore Pressure*, Louisiana State Univ., Baton Rouge, La., May.

Lowry, W.D., 1956. Factors in loss of porosity by quartzose sandstones of Virginia. *Bull. Am. Assoc. Pet. Geol.*, 40: 489.

MacKay, J.R., 1963. The Mackenzie delta area, N.W.T. *Can. Dep. Min. Tech. Sur. Geogr. Br., Mem.*, 8, 202 pp.

Magara, K., 1971. Permeability considerations in generation of abnormal pressures. *Soc. Pet. Eng. J.*, 11:236—242.

Marenko, Y.I. and Postnikov, V.G., 1967. Causes of abnormally high rock pressure in the Osinskii zone of Markovo oil field. *Neftegazov. Geol. Geofiz.*, 10: 10—12.

Marsden, S.S. and Davis, S.N., 1967. Geological subsidence. *Sci. Am.*, 216: 93—100.

Martin, G.B., 1972. Abnormal high pressure and environment of deposition. *SPE 3846, 3rd Symp. on Abnormal Subsurface Pore Pressure*, Louisiana State Univ., Baton Rouge, La., May.

Mayuga, M.N. and Allen, D.R., 1969. Subsidence in the Wilmington oil field, Long Beach, California, U.S.A. In: *Land Subsidence*, Publ. No. 88. IASH-UNESCO, Gentbrugge.

McGinnis, L.D., Clark, C.C. and Nakao, K., 1973. Geophysical identification of frozen and unfrozen ground, Antarctica. *Proc. 2nd Int. Can. Organ. Comm. U.S. Planning Comm. Permafrost Conf.*, Yakutsk, U.S.S.R. U.S. Academy of National Sciences, Washington, D.C., pp.136—146.

McNeal, R.P., 1965. Hydrodynamics of the Permian Basin. In: A. Young and J.E. Galley (Editors), *Fluids in Subsurface Environments*. American Association of Petroleum Geologists, Tulsa, Okla., pp.308—326.

Meinzer, O.E., 1928. Compressibility and elasticity of artesian aquifers. *Econ. Geol.*, 23: 263—291.

Mellor, M., 1973. Mechanical properties of rocks at low temperatures. *Proc. 2nd Int. Can. Organ. Comm. U.S. Planning Comm. Permafrost Conf.*, Yakutsk, U.S.S.R. U.S. Academy of National Sciences, Washington, D.C., pp.334—344.

Meyers, J.D., 1968. Differential pressures: a trapping mechanism in Gulf Coast oil and gas fields. *Trans. Gulf Coast Assoc. Geol. Soc.*, 18: 56—80.

Moore, C.A., 1971. Geologic factors which may affect gas occurrence in the Anadarko Basin, Oklahoma. *Annu. AAPG-SEPM Meet.*, Houston, Texas.

Morgan, J.P., 1952. Mud lumps at the mouth of the Mississippi River. In: J.W. Johnson et al. (Editors), *Coastal Engineering*. Eng. Foundation, Univ. of California Press, Los Angeles, Calif., pp.130—144.

Muehlberger, W.R., 1959. Internal structure of the Grand Saline salt dome, Van Zandt County, Texas. *Texas Univ. Bur. Econ. Geol. Rep. Invest.*, 38, 18 pp.

Mueller, F., 1959. Beobachtungen ueber Pingos. *Medd. Groenl.*, 153: 127.

Muller, S.W., 1947. *Permafrost or Permanently Frozen Ground and Related Engineering Problems*. J.W. Edwards, Ann Arbor, Mich., 231 pp.

REFERENCES

Murray, G.E., 1961. *Geology of the Atlantic and Gulf Coastal Province of North America.* Harper and Brothers, New York, N.Y., 692 pp.
Nakano, Y. and Froula, N.H., 1973. Sound and shock transmission in frozen soils. *Proc. 2nd Int. Can. Organ. Comm. U.S. Planning Comm. Permafrost Conf.*, Yakutsk, U.S.S.R., pp.359—369.
Ocamb, R.D., 1961. Growth faults of South Louisiana. *Trans. Gulf Coast Assoc. Geol. Soc.*, 11: 139—175.
Oomkens, E., 1966. Environmental significance of sand dikes. *Sedimentology*, 7: 145—148.
Perkins, T.K., Rochon, J.A. and Knowles, C.R., 1973. Studies of pressures generated upon refreezing of thawed permafrost around a wellbore. *SPE 4588, 48th AIME Fall Meet.*, Las Vegas, Nev., October.
Perry, E. and Hower, J., 1970. Burial diagenesis in Gulf Coast pelitic sediments. *Clays Clay Miner.*, 18: 165—177.
Philipp, W., 1963. Zur Geschichte der Migration im Gifhorner Trog. *Erdoel Kohle*, 16: 456—468.
Pirson, S.J., 1967. How to use well logs to seek hydrodynamically trapped oil. *World Oil*, 164: 100—106, April.
Powers, M.C., 1959. Adjustment of clays to chemical change and the concept of the equivalent level. *Clays Clay Miner.*, 6: 309—326.
Powers, M.C., 1967. Fluid-release mechanisms in compacting marine mudrocks and their importance in oil exploration. *Bull. Am. Assoc. Pet. Geol.*, 51: 1240—1254.
Raleigh, C.B., Bredehoeft, J., Healy, J.H. and Bohn, J., 1970. Earthquakes and water flooding in the Rangely oil field. *Annu. Jt. Geol. Soc. Am. Allied Soc. Meet.*, 2: 600—661.
Richard, J.J., 1945. The mud volcanoes of Moa near Tanga. *Tanganyika Notes Rec.*, 19: 3—8.
Ridd, M.F., 1970. Mud volcanoes in New Zealand. *Bull. Am. Assoc. Pet. Geol.*, 54: 601—616.
Roberts, J.L., 1972. The mechanism of overthrust faulting: a critical review. *Proc. 29th Int. Geol. Congr.*, Montreal, Que., Sect. 3, pp.593—598.
Rochon, R.W., 1967. Relationship of mineral composition of shales to density. *Trans. Gulf Coast Assoc. Geol. Soc.*, 17: 135—142.
Rubey, W.W. and Hubbert, M.K., 1959. Role of fluid pressure in mechanics of overthrust faulting, 2. Overthrust belt in geosynclinal area of western Wyoming in light of fluid pressure hypothesis. *Bull. Geol. Soc. Am.*, 70: 167—206.
Russel, W.L., 1972. Pressure—depth relations in Appalachian region. *Bull. Am. Assoc. Pet. Geol.*, 56: 528—536.
Shelton, J.W., 1967. Deformational pattern in the Springer Group of southeastern Oklahoma. *Geol. Soc. Am., South Cent. Sec. Annu. Meet. Program*, Norman, Okla. p.25.
Slatyer, R.O. and Mabutt, J.A., 1964. In: V.T. Chow (Editor), *Handbook of Applied Hydrology.* McGraw-Hill, New York, N.Y., pp.24—32.
Smith, J.E., 1973. Shale compaction. *Soc. Pet. Eng. J.*, 12: 12—22.
Suter, H.H., 1960. *The General and Economic Geology of Trinidad.* H.M. Stationary Office, London, 145 pp.
Teodorovich, G.I. and Konyukhov, A.I., 1970. Mixed-layer minerals in sedimentary rocks as indicators of the depth of their categenetic alterations. *Dokl. Akad. Nauk S.S.S.R.*, 191: 1123—1126.
Thomeer, J.H.M.A. and Bottema, J.A., 1961. Increasing occurrence of abnormally high reservoir pressures in boreholes, and drilling problems resulting therefrom. *Bull. Am. Assoc. Pet. Geol.*, 45: 1721—1730.

Thorsen, C.E., 1963. Age of growth faulting in southeast Louisiana. *Trans. Gulf Coast Assoc. Geol. Soc.*, 13: 103—110.

Trusheim, F., 1960. Mechanism of salt migration in northern Germany. *Bull. Am. Assoc. Pet. Geol.*, 44: 1519—1540.

Van der Knaap, W. and Van der Vlies, A.C., 1967. On the cause of subsidence in oil-producing areas. *Proc. 7th World Pet. Congr.*, Mexico City, 3: 85—97.

Van Everdingen, R.O., 1968. Studies of formation waters in western Canada: geochemistry and hydrodynamics. *Can. J. Earth Sci.*, 5: 523—543.

Van Moort, J.C., 1971. A comparative study of the diagenetic alteration of clay minerals in Mesozoic shales from Papua, New Guinea, and in Tertiary shales from Louisiana. *Clays Clay Miner.*, 19: 1—20.

Voroshilov, E.A., Dzhalilov, T.I. and Samedov, N.K., 1968. Precipitation of salts during a change in the chemical composition of stratal waters during a reaction. *Azerb. Neft. Khoz.*, 47: 28—30.

Walton, M.S. and O'Sullivan, R.B., 1950. The intrusive mechanics of a clastic dike. *Am. J. Sci.*, 248: 1—21.

Weaver, C.E., 1960. Possible uses of clay minerals in search for oil. *Bull. Am. Assoc. Pet. Geol.*, 44: 1505—1518.

Weaver, C.E., 1961. Clay mineralogy of the late Cretaceous rocks of the Washakie Basin. *Wyo. Geol. Assoc., Guideb., 16th Annu. Field Conf.*, pp.148—154.

Weaver, C.E. and Beck, K.C., 1971. Clay water diagenesis during burial: how mud becomes gneiss. *Geol. Soc. Am., Spec. Paper*, 134, 96 pp.

Wilson, C.C. and Birchwood, K.M., 1965. The Trinidad sand volcano island of 1964. *Proc. Geol. Soc. Lond.*, 1626: 169—174.

Winslow, A.G. and Wood, L.A., 1959. Relation of land subsidence to groundwater withdrawals in the upper Gulf Coast region, Texas. *Trans. AIME*, 214: 1030—1034.

Yakubov, A.A., Ali-Zade, A.A. and Kastryulin, N.S., 1973. A new eruption of Lokbatan mud volcano. *Azerb. Neft. Khoz.*, 3: 5—7.

Young, A. and Low, P.F., 1965. Osmosis in argillaceous rocks. *Bull. Am. Assoc. Pet. Geol.*, 49: 1005—1007.

Zen, E.A. and Hanshaw, B.B., 1964. Osmotic equilibrium and mechanics of overthrust faultings. *Geol. Soc. Am., Annu. Meet. Program*, pp.232—233.

Chapter 2. COMPACTION OF ARGILLACEOUS SEDIMENTS

GEORGE V. CHILINGARIAN and HERMAN H. RIEKE, III

Introduction

During the past 25 years, the exploitation of oil and natural gas reserves, associated with thick sequences of very fine-grained rocks in Tertiary basins, has become increasingly important for fulfilling the world's energy needs. Many reservoir development problems have arisen which demand an engineering solution. Successful drilling to depths greater than 20,000 ft [6096 m] in these sedimentary basins and the amounts of hydrocarbons discovered depend to a great extent on knowledge of the fundamental physical and mechanical properties and deformation characteristics of the encountered formations and the interrelationship between various rock properties. It is very important, therefore, to have a thorough understanding of how abnormal fluid pressures are created and maintained in these geologically young sedimentary rocks.

Considerable disagreement exists among engineers and geologists as to the mechanism responsible for the origin of abnormally high formation pressures in the Tertiary basins. A number of causes for the creation and maintenance of abnormal fluid pressures in argillaceous sediments have been proposed: (1) continuous loading and incomplete gravitational compaction of sediments, (2) tectonic compression, (3) faulting, (4) salt and shale diapirism, (5) phase changes in minerals during compaction and diagenesis, (6) geothermal temperature changes creating fluid-volume expansion and/or cracking of organic compounds, (7) osmotic and diffusion pressures, and (8) invasion of water derived from magmatic intrusions (see Rieke and Chilingarian, 1974).

Compaction of sediments under the influence of a vertical monotonic loading has long been a well-documented geologic phenomenon (Sorby, 1908; Hedberg, 1926, 1936). This mechanism appears to be a likely candidate for creating an environment conducive to the formation of high fluid potentials in sediments having low permeability. Knowledge of both the vertical and lateral orogenic stress patterns in a depositional basin is of the utmost importance in interpreting abnormal fluid pressure environments and anticipating the location of oil and natural-gas reservoirs. The influence of tectonic forces on compaction mechanism has been discussed by Hubbert and Rubey (1959), Berry (1969), Thompson (1973), Hergert (1973), and Rieke and Chilingarian (1974).

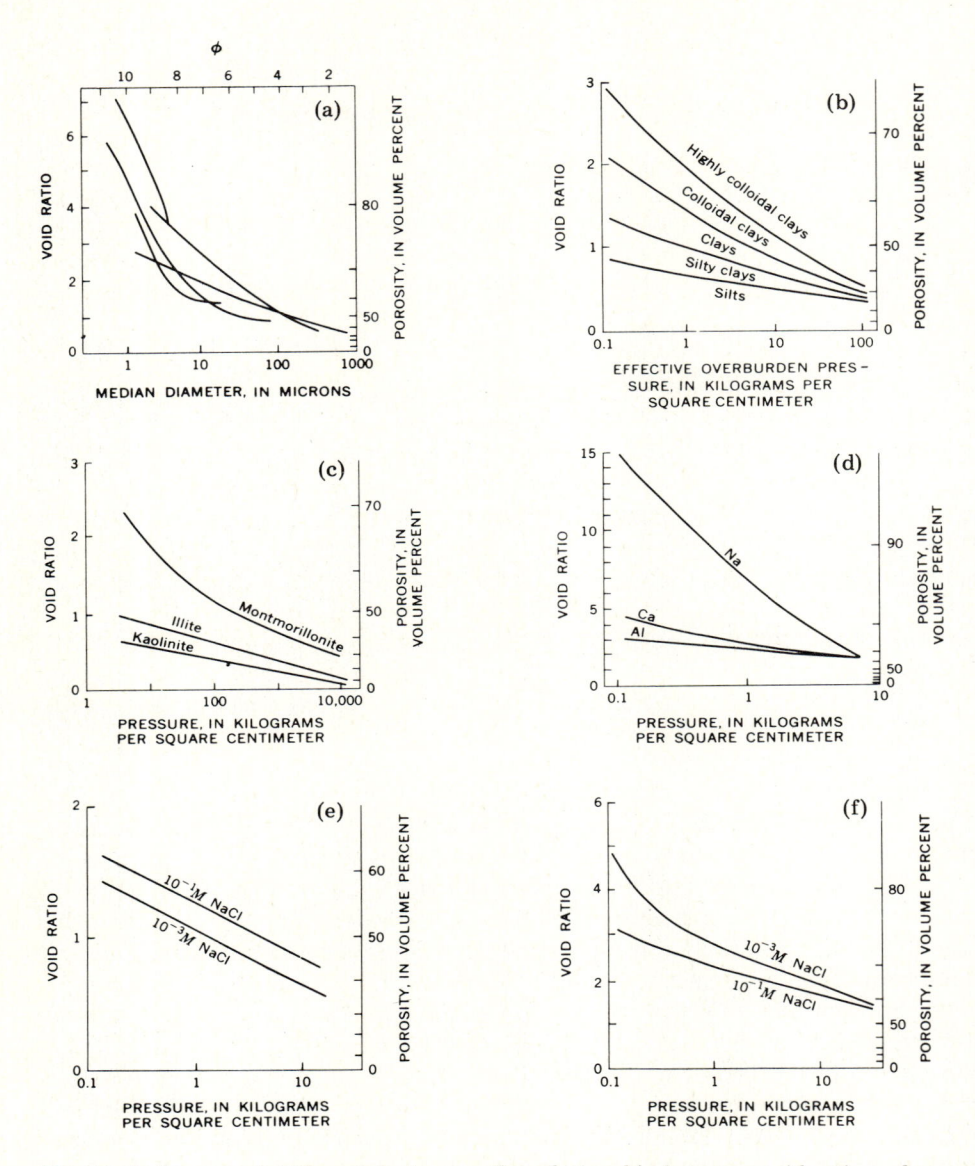

Fig.2.1. Influence of different factors on the relationship between void ratio and pressure in clayey materials. (a) Relationship between void ratio and median particle diameter at overburden pressures less than 1 kg/cm² (after Meade, 1964, p.B6). (b) Generalized influence of particle size (modified from Skempton, 1953, p.55). (c) Influence of clay-mineral species (modified from Chilingar and Knight, 1960, p.104). (d) Influence of cations adsorbed by montmorillonite (modified from Samuels, 1950). (e) Influence of NaCl concentrations in unfractionated illite, about 60% of which was coarser than 2 μm in size (modified from Mitchell, 1960, fig.M3). (f) Influence of NaCl concentration in illite finer than 0.2 μm (modified from Bolt, 1956, p.92). (After Meade, 1968, fig.1, p.D4.)

INTRODUCTION

Because of the intimacy of the relations between clay-sized mineral grains and water, a reduction of pore volume in sediments under increasing loads can best be considered in terms of the removal of the pore fluids. The factors that are known to influence the water content of argillaceous sediments under applied loads are the type of clay minerals, particle size, adsorbed cations, temperature, pH, Eh, and type of interstitial electrolyte solutions. The general effects of all but pH, Eh, and temperature are shown in Fig.2.1. With the exception of particle size, the influence of these factors is deduced mainly from laboratory compaction experiments, i.e. squeezing of monomineralic clays mixed with simple electrolytes.

Overpressured, undercompacted formations in many parts of the world probably owe their origin to rapid sedimentation, which trapped water in compacting clays with practically no permeability. As mentioned previously, laterally directed stresses of tectonic origin could also give rise to overpressured formations. Dickey (1972), however, stated: "It seems improbable that the comparatively unconsolidated shales could transmit horizontal stress." Anikiev (1964), on the other hand, attributed the origin of high pressures in numerous overpressured formations to Recent and Quaternary tectonic movements. He (p.108) stated that gravitational compaction is not a probable cause of overpressures, because compaction occurred in many places where one does not find overpressured formations. Gravitational compaction also could not explain the origin of high pressures present in formations surrounded by halogenous deposits, in fractured limestones and tuffs, and in well-consolidated Paleozoic rocks.

The forced influx of additional volumes of oil and gas under high pressures into reservoirs having a definite volume could cause overpressures.

Hedberg (1974) discussed the relation of methane generation to undercompacted shales, shale diapirs, and mud volcanoes. The organic matter, which constitutes a substantial part of the freshly deposited muds, decomposes during diagenesis as a result of biochemical and thermochemical processes. The resulting methane gas could create, or accentuate, the overpressured undercompacted state of the compacting mud sediments in two ways: (1) by building up additional internal pore pressure, and (2) by further impeding the expulsion of interstitial pore water due to the development of a second phase (gas) in the fluid (Hedberg, 1974, p.661). The presence of gas bubbles dispersed in a liquid phase will reduce the permeability of the rock to either phase.

The origin of abnormal subsurface pressures and factors causing these pressures to persist for many millions of years in sedimentary rocks are discussed in this chapter. A mathematical description of the sedimentation and the compaction of argillaceous sediments is also presented. A quantitative grasp of the compaction mechanism is based on the relationship between overburden stress, effective stress, pore fluid stress, and the physical properties of the competent and incompetent rocks. One continuum model, which represents the mechanical behavior of compacting sediments, treats the gravitational compaction process using a deformable one-dimensional porous medium (see Raghavan and Miller, 1975, for a detailed mathematical analysis).

In most sedimentary rocks, connate water is the predominant interstitial fluid. The degree of expulsion of water from the pore space by compaction provides a foundation for developing interrelationships between abnormal formation pressures, velocity of fluid expulsion, and pressure profiles through the sedimentary column. Some theories on how free and bound interstitial waters, along with hydrocarbons, are expelled and transported through the porous rocks in the form of an emulsion and/or a gaseous phase have been discussed by Chilingar and Adamson (1964), Smith et al. (1971), and Pandey et al. (1974).

Compaction model development

Dickinson (1951) performed a very thorough study on the geologic aspects related to abnormally high fluid pressures in the U.S. Gulf Coast Tertiary Basin. These high-pressure zones occur frequently in isolated Miocene and Pliocene porous sand beds surrounded by thick shale sections, which are located below the main deltaic sand series. Location of the high fluid potentials is controlled by the regional facies changes in the basin and appears to be independent of the depth and geologic age of the formation.

In sedimentary environments where sedimentation has been rapid, the thick accumulation of shales and mudstones of low permeability has retarded the expulsion of water and hydrocarbons. This excess fluid is trapped as pore fluid, which must bear a portion of the load that would normally be carried by the grain-to-grain contacts.

First basic premise in a model development is that the rate of gravitational compaction of sediments of a depositional basin is limited by the low permeability of the argillaceous members. In areas of rapid deposition, the expulsion of water from fine-grained sediments has been shown to lag behind the loss of water from sands (see Rieke and Chilingarian, 1974).

One can visualize the sediments as a two-phase continuum, with sediment grains in mechanical contact with each other and fluid filling every pore space not occupied by solids and wetting everything except the areas of the interparticle contacts. A diagrammatic sketch of stresses in a pore space within a sediment is presented in Fig.2.2. Each phase, i.e. solid and liquid, is assumed to occupy continuously a portion of the entire space, somewhat analogously to two vapors sharing a space in which they are assumed to exert their own partial pressure. The interstitial fluid is considered to be a slightly compressible homogeneous liquid and the pore pressure may vary from point to point. Pore fluids will flow through the void space under the influence of excess pore pressures. If the structure of the sediment remains rigid during

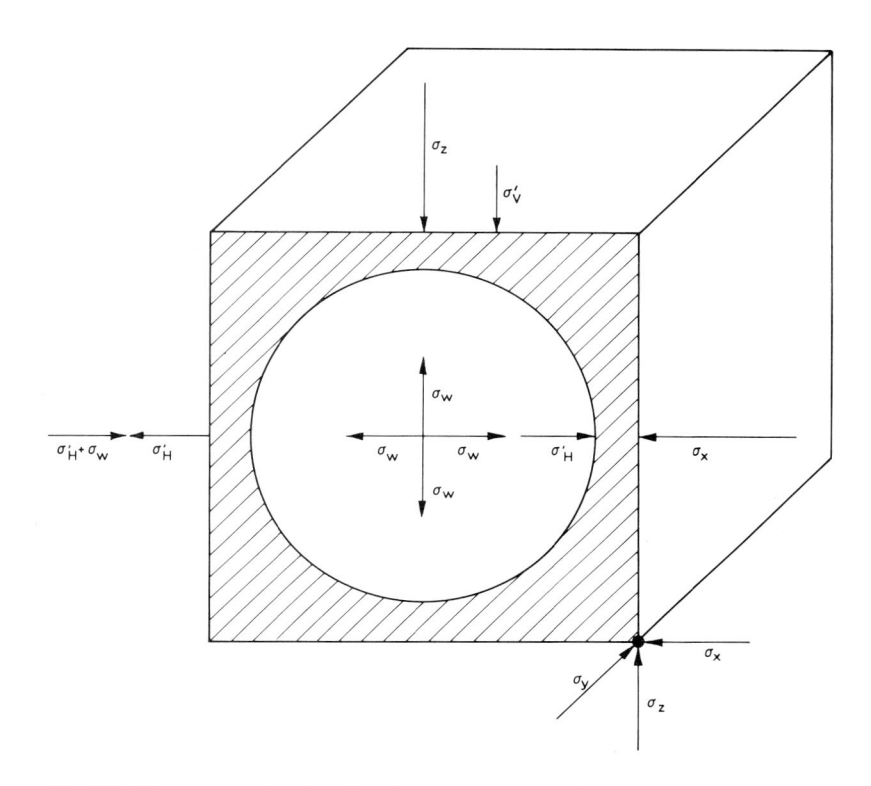

Fig.2.2. Diagrammatic sketch of a pore space within a sedimentary framework and the corresponding stress state in the system. σ'_V is the effective (intergranular) stress in the vertical direction, σ'_H is the horizontal effective stress, σ_w is the pore water stress, and σ_z is the total vertical stress component. The total horizontal stress component in the x direction, σ_x, is equal to $\sigma'_H + \sigma_w$.

the flow of fluids, a steady-state seepage will occur. On the other hand, if the grain arrangement alters to a different packing order, then an unsteady-state flow will arise. The steady-state flow, which requires a rigid stationary framework, can be easily developed from a mathematical viewpoint. Transient flow implies change in the effective stress which results in deformation of the solid matrix.

The Katz and Ibrahim compaction model

Katz and Ibrahim (1971) presented a mechanical model for explaining compaction and fluid expulsion from shales (Fig.2.3). Their model is based on Terzaghi's simple piston and spring analogy presented in Fig.2.4. The Katz and Ibrahim model is based on the compaction of an argillaceous layer

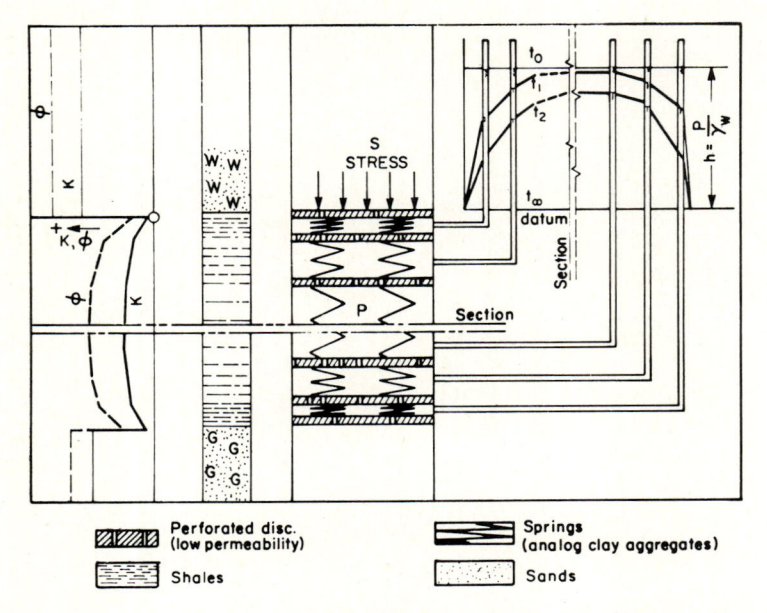

Fig.2.3. Schematic representation of clay compaction, porosity and permeability relationships, and creation of abnormally high formation fluid pressures (after Katz and Ibrahim, 1971, fig.12 — Courtesy of the Society of Petroleum Engineers of AIME). k = permeability; ϕ = porosity; t = time; p = pore pressure; γ_w = specific weight of water; h = height to which fluid will rise in the tubes, which represents the pressure head (p/γ_w); W = water; and G = gas.

between two permeable sand layers. As proposed by Terzaghi, the argillaceous sediment is represented by a series of springs and perforated disks. The perforated disks represent low-permeability clays, which restrict the escape of fluids, whereas the springs represent the deformable clay matrix. Sudden loading on the model corresponds to a rapid sedimentation rate. Water contained in the spaces between the perforated disks represents the interstitial fluid. If a stress is applied suddenly to the system, the water between the disks initially will support the entire load. After a brief period of time, the water will be forced through the perforations in the disks either in an upward or a downward direction, depending on the relative magnitudes of pressure in the compacting system, without lateral flow.

As the top and bottom disks move closer to the internal disks, the springs will begin to carry part of the applied load (Fig.2.3). Consequently, the fluid pressure between the external disks will decrease. When these disks approach each other, it will become difficult for the pore fluid to escape from inside the system. Katz and Ibrahim (1971) mentioned that the gradual decrease of permeability from the center toward the top and bottom of the model could be represented either by a decrease in the number of openings in the disks or

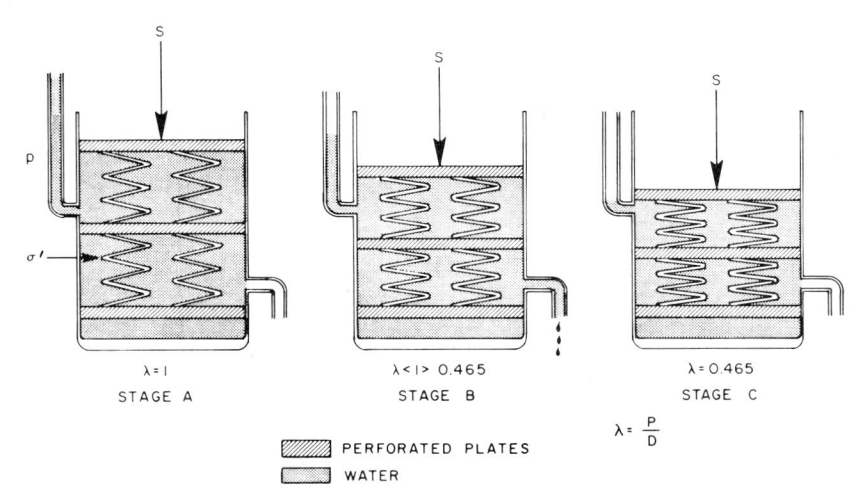

Fig.2.4. Simple schematic representation of clay compaction (after Terzaghi and Peck, 1948, in: Hottman and Johnson, 1965, p.718). σ' = grain-to-grain bearing strength, S = axial component of total stress (overburden pressure), p = fluid pressure, and λ = ratio of the pore stress to the total stress (e.g. 0.465 psi/ft [0.1074 kg cm^{-2} m^{-1}] of depth); σ' = $S - p$. *Stage A:* overpressured system; water is not allowed to escape. *Stage B:* water is allowed to escape; springs carry part of the applied load. *Stage C:* compaction equilibrium; load is supported jointly by the springs and the water (water pressure is simply hydrostatic).

by an increase of the number of disks per unit length in the model. Fig.2.3 also shows the fluid pressure distribution as a function of time. Higher fluid potential is shown to exist in the central portion rather than in the upper or lower portions of the model. This means that it takes more time for the fluid in the center of the model to escape than at the outer boundaries. The permeability and porosity distribution in the model is given on the left-hand side of Fig.2.3.

The behavior of the Katz and Ibrahim mechanical model is in general agreement with the observed performance of the overpressured formations in the U.S. Gulf Coast area. This model illustrates the reasons for the higher porosity of undercompacted shales, the extreme drop in permeability with increasing lithostatic pressure, and the entrapment of high interstitial fluid pressure in the shales.

Mathematical description of compaction

Mathematical description of sediments undergoing compaction is extremely difficult. In addition to having elastic characteristics, the sediments may possess plastic, viscoelastic, or thixotropic properties under certain conditions. The development of a mathematical model which encompasses all these compatibility characteristics, even as an approximation, has not been rigorously accomplished. The acceptable mathematical complexity of any sediment description is limited by the ability to solve boundary conditions. There are various phenomena related to sediment compaction, such as mineral and petroleum genesis, osmosis, reverse osmosis, variation in the interstitial fluid chemistry, diagenesis, release of bound water from minerals, and changes in mineral stability, that are still beyond the capabilities of finite-element analysis and digital computers.

Bulk (total) volume, V_b, of the element in Fig.2.2 is expressed as:

$$V_b = V_s + V_p \tag{2.1}$$

where V_s is the volume of solid grains comprising the rock framework, and V_p is the volume of pores (voids) in the rock.

The porosity, ϕ, of the compacting sediments is defined as:

$$\phi = \frac{V_p}{V_b} \tag{2.2}$$

Another parameter frequently used in the measurement of the degree of compaction is the void ratio, e, which is defined as:

$$e = \frac{V_p}{V_s} \tag{2.3}$$

The void ratio can be related to porosity by using the previous relationships (eqs. 2.1, 2.2, 2.3):

$$e = \frac{V_p}{V_b - V_p} \tag{2.4}$$

Multiplying the numerator and the denominator by V_b and rearranging the terms:

$$e = \frac{V_p}{V_b} \left(\frac{V_b}{V_b - V_p} \right) \tag{2.5}$$

$$e = \frac{V_p}{V_b} \bigg/ \left(\frac{V_b - V_p}{V_b} \right) \tag{2.6}$$

$$e = \frac{V_p}{V_b} \bigg/ \left(\frac{V_b}{V_b} - \frac{V_p}{V_b} \right) \tag{2.7}$$

$$e = \frac{\phi}{1 - \phi} \tag{2.8}$$

Solving for the porosity, ϕ:

$$\phi = \frac{e}{1 + e} \tag{2.9}$$

Robertson (1967) proposed a new compaction parameter called the solid-grain proportion, G, which can be expressed in terms of void ratio, e, and porosity, ϕ, as follows:

$$G = \frac{\gamma_{bd}}{\gamma_s} = \frac{V_s}{V_b} = \frac{1}{1+e} = 1 - \phi \qquad (2.10)$$

The solid-grain proportion, which is the ratio of dry-bulk specific weight, γ_{bd}, to grain specific weight, γ_s, is a linear measure of the approach of the sediment's dry-bulk specific weight to its solid-grain specific weight at any stage of compaction. The solid-grain proportion is an index of change in mass per unit volume, whereas the void ratio is an index of volume change in the sediment. Robertson (1967) prepared a graph illustrating the relationship between G, ϕ, and e (Fig.2.5).

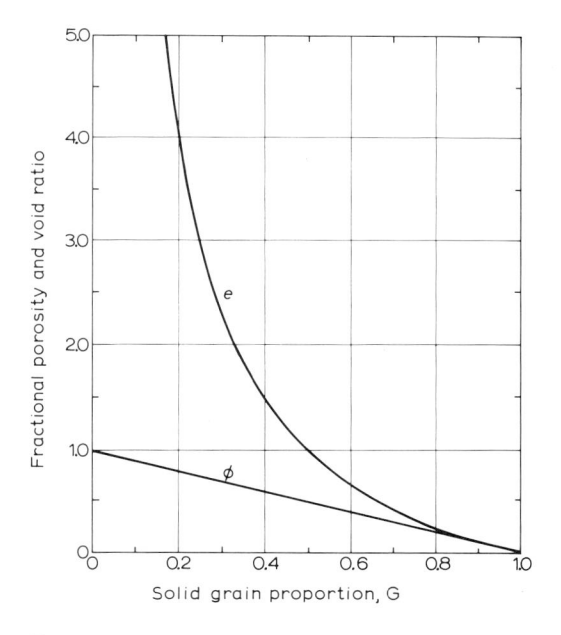

Fig.2.5. Relationship between void ratio, fractional porosity, and solid-grain proportion (adapted after Robertson, 1967, fig.1, p.124).

Berner (1971) presented a simplified and very elegant mathematical analysis of diagenesis and compaction. The following method of calculating compaction rate and total compaction of sediments saturated with water, is a modified Berner's approach.

Inasmuch as void ratio, e, is defined as the ratio of pore volume, V_p, to volume of solids, V_s, and the fractional moisture content, M, on a dry-weight basis is equal to the weight of water, W_w, divided by the weight of solids, W_s, the following relation can be developed:

$$e = \frac{V_p}{V_s} = \frac{W_w/\gamma_w}{W_s/\gamma_s} = M \cdot \frac{\gamma_s}{\gamma_w} \tag{2.11}$$

where γ_w and γ_s are the specific weights of water and solids, respectively. The fractional moisture content, M, is usually determined by measuring the loss of weight of water-saturated sediments upon drying at 100°C (see Rieke and Chilingarian, 1974, pp.33, 40, for the effect of drying temperature on solid-grain density determined). The volume of solids within a sediment layer per unit of horizontal cross-sectional area, $\overline{V}_s$, can be related to the rate of deposition, R_d, in weight of sediment per unit area per year, as follows:

$$\overline{V}_s = \frac{R_d \cdot t}{\gamma_s} \tag{2.12}$$

where t is equal to time in years necessary to deposit a volume of solids $\overline{V}_s$. The bulk volume of a sediment per unit area, $\overline{V}_b$, deposited annually is equal to the thickness of an annual sediment layer, h_a. Combining eqs. 2.9, 2.11, and 2.12, the following relationship can be obtained:

$$h_a = \frac{R_d}{\gamma_s} \frac{1}{1-\phi} \tag{2.13}$$

Fractional compaction, C_f, can be expressed as:

$$C_f = \frac{h_i - h}{h_i} \tag{2.14}$$

where h is the present thickness of the sediment layer and h_i is the initial thickness of this sediment layer at the time of deposition. If the initial porosity is ϕ_i and the present porosity is ϕ, then:

$$h(1-\phi) = h_i(1-\phi_i) \tag{2.15}$$

because a volume of solids for a given layer is assumed to be constant. Thus:

$$h_i = \left(\frac{1-\phi}{1-\phi_i}\right) h \tag{2.16}$$

and:

$$C_f = \frac{h_i - h}{h_i} = \frac{\phi_i - \phi}{1-\phi} \tag{2.17}$$

The rate of compaction, $d\phi/dt$, can be calculated using a partial differential equation as follows, if porosity, ϕ, is a function of depth of burial, D, and time, t:

$$d\phi = \left(\frac{\partial \phi}{\partial t}\right)_D dt + \left(\frac{\partial \phi}{\partial D}\right)_t dD \tag{2.18}$$

Dividing through by dt:

$$\frac{d\phi}{dt} = \left(\frac{\partial \phi}{\partial t}\right)_D + \left(\frac{\partial \phi}{\partial D}\right)_t \left(\frac{dD}{dt}\right) \quad (2.19)$$

and inasmuch as dD/dt is equal to the thickness of an annual layer, h_a:

$$\frac{d\phi}{dt} = \left(\frac{\partial \phi}{\partial t}\right)_D + \left(\frac{\partial \phi}{\partial D}\right)_t h_a \quad (2.20)$$

In the case of steady-state compaction, $(\partial \phi/\partial t)_D$ is equal to zero and, consequently:

$$\frac{d\phi}{dt} = \left(\frac{\partial \phi}{\partial D}\right)_t h_a \quad (2.21)$$

or combining with eq. 2.13:

$$\frac{d\phi}{dt} = \left(\frac{\partial \phi}{\partial D}\right)_t \frac{R_d}{\gamma_s} \left(\frac{1}{1-\phi}\right) \quad (2.22)$$

In the above equation, the $(\partial \phi/\partial D)_t$ term can be determined from depth-versus-porosity curves (e.g. see Rieke and Chilingarian, 1974).

Differential equations have been derived by several investigators (e.g. Cooper, 1966) to describe the changes in a space-time continuum for a conceptual model. The treatment is usually limited to the upward and downward movement of fluids out of one-dimensionalized, horizontal, elastic porous layers. In order to obtain a solution to the compaction process, involving the components of the overburden and tectonic stresses and particle and fluid displacements in the sediment system, one must consider the following equations and conditions: (1) continuity equations, (2) constitutive equations, (3) compatibility equations, (4) boundary conditions, and (5) equilibrium conditions.

In addition, the physicochemical factors and processes (e.g. osmosis, reverse osmosis, chemical composition of interstitial solutions, and temperature) must be evaluated. Bredehoeft and Hanshaw (1968) included thermodynamic considerations in their analysis of anomalous fluid potentials. The depositional environment, which can be visualized for the model, is a level surface of infinite extent. Water depth is constant over the entire area and may vary with time. When the upper layer of a freshly deposited, fine-grained sediment is undergoing initial compaction at the bottom of the sea, the interstitial fluid is in continuous communication with the overlying seawater and the pore pressure is essentially hydrostatic (see Fig.2.6). Beginning at time zero, the sediment is deposited at a uniform constant rate, e.g. 0.1 m/1000 yr to 0.5 m/1000 yr over the entire area. The physical characteristics of the freshly deposited sediment, such as the mineralogy, porosity, packing geometry and grain size are assumed to be constant throughout the entire area. As each sediment layer is buried under subsequently deposited

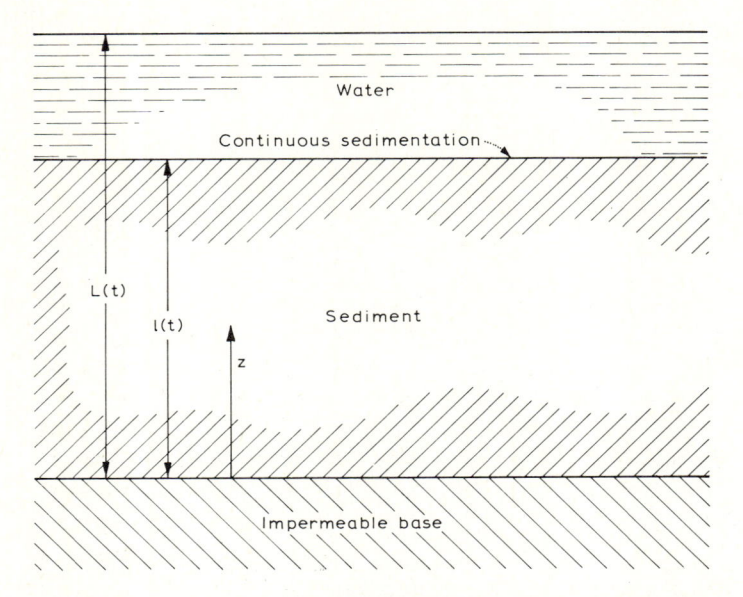

Fig.2.6. Schematic diagram of continuous sedimentation in water (after Bredehoeft and Hanshaw, 1968, fig.4, p.1103 — Courtesy of the Geological Society of America Bulletin.) l = thickness of the sediment, t = time, z = vertical coordinate, and L = depth of water (interstitial fluid is in continuous communication with the overlying seawater).

mud and sand layers, a gradual compaction takes place. If the sedimentation rate is slow, the compacting sediment will gradually adjust to the additional load imposed by the overlying sediments, and as the mineral grains are pressed together, pore fluids are expelled.

Inasmuch as the argillaceous layers have high porosity and are relatively more permeable in their initial state (see Rieke and Chilingarian, 1974), the expelled fluids will flow in the direction of least resistance, usually upwards and into a porous sand layer. As long as the fluid can escape under normal loading conditions and porosity is intercommunicating, hydrostatic pressures will be encountered. If the gravitational loading rate of muds and sand layers is high, the permeability of muds decreases rapidly and, as a result, the pore fluids cannot escape from the sands through the argillaceous layers. These interstitial fluids would help to support the increasing overburden weight and further compaction of the formation is retarded or stopped. Thus, the formation becomes overpressured, because the contained fluids are subjected not only to the hydrostatic force, but also to the weight of newly deposited sediments. Bredehoeft and Hanshaw (1968) presented a hydrodynamic compaction model based on a continuous sedimentation rate, which possibly describes the creation of high pressures (see also Rieke and Chilingarian, 1974). Gibson (1958, p.175) presented the following equation for continuous sedimentation in which the rate of sedimentation, $w = \partial l/\partial t$, is constant:

$$\frac{h' \rho_w}{l \rho'} = 1 - \left(\frac{\pi K t}{S_s l^2}\right)^{-\frac{1}{2}} \exp\left[-\left(\frac{z}{l}\right)^2 \frac{S_s l^2}{4 K t}\right]$$

$$\times \int_0^\infty \xi' \tanh\left(\frac{\xi' S_s l^2}{2 K t}\right) \cosh\left[\left(\frac{z}{l}\right) \frac{\xi' S_s l^2}{2 K t}\right] \exp\left(-\frac{\xi' S_s l^2}{4 K t}\right) d\xi' \quad (2.23)$$

where h' is the excess fluid head, which is equal to $p_a/\rho_w g$ (p_a is the transient pore pressure in excess over the normal hydrostatic pressure, originally present); h is the head, which is equal to $[h' + L(t)]$, where L = depth of water and t = time; z is the vertical coordinate; S_s is the specific storage and is defined as the volume of water taken into storage or discharged per unit volume per unit change in head; S_s is equal to $\rho_w g[(1/E_s) + (\phi/E_w)]$, where $\rho_w g$ = weight of unit volume of water, E_s = modulus of compression of the sediment matrix confined in situ (= σ_z/ϵ_z, i.e. ratio of the vertical stress to the vertical strain), ϕ = fractional porosity, and E_w = bulk modulus of elasticity of water; K is the hydraulic conductivity, which is equal to $(k\rho_w/\mu)g$, where k = permeability, ρ_w = density of fluid, g = acceleration of gravity, and μ = viscosity of fluid; t is the time; $\rho' = \rho_b - \rho_w$, i.e. the difference between the bulk density, ρ_b, of the sediment and the density of the pore fluid, ρ_w; l is the thickness of the sediment; and $\xi' = z/l(t)$. Bredehoeft and Hanshaw (1968, fig.5, p.1104) presented a figure which enables solution (dimensionless form) of the above equation (see also Rieke and Chilingarian, 1974, fig.168, p.317). Gibson (1958, p.171) stated, however, that it is unlikely that closed solutions can be obtained for arbitrary rates of sediment deposition. In addition, Gibson's equation does not take into consideration the deformable coordinates.

The development of a compaction model depends on certain fundamental concepts and several basic relationships, which must be stated at the onset so that any mathematical derivation can proceed with minimum interruption. Inasmuch as constitutive equations can define an ideal geological material, no advance can be made in the description of the sediment's properties without a prior knowledge of its behavior under an external stress. Constitutive theory is intended only to describe a limited number of physical properties decided at the outset for a given material.

It is postulated that compaction is directly related to the following parameters and is functionally represented in the following form:

$$C = f(\sigma, v, \rho, V, \phi, k, D, t, c) \quad (2.24)$$

where C is the compaction, σ is the stress on the system, v is the velocity parameter for solids and interstitial fluids in the system, ρ is the density, V is the volume relationships, ϕ is the porosity, k is the permeability of the system, D is the burial depth, t is the time, and c is the compressibility relationship.

Fig.2.7 illustrates the spatial relationships of a compacting sediment and the coordinate system adopted by Smith (1971b). In this system the position of the differential element of solids or pore fluid is measured as the vertical distance z upward (positive) from the base of the argillaceous sediment. A differential element has a negative velocity if the material is moving downward and a positive velocity if it is moving upwards relative to the base of the sediment unit. The interstitial water in the unit cross section between z and $z + dz$ (Fig.2.7) has an average velocity, v_w, which is a function of the time, t, distance z, and the depositional history and boundary conditions:

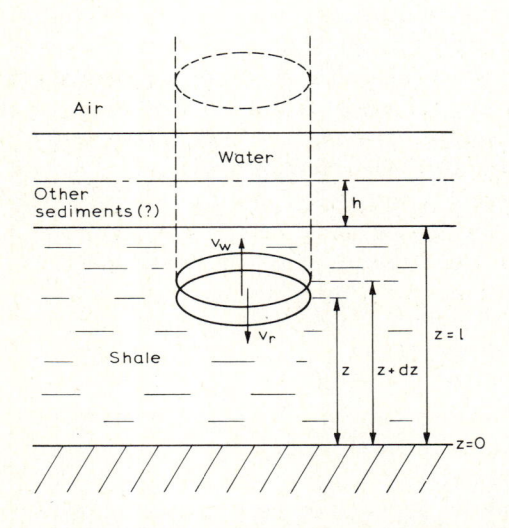

$$v_w = v_w(z, t) \tag{2.25}$$

Fig.2.7. Spatial relationships and the coordinate system of a compacting sediment (modified after Smith, 1971b, fig.1, p.242). v_w = average velocity of the water, v_r = average velocity of the matrix.

Similarly, the average velocity of the matrix material between z and $z + dz$ will be:

$$v_r = v_r(z, t) \tag{2.26}$$

Based on the law of conservation of mass, the increase in mass within the volume increment between z and $z + dz$, during the differential time increment dt, is equal to:

(mass *in* at z) — [mass *out* at $(z + dz)$]

$$= (\rho_w v_w \phi)_z \, dt - (\rho_w v_w \phi)_{z+dz} \, dt = \left(\frac{\partial \rho_w \phi}{\partial t}\right) dz\, dt \tag{2.27}$$

or:

$$\frac{\partial \rho_w v_w \phi}{\partial z} = -\frac{\partial \rho_w \phi}{\partial t} \tag{2.28}$$

where ϕ is the porosity and ρ_w is the density of water. Similarly, Smith (1971b, p.243) derived an equation for solid matrix:

$$\frac{\partial \rho_r v_r (1-\phi)}{\partial z} = \frac{-\partial \rho_r (1-\phi)}{\partial t} \tag{2.29}$$

where ρ_r is the density of solid grains. For a detailed derivation which applies to fine-grained sediments the reader may consult Smith (1971b).

Flow of fluids through porous media

Bear et al. (1968) defined a porous medium as a space occupied by multi-phase matter in which one or more of the phases is not solid. The solid phase is termed the solid matrix and the space within the porous medium, which is not a part of the solid matrix, is referred to as the pore space. The solid matrix has to be distributed throughout the porous medium within the domain occupied by the porous medium. The majority of the pores should be interconnected if fluid flow is to occur; however, unconnected pores can be considered as a part of the solid matrix.

Henry Darcy investigated the flow of water in vertical homogeneous sand filters in 1856. From these investigations, Darcy concluded that the volumetric rate of flow, q, can be expressed as:

$$q = \frac{kA\Delta p}{L} \tag{2.30}$$

where k is the coefficient of proportionality, termed permeability; A is the cross-sectional area; and Δp is the difference in fluid pressure across a sediment column having length L. Eq. 2.30 has been modified into a general form that considers the viscosity, μ, of any fluid flowing in a nondeforming porous medium:

$$q = \frac{kA\Delta p}{L\mu} \tag{2.31}$$

The potential, Φ, governing the fluid flow is defined by:

$$\Phi = g(z - z_0) + \int_{p_0}^{p} \frac{dp}{\rho_f} \tag{2.32}$$

where g is the acceleration due to gravity; z is the elevation above the arbitrarily selected datum z_0; z_0 is the elevation of the point where $\Phi = 0$; p_0 is the pressure at the point where $z = z_0$ ($\Phi = 0$); and ρ_f is the density of the fluid.

The velocity of fluid flow can then be generalized as the macroscopic flow velocity, v_{ma} (Hubbert, 1940):

$$v_{ma} = -\left(\frac{k\rho_f}{\mu}\right)\nabla\Phi \tag{2.33}$$

The macroscopic velocity, however, is not the true velocity of the liquid at any particular point. Bear (1972) pointed out that the complexity of the geometry of the porous medium precludes the use of the true velocity of the migrating fluid at any particular point. A statistically averaged microscopic velocity, v_{mi}, however, can be determined if the average porosity, ϕ, of the formation is known:

$$v_{mi} = -\left(\frac{k\rho_f}{\mu\phi}\right)\nabla\Phi \tag{2.34}$$

The microscopic and macroscopic velocity concepts presented above are valid in a stationary coordinate system. Hubbert (1940) discussed these relationships for nondeformable media in detail and the reader is referred to his work.

Equation 2.34 does not reveal that a porous medium undergoing compaction is compressible and that there is a relative motion of the solid particles in the system. Darcy's law supposes that the fluid velocity is measured with respect to a fixed coordinate system outside the solid matrix, and that the matrix does not move or distort. Only under these conditions would eq. 2.33 give a correct answer. Thus, this equation must be generalized, because Darcy's law applies to flow relative to the solid matrix only:

$$v_{ma,r} = -\left(\frac{k\rho_f}{\mu}\right)\nabla\Phi \tag{2.35}$$

where $v_{ma,r}$ is the macroscopic velocity relative to the motion of the solid particles. The value of the calculated permeability would be meaningless if the flow velocity is not measured relative to the solid matrix. Raghavan (1974) discussed this problem in detail.

The differential equation governing the transient three-dimensional flow of water in an elastic aquifier was derived several years ago by Jacob (1940; 1950). De Wiest (1966) questioned the validity of Jacob's derivations (see eq. 2.40) based on the fact that in one side of the equation the net inward mass flux was calculated for a volume element without deformation, whereas in the other side of the equation, in order to compute the rate of change of the mass inside the volume element, the element itself was deformed. The solution by De Wiest (1966) did not distinguish between the rate of flow relative to the moving grains of the medium and the rate of flow across the fixed boundaries of his control volume as long as the material was being deformed:

$$\nabla^2 h - 2\rho g\beta \frac{\partial h}{\partial z} = -S^* \frac{\partial h}{\partial t} \tag{2.36}$$

FLOW OF FLUIDS THROUGH POROUS MEDIA

where specific storage $S^* = \rho g[(1-\phi)\alpha + \phi\beta]$ and has dimensions of $1/L$; ρ is the fluid density (M/L^3); β is the compressibility of the liquid (LT^2/M); g is the gravity field strength (L/T^2); $h = z + 1/g \int_{p_0}^{p} dp/\rho(p)$, i.e. head above common datum (L); z is the fixed spatial coordinate (L); t is the time (T); α is the compressibility of the medium (LT^2/M); and ϕ is the fractional porosity of the medium (dimensionless).

Cooper (1966) pointed out that the first one of these two flows obeys Darcy's law, whereas the second does not. Cooper (1966) derived two different forms of the flow equation for a compressible liquid in an elastic porous medium by considering mass conservation in: (1) a control volume, the boundaries of which are fixed in space (Eulerian stationary coordinates), and (2) a control volume that deforms and moves through space (Lagrangian moving coordinates) when the material deforms:

$$\nabla^2 h - 2\rho g\beta \frac{\partial h}{\partial z} = \frac{\rho g}{K}(\alpha + \phi\beta)\left(\frac{\partial h}{\partial t} - w_g\right) \tag{2.37}$$

in the case of fixed coordinates [w_g = vertical component of velocity of grains of medium (L/T)], and:

$$\nabla'^2 h - 2\rho g\beta \frac{\partial h}{\partial z'} = \frac{\rho g}{K}(\alpha + \phi\beta)\frac{\partial h}{\partial t} \tag{2.38}$$

in the case of deforming coordinates. K = fluid conductivity (L/T) and z' = deforming spatial coordinate (L). Usually, the second term in eq. 2.38 is negligible and, therefore, the equation is very closely approximated by:

$$\nabla'^2 h = \frac{S_s}{K}\frac{\partial h}{\partial t} \tag{2.39}$$

where $S_s = \rho g(\alpha + \phi\beta)$. The latter equation (eq. 2.39) shows that Jacob's expression (eq. 2.40), for what is referred to in the terminology on groundwater hydrology as *specific storage*, is essentially correct for the use in deforming coordinates. In Jacob's equation:

$$\frac{\partial^2 h}{\partial x^2} + \frac{\partial^2 h}{\partial y^2} + \frac{\partial^2 h}{\partial z^2} = \frac{S_s}{K}\frac{\partial h}{\partial t} \tag{2.40}$$

h is the potential head above a common datum, K is the fluid conductivity, t is the time, and S_s is the specific storage (see Rieke and Chilingarian, 1974, p.144). Specific storage, S_s, is the volume of water which a unit volume of the formation releases from storage under a unit head of decline. It is related to Theis's storage coefficient, S, which is dimensionless, as follows:

$$S = bS_s \tag{2.41}$$

where b is the thickness of the formation. As pointed out by Cooper (1966), aside from the advantage of eliminating the grain velocity from the flow equation, several other factors point in favor of deforming coordinates in the case of nonsteady flow in elastic media, e.g. it gives a somewhat closer approximation of Darcy's law if z' is used rather than z. Statistically speaking, a vertical streamline through the deforming elemental volume $\Delta x \Delta y \Delta z'$ would traverse the same number of pores regardless of the deformation. Streamlines through the fixed element $\Delta x \Delta y \Delta z$ traverse a greater or lesser number of pores when the sediment deforms.

The vertical component of flow per unit area relative to grains of the medium, $w' (= K \, \partial h/\partial z')$, therefore, is more nearly proportional to the head differential across $\Delta z'$ than to the one across a fixed element Δz (Cooper, 1966, p.4790). Thus, K is more nearly constant when Darcy's law is approximated in terms of the deforming coordinate z'.

Tortuosity in argillaceous sediments plays an important role in compaction, more so than in the coarser-grained clastics. An important question arises here as to whether Darcy's law is generally valid for fluid flow in saturated clays and clayey sediments or not. Two main criteria must be met for Darcy's equation to be valid for clays: (1) the interstitial fluid in the pores must exhibit Newtonian behavior, and (2) the clay particles must be arranged in a rigid manner so that forces due to fluid movements do not modify the pore geometry. A deforming coordinate system violates the second premise owing to the change in tortuosity by a mobile matrix. In addition, various electrokinetic forces become operative during the flow of water through the clayey sediments, because of the presence of fixed and mobile double layers, and invalidate Darcy's equation. The occurrence of phenomena such as osmosis, reverse osmosis, and filtration through clay membranes further complicates the picture.

Overburden potential

Philip (1969a) viewed the overburden potential as consisting of two components: (1) the mechanical stress σ on the rock column, against which work has to be done if the column deforms, and (2) the work which has to be performed at any point where liquid is added to the porous medium, inasmuch as the porous matrix is capable of being deformed and the addition of fluid results in an increase in bulk volume.

Philip (1969a), in studying swelling soils, formulated a quantity, θ, which is equal to the volume of liquid water per unit volume of sediment particles. This quantity, called the *moisture ratio* by Philip (1969a), is apparently equal to:

$$\theta = (1 + e)\phi \tag{2.42}$$

The volume of voids, V_p, is equal to:

$$V_p = V_w + V_g, \tag{2.43}$$

where V_w is the volume of interstitial fluid in the pores and V_g is the volume of free gas in the pores. If the pores are assumed to be fully saturated with an interstitial liquid, then eq. 2.42 reduces to:

$$\theta = e \tag{2.44}$$

by substituting $e/(1 + e)$ for ϕ.

In a vertical column of argillaceous sediments with the surface unconstrained and the mineral grains constrained from movement only at the base, the necessary upward movement of both the interstitial fluid and the solids above the point where interstitial fluid is added to the system (lateral or vertical migration from adjacent formations) requires that work be done against the gravitational field. A load on the surface of the vertical sediment column also requires that work must be done against this load. The component of the total potential which arises in this manner is termed the overburden potential. The overburden potential, Ω, in swelling soils can be defined as (Philip, 1969a):

$$\Omega = \frac{de}{d\theta}\left[\sigma(z) + \int_{z_0}^{z} \rho_a dz\right] \tag{2.45}$$

where ρ_a is the apparent wet density (= specific gravity) of the sediment and $\sigma(z)$ is the total load carried by the sediment column at some convenient plane z which could be the upper surface of the column.

Philip (1969a) defined ρ_a as:

$$\rho_a = \frac{\theta + (\rho_g/\rho_f)}{(1 + e)} \tag{2.46}$$

where ρ_g is the density of the matrix and ρ_f is the density of the interstitial fluid. Substituting $(1 + e)\phi$ for θ (eq. 2.42) in eq. 2.46, the following equation would result:

$$\rho_a = \phi + \left(\frac{\rho_g}{\rho_f}\right)\left(\frac{1}{1 + e}\right) \tag{2.47}$$

Utilizing Robertson's solid-grain proportion relationships in eq. 2.10:

$$\rho_a = [\phi + (\rho_g/\rho_f)(1 - \phi)] \tag{2.48}$$

or:

$$\rho_a = [\phi + (\rho_g/\rho_f)(G)] \tag{2.49}$$

Effect of compaction on the chemistry of interstitial solutions

Chemistry of interstitial fluids

The chemical properties of interstitial fluids, associated with abnormally and normally pressured sediments, have been considered in very few abnormal-pressure investigations. Most investigators have ignored the pore fluids associated with less permeable shales. One reason for this was that the permeability of shales is so low that the exposed shale intervals in oil wells rarely produce fluids in measurable quantities. In addition, the produced water for the most part is not representative of the interstitial fluid because of contamination. The concentration of the in-situ fluids may be due to associated salt deposits, or contamination may occur during drilling and production of the oil well. Drilling fluids may cause dilution of interstitial fluids and mixing of fluids from several different horizons is common. Hottman and Johnson (1965) attempted to determine the salinity of interstitial fluids using electric log calculations. Analysis of water-soluble constituents, leached or squeezed from shale and clay samples, was performed by Von Engelhardt and Gaida (1963), Hedberg (1967), Weaver and Beck (1969), Chilingarian and Rieke (1968), Manheim and Sayles (1970), Long et al. (1970), and Schmidt (1973).

One way in which data on interstitial fluids in shales can be obtained, is by core analysis. The pore solution can be flushed or distilled out of the core, permeability permitting, or the core can be pulverized and then leached with distilled water in order to obtain the soluble salts. The latter technique, however, will not give the true composition of the salts dissolved in the pore solution, because any soluble minerals present in the rock will also be dissolved along with the salts. Von Engelhardt and Gaida (1963) leached interstitial salts from Jurassic claystone and shale core samples. They felt that analysis for the chloride content of the leached solution alone would be sufficient, because it is very likely that it would be derived from the pore fluid only. The proportions of calcium, magnesium and sodium ions may be changed by the base exchange of the clays. They also concluded that the chloride content in shale pore waters was of the same magnitude as that in the associated permeable strata, which is questioned by other investigators as discussed later. Possibly, the formation-logging techniques (electric, radioactive, etc.) may be improved to such an extent in the future that the chemistry of fluids in shales could be determined directly without using leaching or other procedures.

Changes in concentration of interstitial fluids during the process of compaction, as reported by different investigators, are presented in this chapter and related to field data obtained from various sources. In order to obtain reliable new data concerning the composition of interstitial water in shales, Schmidt (1973) analyzed sidewall cores from a shale section of a well in Calcasieu Parish, Louisiana. These cores were analyzed to determine (1) the concentration of various cations and anions, (2) base-exchange capacity of clays, (3) exchangeable cations of clays, and (4) clay mineralogy. Additional information was provided by (1) electric logs from selected wells, (2) subsurface temperature and pressure measurements, (3) shale density measurements, and (4) analyses of produced water.

To date, the formation of sodium chloride solutions in rocks at depth is not well understood. Problems associated with the origin of subsurface fluids are still in an area of scientific controversy. Several explanations have been proposed for the generation of subsurface salinities, which are greater than that of seawater. These include (1) ion filtration, (2) evaporation of water, (3) salt diffusion from nearby evaporite deposits, and (4) gravitational segregation (Dickey et al., 1968). In the presence of salt domes or salt beds, the high-salinity brines in the surrounding strata owe their origin in large measure to solution of the available salt. Formation waters with high salinity, however, also occur in sedimentary rocks which are not associated with salt deposits. It seems probable that under certain conditions gravitational compaction acts as the main mechanism producing concentrated salt solutions.

In geosynclines, where the depositional rate is rapid, large quantities of water are continuously extracted from the hydrosphere during sedimentation. Recent muds, for example, may contain up to about 80% (and higher) water by volume (Degens and Chilingar, 1967, p.478). The interstitial fluids occupy the pore space of most of the buried sediments. Upon compaction of the sediments, the connate waters, as the interstitial fluids are sometimes termed, are expelled into the associated sandstones. The speed with which the water is expelled from the original argillaceous sediments depends not only on the overburden pressure and the physical and chemical properties of fluids, but also on the texture, structure, and mineral composition of the sediment (see Fig.2.1).

The composition of fluids contained in the pore space of permeable sediments at depth is known from the many analyses of formation waters obtained from drill holes and producing oil wells. In many sedimentary basins the salinity of fluids increases with increasing depth, especially in the case of those fluids that are associated with basins older than the Tertiary. These fluids differ greatly from seawater in their chemical composition. All trapped fluids in the sediments are believed to have undergone chemical alteration, which is a function of time, temperature, and pressure. In studying these fluids, the basic assumption is made that the salinity and ionic composition of seawater, especially since Mesozoic times, was the same as today. The original water, expelled by compaction, has been gradually replaced by, or mingled with, meteoric water, especially if the sediments were uplifted or exposed at the earth's surface by tectonic activity or erosional processes. The majority of analyzed fluids from these sediments are derived from permeable strata.

Composition of interstitial solutions related to seawater

Interstitial solutions can be classified as (1) syngenetic (formed at the same time as the enclosing rocks), and (2) epigenetic (owing their origin to subsequent infiltration of meteoric and other waters into already formed rocks). The main processes which alter the chemistry of buried waters are: (1) physical (compaction), (2) chemical (reactions involving rock minerals, organic matter, and interstitial solutions, etc.), (3) physicochemical (filtration through charged-net clay membranes, adsorption and base exchange, etc.), (4) electrochemical, and (5) biochemical.

Degens et al. (1964) analyzed the oxygen isotope composition of a number of connate waters ranging in age from the Cambrian to the Tertiary, and reported that the $\delta^{18}O$ values of the highly saline oilfield brines do not deviate appreciably from the $\delta^{18}O$ of modern seawater (also see Degens and Chilingar, 1967). Deviation from the mean value in some of the samples into the negative range of $\delta^{18}O$ are always well correlated to a decrease in salinity. This feature can be easily explained by effects of dilution with meteoric waters during the migration of brine, or by subsequent infiltration caused by a change in the geological setting through uplift, denudation, and faulting. The similarity between the isotope characteristics of brines and modern seawater suggests that the concentration of inorganic salts has not been accomplished by syngenetic evaporation in most cases studied. Slight deviations into the positive range of $\delta^{18}O$ values in some samples studied may have been caused by original evaporation in a surface environment, or by isotope equilibration with the surrounding mineral matter for millions of years (Degens and Epstein, 1962).

The concentration of amino acids in the oilfield brine waters is a function of salinity (Degens et al., 1964), i.e. the content of amino acids increases with increasing salt concentration. On adjusting the salinity of brine waters to that of present-day oceans and applying the same calculation factors to the original amino acid values, the similarity between the amino acid spectra in the Recent seawater and fossil brines is pronounced (see fig.116, Rieke and Chilingarian, 1974, p.222).

There are systematic chemical differences (both qualitative and quantitative) between the ancient and modern connate waters. Magnesium, which is abundant in seawater, is present only in minor amounts in oilfield waters and the opposite is true for calcium. Calcium chloride waters, which are not formed in surface environment, are widespread among oilfield brines. This feature can possibly be linked to the dolomitization process.

Magnesium may also replace various cations in chlorites and clay minerals. Adsorption and exchange phenomena may explain variations in K/Na and Ca/Na ratios in fossil and modern interstitial solutions; the former ratio also appears to be temperature dependent (White, 1965, p.359). It should be remembered, however, that the significance of base exchange decreases with depth, and therefore, could not account for the formation of deeply buried calcium chloride waters. Graf et al. (1966) noted that the calcium content of the Michigan Basin brines is very high, greater than that of sodium in some fluid samples, and it increases proportionately with increase in total solids content. A definite relationship appears to exist between calcium and total solids content in oilfield brines of various geological ages (Figs.2.8 and 2.9). Graf et al. (1966) felt that the more obvious geologic processes are inadequate as explanations of the origin of concentrated calcium chloride brines, which occur in geosynclines free of major orogeny. Two simple processes were proposed by Graf et al. (1966) for deriving such a composition: (1) shale ultrafiltration of the dissolved solids contained in the original seawater, and (2) mixing various proportions of fresh water and seawater.

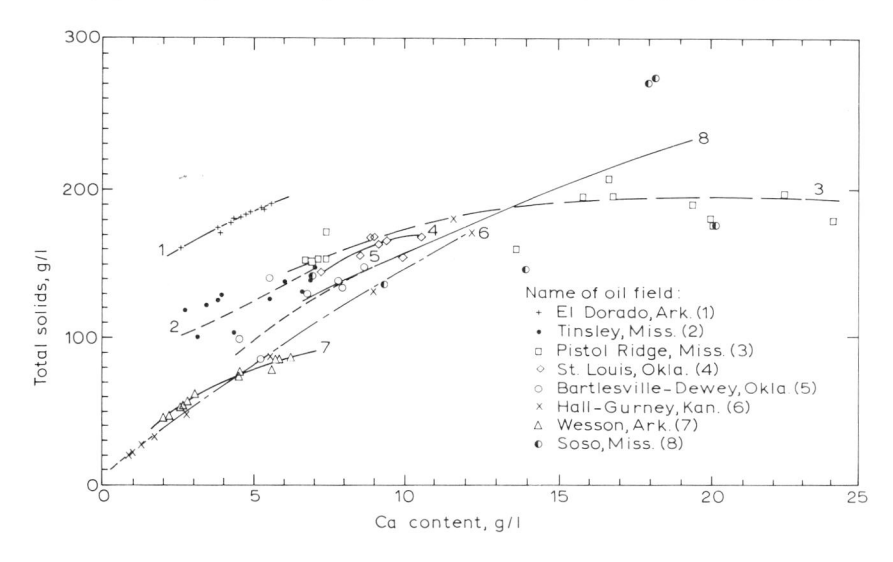

Fig.2.8. Relationship between calcium content and total solids in various oilfield waters (data from Rall and Wright, 1953; Wright et al., 1957; Hawkins et al., 1963a,b). Name of field, sampled interval and geologic age: 1 = El Dorado East, 2150—3172 ft [655—967 m], Upper Cretaceous; 2 = Tinsley, 4797—5770 ft [1462—1759 m], Upper Cretaceous; 3 = Pistol Ridge, 7423—10,964 ft [2263—3342 m], Upper Cretaceous; 4 = St. Louis, 2610—3174 ft [796—967 m], Ordovician/Pennsylvanian; 5 = Bartlesville—Dewey, 1241—2557 ft [378—779 m], Cambrian/Pennsylvanian; 6 = Hall—Gurney, 2610—3293 ft [796—1004 m], Ordovician/Pennsylvanian; 7 = Wesson, 2132—3598 ft [650—1099 m], Lower Cretaceous/Upper Cretaceous; 8 = Soso, 6498—12,045 ft [1981—3671 m], Lower Cretaceous/Upper Cretaceous.

Field data from various formations showing relationships between the Na/Ca ratio and the total solids are presented in Fig.2.10. In this figure brines from formations older than Pliocene lie generally parallel to the y-axis (total solids), whereas brines from younger formations or marine sediments lie more or less parallel to the x-axis (Na/Ca ratio). In some cases there is a tendency for the Na/Ca ratio to increase with increasing total solids (generally older formations), whereas in others (generally younger formations) the reverse is true. It is of interest to note that the interstitial waters

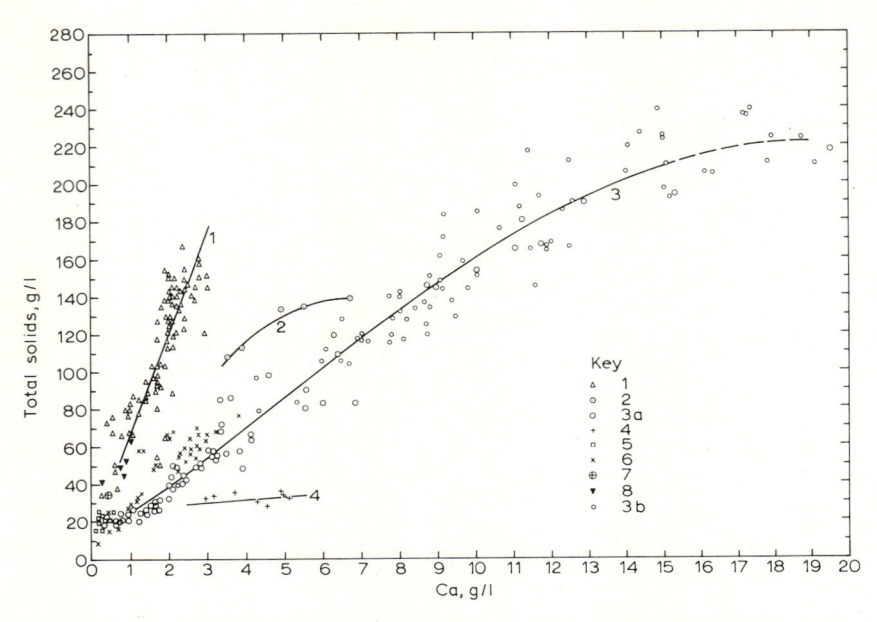

Fig.2.9. Relationship between calcium and total solids content in various oilfield brines, each curve representing a different formation, except curve 3 (data from Rall and Wright, 1953; Wright et al., 1957; Gullikson et al., 1961; Hawkins et al., 1963a,b, 1964; and Graf et al., 1966). *1* = Wilcox Formation, Eocene, Gulf Coast; *2* = Ste. Genevieve Limestone, Mississippian, Illinois Basin; *3a* = Arbuckle Limestone, Cambrian/Ordovician; *3b* = Bartlesville Sandstone, Pennsylvanian, Oklahoma; *4* = Sespe Formation, Oligocene, California; *5* = Pico Formation, Pliocene, California; *6* = Nacatoch Sandstone, Cretaceous, Texas-Louisiana; *7* = seawater; *8* = laboratory data obtained by the writers during compaction experiments.

of the Sespe Formation have a closer affinity with the more "stabilized" brines than the interstitial waters from the younger Pico Formation; the stratigraphic distance between the Sespe and Pico Formations in Ventura County, California, is approximately 20,000 ft [6096 m].

The distribution of Cl^- with depth may be affected by the proximity of salt, which should promote higher salt concentrations in pore fluids. For example, Manheim and Sayles (1970) observed marked increases in interstitial water salinity in two drill holes located in the Gulf of Mexico at a water depth of more than 3500 m. They attributed these increases to diffusion of salt from buried evaporites. In one hole, however, on penetrating the oil-permeated cap rock of a salt dome, they encountered fresh water, which could have originated during oxidation of petroleum hydrocarbons and decomposition of gypsum giving rise to native sulfur.

Posokhov (1966) discussed in detail many of the factors which affect the chemical composition of underground waters: (1) physiogeographic, (2)

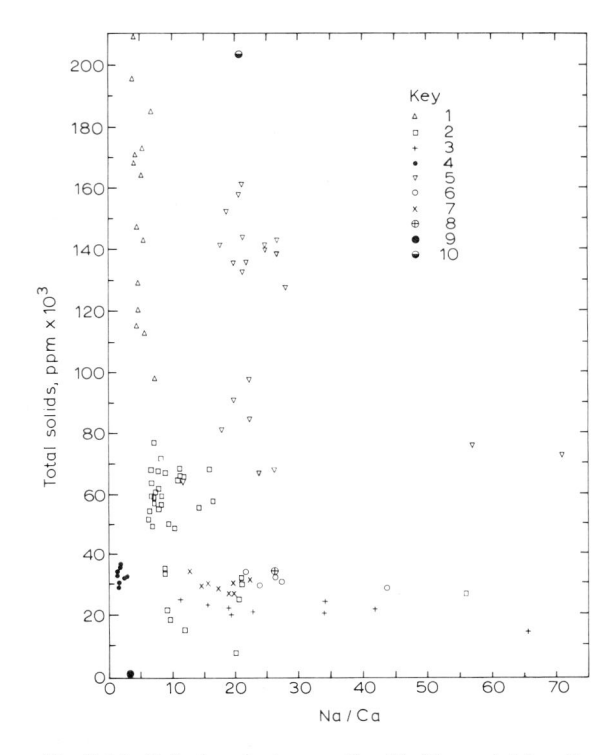

Fig. 2.10. Relation between the Na/Ca weight ratio and total solids in oilfield brines. *1* = Bartlesville Sandstone, Pennsylvanian, Oklahoma; *2* = Nacatoch Formation, Upper Cretaceous, Texas—Louisiana; *3* = Pico Formation, Pliocene, California; *4* = Sespe Formation, Oligocene, California; *5* = Wilcox Formation, Eocene, Gulf Coast; *6* = Recent marine sediments (Siever et al., 1965); *7* = Wilmington oil field, California. (Data after Wright et al., 1957; Gullikson et al., 1961; Hawkins et al., 1963a, 1964.) *8* = seawater; *9* = river water; and *10* = Great Salt Lake (Bentor, 1961).

geologic, (3) hydrogeologic, (4) biologic, (5) physical, and (6) physicochemical. Under physicochemical factors Posokhov discussed: (a) oxidation-reduction conditions of underground waters, (b) solubility of salts, (c) diffusion, (d) osmosis, (e) gravitational differentiation, (f) mixing of different waters, and (g) base exchange.

Changes in the chemistry of solutions squeezed out at different overburden pressures

Most of the dissolved salts present in the interstitial fluids, which are trapped during sedimentation, are squeezed out during the initial stages of compaction. Laboratory results (Kryukov, 1971) showed that mineralization of expelled solutions progressively decreases with increasing overburden pressure. These experimental results led researchers to conclude that the concentrations of interstitial solutions in shales should be lower than those in associated sandstones (see Chilingar et al., 1969).

A corollary of this premise suggests that solutions squeezed out at the beginning of compaction should have higher concentration than the interstitial solutions initially present in argillaceous sediments.

Rieke et al. (1964) determined the percentage increase in the resistivity of expelled solutions from a marine mud with increasing overburden pressure. The mud was obtained from the Santa Cruz Basin, off the coast of southern California. Their results indicate that the mineralization of squeezed-out solutions decreases with pressure.

In some of the experiments conducted by Rieke et al. (1964), the percentage decrease in concentrations of the principal cations and anions with increasing pressure was about the same. These results suggested that (1) the ions being removed represent interstitial electrolyte solution and do not include the adsorbed cations, and (2) the analysis for a single ion in the effluent (for example, Cl^-) might reveal as much insight into the problem as the analysis for all of the ions.

The results of Kryukov and Zhuchkova (1963) demonstrated that the last portions of water (adsorbed?) squeezed out of sediments are poor in electrolytes (Fig.2.11a, b). According to Manheim (1966), the threshold pressure

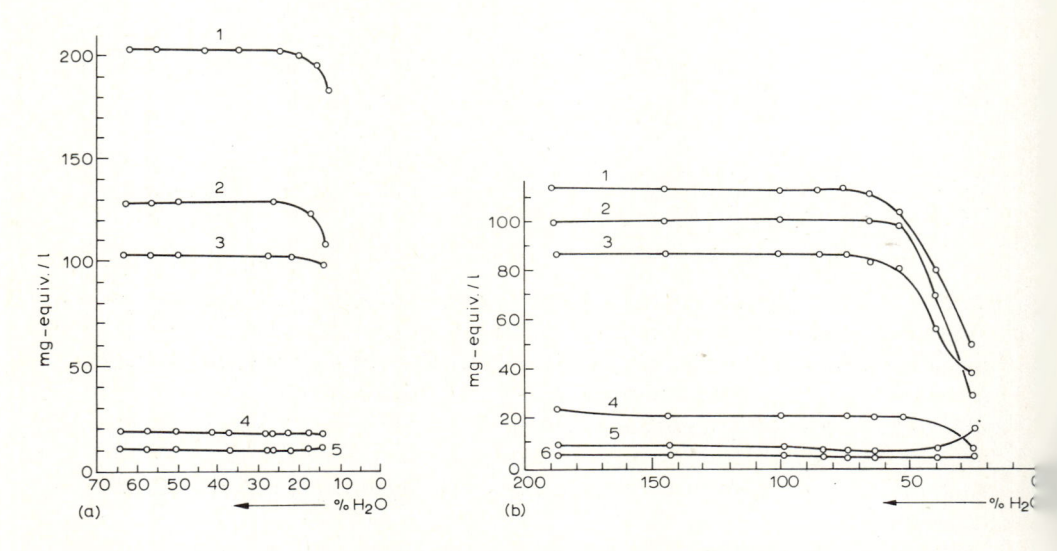

Fig.2.11. (a) Changes in composition of solutions squeezed out of kaolinite clay. $1 = Na^+$; $2 = SO_4^{2-}$; $3 = Cl^-$; $4 = Ca^{2+}$; $5 = Mg^{2+}$ (after Kryukov and Zhuchkova, 1963, p.97). (b) Changes in composition of solutions squeezed out of bentonite. $1 = \kappa \times 10^4$, specific conductivity of solution; $2 = Na^+$; $3 = Cl^-$; $4 = SO_4^{2-}$; $5 = Mg^{2+}$; $6 = Ca^{2+}$ (after Kryukov and Zhuchkova, 1963, p.38).

for the chloride anion in 0.86N NaCl solution was about 1410 kg/cm² (also for Na-bentonite), whereas for fresher waters the influence of pressure on composition was noted at lower pressures. (Threshold pressure is the compaction pressure at which the composition of squeezed-out interstitial solutions starts to change.) Kryukov and Zhuchkova (1963) pointed out that for ordinary sediments, the pressure threshold for influence on the composition of water is shifted to higher pressures. According to Chilingarian and Rieke (1968), the chemistry of squeezed-out solutions begins to change appreciably when the remaining moisture content is about 20—25% for kaolinite and about 50—70% for montmorillonite clay.

Kazintsev (1968, p.186) observed in laboratory experiments a gradual decrease in chlorine concentration on squeezing samples of Maykop clay (eastern Pre-Caucasus) having an initial moisture content of 20 and 25% (Fig.2.12). The final moisture contents after compaction constituted 8.83

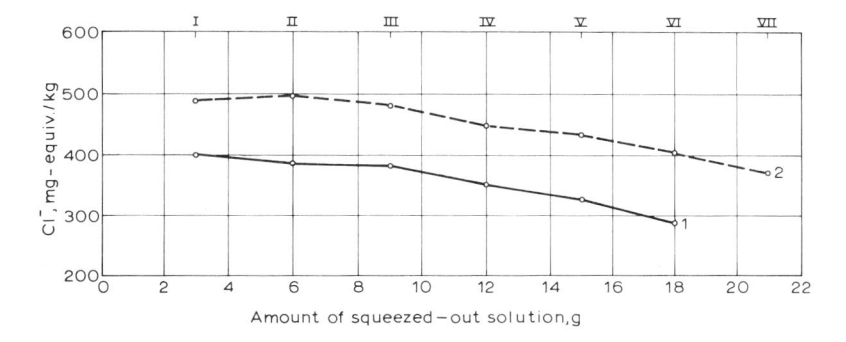

Fig.2.12. Variation in chlorine content in subsequent fractions (I—VII) of squeezed-out interstitial solutions of Maykop clay, eastern Pre-Caucasus (after Kazintsev, 1968, fig.1, p.186). 1 = depth of 42 m, Divnoe area, 2 = depth of 158 m, Divnoe area, U.S.S.R.

and 10.88%, respectively. He also determined the effect of temperature (heating to 80°C) on the concentration of various ions in squeezed-out solutions (Fig.2.13). The concentration of Cl^- and Na^+ decreases with increasing pressure and temperature does not seem to have any appreciable effect. The Mg^{2+} ion concentration increases (about 1.5 times) with increasing pressure, but the absolute values are lower at high temperatures than at low temperatures. The concentration of K^+ decreases with pressure. Concentrations of K^+, Li^+, and I^- are higher in solutions expelled at higher temperatures, whereas that of SO_4^{2-} is lower.

Krasintseva and Korunova (1968, p.191) studied the variations in chemistry of solutions expelled from unlithified marine muds from the Black Sea. At room temperature, the chlorine concentration definitely decreases with increasing pressure, whereas the concentrations of some components go

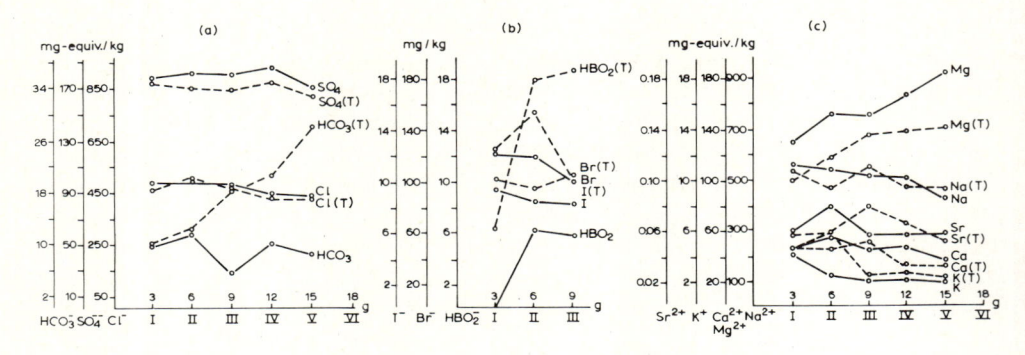

Fig.2.13. Changes in concentration of anions, cations and microcomponents with increasing compaction in subsequent fractions (I—VII) of extruded interstitial solutions. Maykop clay, depth of 158 m, Divnoe area, eastern Pre-Caucasus, U.S.S.R. (after Kazintsev, 1968, fig.2, p.188). Solid curves = room temperatures, dashed curves = heated to 80°C. The amount of extruded solutions in grams is plotted on the abscissa.

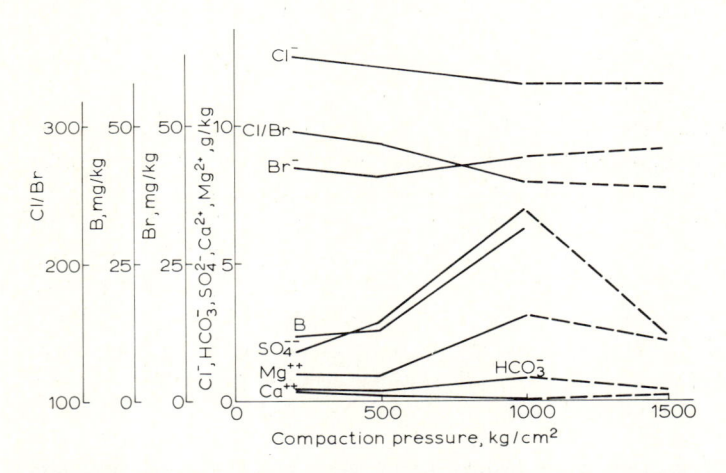

Fig.2.14. Relationship between concentration of various ions in interstitial solutions squeezed out of marine mud and compaction pressure at room temperature (after Krasintseva and Korunova, 1968, fig.2, p.195).

through a maximum at pressures of 500—1000 kg/cm² (Fig.2.14). The Br⁻ and B^{3+} contents increase with increasing compaction pressure. Krasintseva and Korunova also presented the relationship between the concentration of various ions and compaction pressure at 80°C. Increasing temperature seems to decrease the amount of Mg^{2+} cation in expelled solutions (Fig.2.15).

Some investigators, however, disagree with the above-described findings. For example, the study by Manheim (1966), who used pressures ranging from 41 to 844 kg/cm², indicates that pressure does not appreciably affect

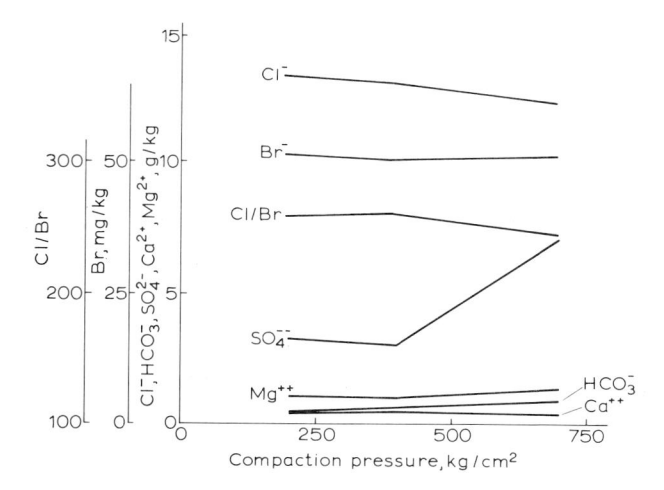

Fig.2.15. Variation in concentration of various ions in interstitial solutions expelled from marine mud with increasing pressure at 80°C (after Krasintseva and Korunova, 1968, fig.3, p.196).

the composition of extracted waters. Shishkina (1968, p.167), on investigating interstitial solutions in marine muds from the Atlantic and the Pacific Oceans and from the Black Sea, did not observe any appreciable changes in the chemistry of squeezed-out solutions up to a pressure of 1260 kg/cm² in some samples and up to a pressure of 3000 kg/cm² in others. There was some increase in Ca^{2+} concentration at a pressure range of 675—1080 kg/cm² and then there was a decrease at higher pressures. Shishkina (1968) stated that at compaction pressures, at which 80—85% of interstitial water is squeezed out, there are no changes in concentration. Obviously, the chemistry of the remaining 20—15% interstitial fluid is also of great interest, and should be determined in most studies.

Von Engelhardt and Gaida (1963) compacted pure montmorillonite and kaolinite clay muds saturated with solutions having different concentrations of NaCl and $CaCl_2$ at different pressures ranging from 30 to 3200 atmospheres. Their results show that for a given clay, the equilibrium porosity which is reached at a distinct overburden pressure does not depend on electrolyte concentration. For pressures between 30 and 800 atmospheres, the concentration of electrolyte in pore fluids in montmorillonite clay diminishes with increasing compaction. This was explained by Von Engelhardt and Gaida (1963) as due to the electrochemical properties of base-exchanging clays. If the pore fluids contain an electrolyte, the liquid immediately surrounding the clay particle will contain less electrolyte than fluids farther away from the double layer. Base-exchanging clays suspended in electrolyte solution adsorb a certain amount of pure water which is bound in double layers around each clay particle (Von Engelhardt and Gaida, 1963, p.929). During compression, the electrolyte-rich solution is removed and the fluid of the double layers, poor in electrolyte content, is left behind. At higher overburden pressures (from 800 to 3200 atmospheres), an increase of salt concentration within the remaining pore water may be caused by the inclusion of small droplets of fluid in the highly compressed clay, acting as a barrier to movement of ions. The passage of anions through the double layer is retarded by the fixed negative surface charges on the clay particles. Ion blocking increases ion-exchange capacity and compression of the clay. Apparently, ion blocking is greater for dilute solutions than for concentrated ones.

Chilingarian et al. (1973b) experimented with a sample of montmorillonite clay that was saturated in seawater for a period of seven days (the volume of seawater was in excess of that of clay solids). The sample was shaken vigorously twice a day. Then the supernatant liquid, which was assumed to have the same composition as the free interstitial water, was removed and analyzed. The remaining saturated sample was placed in a hydrostatic compaction apparatus and the successive portions of the expelled solutions were analyzed. The final remaining moisture content was equal to 62%, which corresponded to an overburden pressure of about 35 kg/cm^2. Table 2.I shows the variation in concentration of various ions. Variation in total dissolved solids in subsequent fractions of expelled solutions indicates that the concentrations of the solutions squeezed out at the

TABLE 2.I

Variation in concentration of various components in solutions expelled from montmorillonite clay saturated with seawater (Chilingarian et al., 1973b)

Ion	Concentration (mg/l)						
	seawater	supernatant fluid	expelled solutions				
			fraction no.	I	II	III	IV	V
			cumulative volume (cm^3) 16.5	25.5	34.5	49.5	59.0
Ca^{2+}	690	560		460	560	580	620	560
Mg^{2+}	1,189	572		644	557	669	657	754
Na^+	10,116	13,400		13,300	13,200	13,400	13,200	12,800
K^+	400	210		226	216	206	206	160
HCO_3^-	520	165		262	165	189	165	128
SO_4^{2-}	2,759	4,610		5,350	5,840	5,270	4,610	5,180
Cl^-	18,929	19,310		19,030	19,200	19,170	19,170	18,990
F^-	3	20		20	20	<20	<20	40
NO_3^-	34	3		0	0	0.5	0	0
$CaCO_3$	6,612	3,750		3,800	4,100	4,200	4,250	4,500
Fe^{2+}	—	24		36	56	28	28	64
Mn^{2+}	—	5		<5	<5	<5	<5	<5
SiO_2	43	15		5	0	5	10	20
B^{3+}	4	14		21	14	17	4	0
Total dissolved solids	34,423	38,804		39,216	39,844	39,433	38,576	38,611

initial stage of compaction are slightly higher than that of the interstitial solution initially present in montmorillonite clay saturated in seawater. Concentrations of expelled solutions, therefore, go through a maximum (peak), or at least remain constant, before starting to decrease with increasing overburden pressure.

As discussed before, according to many investigators, the salinity of squeezed-out solutions progressively decreases with increasing overburden pressure. Consequently, the salinity of interstitial solutions in shales is possibly less than that of waters in associated sandstones, because practically all of the interstitial fluids were expelled in many of the laboratory experiments. It has been also observed that during production of crude oil from sandstones, surrounded by thick shale sequences, the salinity of produced water gradually decreases with time, possibly owing to the influx of fresher water from the associated shales.

The mineralization of solutions moving upward through a thick shale sequence as a result of compaction probably will progressively increase in salinity. It should be remembered, however, that if water from a sandstone bed moves through a shale layer into another sandstone bed, the water in the latter bed may be less mineralized because of filtration through a charged-net membrane.

Overton and Timko (1969) studied the chemistry of interstitial solutions in the subsurface formations (sands and shales) of the Gulf Coast area. Figs.2.16 and 2.17, which are based on their data, show on semilog paper the relationship between (1) pore water salinity in sandstones and depth, and (2) porosity of shale and depth. Overton and Timko calculated the pore water salinity from an SP log, assuming that the salinity in sandstones is in equilibrium with that in the nearby shales. Figs.2.16 and 2.17 suggest that in the normally compacted zones the shale porosity and pore water salinity are inversely related. In these examples the normally compacted zones are the intervals above a depth of about 9000 ft [2743 m]. Overton and Timko (1969) stated: "as the shale is compressed to one-half its original pore fraction, the water is pressed out, leaving salt behind to concentrate itself by a factor of two." W.H. Fertl (personal communication, 1974), however, observed that in many instances undercompacted shales contain more saline water than comparable well-compacted ones (also see Rieke and Chilingarian, 1974, p.25).

Magara (1974, p.283) stated that the salinity distribution in shales should be a reciprocal of the porosity, due to the ion-filtration effect of the clays, and that the salinity in the shales should tend to increase towards the sand beds, because of the higher porosity of the central zones of the clays undergoing compaction. These conclusions, however, do not seem to be substantiated by the laboratory experiments discussed above. In addition, as pointed out by Rieke and Chilingarian (1974, p.234), a semipermeable membrane effect becomes operative only at a certain minimum void ratio, at some high overburden pressure, e.g. 11,000 psi [773 kg/cm^2].

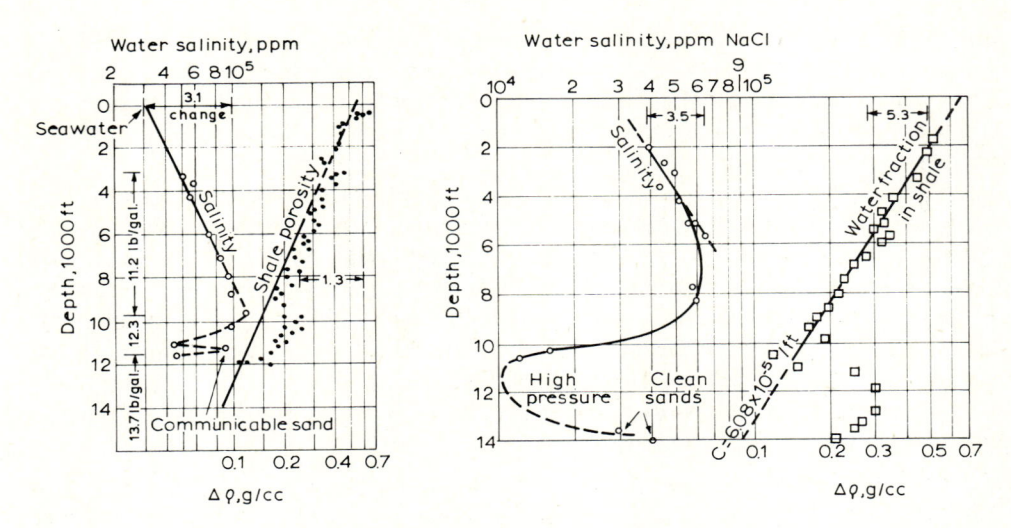

Fig.2.16. Pore water salinity—depth and shale porosity—depth relationships in a Gulf Coast well (after Overton and Timko, 1969, fig.2, p.116). $\Delta\rho = 2.55 - \rho_{sh}$, where 2.55 is the matrix density in g/cm³ and ρ_{sh} is the shale density. Shale porosity ϕ_{sh} is equal to $(2.55 - \rho_{sh})/(2.55 - \rho_w)$ or $(2.55 - \rho_{sh})/1.5$, inasmuch as water density, ρ_w, can be assumed to be equal to 1. Values of 11.2, 12.3, and 13.7 lb/gal indicate specific weight of drilling fluid ("mud") used in drilling. Location: Terrebonne Parish, Ship Shoal area, Gulf Coast, U.S.A.

Fig.2.17. Pore water salinity—depth and shale porosity—depth relationships in a Gulf Coast well (after Overton and Timko, 1969, fig.3, p.116). $\Delta\rho = 2.55 - \rho_{sh}$ (see explanation for Fig.2.16); C = compaction coefficient in psi⁻¹. Location: Grand Isle area, Gulf Coast, U.S.A.

Salinity distribution in sandstones and associated shales

In 1947, De Sitter reported that the salinity of formation waters in sandstones varies from that of fresh water to ten times the salinity of seawater. The distribution of the salinity of interstitial waters present in young geosynclinal sediments along the U.S. Gulf Coast has been well documented by Timm and Maricelli (1953), Myers (1963), and Fowler (1968).

Timm and Maricelli (1953, p.394) stated that high salinities up to 4.5 times that of normal seawater characterize the interstitial solutions in Miocene/Pliocene sediments, where the relative quantity of undercompacted shale is small. In Eocene/Oligocene sediments, where the relative quantity of shale is large and the degree of compaction is high, interstitial solutions have salinities as low as one-half that of normal seawater. Fig.2.18 illustrates their concept that the formation waters in downdip, interfingering, marine sandstone members, which have proportionately less volume than the associated

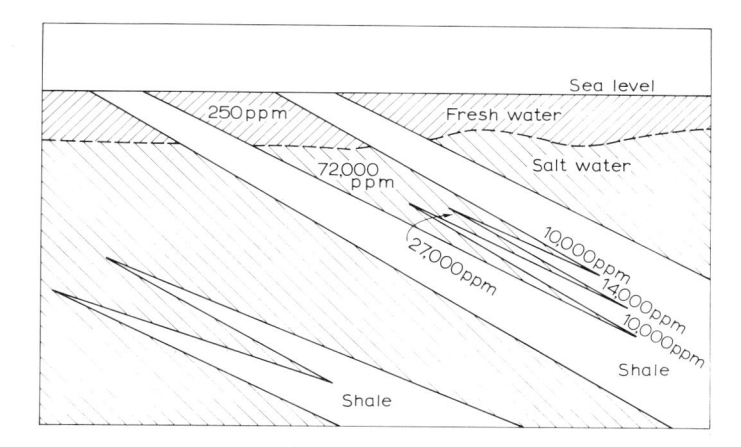

Fig.2.18. Idealized typical cross section of some sands and shales in southwest Louisiana showing generalized salinity relationships (modified after Timm and Maricelli, 1953, p.396, 397, and 408 — Courtesy of the American Association of Petroleum Geologists).

massive shales, have lower salinities than that of seawater. More massive sands updip have salinities greater than that of seawater, because of lack of influx of fresher waters from shales. Salinity was determined by using the following techniques: electrical resistivity, complete chemical analysis, and titration (see Gullikson et al., 1961). Calculations showed that all water samples, of which complete mineral analyses were made, are secondary saline according to the Palmer's system of water-analyses interpretation.

Myers (1963) studied the chemical properties of formation waters, down to a depth of 12,400 ft [3780 m], in four producing oil wells in Matagorda County, Texas. Salinities of interstitial waters ranging from 5000 ppm to 12,500 ppm were found below 10,000 ft [3048 m] in each of the four wells, as compared to salinities of about 70,000 ppm above this depth. Myers commented that in this deeper section, the proportion of massive shale is large and the sands are near their downdip limits (become thinner). These results were in close accord with those of Timm and Maricelli (1953).

Some investigators, including Hottman and Johnson (1965), observed that the sands with abnormally high pore water pressures are associated with undercompacted shales having very high porosity. In an excellent paper, Dickey et al. (1968) observed that faults which transect oil reservoirs form pressure discontinuities and act as seals for zones of high fluid pressure for long periods of time. The high porosity of shales in such zones is reflected by the high values of conductivity. The depth marking the beginning of the abnormally high fluid pressures in the sandstones coincides with the abnormal increase in conductivity of associated undercompacted shales (Wallace, 1965; Williams et al., 1965). Yet, calculations by the writers indicate that possibly high porosity of shales alone could not account for this abnormal increase in conductivity; the salinity of interstitial waters also appears to be important.

Although the process of clay compaction is continuous until the complete (?) lithification of clays, the volume of expelled fluids into the reservoir rocks gradually decreases. Consequently, there is a gradual decrease in excessive pore water pressure in highly permeable horizons in deep parts of the basin until, finally, it becomes equal to the hydrostatic pressure. After that, theoretically the movement of waters occurs in the opposite direction, from the periphery of the basin (recharge areas). The entire hydrodynamic system strives to attain equilibrium, which is controlled by the areal distribution and size of recharge and discharge areas.

The findings of Fowler (1968) for the Chocolate Bayou field, Brazoria County, Texas, seem to suggest to the writers that the salinity of water in undercompacted shales is higher than in well-compacted ones. He discovered a definite correlation between the high salinity of interstitial fluids and abnormally high pressures. This is possibly owing to the fact that undercompacted shales did not have a chance to contribute their fresher water to the associated sandstones. In addition, he studied the variation in salinity of produced water with time. The typical pattern is one of decreasing salinity with time, and the freshest water is found in sands receiving most of this water from associated shales. This is in agreement with the experimental results of several investigators, as discussed earlier, which indicate that the salinity of waters in shales should be less than that in associated sands.

The pore water salinities in shales and in associated sandstones are compared in Fig.2.19. The examples include field case studies from the

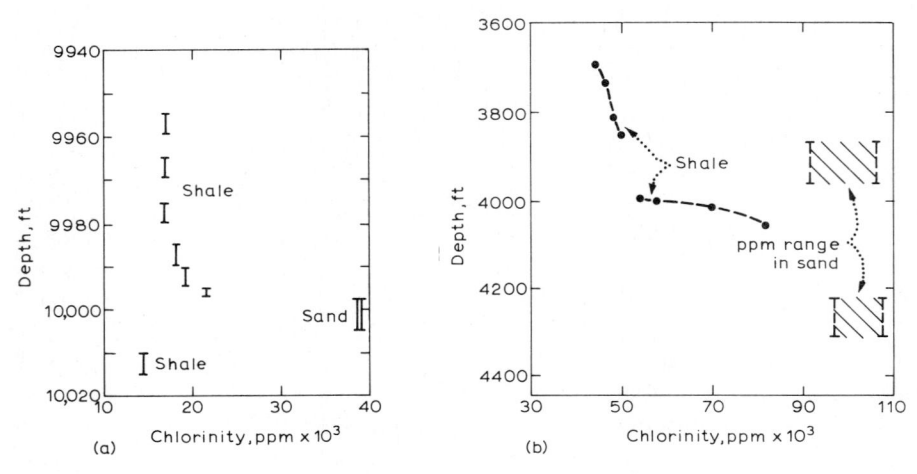

Fig.2.19. Chloride concentrations in shales and sands (in: Fertl and Timko, 1970c, fig.4, p.15, based on data by Hedberg, 1967).

Middle East and Texas. Some data from offshore wells in Louisiana have been given by Fertl and Timko (1970c).

Schmidt's (1973) study showed that the compositions of interstitial water in shales and those in sandstones are different. Sidewall cores of shales, analyzed by him, were taken at intervals of every 500 ft [152 m] between depths of 3000 ft [914 m] and 14,000 ft [4257 m] in a well in Calcasieu Parish, Louisiana. Abnormally high fluid pressures were encountered just below a depth of 10,000 ft [3048 m]. He noted significant differences between the total dissolved solids concentrations in waters from the normally pressured sandstones (600—180,000 mg/l) and the highly pressured sandstones (16,000—26,000 mg/l). The salinity of the water in shales is lower than that in the adjacent normally pressured sandstones. The concentrations, however, were found to be more similar in the high-pressure zone. Schmidt found that in shale pore water, the concentration order is generally $SO_4^{2-} > HCO_3^- > Cl^-$, whereas water in normally pressured sandstones has an opposite concentration order. In all cases, the salinities of interstitial fluids in shales were found to be considerably lower than those in associated sandstones. Rieke et al. pointed out this relationship in 1964.

Stresses in sediments

Hydrostatic stress

Jürgenson (1973) recommended that the term hydrostatic stress be clarified. This is necessary in order to apply the term properly to the constitutive relationships.

The term hydrostatic stress is usually defined as the fluid pressure exerted by a liquid at rest, and at a given point it acts with equal intensity in all directions. A submerged solid is subjected to equal pressure from all sides. For all practical purposes, the hydrostatic stress caused by water does not affect the friction between mineral grains and does not contribute to the strength (shearing resistance) of the sediment body.

Hubbert and Rubey (1959, p.129) stated that within depths of 1 or 2 km, the pressure of the water as a function of the depth, D, can be closely approximated by the equation:

$$p_w = \rho_w g D \qquad (2.50)$$

where p_w is the hydrostatic pressure of a column of water extending from the surface of the ground to a depth of D within the porous column; ρ_w is the density of the water; and g is the acceleration of gravity.

The specific weight of any fluid, γ_f, can be expressed as:

$$\gamma_f = \rho_f g \qquad (2.51)$$

or:

$$\rho_f = \gamma_f/g \tag{2.52}$$

Thus, the hydrostatic pressure, p_f, at a depth D is equal to:

$$p_f = \gamma_f \cdot D \tag{2.53}$$

The specific gravity, SG, is equal to:

$$SG = \gamma_f/\gamma_w \tag{2.54}$$

and:

$$\gamma_f = SG \cdot \gamma_w \tag{2.55}$$

where γ_w is the specific weight of water. On combining eqs. 2.53 and 2.55:

$$p_f = SG \cdot \gamma_w \cdot D \tag{2.56}$$

If γ_w is in pounds per cubic foot and D is in feet, then p_f is in pounds per square foot. The pressure gradient (p_f/D) for pure water ($\gamma_w = 62.4$ lb/ft^3) is equal to 0.433 psi/ft $[= (62.4 \text{ lb/ft}^3)/(144 \text{ in}^2/\text{ft}^2)]$.

Resolution of the total stress field

The stress tensor for a porous, homogeneous, isotropic sediment body can be written in the conventional way:

$$\mathbf{S} \equiv \begin{bmatrix} \sigma_x & \tau_{xy} & \tau_{xz} \\ \tau_{yx} & \sigma_y & \tau_{yz} \\ \tau_{zx} & \tau_{zy} & \sigma_z \end{bmatrix} \tag{2.57}$$

where S signifies the symmetrical tensor of the total stress; σ_i and τ_{ij} represent the normal and shear forces, respectively, acting on the faces of a unit volume of sediment (Fig.2.20). Moments can be taken about point O in Fig.2.20. The tangential stress, τ_{xy}, multiplied by the area in which it acts, gives the force τ_{xy} $dzdy$, and this times dx gives a clockwise moment about O. The stress τ_{yx} times the area gives τ_{yx} $dxdz$, and the latter times dy results in a counterclockwise moment τ_{yx} $dxdzdy$. At equilibrium, the two moments balance each other:

$$\tau_{xy}\, dzdydx = \tau_{yx}\, dxdzdy \tag{2.58}$$

or:

$$\tau_{xy} = \tau_{yx} \tag{2.59}$$

Then it follows that:

$$\tau_{xz} = \tau_{zx} \tag{2.60}$$

and:

$$\tau_{yz} = \tau_{zy} \tag{2.61}$$

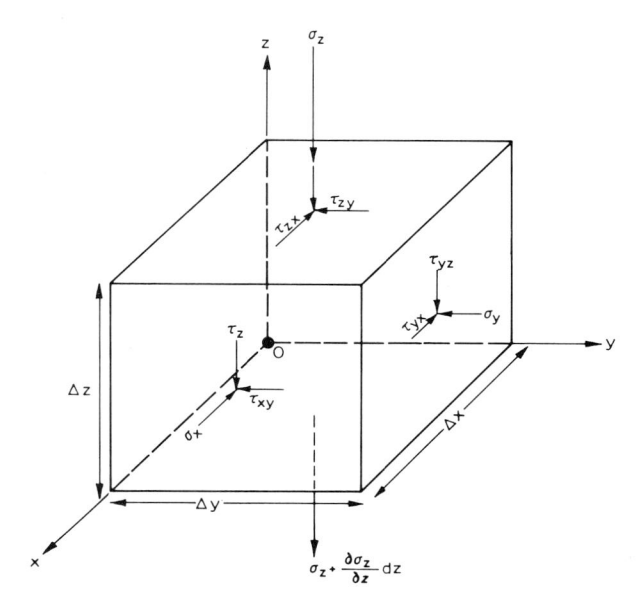

Fig.2.20. Stress notation in a cubic argillaceous rock slice. Stress notation of the normal component of stress, σ_z, on the plane normal to the z-axis; τ_{zx} and τ_{zy} refer to the shear stress components in the plane normal to the z-axis and acting in the x and y directions, respectively. Point O is in equilibrium with respect to the forces in the x and y directions. $\sigma_z + (\delta\sigma_z/\delta z)dz$ is the incremental change in the vertical stress through the free body.

The total stress array for a point in a cylindrical body under compaction, such as the ones studied in the laboratory by numerous investigators, can be expressed in cylindrical coordinates r, θ, and z:

$$\mathbf{S} \equiv \begin{bmatrix} \sigma_r & \tau_{r\theta} & \tau_{rz} \\ \tau_{\theta r} & \sigma_\theta & \tau_{\theta z} \\ \tau_{zr} & \tau_{z\theta} & \sigma_z \end{bmatrix} \tag{2.62}$$

In order to visualize more clearly the ability of sedimentary deposits to bear an overburden load, the total stress tensor is decomposed into two distinct parts for a body in equilibrium: (1) the "hydrostatic stress" tensor, which in analogy with the pressure in water, is acting with equal intensity in all directions, and (2) the deviator stress component which is not equal in all directions. The term "hydrostatic stress" above does not mean a pressure acting in the water.

Jürgenson (1973) pointed out that in the theory of elasticity this "hydrostatic stress", which would produce the same volume stress as the actual state of stress considered, is numerically equal to the octahedral normal stress $[\frac{1}{3}(\sigma_1 + \sigma_2 + \sigma_3)]$. In the case of a sediment, it is understood to include the pore pressure, i.e. the *true hydrostatic pressure*. Thus, a double perplexity exists. The "hydrostatic stress" exerts an intergranular pressure and contributes to strength, which a true hydrostatic stress cannot do. In addition, one of the components of the "hydrostatic stress" is a true hydrostatic stress. The "hydrostatic stress", therefore, is merely a mathematical quantity and does not exist physically as an all-sided pressure (Jürgenson, 1973, p.448).

It seems that the term hydrostatic has to be retained in its true sense of a stress acting in a liquid at rest. An entirely different term must be used for the all-sided stress producing an equivalent state of volume stress. It was suggested by Jürgenson (1973) that the term most appropriate would be *volumetric stress*. The magnitude of the volumetric stress varies between the limits of $\frac{1}{3}(\sigma_1 + 2\sigma_2)$ and $\frac{1}{2}(2\sigma_1 + \sigma_3)$. The deviator component, which represents a rather complicated concept, is defined as $(\sigma - \sigma_3)$, where σ_3 is the spheric stress that is equal in all directions.

The above division into components is very useful in explaining the strength behavior of sediments undergoing compaction, i.e. the ability of the sediments to resist shear. The spheric component, being equal in all directions within the sediment body, causes no shearing stresses. When the mineral grains are pressed against each other, the frictional resistance develops, which is equal in all directions. As pointed out by Jürgenson (1973, p.450), the intergranular pressure and, thus, the material's strength, is raised by the deviator pressure as well, but in unequal degrees in different planes.

Spheric stress state

The component attributable to the interstitial fluid is the hydrostatic stress (pressure), σ_w, which can be regarded as being continuous throughout the medium. The normal and shear stress components are given by:

$$\mathbf{P} \equiv \begin{bmatrix} \sigma_{wx} & \tau_{wxy} & \tau_{wxz} \\ \tau_{wyx} & \sigma_{wy} & \tau_{wyz} \\ \tau_{wzx} & \tau_{wzy} & \sigma_{wz} \end{bmatrix} \quad (2.63)$$

where **P** is the spheric stress tensor. It can be safely assumed that under hydrostatic conditions no shearing stresses exist in the interstitial fluid. As a matter of fact and by definition, a fluid is a substance which cannot sustain tangential or shear forces when in static equilibrium. This may not hold true for the adsorbed water because of its probable quasicrystalline nature. Hubbert and Rubey (1959, p.138) stated that if a viscous liquid occupies the pore space, there are then microscopic shear stresses which are expended locally against the fluid/solid boundaries. Thus, their only macroscopic effect is to transmit to the solid skeleton by viscous coupling whatever net impelling force may be applied to the interstitial fluid.

STRESSES IN SEDIMENTS

In any stress system with the principal stresses, σ_x, σ_y, and σ_z, one can define the local mean value of the quantity for the spheric stress, $\bar{\sigma}_w$, as:

$$\bar{\sigma}_w = \tfrac{1}{3}(\sigma_{wx} + \sigma_{wy} + \sigma_{wz}) \tag{2.64}$$

Now, the spheric stress tensor, **P**, can be represented by:

$$\mathbf{P} \equiv \begin{bmatrix} \bar{\sigma}_w & 0 & 0 \\ 0 & \bar{\sigma}_w & 0 \\ 0 & 0 & \bar{\sigma}_w \end{bmatrix} \tag{2.65}$$

and:

$$\mathbf{P} = \tfrac{1}{3}(3\bar{\sigma}_w) = \bar{\sigma}_w \tag{2.66}$$

The above expression represents the hydrostatic pressure of a fluid, whether it is flowing or is stationary in the porous system of a shale. Thus, $\sigma_{wx} = \sigma_{wy} = \sigma_{wz} = \bar{\sigma}_w$ and the hydrostatic portion of the total stress system causes only volume changes in the deformed material.

Deviatoric stress state

The second component is the stress deviator from the hydrostatic state. It is expressed as the difference between the total stress and the hydrostatic stress which resists deformation:

$$\mathbf{D} \equiv \begin{bmatrix} (\sigma_x - \sigma_{wx}) & \tau_{xy} & \tau_{xz} \\ \tau_{yx} & (\sigma_y - \sigma_{wy}) & \tau_{yz} \\ \tau_{zx} & \tau_{zy} & (\sigma_z - \sigma_{wz}) \end{bmatrix} \tag{2.67}$$

where **D** is the deviatoric part of the total stress tensor. The effect of the deviator stress is to produce a distortion which is elastic or plastic in nature and is introduced into the sediment body.

Total stress tensor

If the sediment body is not in equilibrium, the second component will not be a symmetric tensor for $\tau_{xy} \neq \tau_{yx}$. This asymmetric tensor can be subdivided into symmetric and skew-symmetric parts (Ramsay, 1967, p.282). The hydrostatic stress component is the same as in eq. 2.64. The second symmetrical part is the deviatoric stress component which can be expressed as:

$$\mathbf{D} \equiv \begin{bmatrix} (\sigma_x - \bar{\sigma}_w) & \tfrac{1}{2}(\tau_{xy} + \tau_{yx}) & \tfrac{1}{2}(\tau_{xz} + \tau_{zx}) \\ \tfrac{1}{2}(\tau_{xy} + \tau_{yx}) & (\sigma_y - \bar{\sigma}_w) & \tfrac{1}{2}(\tau_{yz} + \tau_{zy}) \\ \tfrac{1}{2}(\tau_{xz} + \tau_{zx}) & \tfrac{1}{2}(\tau_{yz} + \tau_{zy}) & (\sigma_z - \bar{\sigma}_w) \end{bmatrix} \tag{2.68}$$

The skew-symmetric part is termed the disequilibrium component, which causes the sediment to undergo a rotation in space and is expressed as:

$$\mathbf{R} \equiv \begin{bmatrix} 0 & \frac{1}{2}(\tau_{xy} - \tau_{yx}) & \frac{1}{2}(\tau_{xz} - \tau_{zx}) \\ \frac{1}{2}(\tau_{yx} - \tau_{xy}) & 0 & \frac{1}{2}(\tau_{yz} - \tau_{zy}) \\ \frac{1}{2}(\tau_{zx} - \tau_{xz}) & \frac{1}{2}(\tau_{zy} - \tau_{yz}) & 0 \end{bmatrix} \quad (2.69)$$

where **R** is the disequilibrium component. Such a stress state would be expected if tectonic forces were acting on the sediment mass in a basin within a geosyncline. The total stress tensor for a sediment body not in equilibrium is expressed as the sum of the above-described parts:

$$\mathbf{S} = \mathbf{P} + \mathbf{D} + \mathbf{R} \quad (2.70)$$

(total stress = spheric stress + deviatoric stress + disequilibrium component)

Each one of the three components making up the state of stress is directly related to the respective component of the strain tensor. The spheric part of the stress system causes changes in volume, the deviatoric stress components cause distortion, and the disequilibrium components cause the material to undergo rotation in space (Ramsay, 1967).

Lo (1969) showed mathematically that the pore pressure induced by shear may be expressed as a sole function of the major-principal strain. According to him, the only unambiguous and correct principle of superposition of pore pressure is to consider an isotropic stress system and a deviatoric stress system, namely:

$$\begin{bmatrix} \Delta\sigma_1 & 0 & 0 \\ 0 & \Delta\sigma_2 & 0 \\ 0 & 0 & \Delta\sigma_3 \end{bmatrix} = \begin{bmatrix} \Delta\sigma_3 & 0 & 0 \\ 0 & \Delta\sigma_3 & 0 \\ 0 & 0 & \Delta\sigma_3 \end{bmatrix} + \begin{bmatrix} \Delta\sigma_1 - \Delta\sigma_3 & 0 & 0 \\ 0 & \Delta\sigma_2 - \Delta\sigma_3 & 0 \\ 0 & 0 & 0 \end{bmatrix} \quad (2.71)$$

where σ_1 is the total major stress, σ_2 is total intermediate stress and σ_3 is total minor stress.

According to Lo (1969), the physical justification for eq. 2.71 lies in the fact that under ambient stress, the induced pore pressure corresponds almost exactly to the applied pressure, because the compressibilities of pore water and sediment grains are much lower than that of the sediment structure. In the experience of the writers, however, most of the pore pressure equations presented in the literature give almost identical results, provided they are properly used.

According to Jürgenson (1973): "Defenders of the old terminology adopted in the theory of elasticity use as their main argument the longevity of their 'hydrostatic' and 'deviatoric' stresses." Although this is perfectly true, it does not take into account that mechanics of sediments deals with a type of material quite different from metals which have been the main concern in the theory of elasticity. Jürgenson (1973) continued to state that "the difference in the physics of the strength of these materials can well be compared considering that in metals the intergranular pressure is estimated to reach several hundred thousand atmospheres and the angle of internal friction is measured in decimals of a degree, while in soils the intergranular pressure and the pore pressure (which as a term is a comparative newcomer) usually do not exceed a few atmospheres, but the angle of internal friction may exceed 40 degrees."

As pointed out by Jürgenson (1973), the insufficiency of the old terminology is caused by the peculiarities of the mechanics of sediments and soils, which require the consideration of widely different factors of strength. The pressure caused by water is included here (Table 2.II).

TABLE 2.II

Terminology recommended by Jürgenson (1973, p.450)

Term	Symbol*
Hydrostatic stress	σ_w
Spheric	σ_3
Deviatoric	$\sigma - \sigma_3$
Main deviator	$\sigma_1 - \sigma_3 = 2\tau_{max}$
Compressive strength	$\sigma_1 - \sigma_3 = 2\tau_{max}$
Volumetric stress	$\frac{1}{3}(\sigma_1 + \sigma_2 + \sigma_3)$
Mean normal stress	$\frac{1}{3}(\sigma_1 + \sigma_2 + \sigma_3)$
Normal octahedral	$\frac{1}{3}(\sigma_1 + \sigma_2 + \sigma_3)$
Octahedral deviatoric	$\sigma - \frac{1}{3}(\sigma_1 + \sigma_2 + \sigma_3)$

*One should consult the state of stress in Mohr's representation and the state of stress in polar coordinates.

Compressibilities of sand and clayey sediments

Compressibility, c, can be defined as the rate of change of volume, ∂V, with respect to the applied stress, σ, per unit of volume, V:

$$c = -\frac{1}{V}\left(\frac{\partial V}{\partial \sigma}\right) \tag{2.72}$$

There are several different usages of the term compressibility, which appear in the literature depending on the method of determination: (1) bulk compressibility, (2) pore compressibility, (3) formation compressibility, (4) rock solids compressibility, and (5) pseudo-bulk compressibility (Table 2.III).

Although many consolidation (compressibility) studies on clays and shales have been performed in soil-mechanics laboratories for more than 50 years, these tests have been limited largely to a low-pressure range (<1000 psi

TABLE 2.III

Various compressibility formulas used in the literature

Formula	
Bulk compressibility: $$c_b = -\frac{1}{V_b}\left(\frac{\partial V_b}{\partial \bar{\sigma}}\right)_{p_p, T}$$	Change in the bulk volume (∂V_b) per unit of bulk volume (V_b), per unit change in total external stress ($\bar{\sigma}$), while keeping the pore pressure (p_p) and temperature (T) constant.
$$c_b = -\frac{1}{V_b}\left(\frac{\partial V_b}{\partial p_e}\right)_{\bar{\sigma}, T}$$	Change in the bulk volume (∂V_b) per unit of bulk volume (V_b), per unit change in effective pressure ($p_e = \bar{\sigma} - p_p$), while keeping the total external stress ($\bar{\sigma}$) and temperature (T) constant.
$$c_b = -\frac{1}{e+1}\left(\frac{\partial e}{\partial p_e}\right)$$ or: $$c_b = -\frac{1}{h}\left(\frac{\partial h}{\partial p_e}\right)$$	Determined in the uniaxial compaction apparatus, if pore pressure $p_p = 0$, i.e. $\bar{\sigma} = p_e$. Void ratio, e, is equal to the volume of voids (V_p) divided by the volume of solids (V_s): $e = V_p/V_s = \phi/(1-\phi)$, where ϕ is fractional porosity. h is a sample thickness in a uniaxial compaction apparatus, i.e. thick-walled cylinder.
Pore compressibility: $$c_p = -\frac{1}{V_p}\left(\frac{\partial V_p}{\partial \bar{\sigma}}\right)_{p_p, T}$$ or: $$c_p = -\frac{1}{V_p}\left(\frac{\partial V_p}{\partial p_e}\right)_{\bar{\sigma}, T}$$	Change in the pore volume (∂V_p) per unit of pore volume (V_p), per unit change of external stress ($\bar{\sigma}$), keeping the pore pressure (p_p) and temperature (T) constant. Change in the pore volume (∂V_p) per unit of pore volume (V_p), per unit change in effective pressure (p_e), while keeping the total external stress ($\bar{\sigma}$) and temperature (T) constant.
Formation compressibility (pore compressibility of some authors): $$c_f = -\frac{1}{V_p}\left(\frac{\partial V_p}{\partial p_p}\right)_{\bar{\sigma}, T}$$	Change in the pore volume (∂V_p) per unit of pore volume (V_p), per unit change of pore pressure (p_p), while keeping the total external stress ($\bar{\sigma}$) and temperature (T) constant.
Rock solids compressibility: $$c_r = -\frac{1}{V_s}\left(\frac{\partial V_s}{\partial p_p}\right)_{\bar{\sigma}=p_p, T}$$ $$c_r = -\frac{1}{V_s}\left(\frac{\partial V_s}{\partial \bar{\sigma}}\right)_{\bar{\sigma}=p_p, T}$$	Change in the rock solids volume (∂V_s) per unit of rock solids volume (V_s), per unit of external stress ($\bar{\sigma}$), at constant temperature. If a rock sample is tested without a jacket, external stress will be equal to the pore pressure ($\bar{\sigma} = p_p$).
Pseudo-bulk compressibility: $$c_{bt} = -\frac{1}{V_b}\left(\frac{\partial V_b}{\partial p_p}\right)_{\bar{\sigma}, T}$$	Change in the bulk volume (V_b), per unit of bulk volume (V_b), per unit change of pore pressure (p_p), at constant external stress ($\bar{\sigma}$) and temperature (T).
Coefficient of compressibility: $$a_v = -\frac{\partial e}{\partial p}$$	Change in void ratio (∂e) per unit change of net confining pressure (∂p).

Note: In calculating bulk and pore compressibility, one can use either the *initial* bulk (V_b) or pore (V_p) volume in all cases, or measure volumes at each particular pressure for which compressibility is being calculated. The results appear to plot better in the former case.

[70.3 kg/cm^2]). During the same period, high-pressure confining tests on consolidated sedimentary rocks have exceeded 15,000 psi [1055 kg/cm^2]. Most investigators used mainly well-indurated sandstones or limestones in their laboratory experiments. Knutson and Bohor (1963) tested the oil-reservoir rocks typical of the Texas—Louisiana Gulf Coast region (orthoquartzites to calcareous subgraywackes). Van der Knaap and Van der Vlis (1967) determined the compressibilities of unconsolidated clays and sands from the Bolivar Coast in Venezuela.

Carpenter and Spencer (1940) measured the "pseudo-bulk" compressibility of various consolidated sandstones in an attempt to investigate whether or not fluid withdrawal from U.S. Gulf Coast oil reservoirs and the resulting volume reduction could account for ground subsidence. They defined "pseudo-bulk" compressibility as:

$$\beta = \frac{1}{V_b} \left(\frac{\partial V_p}{\partial p_t} \right) \tag{2.73}$$

where β is the pseudo-bulk compressibility in psi^{-1}, V_b is the original bulk volume in cm^3, ∂V_p is the change in void volume in cm^3, and ∂p_t is the change in the applied pressure in psi. Their experiments showed that sediments compact owing to fluid withdrawal from the pore space.

Fatt (1958a, b) studied the relationship between compressibility and rock composition. He reported that unconsolidated sediments, which are poorly sorted and contain clay, have higher compressibilities than do consolidated and well-sorted sands. Fatt (1958b) found that the bulk compressibilities of sandstones are a function of rock composition for a given grain shape and sorting. If sandstones are divided into two groups (one with well-sorted, well-rounded grains and the other with poorly sorted, angular grains), then for each group the compressibility is a linear function of the amount of intergranular material.

The procedure used in the laboratory by Fatt (1958b) was similar to that of Carpenter and Spencer (1940), but in the former case the fluid was expelled under constant external pressure with a reduction in pore pressure rather than an increase in the external stress. This is believed to closely duplicate petroleum-reservoir producing conditions. Fatt's (1958b) procedure was to apply a constant external stress to the core and decrease or increase the pore pressure. His apparatus simultaneously measured both the bulk and pore-volume changes at room temperature. Volume changes of the core in the pressure cell were measured through the use of a linear potentiometer that could resolve a movement of 1×10^{-3} inch [2.54×10^{-3} cm].

Van der Knaap (1959) noted that pore compressibility increases with decreasing porosity. It has been suggested by some investigators that between certain minimum and maximum pressures, the relationship between pore compressibility and the logarithm of pressure can be approximated by a straight line. A straight-line relationship has been found to exist between the log of the bulk compressibility and the log of the "effective" pressure, which in this case was equal to the direct applied axial load, because pore pressure was atmospheric (Van der Knaap and Van der Vlis, 1967). Bulk compressibilities of unconsolidated clays and sands decreased with increasing overburden pressure (Fig.2.21). From their studies, Van der Knaap and Van der Vlis concluded that clay and sand layers compact almost to the same extent, the main difference being that the low permeability to water of the clay prevents instantaneous compaction and time effects become important.

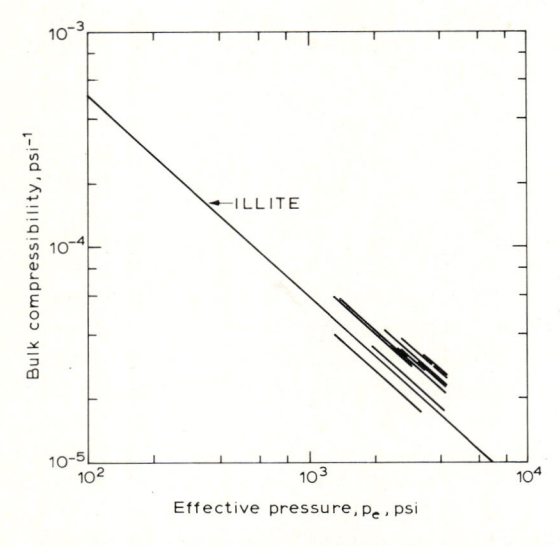

Fig.2.21. Compressibilities of a number of sand samples from the post-Eocene formations of the Bolivar Coast, Venezuela. The illite curve is after Chilingar and Knight, 1960, fig.2, p.104. (After Van der Knaap and Van der Vlis, 1967, fig.5, p.90.)

In his classical paper on the compaction of freshwater-bearing alluvial clays, silts, and silty sands in California, Meade (1968) found that the loss in pore volume that results from compaction by effective overburden pressures in the range between 3 and 70 kg/cm^2 averages about 0.3 void ratio ($e = V_p/V_s$) units or about 15% of the bulk volume of the fine-grained sediments. Meade stated that when one allows for the lesser compaction of the interbedded coarser sands and gravels, the reduction of the total volume of the alluvial sediments amounts to about 12% in the pressure range of 10—70 kg/cm^2 on the west side of the San Joaquin Valley and about 10% in the 3—33 kg/cm^2 pressure range in the Santa Clara Valley, California. The factors that directly influence the compressibility of shallow marine and alluvial sediments are average particle size, the particle sorting, the amount of montmorillonite clay, the proportion of exchangeable sodium cation relative to the calcium and magnesium cations in the clay minerals, presence of diatom skeletons, and, probably, the mica content.

Compressibilities of unconsolidated sands and clays are of the order of 10^{-3} to 10^{-5} psi^{-1} in the 100—10,000-psi* pressure range. Chilingarian et al. (1973a) found that the bulk compressibilities $[c_b = (-1/V_b)(\partial V_b/\partial p_e)_{p_t}]$ of unconsolidated sands range from 7.4×10^{-4} to 3×10^{-5} psi^{-1} at an effective pressure range of 0—3000 psi, whereas the pore-volume compressibilities $[c_p = (-1/V_p)(\partial V_p/\partial p_e)_{p_t}]$ range from 1×10^{-3} to 1×10^{-4} in the same pressure range using a hydrostatic compaction apparatus (pressures being equal in x, y, and z directions). These values are greater by about 55—100% than those obtained on using uniaxial compaction equipment. Different possible loading conditions on sediments are presented in Fig.2.22.

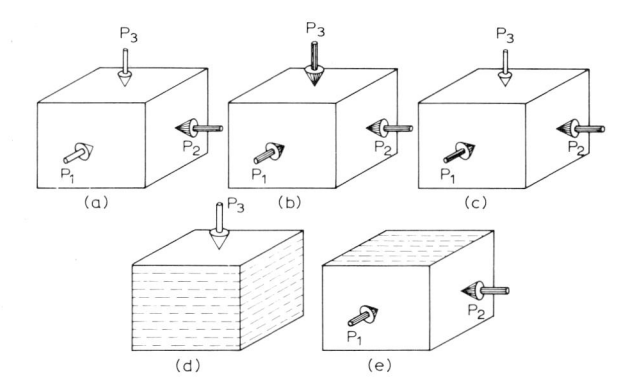

Fig.2.22. Compaction loading classification (after Sawabini et al., 1974, fig.1, p.133). (a) *Polyaxial loading* ($p_1 \neq p_2 \neq p_3$), called triaxial loading by some investigators. (b) *Hydrostatic loading* ($p_1 = p_2 = p_3$). (c) *Triaxial loading* ($p_1 = p_2 \neq p_3$), called biaxial loading by some investigators. (d) *Uniaxial loading* (four sides parallel to the stress p_3 are kept stationary), referred to as biaxial loading by some investigators. (e) *Biaxial loading* ($p_1 = p_2$ and sides parallel to these two stresses are kept stationary).

The bulk compressibilities obtained on compacting montmorillonite clay saturated in seawater, using the hydrostatic compaction apparatus (5×10^{-4} to 2.9×10^{-5} psi^{-1}), were found to be about 300—500% higher than those obtained on using uniaxial loading (1.85×10^{-4} to 5.4×10^{-6} psi^{-1}) in the applied pressure range of 400—20,000 psig (pounds per square inch, gage).

Compressibilities of the consolidated sandstones, shales, and carbonates are lower, and range from 10^{-5} to 10^{-7} psi^{-1} in the 500—15,000-psi pressure range (see Fig.2.23).

---

*1 kg/cm^2 = 14.223 psi and 1 psi^{-1} = 14.223 cm^2/kg.

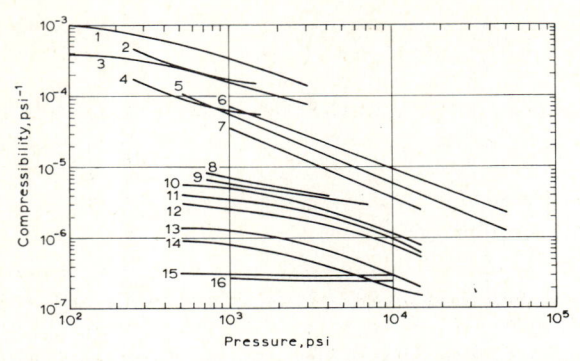

Fig.2.23. Relationship between compressibility (psi^{-1}) and applied pressure (psi) for unconsolidated sands, illite clay, limestone, sandstones, and shale (after Chilingarian et al., 1973a, fig.1).

No.	Investigator	Rock type	Type of applied pressure	Compressibility
1	Chilingarian et al. (1973a)	California unconsolidated arkosic sands[1]	hydrostatic	pore $[-(1/V_p)(\partial V_p/\partial p_e)_{\bar{\sigma}}]$
2	Kohlhaas and Miller (1969)	California unconsolidated sands	uniaxial	pore
3	Chilingarian et al. (1973a)	California unconsolidated arkosic sands[1]	hydrostatic	bulk $[-(1/V_b)(\partial V_b/\partial p_e)_{\bar{\sigma}}]$
4	Kohlhaas and Miller (1969)	California unconsolidated sands	uniaxial	bulk
5	Chilingarian et al. (1973a)	Illite clay (API No. 35) (wet)[2]	uniaxial	bulk $[-(1/e+1)(de/dp_e)]$
6	Chilingarian et al. (1973a)	Illite clay (API No. 35) (dry)	uniaxial	bulk $[-(1/h)(dh/dp_e)]$
7	Knutson and Bohor (1963)	Repetto Formation (Grubb Zone) (wet)[1]	net confining	"pore" $\bigg\} [-(1/V_p)(\partial V_p/\partial \bar{\sigma})_{p_p}]$
8	Knutson and Bohor (1963)	Lansing–Kansas City Limestone (wet)[1]	net confining	"pore"
9	Carpenter and Spencer (1940)	Woodbine Sandstone (wet)	net confining	"pseudo-bulk" $[-(1/V_b)(\partial V_p/\partial p)]$
10	Fatt (1958b)	feldspathic graywacke (No. 10) (wet)[3]	net confining[4]	bulk
11	Fatt (1958b)	graywacke (No. 7) (wet)[3]	net confining	bulk
12	Fatt (1958b)	feldspathic graywacke (No. 11) (wet)[3]	net confining	bulk $\bigg\} [-(1/V_b)(\partial V_b/\partial p_t)_{p_p}]$
13	Fatt (1958b)	lithic graywacke (No. 12) (wet)[3]	net confining	bulk
14	Fatt (1958b)	feldspathic quartzite (No. 20) (wet)[3]	net confining	bulk
15	Podio et al. (1968)	Green River Shale (dry)	net confining	bulk
16	Podio et al. (1968)	Green River Shale (wet)[2]	net confining	bulk

[1] Saturated with formation water.
[2] Saturated with distilled water.
[3] Saturated with kerosene.
[4] Net confining pressure = $p_e = (\bar{\sigma} - 0.85 p_p)$, where $\bar{\sigma}$ is the total overburden stress and p_p is the pore pressure.

The compressibilities of unconsolidated sands appear to be very close to those of clays. Unconsolidated sands are as compressible as clays, or even more so. Compressibility values for sands obtained in a hydrostatic compaction apparatus are usually about twice as high as those determined on using uniaxial compaction equipment. Sawabini et al. (1974) found that compressibility increases with increasing feldspar content.

Effect of rock compressibility on the estimation of petroleum reserves

The material balance equation for a finite reservoir containing undersaturated oil can be presented as follows:

$$\frac{N_p}{N} = \frac{B_{oi}}{B_0} \left[\frac{S_{oi}c_o + S_{wi}c_w + c_f}{(1-S_w)} \right] (p_i - p) \qquad (2.74)$$

where N = initial oil-in-place in the reservoir, N_p = volume of stock-tank oil produced, p = reservoir fluid pressure $(p_i > p > p_b)$, p_i = initial reservoir fluid pressure, p_b = bubble-point pressure, B_0 = formation volume factor (= volume at reservoir conditions/volume at standard conditions), B_{oi} = initial formation volume factor, c_f = pore compressibility $[= (1/V_p)(\partial V_p/\partial p_p)_{\bar{\sigma}, T}$, where V_p is the pore volume, p_p is the reservoir fluid pressure, $\bar{\sigma}$ is the average external pressure, and T is the temperature], c_o = compressibility of oil, c_w = compressibility of water, S_{wi} = initial water saturation, and S_{oi} = initial oil saturation. Thus, the significance of pore compressibility, c_f, in eq. 2.74 depends on its numerical value compared with $(S_{oi} c_o + S_{wi} c_w)$ (Scorer and Miller, 1974). At bubble-point pressures, oil compressibility varies from 5×10^{-6} to 25×10^{-6} psi^{-1}, whereas water compressibility is about 3×10^{-6} psi^{-1}.

The effective gas compressibility in gas reservoirs can be estimated by using the following equation:

$$(c_g)_{\text{effective}} = c_g + \frac{c_w S_w + c_f}{S_g} \qquad (2.75)$$

Inasmuch as c_g usually is much larger than c_f, the latter can be neglected. In deep formations and, especially, unconsolidated ones, c_f is of the same order of magnitude as c_g, which decreases with increasing pressure. As pointed out by Scorer and Miller (1974, p.16), the above expression (eq. 2.75) must be continuously evaluated as the pressure changes, and any attempt to use an average gas compressibility in flow equations is likely to lead to serious errors.

References

Anikiev, K.A., 1964. *Anomalously High Formation Pressures in Oil and Gas Deposits*. Tr. VNIGRI, No. 233. Izd. Nedra, Leningrad, 167 pp.

Bakhtin, V.V. and Martirosova, A.O., 1972. Variation in density and porosity of clayey rocks as related to their depth of burial. *Geol. Nefti Gaza*, 10: 57—60.

Bear, J., 1972. *Dynamics of Fluids in Porous Media*. American Elsevier, New York, N.Y., 764 pp.

Bear, J., Zaslavsky, D. and Irmay, S., 1968. *Physical Principles of Water Percolation and Seepage*. UNESCO, Paris.

Bentor, Y.K., 1961. Some geochemical aspects of the Dead Sea and the question of its age. *Geochim. Cosmochim. Acta*, 25: 239—260.

Berner, R.A., 1971. *Principles of Chemical Sedimentology*. McGraw-Hill, New York, N.Y., 240 pp.

Berry, F.A.F., 1969. Origin and tectonic significance of high fluid pressures in California coast ranges. *J. Pet. Technol.*, 21: 13—14.

Berry, F.A.F., 1973. High fluid potentials in California coast ranges and their tectonic significance. *Bull. Am. Assoc. Pet. Geol.*, 57: 1219—1249.

Bogomolova, G.V. et al. (Editors), 1973. *Problem of Interstitial Solutions in Geology*. Izd. Nauka i Tekhnika, Minsk, 254 pp.

Bolt, G.H., 1956. Physico-chemical analysis of the compressibility of pure clays. *Geotechnique*, 6: 86—93.

Bredehoeft, J.D. and Hanshaw, B.B., 1968. On the maintenance of anomalous fluid pressure, I. Thick sedimentary sequences. *Geol. Soc. Am. Bull.*, 79: 1097—1106.

Bryant, W.R., 1973. Consolidation of marine clays. Paper presented at ONR Symposium and Workshop on the Physical and Engineering Properties of Deep-Sea Sediments, April 24—27, 23 pp.

Carpenter, C.B. and Spencer, G.B., 1940. Measurements of compressibility of consolidated oil-bearing sandstones. *U.S. Bur. Min. Rep. Invest.*, No. 3540, 20 pp.

Cheng, J.T., 1973. *The effect of pressure and temperature on pore volume compressibility of reservoir rock*. M.S. Thesis, Texas A & M Univ., College Station, Texas, 44 pp.

Chilingar, G.V. and Adamson, L.G., 1964. Does some migration of oil occur in a gaseous form? *Proc. 22nd Int. Geol. Congr.*, New Delhi, Sec. 1, Part 1: 64—70.

Chilingar, G.V. and Knight, L., 1960. Relationship between pressure and moisture content of kaolinite, illite and montmorillonite clays. *Bull. Am. Assoc. Pet. Geol.*, 44: 101—106.

Chilingar, G.V., Rieke, III, H.H. and Robertson, Jr., J.O., 1963. Relationship between high overburden pressure and moisture content of halloysite and dickite clays. *Geol. Soc. Am. Bull.*, 74: 1041—1048.

Chilingar, G.V., Rieke, III, H.H., Sawabini, C.T. and Ershaghi, I., 1969. Chemistry of interstitial solutions in shales versus that in associated sandstones. *SPE 2527, 44th AIME Fall. Meet.*, Denver, Colo., 8 pp.

Chilingarian, G.V. and Rieke, III, H.H., 1968. Data on consolidation of fine-grained sediments. *J. Sediment. Petrol.*, 38: 811—816.

Chilingarian, G.V., Sawabini, C.T. and Rieke, III, H.H., 1973a. Comparison between compressibilities of sands and clays. *J. Sediment. Petrol.*, 43: 529—536.

Chilingarian, G.V., Sawabini, C.T. and Rieke, III, H.H., 1973b. Effect of compaction on chemistry of solutions expelled from montmorillonite clay saturated in sea water. *Sedimentology*, 20: 391—398.

Cooper, Jr., H.H., 1966. The equation of ground water flow in fixed and deforming coordinates. *J. Geophys. Res.*, 71: 4785—4790.

REFERENCES

Darcy, H., 1856. *Les Fontaines Publiques de la Ville de Dijon.* Victor Dalmont, Paris.
Degens, E.T., 1964. *Geochemistry of Sediments.* Prentice-Hall, Englewood Cliffs, N.J., 342 pp.
Degens, E.T. and Chilingar, G.V., 1967. Diagenesis of subsurface waters. In: G. Larsen and G.V. Chilingar (Editors), *Diagenesis in Sediments.* Elsevier, Amsterdam, pp.477—502.
Degens, E.T. and Epstein, S., 1962. Relationship between $^{18}O/^{16}O$ ratios in coexisting carbonates, cherts, and diatomites. *Bull. Am. Assoc. Pet. Geol.*, 46: 534—542.
Degens, E.T., Hunt, J.M., Reuter, J.H. and Reed, W.E., 1964. Data on the distribution of amino acids and oxygen isotopes in petroleum brine waters of various geologic ages. *Sedimentology*, 3: 199—225.
De Sitter, L.U., 1947. Diagenesis of oil-field brines. *Bull. Am. Assoc. Pet. Geol.*, 31: 2030—2040.
De Wiest, R.J.M., 1966. On the storage coefficient and the equation of groundwater flow. *J. Geophys. Res.*, 71: 1117—1122.
Dickey, P.A., 1972. Migration of interstitial water in sediments and the concentration of petroleum and useful minerals. *Proc. 29th Int. Geol. Congr.*, Montreal, Que., Sect. 5, pp.3—16.
Dickey, P.A., Shriram, C.R. and Paine, W.R., 1968. Abnormal pressures in deep wells of southwestern Louisiana. *Science*, 160: 609—615.
Dickinson, G., 1951. Geological aspects of abnormal reservoir pressures in the Gulf Coast region of Louisiana, U.S.A. *Proc. 3rd World Petrol. Congr.*, 1: 1—17.
Fatt, I., 1958a. Pore structure in sandstones by compressible sphere pack models. *Bull. Am. Assoc. Pet. Geol.*, 42: 1914—1923.
Fatt, I., 1958b. Compressibility of sandstones at low to moderate pressures. *Bull. Am. Assoc. Pet. Geol.*, 42: 1924—1957.
Fertl, W.H. and Timko, D.J., 1970a. Occurrence and significance of abnormal pressure formations. *Oil Gas J.*, 68(1): 97—108.
Fertl, W.H. and Timko, D.J., 1970b. How abnormal pressure detection techniques are applied. *Oil Gas J.*, 68(2): 62—71.
Fertl, W.H. and Timko, D.J., 1970c. Association of salinity variations and geopressures in soft and hard rocks. *11th Prof. Well Log Analysts Symp.*, Los Angeles, Calif., May, pp.1—24.
Fowler, Jr., W.A., 1968. Pressure, hydrocarbon accumulation and salinities — Chocolate Bayou field, Brazoria County. *SPE 2226, 43rd AIME Fall Meet.*, Houston, Texas, September, 9 pp.
Gibson, R.E., 1958. The progress of consolidation in a clay layer increasing in thickness with time. *Geotechnique*, 8: 171—182.
Graf, D.L., Meents, W.F., Friedman, T. and Shimp, N.F., 1966. The origin of saline formation waters, Calcium chloride waters. *Ill. State Geol. Surv. Circ.*, No. 397, 60 pp.
Gullikson, D.M., Caraway, W.H. and Gates, G.L., 1961. Chemical analysis and electrical resistivity of selected California oilfield waters. *U.S. Bur. Min., Rep. Invest.*, No. 5736, 21 pp.
Hawkins, M.E., Dietzman, W.D. and Pearson, C.A., 1964. Chemical analysis and electrical resistivities of oilfield brines from field in east Texas. *U.S. Bur. Min., Rep. Invest.*, No. 6422, 20 pp.
Hawkins, M.E., Dietzman, W.D. and Seward, J.M., 1963a. Analysis of brines from oil-productive formations in South Arkansas and North Louisiana. *U.S. Bur. Min. Rep. Invest.*, No. 6282, 28 pp.
Hawkins, M.E., Jones, C.W. and Pearson, C.A., 1963b. Analysis of brines from oil-productive formations in Mississippi and Alabama. *U.S. Bur. Min., Rep. Invest.*, No. 6167, 22 pp.

Hedberg, H.D., 1926. The effect of gravitational compaction on the structure of sedimentary rocks. *Bull. Am. Assoc. Pet. Geol.*, 10: 1035—1072.
Hedberg, H.D., 1936. Gravitational compaction of clays and shales. *Am. J. Sci.*, 31: 241—287.
Hedberg, H.D., 1974. Relation of methane generation in undercompacted shales, shale diapirs, and mud volcanoes. *Bull. Am. Assoc. Petrol. Geol.*, 58: 661—673.
Hedberg, W.H., 1967. *Pore-water chlorinities of subsurface shales.* Ph.D. Dissertation, Univ. of Wisconsin, Madison, Wisc., 121 pp.
Hergert, F., 1973. Variation of rock stresses with depth at a Canadian iron mine. *Int. J. Rock Mech. Min. Sci.*, 10: 37—51.
Hoshino, K., Koide, H., Inami, K., Iwamura, S. and Mitsui, S., 1972. Mechanical properties of Japanase Tertiary sedimentary rocks under high confining pressures. *Geol. Surv. Japan Rep.*, 244, 200 pp.
Hottman, C.E. and Johnson, R.K., 1965. Estimation of formation pressures from log-derived shale properties. *J. Pet. Technol.*, 16: 717—722.
Hubbert, M.K., 1940. The theory of groundwater motion. *J. Geol.*, 48: 785—944.
Hubbert, M.K. and Rubey, W.W., 1959. Role of fluid pressure in mechanics of overthrust faulting, 1. Mechanics of fluid-filled porous solids and its application to overthrust faulting. *Geol. Soc. Am. Bull.*, 70: 115—166.
Jacob, C.E., 1940. The flow of water in an elastic artesian aquifer. *Trans. Am. Geophys. Union*, 21: 574—586.
Jacob, C.E., 1950. Flow of groundwater. In: H. Rouse (Editor), *Engineering Hydraulics*, John Wiley and Sons, New York, N.Y., pp.321—386.
Jürgenson, L., 1973. Perplexity caused by the term hydrostatic stress. *Geotechnique*, 23: 448—450.
Kalinko, M.K., 1964. *Basic Regularities in Distribution of Oil and Gas in Earth's Crust.* Izd. Nedra, Moscow, 207 pp.
Karpova, G.V., 1972. *Clay Minerals and Their Evolution in Terrigenous Deposits.* Izd. Nedra, Moscow, 172 pp.
Katz, D.L. and Ibrahim, M.A., 1971. Threshold displacement pressure considerations for caprocks of abnormal-pressure reservoirs. *SPE 3222, 5th Conf. on Drilling and Rock Mechanics*, Austin, Texas.
Kazintsev, E.E., 1968. Pore solutions of Maykop Formation of Eastern Pre-Caucasus and methods of squeezing of pore waters at high temperatures. In: G.V. Bogomolov et al. (Editors), *Pore Solutions and Methods of Their Study — A Symposium.* Izd. Nauka i Tekhnika, Minsk, pp.178—190.
Knutson, C.F. and Bohor, B.F., 1963. Reservoir rock behavior under moderate confining pressure. In: C. Fairhurst (Editor), *Rock Mechanics.* Pergamon, New York, N.Y., pp.627—658.
Kohlhaas, C.A. and Miller, F.G., 1969. Rock-compaction and pressure-transient analysis with pressure-dependent rock properties. *SPE 2563, AIME Fall Meet.*, Denver, Colo., 7 pp.
Kozel'skiy, I.T. and Matveev, A.K., 1973. Relationship between lithologic characteristic of sedimentary rocks of coal-bearing deposits and their geophysical parameters. In: V.E. Khain et al. (Editors), *Present-day Problems of Geology and Geochemistry of Combustible Deposits.* Akad. Nauk S.S.S.R., Izd. Nauka, Moscow, pp.184—191.
Krasintseva, V.V. and Korunova, V.V., 1968. Influence of pressure and temperature on composition of extruded solutions during mud compaction. In: G.V. Bogomolov et al. (Editors), *Pore Solutions and Methods of Their Study — A Symposium.* Izd. Nauka i Tekhnika, Minsk, pp.191—204.
Kryukov, P.A., 1971. *Rock, Soil and Mud Solutions.* Akad. Nauk S.S.S.R., Siberian Branch, Izd. Nauka, Novosibirsk, 220 pp.

REFERENCES

Kryukov, P.A. and Zhuchkova, A.A., 1963. Physical-chemical phenomena associated with driving out of solutions from rocks. In: *Present-Day Concept of Bound Water in Rocks*. Izd. Akad. Nauk S.S.S.R., Laboratory of Hydrogeological Problems of F.P. Savarenskiy, Moscow, pp.95—105.

Lo, K.Y., 1969. The pore pressure—strain relationship of normally consolidated undisturbed clays. I. Theoretical considerations. *Can. Geotech. J.*, 6: 383—394.

Long, G., Neglia, S. and Rubino, E., 1970. Pore fluids in shales and their geochemical significance. In: G.D. Hobson and G.C. Speers (Editors), *Advances in Organic Geochemistry*. Pergamon, New York, N.Y., pp.191—217.

Magara, K., 1974. Compaction, ion filtration, and osmosis in shale and their significance in primary migration. *Bull. Am. Assoc. Pet. Geol.*, 58: 283—290.

Manheim, F.T., 1966. A hydraulic squeezer for obtaining interstitial water from consolidated and unconsolidated sediments. *U.S. Geol. Surv. Prof. Paper*, 550C: 256—261.

Manheim, F.T. and Sayles, F.L., 1970. Brines and interstitial brackish water in drill cores from the deep Gulf of Mexico. *Science*, 170: 57—61.

Meade, R.H., 1964. Removal of water and rearrangement of particles during the compaction of clayey sediments — review. *U.S. Geol. Surv. Prof. Paper*, 497B: 1—23.

Meade, R.H., 1968. Compaction of sediments underlying areas of land subsidence in central California. *U.S. Geol. Surv. Prof. Paper*, 497D: 1—39.

Mitchell, J.K., 1960. The application of colloidal theory to the compressibility of clays. In: R.H.G. Parry (Editor), *Interparticle Forces in Clay—Water—Electrolyte Systems*. CSIRO, Melbourne, Vic., pp.2/92—2/97.

Myers, L.L., 1963. *Dynamic Phenomena of Sediment Compaction in Matagorda County, Texas*. Thesis, Univ. of Houston, Houston, Texas, 62 pp.

Overton, H.L. and Timko, D.J., 1969. The salinity factor: a tectonic stress indicator in marine sands. *Oil Gas J.*, 67(40): 115—124.

Pandey, G.N., Tek, M.R. and Katz, D.L., 1974. Diffusion of fluids through porous media with implications in petroleum geology. *Bull. Am. Assoc. Pet. Geol.*, 58: 291—303.

Perrier, R. and Quiblier, J., 1974. Thickness changes in sedimentary layers during compaction history; methods for quantitative evaluation. *Bull. Am. Assoc. Pet. Geol.*, 58: 507—520.

Philip, J.R., 1969a. Moisture equilibrium in swelling soils, 1. Basic theory. *Aust. J. Soil Res.*, 7: 121—141.

Philip, J.R., 1969b. Moisture equilibrium in swelling soils, 2. Applications. *Aust. J. Soil Res.*, 5: 1070—1077.

Podio, A.L., Gregory, A.R. and Gray, K.E., 1968. Dynamic properties of dry and water-saturated Green River Shale under stress. *J. Soc. Pet. Eng.*, 8: 389—404.

Posokhov, E.V., 1966. *Formation of Chemical Composition of Underground Waters (Basic Factors)*. Gidrometeorologicheskoe Izd., Leningrad, 258 pp.; 2nd ed., 1969, 334 pp.

Raghavan, R., 1974. Consolidation and rebound processes in one-dimensional porous columns. *J. Geophys. Res.*, 79: 1687—1698.

Raghavan, R. and Miller, F.G., 1975. Mathematical analysis of sand compaction. In: G.V. Chilingar and K. Wolf (Editors), *Compaction of Coarse-Grained Sediments, 1*. Elsevier, Amsterdam, pp.403—524.

Rall, C.G. and Wright, J., 1953. Analysis of formation brines in Kansas. *U.S. Bur. Min. Rep. Invest.*, No. 4974, 40 pp.

Ramsay, J.G., 1967. *Folding and Fracturing of Rocks*. McGraw-Hill, New York, N.Y., 568 pp.

Rieke, III, H.H., Chilingar, G.V. and Robertson, Jr., J.O., 1964. High-pressure (up to 500,000 psi) compaction studies on various clays. *Proc. 22nd Int. Geol. Congr.*, New Delhi, 15: 22—38.

Rieke, III, H.H. and Chilingarian, G.V., 1974. *Compaction of Argillaceous Sediments*. Elsevier, Amsterdam, 424 pp.

Robertson, E.C., 1967. Laboratory consolidation of carbonate sediments. In: A.F. Richards (Editor), *Marine Geotechnique*. Univ. of Illinois Press, Urbana, Ill., 326 pp.

Samuels, S.G., 1950. The effect of base exchange on the engineering properties of soils. *Build. Res. Stn. (Great Britain) Note* C 176, 16 pp.

Sawabini, C.T., Chilingar, G.V. and Allen, D.R., 1974. Compressibility of unconsolidated, arkosic oil sands. *Soc. Pet. Eng. J.*, 14: 132—138.

Schmidt, G.W., 1971. *Interstitial water composition and geochemistry of deep Gulf Coast shales and sands*. M.S. Thesis, Univ. of Tulsa, Tulsa, Okla.

Schmidt, G.W., 1973. Interstitial water composition and geochemistry of deep Gulf Coast shales and sandstones. *Bull. Am. Assoc. Pet. Geol.*, 57: 321—337.

Scorer, J.D.T. and Miller, F.G., 1974. A review of reservoir rock compressibility, and its relationship to oil and gas recovery. *Inst. Pet. (London)*, 74-003: 25 pp.

Shishkina, O.V., 1968. Methods of investigating marine and ocean mud waters. In: G.V. Bogomolov et al. (Editors), *Pore Solutions and Methods of Their Study — A Symposium*. Izd. Nauka i Tekhnika, Minsk, pp.167—176.

Siever, R., Beck, K.C. and Berner, R.A., 1965. Composition of interstitial waters of modern sediments. *J. Geol.*, 73: 39—73.

Skempton, A.W., 1953. Soil mechanics in relation to geology. *Proc. Yorks. Geol. Soc.*, 29: 33—62.

Smith, J.E., 1971a. Shale compaction. *Soc. Pet. Eng. J.*, 13: 12—22.

Smith, J.E., 1971b. The dynamics of shale compaction and evolution of pore-fluid pressures. *Math. Geol.*, 3: 239—263.

Smith, J.E., Erdman, J.G. and Morris, D.A., 1971. Migration, accumulation and retention of petroleum in the earth. *Proc. 8th World Pet. Congr.*, Moscow, 8: 13—26.

Sorby, H.C., 1908. On the application of quantitative methods to the study of the structure and history of rocks. *J. Geol. Soc. Lond.*, 64: 171—232.

Terzaghi, K. and Peck, R.P., 1968. *Soil Mechanics in Engineering Practice*. John Wiley and Sons, New York, N.Y., 729 pp.

Thompson, T.L., 1973. Application of plate tectonics to petroleum exploration at continental margins. *Bull. Am. Assoc. Pet. Geol.*, (Abst.), 57: 1844—1845.

Timm, B.C. and Maricelli, J.J., 1953. Formation waters in southwest Louisiana. *Bull. Am. Assoc. Pet. Geol.*, 37: 394—409.

Van der Knaap, W., 1959. Non-linear behavior of elastic porous media. *Trans. Soc. Pet. Eng., AIME*, 216: 179—186.

Van der Knaap, W. and Van der Vlis, A.C., 1967. On the cause of subsidence in oil-producing areas. *Proc. 7th World Pet. Congr.*, Mexico City, 3: 85—95.

Von Engelhardt, W. and Gaida, K.H., 1963. Concentration changes of pore solutions during the compaction of clay sediments. *J. Sediment. Petrol.*, 33: 919—930.

Wallace, W.E., 1965. Application of electric log measured pressure to drilling problems and a new simplified chart for well site pressure computation. *Log Analyst*, 6: 4—10.

Weaver, C.E. and Beck, K.C., 1969. *Changes in the Clay—Water System with Depth, Temperature and Time*. OWRR Proj. No. A-008-GA, Soc. Ceram. Eng., Georgia Institute of Technology, Atlanta, Ga., WRC-0769, 95 pp.

White, D.E., 1965. Saline waters of sedimentary rocks. In: A. Young and J.E. Galley (Editors), *Fluids in Subsurface Environments — Am. Assoc. Pet. Geol., Mem.*, 4: 342—366.

Williams, D.G., Brown, W.O. and Wood, J.J., 1965. Cutting drilling costs in high-pressure areas. *Oil Gas J.*, 63(41): 145—152.

Wright, J., Pearson, C., Kurt, E.T. and Watkins, J.W., 1957. Analysis of brines from oil-productive formations in Oklahoma. *U.S. Bur. Min. Rep. Invest.*, No. 5326, 71 pp.

Chapter 3. GEOPHYSICAL METHODS DETECT AND EVALUATE ABNORMAL FORMATION PRESSURES

Introduction

Acoustic characteristics of the upper crust of the earth can be measured for various reasons, e.g. in earthquake seismology, seismic prospecting, and acoustic (sonic) well-logging. The latter methods investigate similar velocity ranges but utilize entirely different frequencies and wave lengths (Thurber, 1959). Velocity ranges to be expected in sedimentary sections are schematically shown in Fig.3.1.

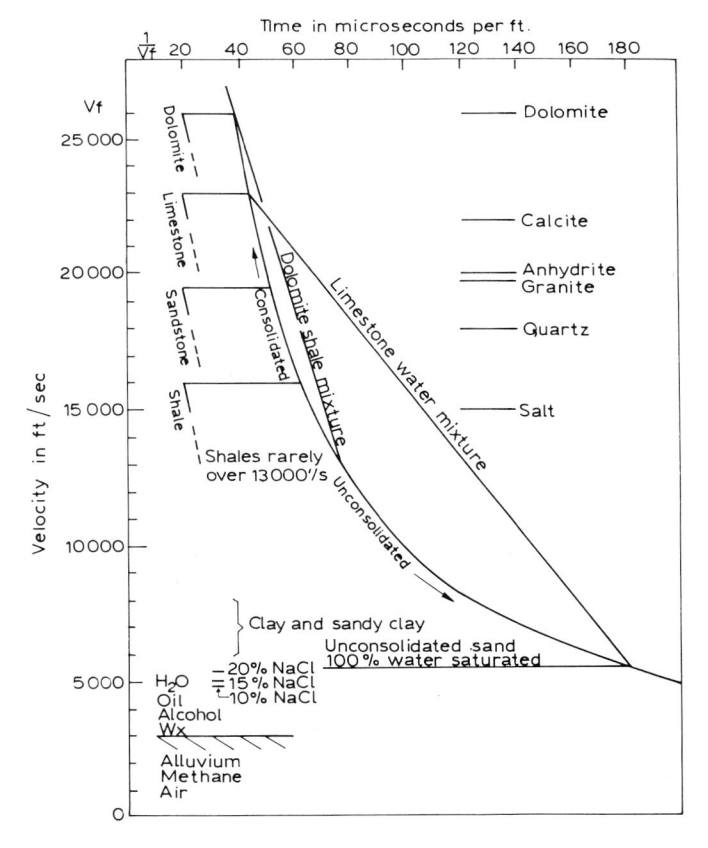

Fig.3.1. Velocity ranges frequently encountered in the sedimentary section (after Thurber, 1959). (Courtesy of *Oil and Gas Journal*.)

Over the years numerous developments and refinements in the acquisition, processing, and interpretation of geophysical data, such as seismic, have made it possible for the oil industry to study not only the structural configurations of formations in sedimentary basins. Today, proper application and interpretation of quality seismic records allow the study of several factors, even though attempts to isolate and measure the influence of individual parameters is not always successful. Such applications include:

Determination of interval velocities.

Study of lithologic and stratigraphic variations in geologic section.

Estimation of geologic age and/or average geothermal gradients.

Study of the effects of lateral folding pressure on a regional scale.

Hydrocarbon recognition, especially gas at shallow depths ("bright-spot" technique).

Detection of natural and/or artificially caused hydrocarbon seepages from the ocean floor.

Investigation of the ocean floor and sub-bottom properties (i.e. buried, ancient glacial channels, etc.) for offshore drilling and production operations.

Detection of the presence and top of abnormally pressured formations and evaluation of the magnitude of pressures.

The present discussion focuses on the application of geophysical, particularly seismic, data to determine *lithological* (formation) *tops*, and both *depth* and *magnitude* of *abnormal formation pressure* environments *prior to spudding* a well. These concepts have a profound impact especially on wildcat well planning (i.e. properly engineered mud and casing programs, predicting penetration rates, etc.) for high-cost, deep-water drilling and/or exploration of remote, yet untested, regions in the arctic, deserts, or jungles.

Seismic data analysis

Determination of formation velocities from seismic data has been known and utilized for many years (Green, 1938; Duerbaum, 1954; Dix, 1955; Musgrave, 1962), and several methods of computing these velocities, frequently using high-speed data handling, exist (Amenson and Darby, 1969; Yancy et al., 1969; Beitzel and Davis, 1973; Boylston and Looney, 1973). All these methods are based on the elementary reflection analysis, summarized by Pennebaker (1968a) as follows:

In Fig.3.2 let SS represent the earth's surface. Assume the shot point O to be at the surface. When explosives at the shot point are detonated, acoustic energy is created in the form of compressional waves. This seismic energy moves equally in all directions. The vertically traveling energy strikes the subsurface plane RR and is reflected back to the surface SS along vertical path OPO. Energy from the shot also propagates along innumerable diagonal paths to the plane RR in the subsurface (i.e. path OT) and is reflected back

SEISMIC DATA ANALYSIS

to the surface along path TW. The time required for the energy to travel the two ray paths is recorded by geophones at points O and W, separated horizontally by distance x.

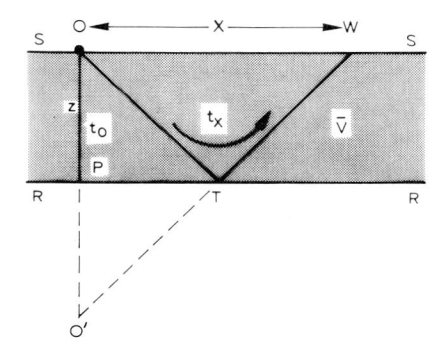

Fig.3.2. Elementary reflection problem (after Pennebaker, 1968a). $OP = O'P$; SS = horizontal ground surface, RR = horizontal reflecting surface, z = vertical distance to reflection point P, x = horizontal distance to geophone W, $\bar{v}$ = average velocity of the medium between surface and reflecting plane, O' = image of shot point O, t_0 = two-way travel time along path OPO, t_x = travel time along path OTW. (Courtesy of *World Oil.*)

With this information, the depth to the reflecting horizon and the average velocity in the medium between the surface and the reflecting horizon can be calculated from the relation distance = average velocity × time:

$$OPO = \bar{v} \cdot t_0 \tag{3.1}$$

$$OTW = \bar{v} \cdot t_x \tag{3.2}$$

where t_0 = travel time along path OPO, t_x = travel time along path OTW, and $\bar{v}$ = apparent average velocity from the surface to the reflecting horizon.

After extending the line OP vertically downward to the image point O', from elementary laws of optics:

$OTW = O'W$

$OPO = O'O$

and:

$$O'TW^2 = O'O^2 + OW^2 \tag{3.3}$$

Substituting eqs. 3.1 and 3.2 in eq. 3.3:

$(\bar{v} t_x)^2 = (\bar{v} t_0)^2 + x^2$

$$\bar{v}^2 (t_x^2 - t_0^2) = x^2 \tag{3.4}$$

$\bar{v} = x^2 / (t_x^2 - t_0^2)$

The depth to the reflecting bed may then be found by the relation:

$$z = \bar{v}(t_0/2) \tag{3.5}$$

In practice, all seismic traces with a common subsurface reflection point can be grouped together to create a *seismic gather*. A convenient approach divides this seismic gather into time windows which are investigated with a series of velocity values. Based on the final value selected, root mean squared velocity versus reflection time can be plotted (Fig.3.3). From this profile one can then construct the plot of *interval velocity* versus depth by simply dividing any desired depth interval, for example 1000-ft [304.8 m] increments, by one-half of the difference in time between bottom and top of the selected depth interval.

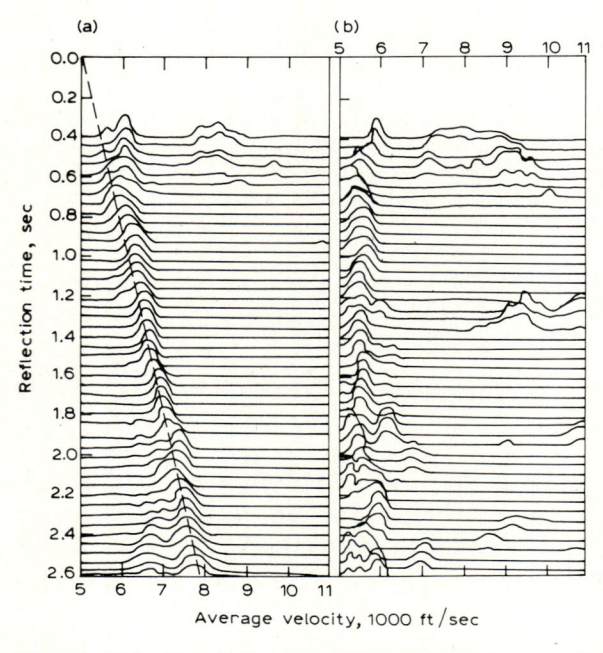

Fig.3.3. Velocity analyses in offshore Louisiana, U.S.A. (after Reynolds, 1970). (a) Good reflection data with few erratic multiples. (b) Useless analysis due to almost total absence of primary reflections and preponderance of multiple energy. (Courtesy of *World Oil*.)

Since *interval velocity*, a term mostly used by geophysicists, is the *reciprocal of interval sonic (acoustic) travel time*, a term more familiar to geologists, well-log analysts, and drilling engineers, one could also replot velocity data as travel-time profiles. This results in a "sonic log-type" presentation in which the data are averaged over rather long sections and not, as in the case of wireline sonic log, over a few feet of vertical distance only (Fig.3.4).

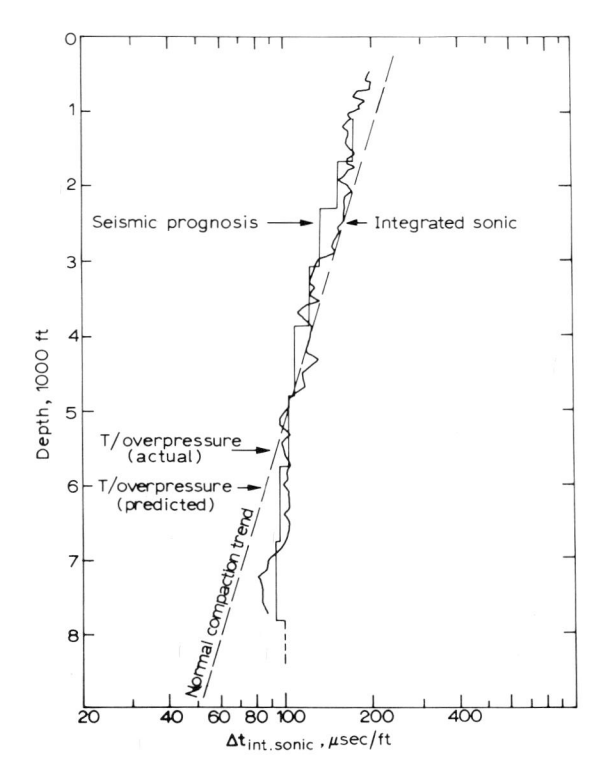

Fig.3.4. Comparison of seismically derived transit travel time and actual velocity data in well (after Reynolds, 1973). (Courtesy of the Society of Petroleum Engineers of A.I.M.E.)

An important application of such velocity analysis is the *prediction of lithology and formation tops* prior to spudding a well. Naturally, such information greatly assists in properly engineered well planning. This will be illustrated by field observations for both soft-rock (shale/sand) and hard-rock sequences (carbonates, etc).

Recently, Woeber and Penhollow (1973) have carried out a statistical study of the accuracy of formation-top predictions from seismic velocity profiles in shale/sand sequences along the Texas Gulf Coast area. Whereas in 80% of field cases an accuracy of 1—2% is found over the depth range between 5000 and 12,000 ft [1524 and 3659 m], the accuracy falls off fairly rapidly at depths below 12,000 ft [3659 m] with only 50% of the cases still showing an accuracy of 2% or more below 15,000 ft [4573 m]. The authors attribute this accuracy to the 2-mile [3.22 km] spread lengths used, the overall good quality and density of the data, and to the moderate velocity range: 7000—11,000 ft/sec [2134—3354 m/sec].

Information in *hard-rock* country on predicted and actual formation tops

TABLE 3.I

Comparison of seismically predicted formation tops and actual lithological tops in well Farrar No. 1, Anadarko Basin, Oklahoma, U.S.A. (after Louden et al., 1971)

Formation tops	Depth (ft)	
	predicted	actual
Brown Dolomite	5,250	4,700
Atoka marker	14,150	13,770
Limestone marker	15,350	15,470
Morrow marker	17,240	16,960
Springer marker	20,500	20,600
Massive carbonates	22,450	22,330

in a Mid-Continent well recently drilled in the Anadarko Basin, Oklahoma, has been reported by Louden et al. (1971). These comparative data are given in Table 3.I with the forecasted detailed geologic column presented in Table 3.II.

In this particular field study, accuracy of the prediction improves with greater depth. This is due to the special seismic design, which was aimed for accurately locating deeper formations only, since in the subject area shallow-well control has been available. Note that below 20,000 ft [6098 m] formation tops are forecasted to within only 100—120 ft [30.5—36.6 m] of the actual depth measured during the drilling and well-logging operations. This corresponds to an accuracy of better than 1%.

Under specific circumstances, the velocity analyses can also be used to predict *presence*, *depth*, and *magnitude* of *abnormal formation pressures*. Basically, in shale/sand sequences the interval velocities decrease with depth (West, 1950; Kaufman, 1953; Dix, 1955; Sarmiento, 1961). Thus, a normal trend line representative of normal, hydrostatic formation pressures can be established. Since overpressured zones exhibit higher than normal porosity at any given depth (see Chapters 1 and 5), this is reflected in a travel-time increase, i.e. divergence from normal to higher values in overpressured formations. Degree of departure from the "normal" then is a function of the magnitude of overpressures.

For *quantitative pore pressure evaluations* empirical correlations of the velocity ratio (Pennebaker, 1968a) or travel time versus expected pressure gradients and/or mud weight requirements can be used (Hottman and Johnson, 1965).

As one would expect, such correlations differ in various Tertiary basins throughout the world (Fig.3.5). Similar concepts have been discussed for well-log-derived shale resistivity correlations in Chapters 5 and 8 (Fig.8.20).

Table 3.III presents results of computations made by Pennebaker (1968a) from seismic records. The velocity values tabulated are the average velocity

TABLE 3.II

Geologic column in well Farrar No. 1 as predicted from seismic analysis (approximate reflection points indicated) plus supplementary data on regional historical geology (after Louden et al., 1971)

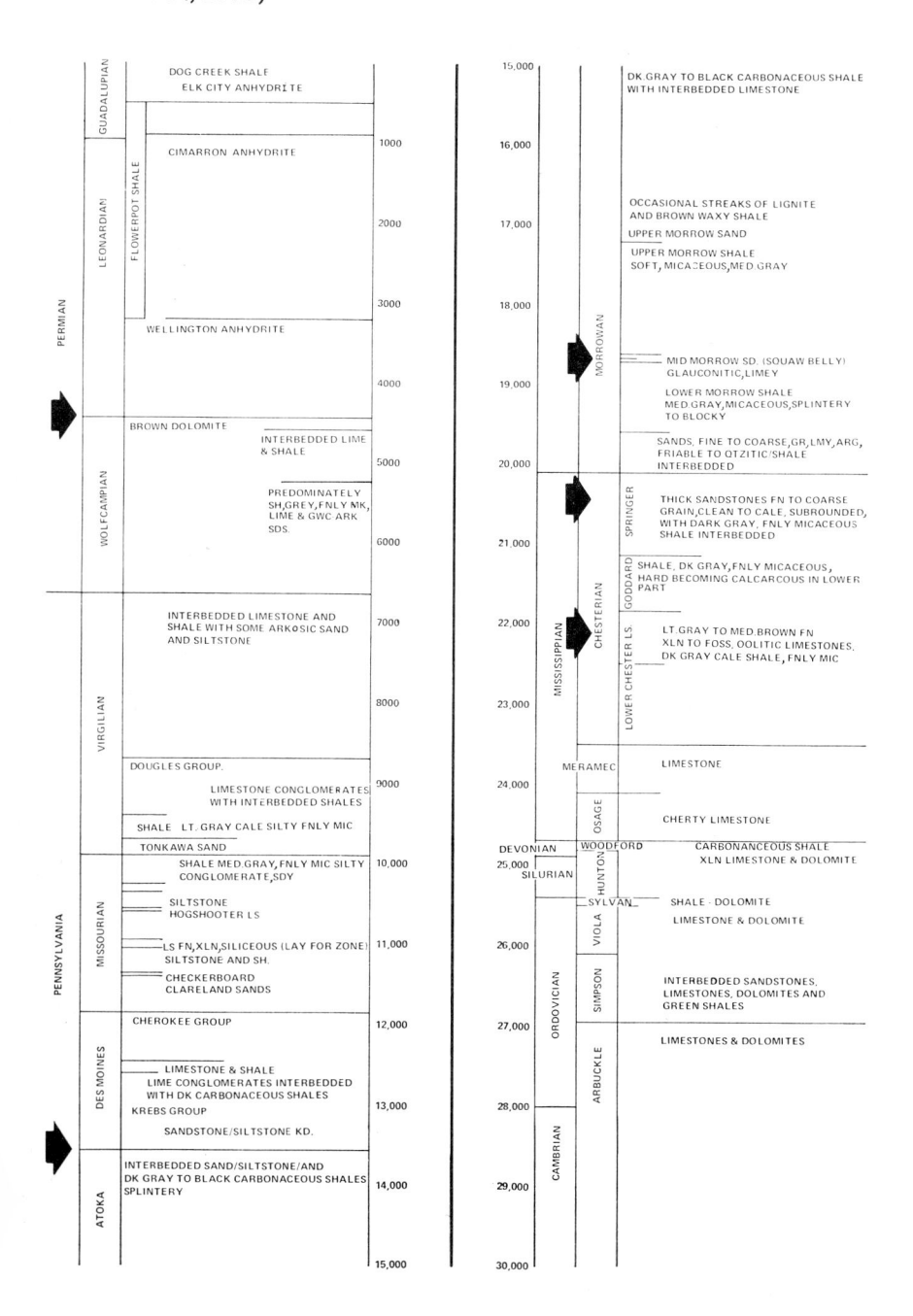

Period	Epoch	Formation	Depth (ft)
PERMIAN	GUADALUPIAN	DOG CREEK SHALE / ELK CITY ANHYDRITE	
	LEONARDIAN	FLOWERPOT SHALE / CIMARRON ANHYDRITE	1000
			2000
		WELLINGTON ANHYDRITE	3000
			4000 ←
	WOLFCAMPIAN	BROWN DOLOMITE / INTERBEDDED LIME & SHALE	5000
		PREDOMINATELY SH, GREY, FNLY MK, LIME & GWC ARK SDS.	6000
	VIRGILIAN	INTERBEDDED LIMESTONE AND SHALE WITH SOME ARKOSIC SAND AND SILTSTONE	7000
			8000
PENNSYLVANIA		DOUGLES GROUP. / LIMESTONE CONGLOMERATES WITH INTERBEDDED SHALES	9000
		SHALE LT. GRAY CALE SILTY FNLY MIC / TONKAWA SAND	
	MISSOURIAN	SHALE MED.GRAY, FNLY MIC SILTY CONGLOMERATE, SDY	10,000
		SILTSTONE / HOGSHOOTER LS	
		LS FN,XLN,SILICEOUS (LAY FOR ZONE) / SILTSTONE AND SH. / CHECKERBOARD / CLARELAND SANDS	11,000
	DES MOINES	CHEROKEE GROUP	12,000
		LIMESTONE & SHALE / LIME CONGLOMERATES INTERBEDDED WITH DK CARBONACEOUS SHALES / KREBS GROUP	13,000
		SANDSTONE/SILTSTONE KD.	
	ATOKA ←	INTERBEDDED SAND/SILTSTONE/AND DK GRAY TO BLACK CARBONACEOUS SHALES SPLINTERY	14,000
			15,000

Period	Epoch	Formation	Depth (ft)
		DK.GRAY TO BLACK CARBONACEOUS SHALE WITH INTERBEDDED LIMESTONE	15,000
			16,000
		OCCASIONAL STREAKS OF LIGNITE AND BROWN WAXY SHALE / UPPER MORROW SAND / UPPER MORROW SHALE SOFT, MICACEOUS, MED GRAY	17,000
	MORROWAN		18,000
	←	MID MORROW SD. (SQUAW BELLY) GLAUCONITIC, LIMEY / LOWER MORROW SHALE MED.GRAY, MICACEOUS, SPLINTERY TO BLOCKY	19,000
		SANDS, FINE TO COARSE, GR, LMY, ARG, FRIABLE TO OTZITIC/SHALE INTERBEDDED	20,000
	SPRINGER ←	THICK SANDSTONES FN TO COARSE GRAIN, CLEAN TO CALE, SUBROUNDED, WITH DARK GRAY, FNLY MICACEOUS SHALE INTERBEDDED	21,000
MISSISSIPPIAN	GODDARD	SHALE, DK GRAY, FNLY MICACEOUS, HARD BECOMING CALCARCOUS IN LOWER PART	
	CHESTERIAN / LOWER CHESTER LS.	LT.GRAY TO MED BROWN FN XLN TO FOSS. OOLITIC LIMESTONES, DK GRAY CALE SHALE, FNLY MIC	22,000
			23,000
	MERAMEC	LIMESTONE	24,000
	OSAGE	CHERTY LIMESTONE	
DEVONIAN / SILURIAN	WOODFORD / HUNTON	CARBONACEOUS SHALE / XLN LIMESTONE & DOLOMITE	25,000
	SYLVAN / VIOLA	SHALE - DOLOMITE / LIMESTONE & DOLOMITE	
ORDOVICIAN	SIMPSON	INTERBEDDED SANDSTONES, LIMESTONES, DOLOMITES AND GREEN SHALES	26,000
	ARBUCKLE	LIMESTONES & DOLOMITES	27,000
			28,000
CAMBRIAN			29,000
			30,000

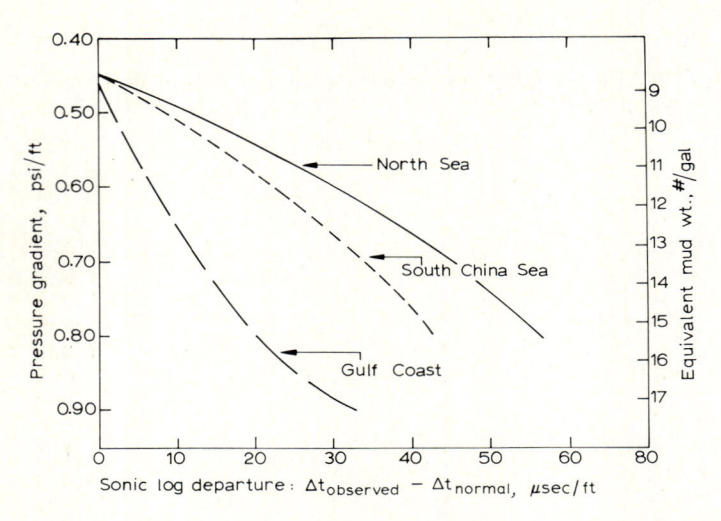

Fig.3.5. Acoustic (sonic) log transit travel time versus pore pressure gradient and equivalent mud weight for several Tertiary basins.

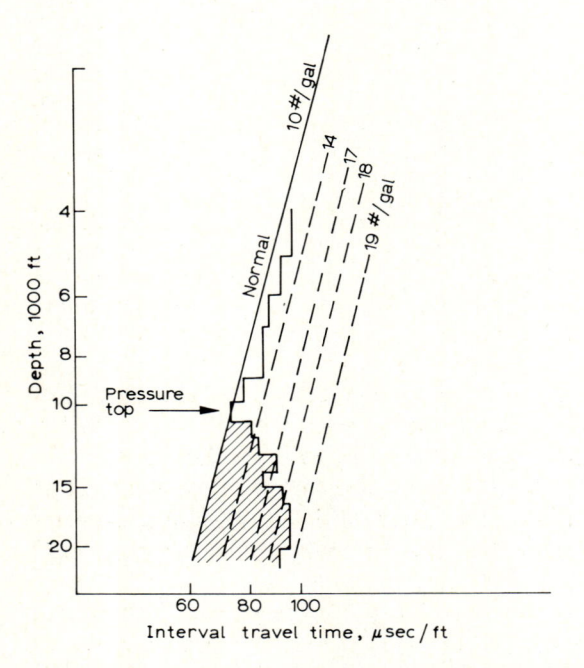

Fig.3.6. Seismically derived interval travel time versus depth, calibrated in mud weight equivalent, in a U.S. Gulf Coast well (after Pennebaker, 1968a). (Courtesy of *World Oil*.)

TABLE 3.III

Two-way reflection time and average velocity determinations in a U.S. Gulf Coast well (after Pennebaker, 1968a)

Event No.	T_0^* (sec)	Average velocity (ft/sec)	Depth (ft)
1	1.060	7,600	4,028
2	1.104	7,400	4,085
3	1.534	8,500	6,520
4	1.651	8,860	7,314
5	1.700	8,850	7,523
6	1.895	9,100	8,622
7	1.935	9,300	8,998
8	2.000	9,200	9,200
9	2.175	9,600	10,440
10	2.235	9,720	10,862
11	2.280	9,800	11,172
12	2.328	9,820	11,430
13	2.439	10,050	12,256
14	2.480	10,100	12,524
15	2.530	10,030	12,688
16	2.675	10,130	13,549
17	2.733	10,120	13,829
18	2.915	10,250	14,939
19	2.968	10,350	15,359
20	3.014	10,300	15,522
21	3.028	10,500	15,897
22	3.137	10,360	16,250
23	3.190	10,220	16,301
24	3.289	10,215	16,799
25	3.505	10,400	18,226
26	3.520	10,400	18,304
27	3.559	10,400	18,507
28	3.614	10,300	18,612

*T_0 is the two-way reflection time from datum (sea level) to a point on a subsurface reflector directly beneath the shot.

computed from datum to the depth of that particular reflecting event. Fig.3.6 shows the resulting plot of travel time versus depth on 1000-ft [304.8 m] vertical increments. Note that the depth scale chosen by Pennebaker for this particular field case is logarithmic. However, it is generally preferable to use a linear depth scale, compatible with the scale used for other pressure indicators. This makes the comparative study of several pressure indicators easier.

Note that in Fig.3.6 travel time decreases with depth in the hydrostatic pressured section, whereas overpressures are designated by the increase in travel time below 11,000 ft [3354 m]. As a matter of fact, geophysicists have for some time been aware of these anomalies which show up on seismic records as regions of rather poor reflections. Since such *velocity inversions* often seemed to coincide with the onset of certain formations, particularly massive shale sections and/or large salt masses, Musgrave and Hicks (1966) proposed the use of seismic data supplemented by gravity information to distinguish between shale masses (<10,000 ft/sec [<3048 m/sec], see Fig.3.1) and salt (approximately 15,000 ft/sec [4573 m/sec], see Fig.3.1).

However, to interpret deviations in interval velocity (or travel time) from the normal compaction trend in a given area or basin as a function of overpressures, one needs some independent knowledge of the lithology in the subject area. In other words, lack of information about major variations in lithology and/or age may cause erroneous interpretations. Reynolds (1970) illustrated such lithology effects for a drilling prospect in southern Mississippi (Fig.3.7), U.S.A., and Fig.3.8 shows the marked change at the top

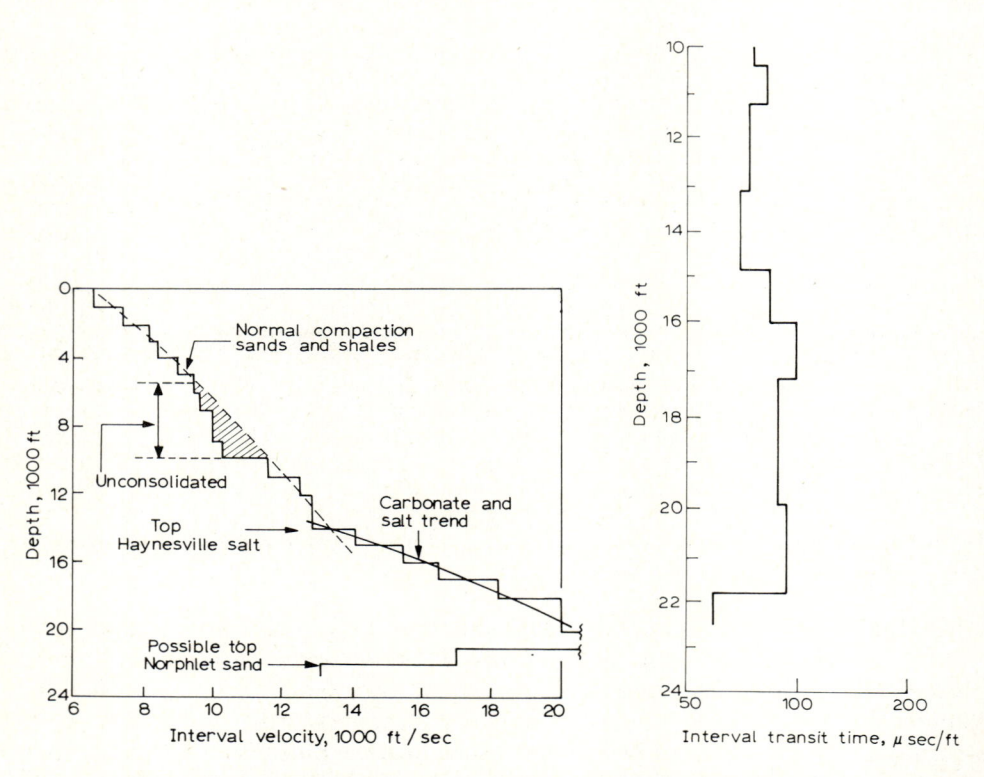

Fig.3.7. Effect of major lithological changes on normal compaction trend, Mississippi, U.S.A. (after Reynolds, 1970). (Courtesy of *World Oil*.)

Fig.3.8. Overpressured Pennsylvanian Atoka shale is defined by a velocity decrease (transit time increase) with depth between 14,000 and 15,000 ft [4268 and 4573 m]. Also note the sharp transit time decrease at the top of Mississippian carbonate at approximately 22,000 ft [6707 m] (after Louden et al., 1971). (Courtesy of *World Oil*.)

of the massive carbonate section at approximately 22,000 ft [6707 m] in the Oklahoma well, which was previously discussed for formation-top predictions (Louden et al., 1971).

Today, proper interpretation of seismic data can predict the top of overpressures to within an average of ±500 ft [±152 m], and pore pressure magnitudes can be estimated to within ±1.0 lb/gal [±0.12 kg/dm^3] mud weight equivalent in more than 50% of field cases studied (Reynolds, 1973). Table 3.IV compares top and magnitude of predicted and actual formation pressures for several wildcats, drilled both in the United States and overseas, while Figs.3.9 and 3.10 illustrate these results graphically.

TABLE 3.IV

Comparison of seismically predicted and actual overpressure environments in 26 wildcat wells, U.S.A. and overseas (after Reynolds, 1973)

Well	Depth of overpressure top (ft)		Pressure magnitude (ppg MWE)*	
	predicted	actual	predicted	actual
1	12,700	12,500	17.5	15
2	4,500	4,300	14	18.4
3	6,200	6,500	12	14.7
4	4,000	5,200	11.5	10.1
5	3,800	4,000	10	9.8
6	4,000	4,000	16.2	15
7	none	2,600	9	<10
8	2,700	none	15	9
9	2,300—2,700	1,800—2,600	<10	9.4
10	4,900	5,600	16.5	15
11	4,100—6,200	5,750	10—11	10
12	9,000	8,800	—	12.1
13	4,500	4,900	—	9.5
14	4,200	4,700	10.3	9.3
15	5,000—8,000	3,000—7,300	<11	11.5
16	6,000	5,900	11.2	12.3
17	8,500	8,250	17	17
18	7,400	3,500—7,600	11.5	12.6
19	4,200	8,200	—	14.5
20	7,800	6,100	11	11.8
21	5,000	4,000	12.7	12.0
22	5,000	4,600	10	9.5
23	6,900	6,500	12	12
24	8,000	5,000	13	17.6
25	6,000	5,780	11	11.2
26	4,500	4,800	11.5	9.7

*MWE = mud weight equivalent.

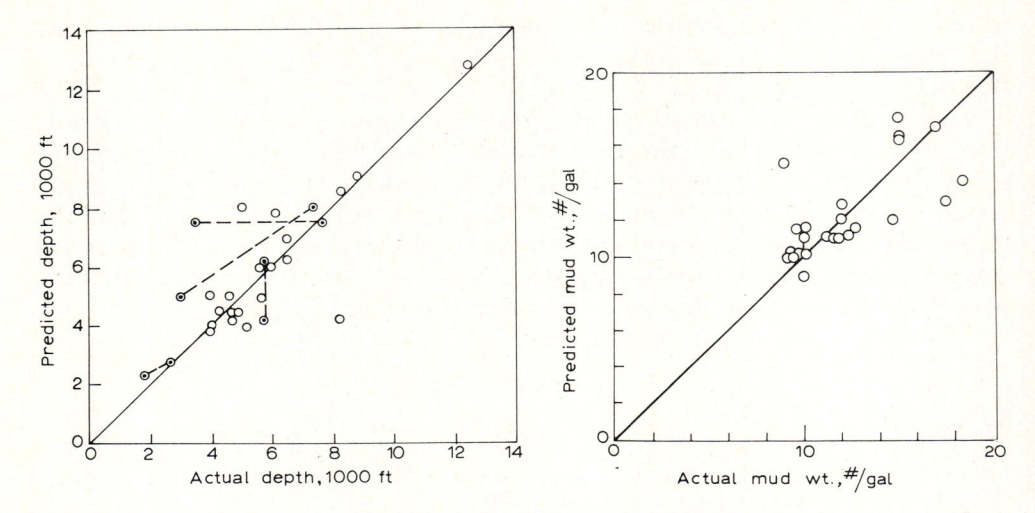

Fig.3.9. Comparison of predicted and actual overpressure tops in United States and overseas wells (after data by Reynolds, 1973). o = predicted values, o — — o = predicted range of values for the subject wells.

Fig.3.10. Comparison of predicted and actual mud weight requirements in United States and overseas wells (after data by Reynolds, 1973). o = predicted values.

Other successful applications of seismic data to pressure evaluation are shown for a North Sea well (Chapter 9, Fig.9.8) and an offshore Louisiana well, U.S.A. (Chapter 9, Fig.9.13).

Prediction of *mud weight requirements* from seismic analysis in a deep Oklahoma well in comparison with determinations from both acoustic (sonic) and induction log, and actual mud weights required in the drilling

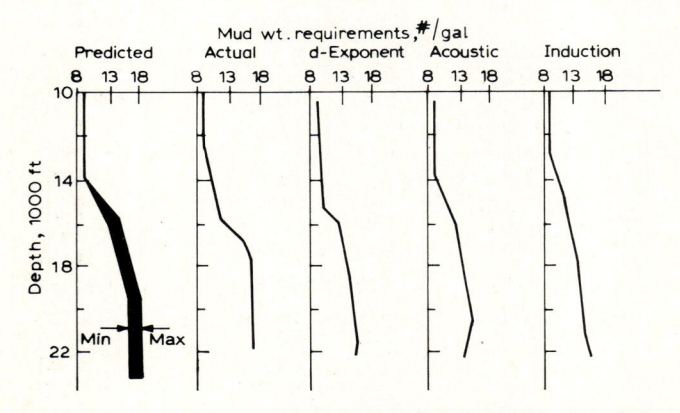

Fig.3.11. Actual mud weights compared with seismically derived data, values determined from drilling parameters (d-exponent, for description see Chapter 4), and data from acoustic (sonic) and induction logs (after Louden et al., 1971). (Courtesy of *World Oil*.)

TABLE 3.V

Comparison of actual and predicted mud-weight requirements in well Farrar No. 1, Anadarko Basin, Oklahoma, U.S.A. (after Louden et al., 1971)

Depth (ft)	Mud weight equivalent (ppg) predicted			Actual (ppg)
	seismic	acoustic log	induction log	
13,000	9.0	8.8	8.8	8.9
14,000	9.0	8.8	8.8	9.6
15,000	10.3	9.6	12.0	11.5
16,000	13.5	13.6	12.8	13.0
17,000	14.6	14.1	13.2	15.3
18,000	15.6	14.7	14.0	16.7
19,000	16.5	15.1	14.4	16.8
20,000	17.1	15.8	14.8	16.8
21,000	17.1	16.1	15.6	16.8
22,000	17.1	15.3	15.8	16.8
Total depth	17.1	15.1	16.5	16.8

TABLE 3.VI

Comparison of predicted and actual drilling mud and casing programs in a U.S. Gulf Coast well (after Pennebaker, 1968b)

	Estimated	Actual
Surface casing set at:	2,500 ft	2,505 ft
Top of abnormal pressure:	10,500 ft	10,200 ft
12.0 ppg pore pressure at:	11,100 ft	11,000 ft
Protective casing set at:	11,100 ft	10,875 ft
Liner set at:	—	11,640 ft
Pore pressure gradients at:		
10,525 ft	—	10.6 ppg
11,100 ft	12.0 ppg	—
11,306 ft	—	14.8 ppg
12,311 ft	—	>16.6 ppg
13,000 ft	17.0 ppg	—
13,965 ft	—	>17.3 ppg
Maximum mud weight:	18.2 ppg	18.1 ppg
Drilling rate at 4,000 ft:	70 ft/hr	76 ft/hr
Drilling rate at 9,000 ft:	14 ft/hr	19.5 ft/hr

operation are listed in Table 3.V and shown in Fig.3.11. Note the excellent quality of the forecasted mud weight requirements all the way down to 24,000 ft [7317 m].

Table 3.VI lists some additional data for a U.S. Gulf Coast well drilled in shale/sand sequences.

The impact of such a high degree of reliability in using these prediction concepts for well planning, such as properly engineered mud and casing programs (see Chapter 6), is obvious from financial, engineering, and safety points of view.

Based upon such formation pressure and lithology data derived from seismic and other supplementary sources, one can go one step further and also attempt to predict *drilling penetration rates* prior to spudding the well. Empirical drilling concepts (equations), such as discussed in Chapter 4, are utilized.

Once more we refer to Table 3.VI which lists one of the first published comparisons of predicted and actual parameters of interest in drilling operations (Pennebaker, 1968b).

In the deep Oklahoma well, drilling rate predictions were found to be within 15% of actual values above 11,000 ft [3354 m] and again below 14,000 ft [4268 m]. From 11,000 to 14,000 ft [3354 to 4268 m] the predicted penetration rate was 50% off (5 ft/hr [1.52 m/hr] versus 7.5 ft/hr [2.29 m/hr]); this was probably caused by a new tooth-shaped insert bit which was used in the operation (Louden et al., 1971).

Although predictions in these field cases cited are quite impressive, seismic velocity analysis is not without possible pitfalls (Reynolds et al., 1971). These include:

Highly dipping formations (Levin, 1969).
Normal move-out error (Shugart, 1969).
Faulting.
Multiple reflections (Dix, 1955; Yancey et al., 1969).
Static errors (Yancey et al., 1969).
Curved ray path (Dix, 1955).
Anisotropy (Postma, 1955; Vander Stoep, 1966).
Interpretation (Amenson and Darby, 1969).

Gravity data analysis

Since overpressured shales are characterized by abnormally low bulk density values (see Chapter 5), gravity data are an additional parameter useful in detecting such formations (Tsimel'zon, 1959; Musgrave and Hicks, 1966). Fig.3.12 schematically shows the apparent density contrast for salt, shale, and normal, hydrostatic shale/sand sequences. However, in overpressure environments the shale density decreases, diminishing the density contrast between overpressured shales and salt masses. This makes differentiation of highly pressured shales and salt by gravity data alone almost impossible.

McCulloh (1967), in an excellent study of sedimentary rocks and gravimetric effects in petroleum and natural-gas reservoirs, stated: "The very large negative density contrast of 0.1—0.3 g/cm^3 between porous reservoir rocks saturated with petroleum fluids and the same rocks saturated with

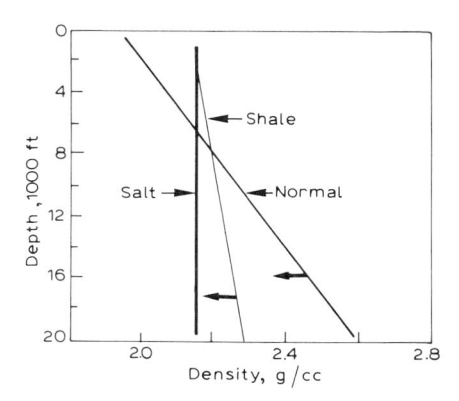

Fig.3.12. Apparent density control of salt and shale masses; normal hydrostatic section and decreasing slopes in overpressured geologic shale/sand sections.

water under reservoir conditions, and the even larger contrast of 0.2—0.6 g/cm^3 between such rocks and interbedded impermeable strata are sufficient to account for those relatively negative gravity anomalies associated with some petroleum and natural-gas fields. In others, *excessive porosity* within the hydrocarbon-bearing part of the reservoir *or other density deficiencies* may be required to account for the gravimetric effects."

Since overpressured formations are characterized by higher than normal (i.e. excessive) porosity, the above statement should also apply to them. This suggests that gravity data can be used as a supplementary data source in the detection and evaluation of overpressure environments.

Other geophysical methods

Besides seismic and gravity data, other geophysical data sources, such as *magnetics* and possibly even *electrical prospecting methods*, should or could be included in the final well plan, provided the investigation is focused on a rather shallow depth (Reynolds et al., 1971).

References

Amenson, H.L. and Darby, E.K., 1969. Evaluation of velocity determination from seismic data. *Soc. Explor. Geophys.*, Preprint G-38.
Beitzel, J.E. and Davis, J.M., 1973. A computer oriented velocity analysis interpretation technique. *SEG Paper, 43rd Annu. Int. SEG Meet.*, Mexico City.
Boylston, L.E. and Looney, J.D., 1973. An application of digital data processing techniques to velocity computations and conversions. *SEG Paper, 43rd Annu. Int. SEG Meet.*, Mexico City.
Dix, C.H., 1955. Seismic velocities from surface measurements. *Geophysics*, 20: 68—86.
Duerbaum, H., 1954. Zur Bestimmung von Wellengeschwindigkeiten aus reflexionsseismischen Messungen. *Geophys. Prospect.*, 2: 151—167.

Green, C.H., 1938. Velocity determinations by means of reflection profiles. *Geophysics*, 3: 295—305.
Hottman, C.E. and Johnson, R.K., 1965. Estimation of formation pressures from log-derived shale properties. *Trans. AIME*, 234: 717—722.
Kaufman, H., 1953. Velocity functions in seismic prospecting. *Geophysics*, 18: 289—297.
Levin, F.K., 1969. Apparent velocity from dipping interface relations. In: *Velocity, Its Measurement and Use — A Symposium*. Houston Geological Society, Houston, Texas.
Louden, L.R., Matthews, W.R., McClendon, R.T., Rehm, W.A. and Aud, B.W., 1971. Ultra-deep drilling guided by seismic data, 1. *World Oil*, 172: 67—72, May; Part 2, 172: 97—105, June.
McCulloh, T.H., 1967. Mass properties of sedimentary rocks and gravimetric effects of petroleum and natural-gas reservoirs. *U.S. Geol. Surv. Prof. Paper*, 528A, 50 pp.
Musgrave, A.W., 1962. Applications of the expanding reflection spread. *Geophysics*, 27: 981—993.
Musgrave, A.W. and Hicks, W.G., 1966. Outlining of shale masses by geophysical methods. *Geophysics*, 31: 711—725.
Pennebaker, E.S., 1968a. Seismic data indicate depth, magnitude of abnormal pressures. *World Oil*, 166: 73—78, June.
Pennebaker, E.S., 1968b. An engineering interpretation of seismic data. *SPE 2165, 43rd AIME Fall Meet.*, Houston, Texas, September.
Postma, G.W., 1955. Wave propagation in a stratified medium. *Geophysics*, 20: 780—806.
Reynolds, E.B., 1970. Predicting overpressured zones with seismic data. *World Oil*, 171: 78—82, October.
Reynolds, E.B., 1973. The application of seismic techniques to drilling techniques. *SPE 4643, 48th AIME Fall Meet.*, Las Vegas, Nev., October.
Reynolds, E.B., May, J.E. and Klaveness, A., 1971. The geophysical aspects of abnormal fluid pressures. In: *Abnormal Subsurface Pressure — A Study Group Report 1969—1971*. Houston Geological Society, Houston, Texas, pp.31—47.
Sarmiento, R., 1961. Geological factors influencing porosity estimates from velocity logs. *Bull. Am. Assoc. Pet. Geol.*, 45: 633—644.
Shugart, T.R., 1969. A critique of delta-T velocity determinations and a method for automating them. In: *Velocity, Its Measurement and Use — A Symposium*. Houston Geophysical Society, Houston, Texas.
Thurber, C.H., 1959. The geo-acoustic spectrum. *Oil Gas J.*, 57: 228—236, October 12.
Tsimel'zon, I.O., 1959. The relationship between local gravity anomalies and the tectonics of oil- and gas-bearing regions. *Izv. Vyss. Uchebn. Zaved., Neft Gaz*, 8: 13—16.
Vander Stoep, D.M., 1966. Velocity anisotropy measurements in wells. *Geophysics*, 30: 900—916.
West, S.S., 1950. Dependence of seismic wave velocity upon depth and lithology. *Geophysics*, 15: 653—662.
Woeber, A.F. and Penhollow, J.O., 1973. The accuracy of depth predictions from velocity profiles on the Texas Gulf Coast. *SEG Paper, 43rd Annu. Int. SEG Meet.*, Mexico City.
Yancy, J.D., Hagen, D.C., Slack, H.A. and Costain, J.K., 1969. On the interpretation of velocity analyses derived from seismic data. *SPE 1122, AIME Offshore Technol. Conf.*, Houston, Texas, 12 pp.

Chapter 4. DRILLING DATA DETECT AND EVALUATE ABNORMAL FORMATION PRESSURES

Introduction

Experience indicates that costly mis-interpretations are best avoided by studying a combination of several measured *pressure indicators.* Many such indicators can be measured or monitored on location, but not all will always be available or usable — or necessarily needed — in any one drilling operation. In Table 4.1 (see overleaf) currently available techniques used to detect and evaluate abnormally high formation pressures are compiled.

All data available, both drilling and logging indicators, have to be interpreted as a group, because if considered individually without relation to other parameters, erroneous interpretation and thus incorrect decisions can easily be made.

This philosophy, which cannot be stressed often enough, is also reflected in a recent statement of an executive of a major oil company: "We used to get good after the fifth or sixth well in the field. Now we have to be good on the very first one. The best way to do this is through a thorough *well plan that incorporates all known information, educated guesses about the unknown and the flexibility to deviate from the plan when necessary.*"

Origins of overpressures are numerous (see Chapter 1), and relate differently or not at all to either logging or drilling data. Furthermore, in shale/sand sequences overpressure environments may have a single cap rock or a sequence of several sealing barriers, which often vary greatly in thickness. Correspondingly, single or multiple transition zones are present, which can be a few to more than several thousand feet thick. The situation is further complicated if faults are present, which may or may not act as overpressure discontinuities. In addition, changes in lithology, such as increasing lime or silt content of the shales, will affect most detection techniques in some manner.

These and similar problems vary with the local geological setting, type of drilling rig equipment, drilling mud and drilling technique, degree of experience of rig personnel, and whether the borehole is straight or deviated.

However, under all these circumstances the ultimate aim is to balance the danger of lost circulation and differential pressure sticking with proper mud weight to contain any encountered abnormally high-pressured pore fluids in permeable formations.

Pressure detection concepts are especially important in balanced pressure

TABLE 4.I

Techniques available to predict, detect, and evaluate overpressures

Source of data	Pressure indicators	Time of recording
Geophysical methods	Seismic (formation velocity) Gravity Magnetics Electrical prospecting methods	Prior to spudding well
Drilling parameters	Drilling rate d-exponent Modified d-exponent Drilling rate equations Drilling porosity and formation pressure logs Logging while drilling Torque Drag	While drilling (no delay time)
Drilling mud parameters	Mud-gas cutting Flow-line mud weight Pressure kicks Flow-line temperature Resistivity, chloride ion, and other novel concepts Pit level and total pit volume Hole fill-up Mud flow rate	While drilling (delayed by the time required for mud return)
Shale cuttings parameters	Bulk density Shale factor Volume, shape, and size Novel, miscellaneous methods	While drilling (delayed by the time required for mud return)

Well logging	Electrical surveys
	  resistivity
	  conductivity
	  shale formation factor
	  salinity variations
	Interval transit time
	Bulk density
	Hydrogen index
	Thermal neutron capture cross section
	Nuclear magnetic resonance
	Downhole gravity data
		After drilling

Direct pressure measuring devices	Pressure bombs Drill-stem test Wire-line formation test	When well is tested or completed

Other methods more or less direct, but generally not as accurate as those indicated above, include surface tubing pressure measurements. The hydrostatic pressure of the fluid in the column must be added to the surface measurements to obtain the formation pressures.

drilling. Not only does the drilling rate decrease with a high overbalance of mud pressure versus formation pressure, but also lost circulation and differential pressure sticking of the drill pipe can readily occur.

A proper mud and casing program is a major cost factor, particularly in high-pressure areas. To design an effective, economical mud program, the fracture gradient as well as pore fluid pressure variations must be known. If pressure exerted by the drilling fluid exceeds the strength limit of the formation, circulation can be lost with a subsequent drop in the hydrostatic pressure head in the wellbore, creating a potential blowout condition.

Data monitoring on present-day drilling rigs ranges from relatively simple equipment used to monitor basic drilling parameters to sophisticated data collection and analysis using on-site computers. In a recent review of the state of the art, Kennedy (1973) commented as follows:

"Over 80 drilling-data collection and analysis units which include the use of a computer are now in operation or available.

These units have been used to help drill more than 1000 wells to date around the world.

Development of hardware has progressed rapidly in recent years to a high degree of sophistication. But in the near future, development emphasis will be shifted to techniques, interpretation methods, and making services more practical and economical.

Though downhole telemetry would make some data available that cannot be collected with surface equipment, the development cost will not be easily justified on the basis of confirming — or even slightly improving — the accuracy of measurements now being made with surface equipment.

Much of today's monitoring equipment is used more to 'keep out of trouble' than to reduce the per-foot cost of drilling. But 'trouble' includes more than just blowouts — it involves stuck pipe, lost circulation, and other costly hazards.

Expenditures of several million dollars are planned in the next 2—3 years by major service firms offering data analysis and gathering equipment to improve existing techniques and develop related items. Those individual estimates that are available are in the $500,000 per company range.

Modern rig instrumentation makes it possible to gather accurate drilling data that are 'untouched by human hands'. Information — flow rate, pit volume, mud weight, flow-line temperature, pipe-running speed, and torque, for example — can be tied to automatic alarms to signal impending danger.

These drilling data can be recorded on strip charts, digital printouts, tape punch, or magnetic tape and transmitted to headquarters computers on a real-time or intermittent basis.

It can be manipulated to update the drilling program and to plan future wells. With an understanding of the relationships involved and the equipment used, data collected can be used to generate *optimum drilling programs, pore pressures at the bit, and logs while drilling.*

Competent personnel must still interpret the data."

Drilling parameters

Drilling rate (penetration rate)

Drilling rate breaks have been used for many years to distinguish sand from shales. However, the apparent relation of penetration rate to variations in pore fluid pressures has been recognized only in the last decade.

Basically, drilling rate is a function of weight on the bit, rotary speed (rpm), bit type and size, hydraulics, drilling fluid, and formation characteristics (Garnier and Van Lingen, 1959; Somerton, 1959; Maurer, 1962, 1966; Galle and Woods, 1963; Gnirk and Cheatham, 1969). Under controlled conditions of constant bit weight, rotary speed, bit type, and hydraulics, the drilling rate in shales decreases uniformly with depth. This is due to compaction increase in shales with depth. However, in pressure transition zones and overpressures the penetration rate increases (Figs.4.1 and 4.2). Slower penetration rate is often observed in the pressure barrier (cap rock) overlying this pressure transition. Also any other major lithological change in the shales

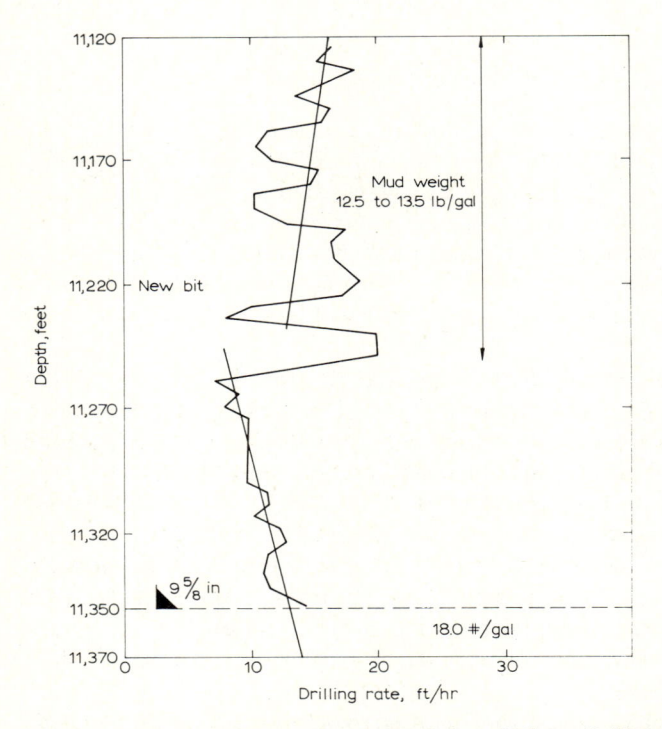

Fig.4.1. Penetration rate versus depth in a U.S. well. Note the decreasing rate of penetration with increasing depth but the reversal below 11,260 ft [3433 m]. $9\frac{5}{8}$-inch casing was set at 11,350 ft [3460 m] and total depth was reached with 18 lb/gal [2.16 kg/dm^3] mud in the hole.

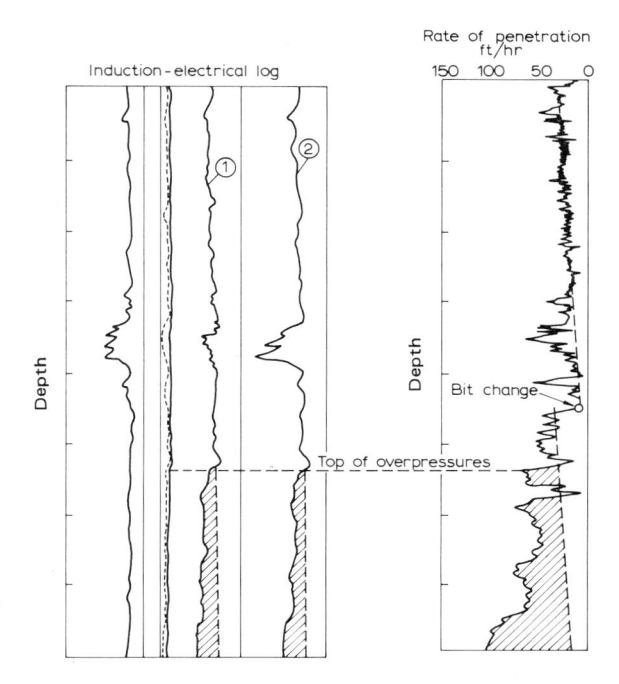

Fig. 4.2. Penetration rate versus depth compared with induction-electric log (after Jordan and Shirley, 1966). Note how in overpressures the amplified short normal resistivity decreases (curve *1*), whereas both the conductivity curve (*2*) and the rate of penetration increase. First and second curve on the left are spontaneous potential and induction curve, respectively. (Courtesy of *Journal of Petroleum Technology*.)

(silty, limey shales, mudstones, etc.) is reflected in penetration rate variations.

Penetration rate should be plotted in 5- to 10-ft [1.5 to 3.05 m] increments in slow-drilling formations or in 30- to 50-ft [9.15 to 15.24 m] increments in fast-drilling intervals. However, plotting such data points should not lag more than twice the plotted depth increment behind the total well depth. Today, drilling rate recorders are available which automatically plot feet per hour versus depth.

Regardless of how the rate of penetration is recorded, one should establish a normal drilling rate trend while drilling shales in normal-pressure environments for comparison with faster-drilling overpressured shales.

According to Forgotson (1969) a penetration rate increase in shales in excess of twice the normal rate indicates proximity to nearby overpressured fluid reservoirs. Observed penetration rate increases are due to reduced differential pressure between the formation pressure and mud-column pressure (Maurer, 1966). However, in practice, simple rules of thumb (i.e. such as a twofold rate increase) do not always apply.

Complications can arise due to bit dulling, which may mask any penetration rate change due to overpressures (Fig.4.3). The penetration rate even may decrease if the rotary torque fluctuates and if the drilling bit action on the bottom of the borehole becomes erratic.

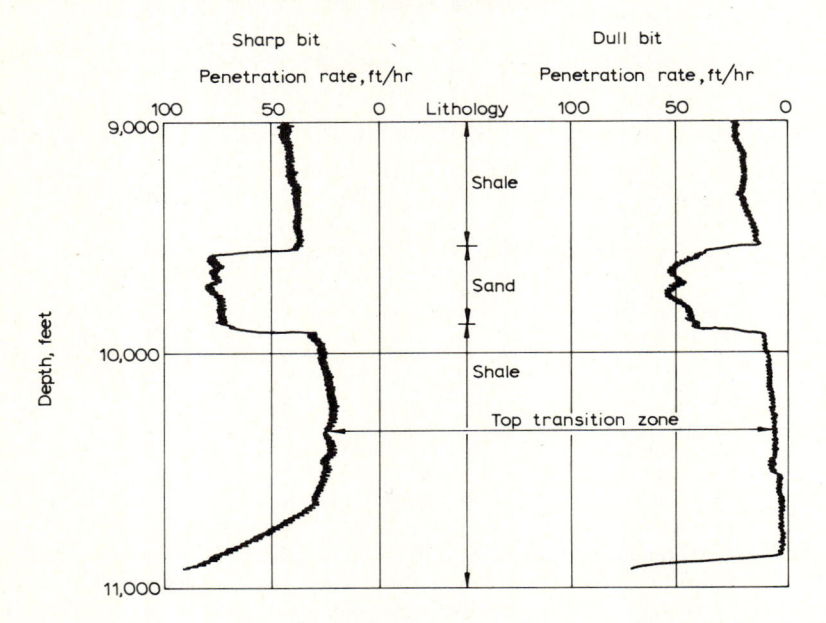

Fig.4.3. Effect of bit-tooth dullness on penetration rate in pressure transition zone (after Matthews, 1969). (Courtesy of *Oil and Gas Journal*.)

Since it is not always possible and/or feasible to maintain bit weight and rotary speed constant, an improved method has been developed which allows plotting of a *normalized penetration rate* (d-exponent) versus depth.

d-Exponent (normalized rate of penetration)

In 1966, Jorden and Shirley proposed a mathematical method to normalize the rate of penetration. Data required to calculate the d-exponent, a dimensionless number, are the penetration rate, bit size (diameter), weight on bit, and rotary speed:

$$d = \frac{\log(R/60N)}{\log(12W/10^6 D)} \tag{4.1}$$

where R = rate of penetration in ft/hr, N = rotary speed, rotations per minute (rpm), W = weight on bit in pounds, and D = bit size (diameter) in inches.

DRILLING PARAMETERS

When $(R/60N) < 1.0$, then the absolute value of $(R/60N)$ varies inversely with R. Therefore, the d-exponent varies inversely with the rate of penetration and, in general, is related to the differential pressure between mud column and formation pressure. Nomographs (Fig.4.4) are available which graphically solve eq. 4.1 quickly.

For increased effectiveness, the d-exponent should be plotted for each 5—10 ft [1.5—3.05 m] drilled. In fast-drilling formation one may have to work with 25- [7.62 m], 50- [15.24 m], or even 100-ft [30.48 m] increments. To make the comparative study of all pressure indicators available easier, the depth scale selected for d-exponent plots should be the same as for any other plotted pressure indicator.

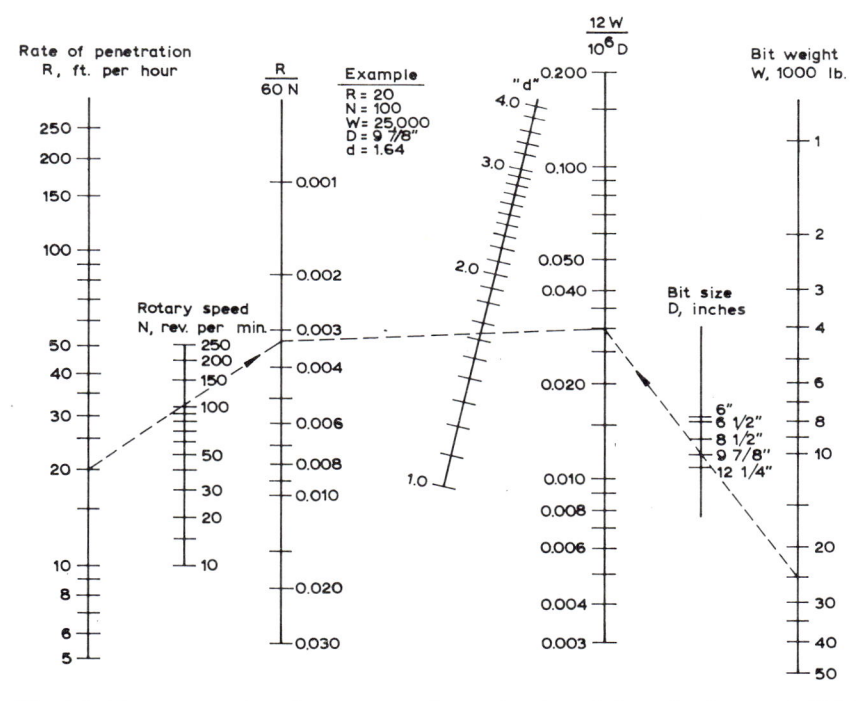

Fig.4.4. Nomogram for d-exponent determination (after Jorden and Shirley, 1966). (Courtesy of *Journal of Petroleum Technology*.)

Basically, plots of d-exponent versus depth show a decreasing trend with depth. In transition zones and overpressure environments, the calculated d-values diverge from the normal trend to lower than normal values.

Quantitative pressure evaluation can then be made using the *equivalent depth method*, such as:

$$P_f = G_o \cdot D_A - D_E \cdot (G_o - G_H) \tag{4.2}$$

where P_f = pore fluid pressure in psi, D_A = depth of interest in overpressured interval in feet, D_E = normal, equivalent depth (ft) corresponding to D_A, G_H = hydrostatic gradient in the subject well in psi/ft, and G_o = overburden gradient in the subject well in psi/ft.

For example, in areas where the overburden gradient equals 1.0 psi/ft [2.31 kg cm^{-2} m^{-1}] and the hydrostatic gradient equals 0.465 psi/ft [0.107 kg cm^{-2} m^{-1}], this relationship simplifies to:

$$P_f = D_A - 0.535 \cdot D_E \qquad (4.3)$$

Eq. 4.3 has been successfully used in the U.S. Gulf Coast area, the North Sea region, and elsewhere. Fig. 4.5 shows its application for a North Sea well. An even faster method is to construct a transparent overlay of parallel, equivalent mud weight lines for the specific depth scale used for the *d*-

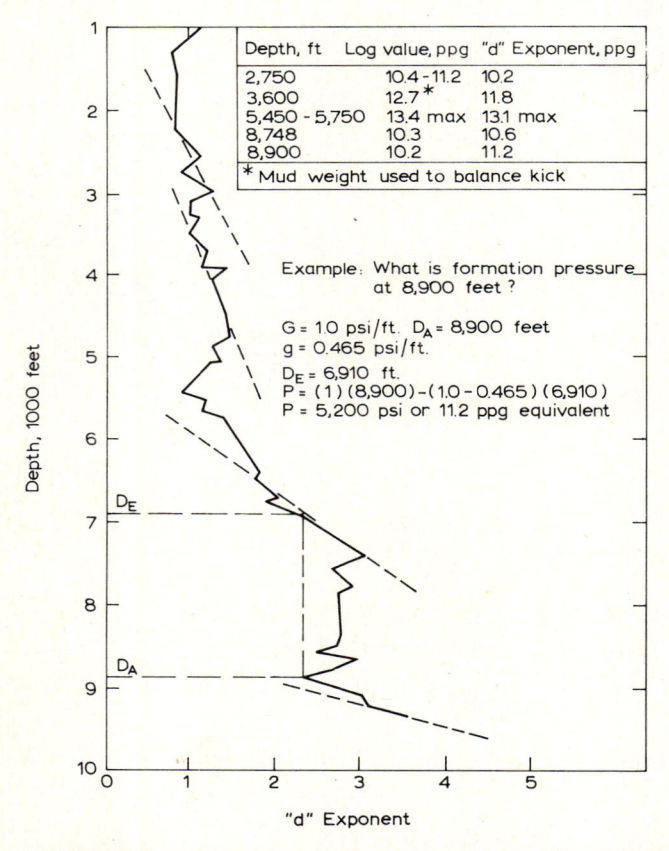

Fig.4.5. *d*-Exponent plot in a North Sea well indicates transition zones at 2600 ft [793 m], 4600 ft [1402 m], and 7400 ft [2256 m] (after Bolt, 1972). Note pressure data comparison from well logs and *d*-exponent. Plot also illustrates application of the equivalent depth method for quantitative pressure evaluation. (Courtesy of *World Oil*.)

exponent. Overlay of this grid allows a fast and direct reading of mud weight requirements in the subject well. This approach is illustrated for the d_c-exponent (i.e. corrected d-exponent) in Figs.4.7 and 4.8.

As one expects, computed d-values are affected by any change in the basic input parameters in eq. 4.1. Thus, changes in bit size and type, bit weight, etc., will affect the d-exponent. In *very soft* formations, such as are localized along the U.S. Gulf Coast, Kalimantan, and elsewhere, reasonable values for bit weight are often difficult to establish. Unfortunately, any major lithological change in the shale section (such as limey, silty shales, mudstones, marls, etc.) will also affect the d-exponent.

Another major problem is the unaccounted effect due to change in mud weight (excessive high mud weight) and/or poorly maintained drilling fluid systems. However, for heavy muds a modification of the standard d-exponent has been developed.

Modified d-exponent

Since the d-exponent is influenced by mud weight variations, a modification has been introduced to normalize the d-exponent for the effective mud weight, such as:

$$d_c = d \cdot (MW_1/MW_2) \tag{4.4}$$

where d_c = modified (corrected) d-exponent, MW_1 = normal mud weight, and MW_2 = actual mud weight used.

d_c-Plots are a substantial improvement over d-plots as far as mud weight corrections are concerned (Fig.4.6). However, other factors and short-

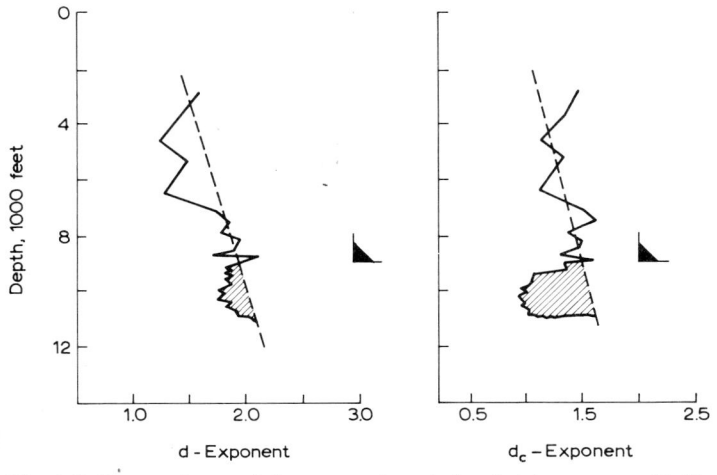

Fig.4.6. Comparison of d-exponent and d_c plot in same well. Protective casing seat is at 8700 ft [2652 m]. Note that d_c-exponent defines overpressures more clearly.

comings — discussed previously — affect both d_c- and d-plots in a similar fashion.

Quantitative pressure methods again utilize the *equivalent depth method* (see eqs. 4.2 and 4.3). Transparent *overlays* of parallel equivalent trend lines for mud weight or pore fluid pressures can be easily constructed for quick quantitative evaluation of d_c-plots. This approach is equally well applicable regardless of whether d_c-values are plotted on a logarithmic scale (Fig.4.7) or on a linear scale (Fig.4.8). Today, slide rules have become available from service companies for quick determination of both d- and d_c-exponents.

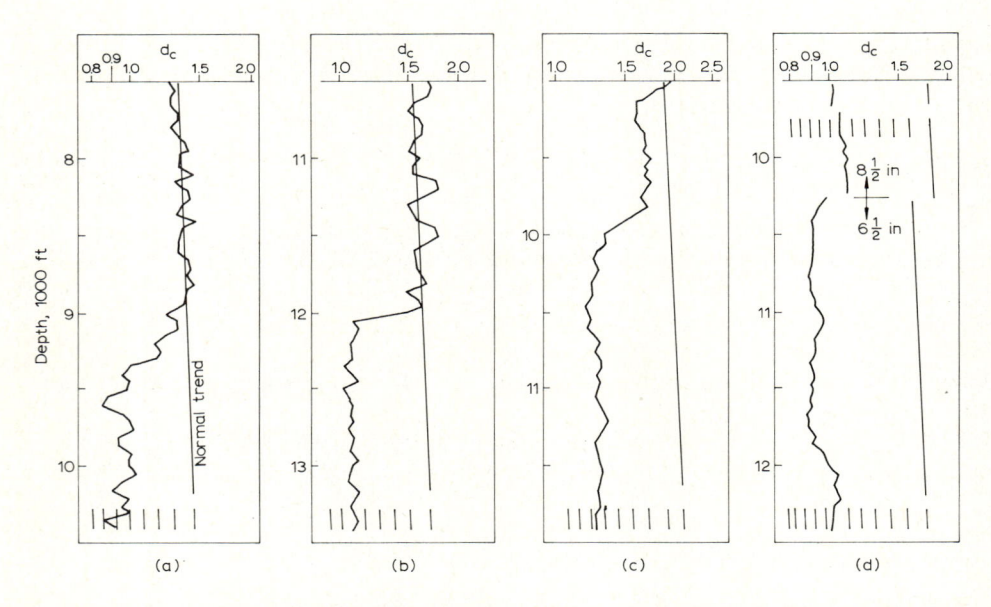

Fig.4.7. d_c-Exponent on logarithmic scale plus use of mud weight overlay technique in (a, d) South Texas, (b) South Louisiana, and (c) Oklahoma (after Zamora, 1972). Note effect of bit size change on normal trend line in the South Texas well (d). |||| = trend lines representing constant mud weights. (Courtesy of *Oil and Gas Journal*.)

Frequency of d_c calculations depends on how fast formations are being drilled. Usually, d_c-exponents are determined every 10 ft [3.05 m]. In fast drilling areas 25- [7.62 m], 50- [15.24 m], or even 100-ft [30.48 m] increments may be adequate. In slow-drilling areas, such as hard-rock intervals, 5-ft [1.5 m] increments may become necessary. Calculations should preferably be made in clean, pure shales. This is easily done in many Tertiary basins throughout the world by simply inspecting drilling rate breaks. It is, however, not as easy in older formations, where drill cuttings analysis (see *Shale factor* discussion) is recommended as a supplementary data source for locating shales.

DRILLING PARAMETERS 127

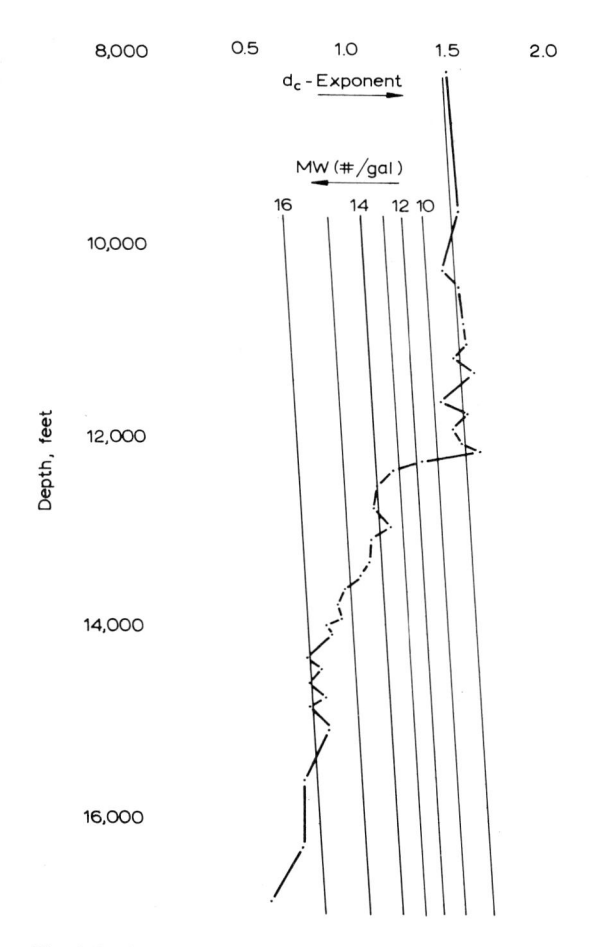

Fig. 4.8. d_c-Plot on linear scale including mud weight overlay technique (after Rehm and McClendon, 1971). (Courtesy of the Society of Petroleum Engineers of A.I.M.E.)

Recently, Rehm and McClendon (1971) went one step further by proposing a modification of the d_c-exponent. Drilling data were collected from over 90 wells throughout the world, and an accuracy approaching 0.2 lb/gal [0.024 kg/dm³] was obtained in all major drilling areas investigated. This was accomplished by an empirical equation of the form:

$$P_f = K \log(D \cdot a + C - d_c) + 16.52 \tag{4.5}$$

where P_f = formation pressure in lb/gal equivalent; K = 7.62, slope of line plotting lb/gal versus d_c; D = depth in feet; and a = slope of normal d_c trend line. According to the authors the value a is fairly constant with geologic age. For example, there is little variation in a between the Miocene of Louisiana, U.S.A., and Indonesia. C = a drillability constant.

TABLE 4.II

Computer solution of eq. 4.5 for formation pressure variations as shown in Fig.4.8 (after Rehm and McClendon, 1971)

Depth (ft)	Mud weight (ppg)	Depth (ft)	Mud weight (ppg)
6,367	9.0	12,946	14.7
8,102	9.5	13,068	14.3
8,858	9.0	13,182	14.2
9,594	9.0	13,354	14.6
10,151	10.2	13,559	15.1
10,409	9.5	13,670	15.4
10,736	9.5	13,873	15.3
10,937	9.0	13,954	15.6
11,104	9.5	14,024	15.6
11,300	9.0	14,208	16.1
11,521	10.4	14,421	15.7
11,701	9.5	14,600	16.0
11,870	9.5	14,678	15.5
11,985	9.6	14,814	15.8
12,063	9.0	14,975	15.8
12,194	11.0	15,115	15.7
12,353	13.6	15,290	15.9
12,523	14.0	15,508	16.1
12,630	13.4	16,893	16.9
12,793	14.2		

Table 4.II shows the results using a computerized version of eq. 4.5 for the well shown in Fig.4.8. Excellent agreement between both methods, the conventional overlay and Rehm's equation, is observed. However, if area experience indicates that d_c plots give the same quantitative evaluation, then there is no reason to use the more tedious eq. 4.5, especially if manual computations are necessary.

Drilling rate equations

A more complex approach is the application of proper drilling rate equations to establish correlations between borehole and formation pressures. However, this requires modern instrumentation to record continuously several drilling parameters and methods to account for the effect of bit wear while drilling.

A general equation for penetration rate in shales has been developed for which constants have been evaluated by a *regression-type analysis* of data from several wells (Combs, 1968). According to the author:

"The correlation assumes that penetration rate is proportional to weight on the bit, rotary speed and a hydraulic term, each raised to a fixed power.

Penetration rate normalized for changes in these variables is found to decrease with increasing differential pressure and with increases in a tooth wear index.

Drillability of shales — defined as the drilling rate at some standard operating condition — decreases with increasing depth, but increases when the pore pressure increases because of the reduced compaction. The equation predicts the penetration rate with a standard deviation of 29%. If the other parameters are known, it predicts the pore pressure with a standard deviation of about 1 lb/gal [0.12 kg/dm^3]. The penetration rate increases when the pore pressure increases because of the reduced differential pressure and the increased drillability of the shale."

Fig.4.9 shows the comparison of observed and predicted rates of penetration using Combs' method.

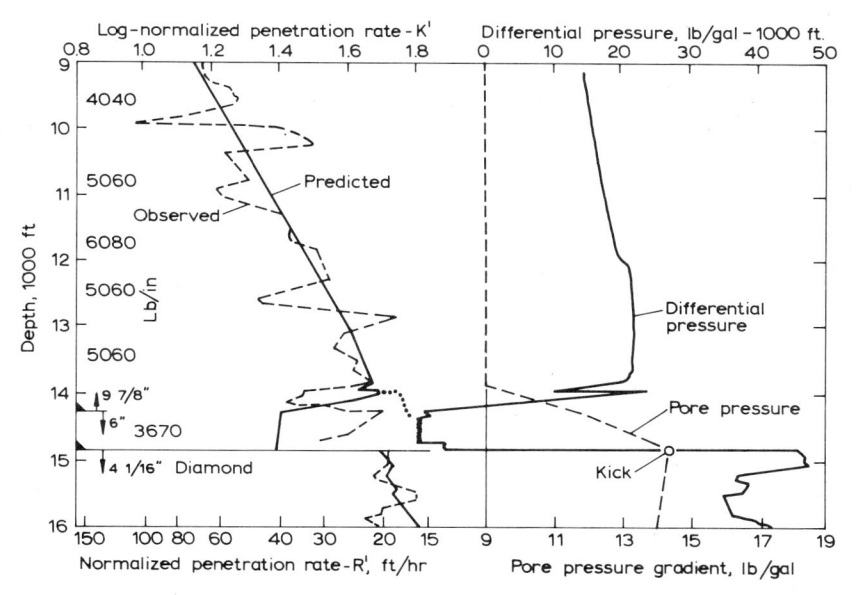

Fig.4.9. Comparison of observed and predicted penetration rate (after Combs, 1968). (Courtesy of the Society of Petroleum Engineers of A.I.M.E.)

In connection with computerized drilling control, Young (1968) expressed penetration rate as:

$$R = G \cdot N^m (W - W_{int}) \cdot K \qquad (4.6)$$

where K, a reciprocal drilling constant, the weight intercept value, W_{int}, and the bit rotary speed exponent, m, are determined by a five-spot drilling test. G is a function of the normalized bit tooth height.

Drilling performance optimization and identification of overpressured formations have been also investigated by Wardlaw (1969).

As mentioned previously, there are several other methods to describe the drilling rate mathematically. Most methods are empirical, and field data for determination of input parameters are usually a prerequisite. These data are obtained from (1) previously drilled and/or adjacent wells, or (2) feedback data from short-interval testing in the subject well.

The main drawback with this type of analysis is that shale lithology (degree of liminess, silt content, etc.) must remain almost constant. Drilling rates will change with lithologic variations, even though the pressure gradient remains constant.

Developing an objective, completely automated pressure detection technique will be difficult, except possibly for local geographic areas where the exact lithology is known. Formation cuttings and wire line and related tools must still be used meticulously.

Drilling porosity and formation pressure logs

Methods have been developed which give a preliminary formation evaluation log while drilling a well, thus reducing and/or eliminating the need to pull the pipe and of running conventional logs.

Using a computerized model describing the relationships between formation characteristics and drilling parameters, an early indication of formation type, porosity, and formation pressure variations is possible (Zoeller, 1970). A review of the method indicates that it can provide drilling optimization, formation evaluation, and pressure control. However, accurate data collection and effective computer processing are necessities (Kennedy, 1970).

We have selected a field example on a deep offshore well in Louisiana, U.S.A. Note the $9\frac{5}{8}$-inch casing seat at 14,401 ft [4392 m], which is followed by a fast pressure increase over the next 600 ft [183 m] (Fig.4.10).

Recently, Boone (1972) discussed the successful application of this specific drilling model in the North Sea area. Sensing devices connected to the drilling rig are used to monitor up to 20 analog and 18 digital signals. As each foot is drilled, the computer scans the data, prints it on teletypewriter, and/or records it on magnetic tape. Further analysis of selected variables, for example, may include rate of penetration, bit weight, rotary speed, porosity, pore fluid pressure, bit tooth dullness, bit bearing wear, cost per foot, etc.

Today, several service companies have made similar data units commercially available. Also, several oil companies have developed their own drilling program models.

For example, since there still is some uncertainty about the exact relations between various controllable drilling variables and the drilling performance, Bourgoyne et al. (1971) have proposed a general drilling model to determine formation porosity, pore fluid pressure, etc. The proposed general equation is expressed as follows:

$$R = K \cdot f_1(W/D) \cdot f_2(N) \cdot f_3(H) \cdot f_4(\Delta P) \tag{4.7}$$

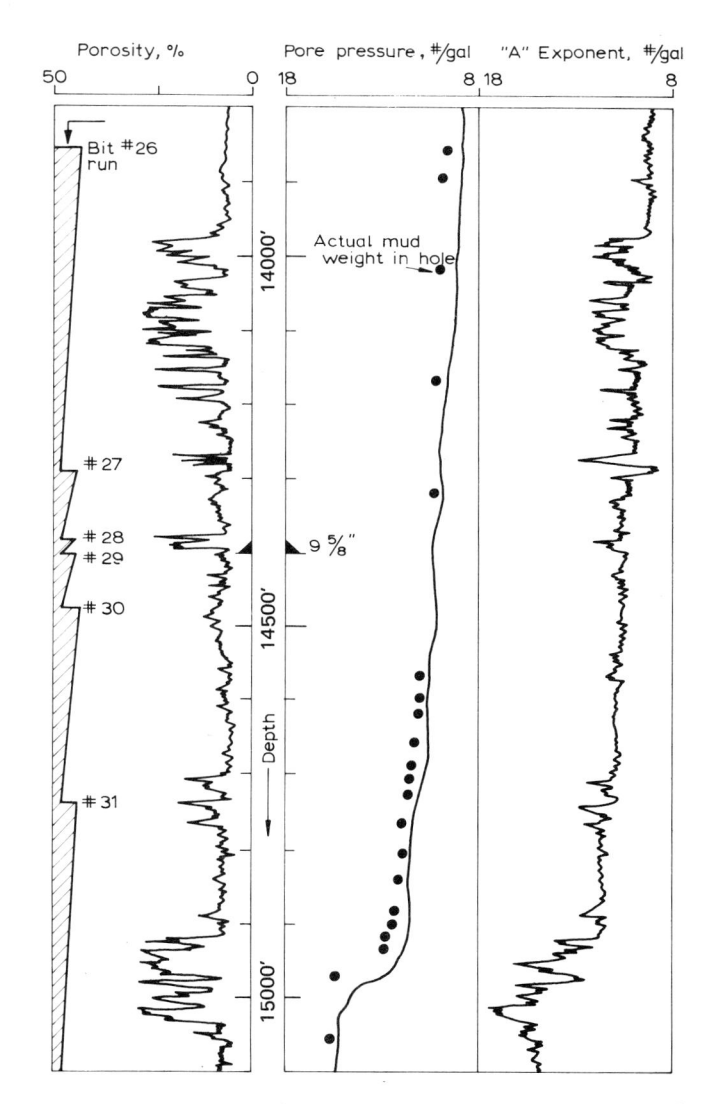

Fig.4.10. Field example of formation porosity—pressure log in an offshore Louisiana wildcat, U.S.A. The curve on the right ("A" exponent) gives the value for pore pressure taking porosity and lithology variations into account.

where R = rate of penetration (ROP); K = drillability constant or normalized ROP; $f_1(W/D)$ = function defining the effect of bit weight, W, per inch of bit diameter, D, on ROP; $f_2(N)$ = function defining the effect of rotary speed, N, on ROP; $f_3(H)$ = function defining the effect of tooth dullness, H, on ROP, and $f_4(\Delta P)$ = function defining the effect of the differential pressure across the hole, ΔP, on ROP.

Normalized rate of penetration, K, is related to bulk density by:

$$\rho_b = 2.65 - 1.65 \frac{S_g + \log K^{-6}}{S_g} \qquad (4.8)$$

where ρ_b = bulk density in g/cm³, and S_g = rock strength parameter (5.2 in the U.S. Gulf Coast area).

Bulk density then is converted to formation pseudo-porosity, which closely agrees with calculated porosity from the density log.

Bourgoyne (1971) then developed a graphical approach for this concept of overpressure detection while drilling, for sites without digital facilities. This approach is indicated on a field example as follows:

"An overpressured formation was anticipated at an approximate depth of 10,000 ft [3048 m]. The normalized penetration rate log was started at a depth of 9000 ft [2744 m]. The calculation was continued to a depth of 10,340 ft [3152 m]. Eleven representative calculations of the 268 calculations required are shown in Table 4.III.

TABLE 4.III

Example calculations for borehole section shown in Fig.4.11 (after Bourgoyne, 1971)

Depth (ft)	Penetration rate (min/5 ft)	Bit weight (1000 lb/in)	Rotary speed (rpm)	Bit rotating time (hr)	Pseudo-density (g/cm³)	Mud weight (ppg)
9,000	5.1	4.1	200	0.00	2.42	11.0
9,005	8.0	4.3	175	0.17	2.46	11.0
9,010	10.5	4.0	180	0.33	2.49	11.0
9,015	9.3	4.2	180	0.50	2.47	11.0
10,050	12.0	4.5	195	8.17	2.50	11.5
10,055	20.0	4.4	195	8.33	2.56	11.5
10,060	13.9	4.5	195	8.50	2.52	11.5
10,155	15.7	3.9	180	12.50	2.49	11.5
10,160	11.5	3.8	180	12.67	2.44	11.5
10,240	13.9	3.6	175	16.83	2.45	11.5
10,245	17.2	3.5	175	17.00	2.48	11.5

Normal shale compaction trends of 5×10^{-5} and 10×10^{-5} g cm⁻³ ft⁻¹ were detected on the first two bit runs. On the third bit run, an abnormal shale trend line of -10×10^{-5} g cm⁻³ ft⁻¹ was detected. At a depth of 10,340 ft [3152 m], the density given by the normal trend line was 2.54 g/cm³ and the density given by the abnormal trend line was 2.48 g/cm³ (Fig.4.11).

As indicated in Fig.4.12, this corresponds to a change in differential pressure of —650 psi [45.7 kg/cm²]. Since the mud density had not increased since establishing the normal trend line, this was the total change in differential pressure. If the mud density had increased 0.1 ppg, the total change in differential pressure would have been (—650 —50 psi) or —700 psi [49.2 kg/

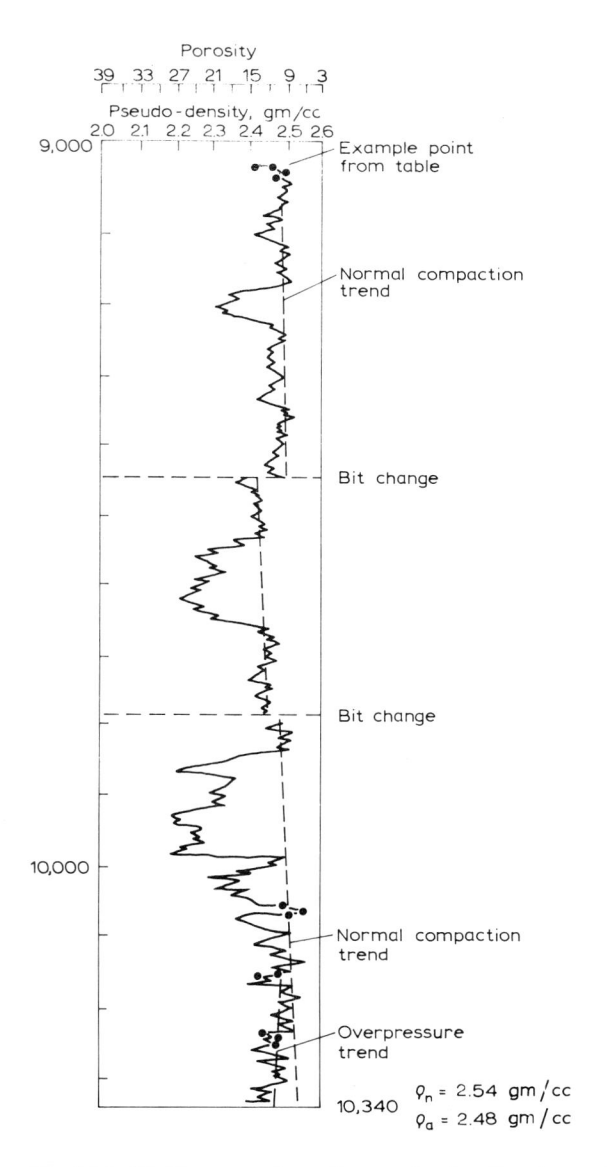

Fig.4.11. Field example of pseudo-density log obtained from drilling data (after Bourgoyne, 1971). Selected example calculations for this depth interval in the subject well are listed in Table 4.III. ρ_n = density value of normally compacted shale; ρ_a = density value of overpressured shale. (Courtesy of *Petroleum Engineer*.)

cm²]. Fig.4.13 shows a change in differential pressure of —650 psi [45.7 kg/cm²] at a vertical depth of 10,340 ft [3152 m] corresponding to an increase in pore pressure of 1.2 ppg [0.145 kg/dm³]."

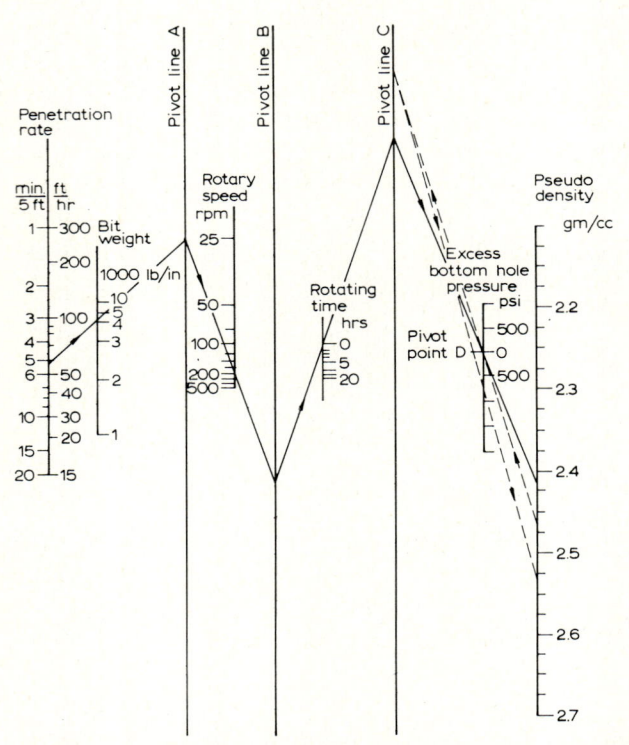

Fig.4.12. Bourgoyne's drilling model nomograph (after Bourgoyne, 1971). (Courtesy of *Petroleum Engineer*.)

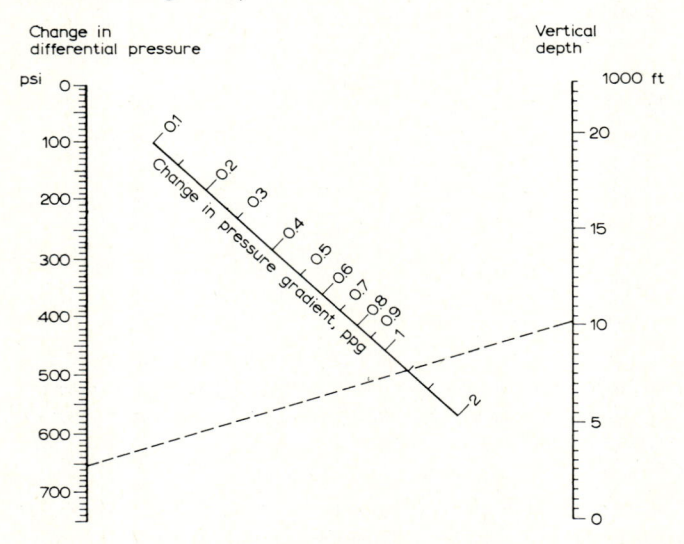

Fig.4.13. Nomograph to detect change in pressure gradient when depth and variation in pressure differential are known (after Bourgoyne, 1971). (Courtesy of *Petroleum Engineer*.)

Three of several other very interesting techniques are also discussed briefly:

Using historical field data from several Louisiana Gulf Coast wells, equations for predicting pore pressure in both normally and abnormally pressured geologic sections penetrated by the borehole were derived by *regression analysis* (Herbert and Young, 1972). When the results of the analysis are applied to drilling data, the transition from normal to overpressures can be predicted, although only on a geographically regional basis.

Correlations between *well log data and rock drillability* have produced encouraging results (Gstalder and Raynal, 1966; El-Hadidi, 1970). Results of these studies indicate that transit time data from sonic (acoustic) logs can be used to predict rock drillability, provided a differentiating lithology parameter is independently available.

Recently, Lutz et al. (1972) have discussed the *SNAP log*. Vibrations from the tricone bit are measured at the kelly and interpreted at the derrick floor to give an instantaneous log of the formation while it is being drilled. In the words of the authors:

"The dynamic theory of drilling presented is based on creation by the

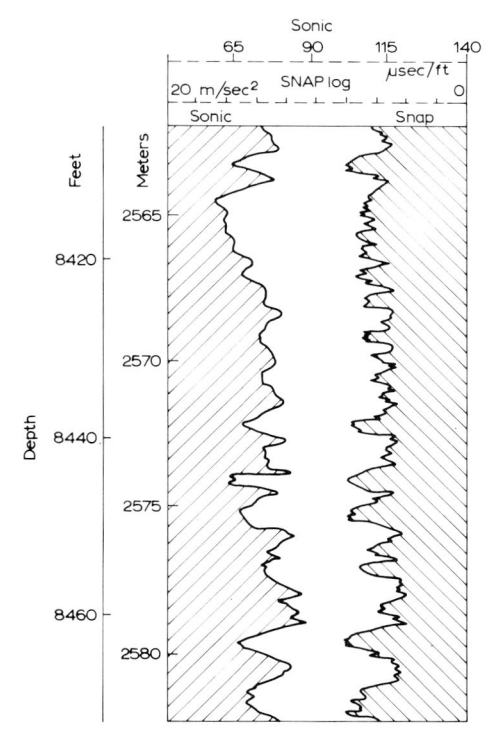

Fig.4.14. Detailed correlation between SNAP log and conventional sonic log in $12\frac{1}{4}$-inch borehole, Aquitanian Basin, France (after Lutz et al., 1972). (Courtesy of *Journal of Petroleum Technology*.)

drilling of longitudinal vibrations and transmissions of these through the drill stem. A study of the entire mechanical drilling system makes it possible to build instrumentation with which the vibrations received can be compared with the signals from a simulated model of the formation. Adjustment of the simulated signals will give instantaneous data on the hardness of the rock being drilled. Today, the most sophisticated model is not yet available, but a simplified treatment unit, associated with sensors and recorders, has been used in the field for several years. An experimental method of instantaneous logging has been developed, which allows one to obtain a log (the SNAP log) related to the hardness of the rock. There is an excellent correlation, as would be expected, between the SNAP log and conventional logs, especially the sonic log.

The information received is of value in formation correlation and in optimizing drilling variables, including making decisions on bit changes. The log is expected to be important in detecting the approach to, and the entrance into, abnormally pressured formations since it reflects the greater hardness of the cap rock above them and the lesser hardness and density of the abnormally pressured shales themselves."

Fig.4.14 shows a detailed correlation between the SNAP log and conventional sonic log in a well located in the Aquitanian Basin, France.

Novel logging-while-drilling concepts

For several years now efforts are under way to record formation, mud, and bit data in situ, i.e. at the bottom of the borehole, and to transmit these data to the surface. The proposed methods and patents granted in subject area are numerous. Future potential of some of these methods appears to be tremendous; however, at the present the *weak link* in all proposed methods is the search for a reliable, accurate, and feasible data transmission system from the drill bit to the surface. Concepts proposed include cables in the drill string, electrical conductors in the pipe itself, and pressure pulses transmitted by the drilling fluid to the surface;

Torque

Because of increased wall-to-wall contact of drill pipe and wellbore, torque usually increases gradually with depth. A drastic increase in torque may serve as an additional abnormal-pressure indicator. In the presence of negative differential pressure in the borehole, overpressured shales will tend to flow or heave into the wellbore.

Today, many drilling rigs continually monitor torque variations.

Drag

Drag is the excess in hook load over free handing load. This excess in load may be caused by bit balling, dog legs, deviated holes, differential sticking, extra cuttings coming into the wellbore when drilling transition zones, etc. The presence of a transition zone may be only one of many causes of an increase in drag. Also in the presence of a gradual pore pressure increase, as in very long transition zones or drilling from floating vessels, this indicator may become questionable.

Furthermore, in deviated holes and severe dog legs, both torque and drag will act as erroneous pressure indicators.

Drilling mud parameters

Mud-gas cutting

Mud logging is being used in an increasing number of wells to aid in abnormal-pressure detection. Experience indicates that, depending on the geological environment and drilling techniques in the subject well, mud-gas cutting may or may not be directly related to increased formation pressures (Fertl, 1973).

As early as 1945, the use of gas measurements in drilling mud was recommended for detection of overpressures and warning of impending blowouts (Pixler, 1945). Similarly, Rochon (1968) proposed mud-gas anomalies as an aid in controlling drilling mud—formation pressure relationships. In a few instances, mud-gas shows are useful to indicate flow into the wellbore if formation permeabilities are extremely low. In such situations, the degree of

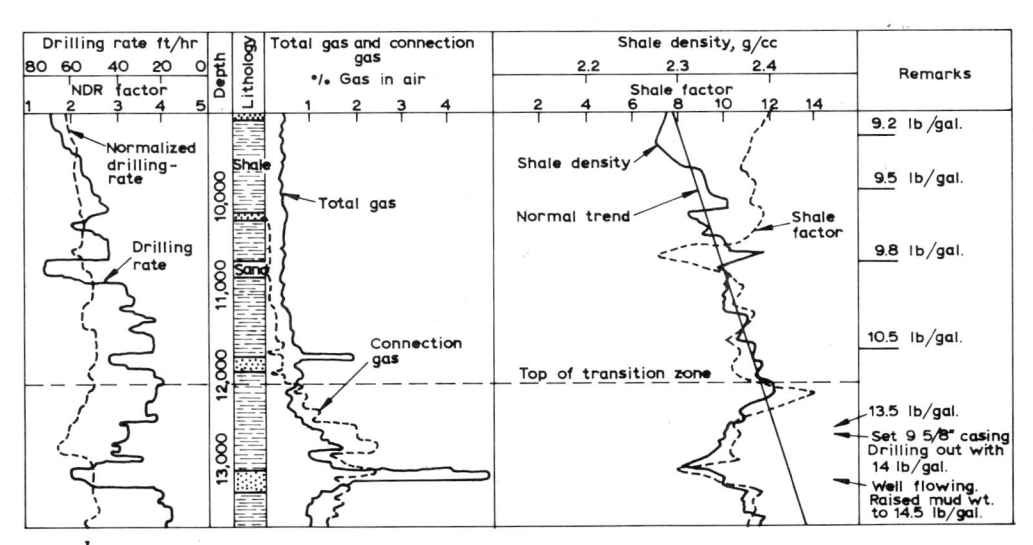

Fig.4.15. Comparison of several pressure indicators in an offshore well, U.S. Gulf Coast. NDR factor is the normalized drilling rate factor; shale factor is the methylene blue test for shale classification (see Table 4.VI).

gas cutting can be roughly correlated with the amount of underbalance (Goldsmith, 1972).

Fig.4.15 shows several pressure indicators, including mud-gas cutting, in a Louisiana well.

Since mud-gas logs are sometimes used to detect overpressures, it is important to be aware of several other factors which may affect and thus complicate their use for pressure detection (Fig.4.16). These include:

(1) *Pay zones.* High gas readings in potential, permeable pay zones — the initial and primary use of mud logs — affect such logs' application for pressure indications. Diffusion of gas into shales immediately above pay zones also complicates interpretation. Such vertical gas migration has been observed above many pools under different geological conditions. In general, a sharp gas increase occurs in these cap rocks, with the thickness of the gas-saturated zone depending on gas composition and reservoir pressure within the pool.

(2) *Connection gas.* Quite frequently, swab effect of raising the kelly when making connections brings small (but often significant) amounts of gas into the borehole.

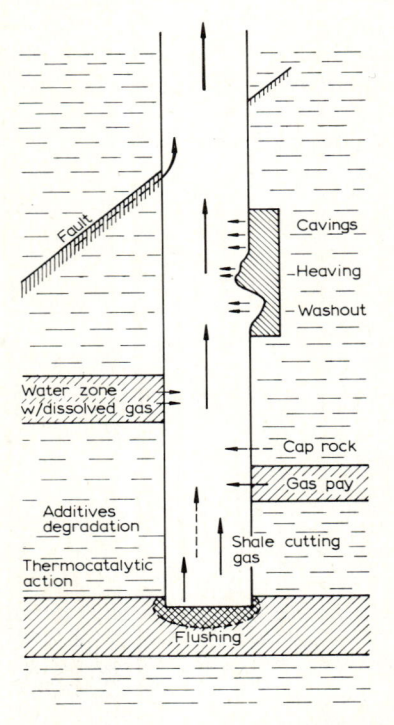

Fig.4.16. Schematic model showing various sources of gas cutting in drilling a well which can complicate detection of abnormal pressures (after Fertl, 1973c). (Courtesy of *World Oil.*)

(3) *Kelly air.* A simple mechanical operation such as unscrewing the kelly for pipe addition can introduce air into the drill pipe. While a continuous mud weight recorder or chromatograph allows its detection at the wellhead after a complete circulation cycle, the kelly-air effect is usually insignificant.

(4) *Downtime.* When mud pumps are shut down for trips, rig repairs, or logging operations, increased gas shows may appear when circulation is re-established.

(5) *Gas composition.* Special consideration has been given to methane-gas cutting. However, H_2S and CO_2 are encountered frequently in modern deep drilling. Both gases are quite soluble in mud, especially oil muds, and their thermodynamic behavior differs from methane. In deep wells, methane kicks usually give sufficient warning time, while H_2S and CO_2 expansion occurs at rather low pressures, i.e. farther up the borehole. Resulting gas kicks give little or no warning.

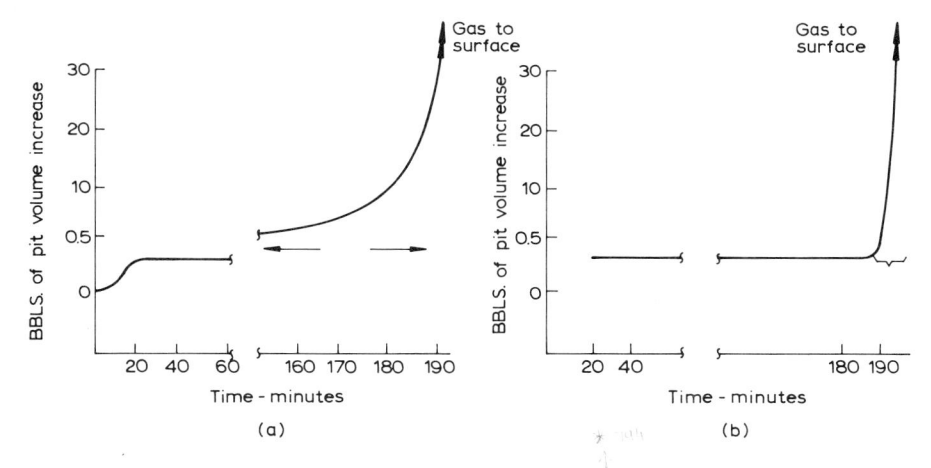

Fig.4.17. Time and pit volume changes for 5 barrels [7.9 liters] of gas swabbed in at a well depth of 20,000 ft [6098 m] in a 6-inch hole with $3\frac{1}{2}$-inch drill pipe. Pumping 3 barrels [4.77 liters] per minute. (a) Typical behavior of methane (CH_4). (b) Typical behavior of hydrogen sulfide (H_2S).

For example, Fig.4.17 compares a five-barrel methane and hydrogen sulfide kick in a 20,000-ft [6098 m] borehole that was swabbed-in during a connection. Methane gives about a 1-hour warning due to high-pressure readings and an increase in both flow and pit volume, since CH_4 starts expanding about halfway up the borehole. Hydrogen sulfide, however, does not show any such pressure increase and gives very little warning (several minutes only) before the hole starts to unload. Thus, not much spare time will be available to close the blowout preventor when necessary.

(6) *Degradation of mud additives.* Gases such as H_2S and CO_2 may also originate from mud additives degrading in hot wells. For example, degrada-

tion of modified lignosulfonates at temperatures exceeding 400°F [204°C] produces significant amounts of H_2S and CO_2 (Skelly and Kjellstrand, 1966).

In a 23,000-ft [7012 m] well in the Caspian area, U.S.S.R., hydrogen was found in mud returns below a certain depth (temperature range). Both well-site and laboratory tests showed considerable hydrogen increase after organic materials were added (Angelopulo et al., 1965). Additional data were recently reported by Ioffe (1972).

Obviously, such gas sources complicate mud-log interpretation. Such gas cutting is also unrelated to any formation pressure variations.

(7) *Lignite*. This brownish-black soft coal of a slightly wooden texture is encountered worldwide in shale/sand sequences. In the U.S. Gulf Coast and elsewhere such lignite zones, and associated high gas readings, are found in thick shales or adjacent to shales.

(8) *Gas flushing*. Total gas readings in drilling muds in western Siberia indicate that productive pay zones sometimes showed background readings, whereas water-bearing zones containing dissolved combustible gas showed high total gas values (Rodionov, 1965). Domestic experience indicates that even with flushing ahead of the bit, residual gas saturation will to some extent cause gas cutting.

(9) *Volcanic material*. Along the northwest coast of the United States, Central America, certain areas of South China Sea (such as on- and offshore Java), and elsewhere, volcanic beds cause high-rate gas flows at very shallow depth. Over geologic time, diagenesis of volcanic ash basically resulted in three components: clay minerals, methane, and carbon dioxide.

Drilling associated shales causes gas cutting without directly reflecting formation pressure variations.

(10) *Deep-seated mud volcanoes* encountered in Central Asia of the U.S.S.R., South America, and New Zealand may be considered as large-scale shale diapirs (for details see Chapter 1). Large amounts of solids, fluids, and gases expelled by such action disturb the equilibrium of rocks and the equilibrium state of liquids and gases they contain. In such environments, gas cutting is frequent, but again it does not necessarily reflect abnormal-pressure variations.

(11) *Faults*. Boreholes often cut faults that may channel gases, resulting in localized gas flow into the wells and gas cutting. Such situations are well documented for Hluk, southeast Moravia, where gas contains up to 30% CO_2 and 50% nitrogen. CO_2 moves along the Hluk fault from rocks underlying the flysch complex at great depth (Kvet, 1970).

(12) *Thermodynamic processes*. Clays exhibiting catalytic activity are present in both formation and drilling mud (Pechorin, 1971). Field tests show that gaseous hydrocarbons also originate as a result of the grinding action by the drill bit and the subsequent temperature increase, in the presence of a catalyst (clay mineral) and organic matter in the rock. This effect may become significant in the interpretation of mud-gas reading.

Flow-line mud weight

Frequent checks for weight reduction at the mud flow line will be an additional indication for gas cutting and possible overpressures.

Issenmann and Lucon (1971) discussed a newly developed continuous mud weight recorder (Densimud), which can be used either alone or in combination with standard gas-logging instruments. Densimud basically consists of a constant-height column through which mud from the wellhead outlet is circulated by a special pump at a constant flow rate. At the bottom of the column, the weight of the mud is measured by a pressure gauge and transmitted to a recorder. The authors also stated that even in the absence of mud gas, a significant decrease in mud weight (up to 0.7 lb/gal [0.084 kg/dm^3]) may occur every time the kelly is disconnected.

Several drilling performance indicators can be deduced if mud density and flow rate in and out of the well and standpipe pressure are accurately known. Accordingly, a high-resolution drilling fluid monitor system consisting of pressure and density sensors has been developed recently (Goddard et al., 1973) and field acceptance of the system has been enthusiastic. According to the authors:

"The density measurements are accomplished by highly accurate specially designed radioactive densometers. The measurement is based on the principle that the degree of absorption of gamma rays by a material is a function of the density of the material. The gamma ray absorption-to-density dependency is generally nonlinear, and it is sensitive to material composition. The design features of these instruments permit highly accurate indications that are relatively unaffected by the material composition of normal mud systems. The measurement is indicated and recorded in a linear fashion and does not require calibration tables. Thus, measured data obtained with these instruments are particularly suitable when incremental and trend information is required. With proper calibration, density measurements accurate to ±0.1 lb/gal [±0.012 kg/dm^3] in the 7—20-lb/gal [0.84—2.4 kg/dm^3] range result."

Today, continuous mud weight indicators have become an integral part of many on-site sophisticated data collection and analysis units.

Pressure kicks

Despite the best engineered approach, a well may kick without forewarning. Current balanced drilling techniques often require only a fine margin between effective pressure control and threatened blowout. Differential pressures on the formation are frequently reduced well below 500 psi [35.2 kg/cm^2]. Therefore, improper pressure balance may cause well kicks of more or less limited size. Such kicks, ranging from 0.5 lb/gal [0.06 kg/dm^3] and higher, can be a direct indication of overpressures. Whereas kicks as high as

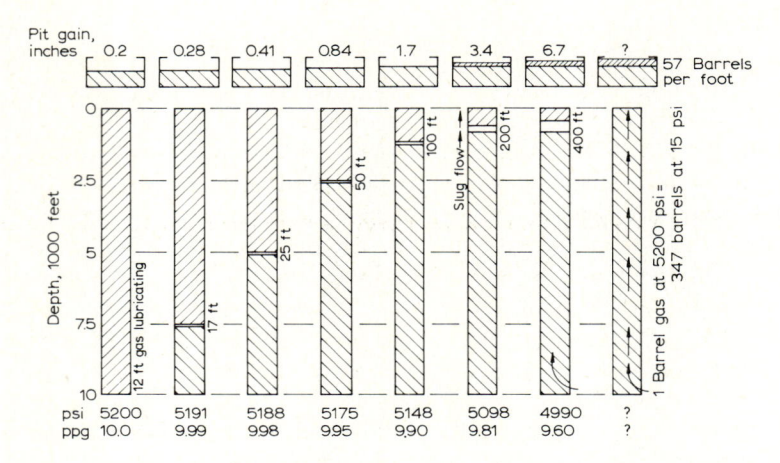

Fig. 4.18. Rise of one barrel of gas with expansion in an $9\frac{7}{8}$-inch hole with a $4\frac{1}{2}$-inch drill pipe (after Goins, 1968). In this illustration, one barrel of gas rises from 10,000 ft [3048 m] to 5000 ft [1524 m], and the volume increases by only one barrel [1.6 liters]. At 2500 ft [762 m], the volume is only four barrels [6.35 liters]. Even though the gas has travelled three-quarters of the distance out of the well, the volume change of mud in the pit is not detectable. Since the rise from 10,000 ft [3048 m] would take about 5 to 10 hours, the pit gain or even the well flow rate is unnoticeable. It is most important to realize that as the gas nears the surface, the rate of expansion will increase rapidly, unloading mud and reducing the bottom-hole pressure to the point that violent bottom flow can begin. At 1250 ft [381 m], the volume is only eight barrels [12.7 liters], and at 625 ft [191 m], it is 16 barrels [25.4 liters], but at surface conditions, the volume of gas, if it originally entered a 10-ppg mud [1.20 kg/dm³], would be 347 barrels [551 liters]. The gas can travel this remaining distance in a relatively short time, and the final expansion can be most rapid. Thus, it is possible for a crew to have checked a well for flow or pit gain while tripping or even when out of the hole and yet be caught by a later, apparently sudden flow from the well. (Courtesy of *World Oil*.)

8 lb/gal [0.96 kg/dm³] are known to have occurred, fortunately, most kicks require less than 2 lb/gal [0.24 kg/dm³] mud weight increase for control.

Kick considerations in high-pressure wells include pressure differential between mud-column and formation pressures, thermodynamic behavior of the gas, its interaction with drilling mud (especially oil base), downhole pressures and temperatures, and time required to circulate the mud system (function of depth) and to record the transmitted pressures. Figs. 4.18 and 4.19 illustrate the rise of one barrel of gas with and without gas expansion (Goins, 1968).

Training programs by oil and service companies, and elsewhere, have largely eliminated the certain amount of confusion associated with well control and pressure detection methods. Nevertheless, questions such as: "Should a well which is kicking be closed in?", cannot be answered by a simple yes or no.

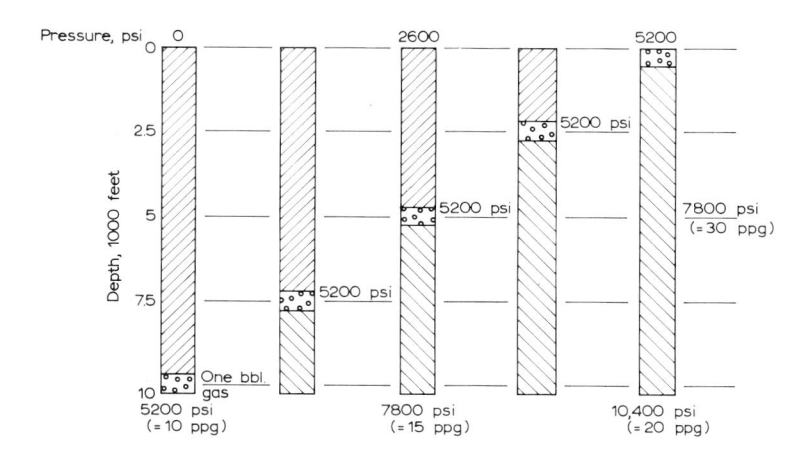

Fig.4.19. Rise of one barrel of gas without expansion (after Goins, 1968). Gas at 10,000 ft [3048 m] in a 10-ppg [1.20 kg/dm³] mud is under 5200 psi [365.7 kg/cm²] pressure, and the borehole at that point is under the same pressure. When the gas rises to the surface and has not expanded, it will still have 5200 psi [365.7 kg/cm²] pressure, but it now exerts the 5200 psi [365.7 kg/cm²] at bottom, and bottom-hole pressure is 10,400 psi [731.4 kg/cm²]. A 20-ppg [2.41 kg/cm³] mud would be necessary to provide this without top-hole pressure. At 5000 ft [1524 m], a 30-ppg [3.16 kg/dm³] mud would be required to equal the combined surface and mud column pressure. A closed-in well that is gas cut will show this effect, and gas lubricating to the surface can either break down the formation or burst the casing. A familiar example occurs in the casing annulus of producers where the pressure is often relieved by bleeding off a small amount of gas. A well that kicks and is uncirculated for a considerable period will undergo this effect. (Courtesy of *World Oil*.)

One concept is the drill pipe (mud pump) pressure technique. This method of controlling a threatening blowout or kick as early as possible is based on a closed "U"-tube system consisting of a borehole and drill pipe, the closed ends being blowout preventer and mud pumps. The basic idea consists of knowing the pressure gradient required to circulate the mud system. By adding the shut-in drill pipe pressure reading, one obtains circulating pressure sufficient to kill the well.

However, closing the well in is not always the best approach since such a procedure could become very dangerous with short surface strings, including the possibility of blowout around the casing.

It is not in the scope of this present work to discuss the advantages, limitations, or mathematical concepts of the "balanced pressure" and "low-choke pressure methods". Excellent discussions and detailed reviews of present-day well control methods are available in the literature: the work of O'Brien and Goins (1960), Goins and O'Brien (1962), Goins (1968, 1969) and Rehm (1969) are notable examples.

In general, however, proper pressure control requires blowout preventer stacks, adjustable surface chokes, accurate and reliable pressure reading equipment, gas separators, sophisticated drilling and mud analysis equipment, and last, but not least, trained rig personnel with a sound understanding of the basic concepts involved, and a planned control program to meet any emergency.

Flow-line temperature

In a pressure transition zone, the formation pressure increases at a rate above the normal increase with depth. The same appears to hold true with temperature. However, a problem exists in determining the formation temperature since the only accurate way is to measure static temperatures in a temperature-stabilized environment after drilling. In drilling operations, interest lies in before-the-fact temperature—pressure related measurements. This can be done by monitoring the flow-line temperature of the mud stream.

Formation heat conductivity is an important parameter for in-situ formation evaluation. As shown previously (Eucken, 1911; Ratcliffe, 1960), heat conductivity decreases hyperbolically with temperature. An increase in pressure promotes heat conduction (Bridgman, 1924; Clark, 1941), whereas convection of pore fluids and radiation hardly seem to affect the heat conductivity (Vries, 1952) provided the grains are smaller than 2 mm.

It is interesting that thermal conductivity of water does not change appreciably with the addition of soluble salts (Nukiyama and Yoshizawa, 1934). Therefore, any salinity variations just above and/or within overpressured sections (see *The salinity principle*, Chapter 5) should have no marked influence on any temperature variations. However, one would still expect to find a decrease in the thermal conductivity due to the increase of porosity in overpressured zones.

With higher pore pressures, porosity is also higher accounting for a higher fluid content. The thermal conductivity of the formation, therefore, decreases. Similar conclusions have been drawn by Langseth (1965) who observed that the thermal conductivity of clays varies inversely with their water content. Additional data show that heat conductivity values of shaly sandstones are less than those of clean sandstone (Zierfuss, 1969). Furthermore, it has been illustrated that the effect of pore fluids on gross conductivity is relatively small for rocks of low to moderate porosity (Zierfuss, 1969), but the reverse is true for high porosity (Bullard et al., 1956).

Jones (1968) also stated that the geothermal gradients are large in the clay beds overlying the reservoirs and very much reduced in the aquifers. There can be no doubt that by checking the upward flow of water from the "waterlogged" sediments beneath them, sealing clay beds of geopressured reservoirs have caused a reduction of the geothermal flux above and overheating of the sediments below.

TABLE 4.IV

Heat conductivity values of various rock and fluid components (after Zierfuss, 1969)

Material	Approx. heat conductivity (mcal cm^{-1} sec^{-1} °C^{-1})
Gas	0.1
Oil	0.3
Water	1.4
Clay	2.4
Quartzite	6.8—18.9

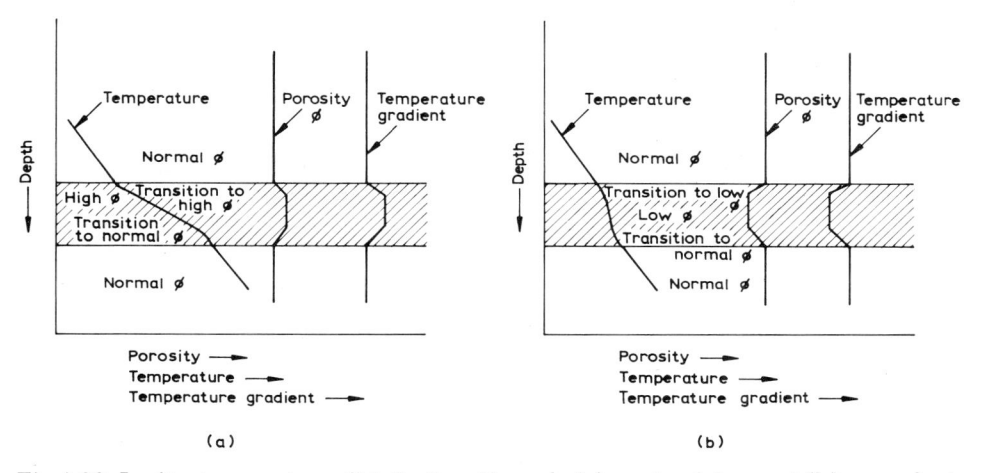

Fig.4.20. In-situ temperature distribution through (a) an insulator and (b) a conductor (after Lewis and Rose, 1970). (Courtesy of *Journal of Petroleum Technology*.)

More recently, Lewis and Rose (1970) proposed a model relating overpressures and high formation temperatures based upon basic heat flow considerations (Guyod, 1946). Inasmuch as heat conductivity varies with rock and fluid components of subsurface formations (Table 4.IV), overpressured, high-porosity shales will act as "thermal barriers" (Fig.4.20), thereby locally increasing the geothermal gradient as has been observed in the field previously (Jones, 1968; Fertl and Timko, 1970). Changes in flow-line temperature gradients of up to 10°F/100 ft [18.2°C/100 m] have been observed prior to and/or when entering overpressured intervals (Fig.4.21).

However, this pressure indicator is also affected by lithology, circulation and penetration rate, tripping the drill string for bit change, etc. Thus, certain precautions and refinements in use and interpretation of such data have to be considered (Wilson and Bush, 1973). Recommendations for plotting the flow-line temperature include (1) using temperature end points

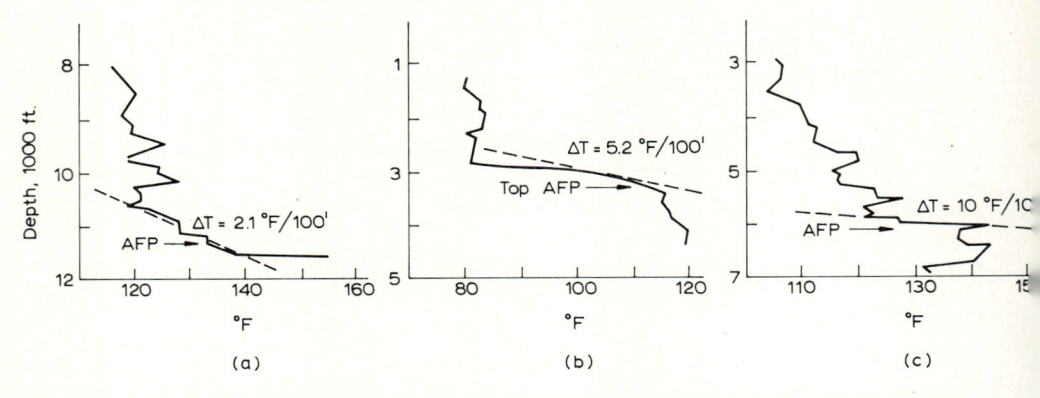

Fig. 4.21. Flow-line temperature data in normal and overpressured environments in (a) a South Texas well, (b) a North Sea well, and (c) a South China Sea well (after Wilson and Bush, 1973). AFP = abnormal formation pressure. (Courtesy of *Journal of Petroleum Technology*.)

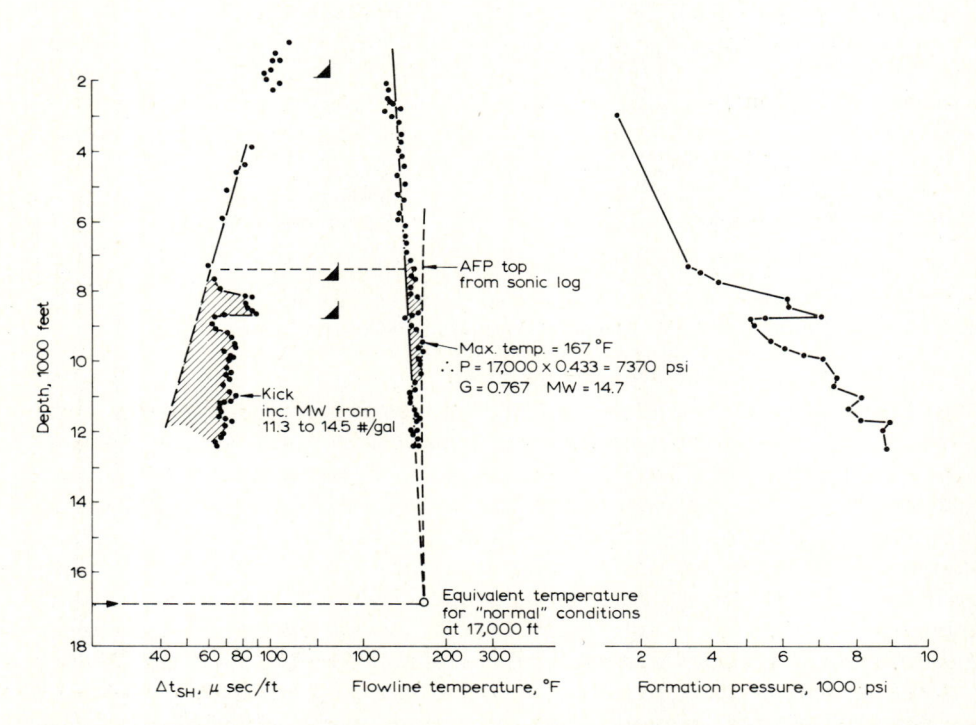

Fig. 4.22. Attempt of quantitative pressure evaluation using flow-line temperature data in a Texan wildcat. For comparison, sonic plot, pore pressure variations, and casing seals are shown (after D. J. Timko, personal communication, 1973). G = geostatic pressure gradient in psi/ft; MW = mud weight in lb/gal.

of each bit run, and (2) replotting segments end-to-end without regard to actual temperature values.

In Fig.4.22 we have gone one step further by attempting to use flow-line temperature data as a semiquantitative pressure indicator in a Texan wildcat. Note the similar, or at least complementary, evaluation of both sonic log and temperature data. As additional information, both the casing seats and log-derived pore pressure profile are also shown.

Finally, reference is made to Fig.9.13 in Chapter 9 showing flow-line temperature changes in comparison with several other pressure indicators in normal and overpressured environments.

Resistivity, chloride ion, and other novel concepts

Association of salinity and formation pressure variations in soft and hard rocks has been discussed in detail previously (Overton and Timko, 1969; Fertl and Timko, 1970). The importance of this observation and several limiting factors were cited by the authors.

Matthews (1969) proposed chloride content measurements of the mud and stated that "this indicator is valid, but is difficult to establish unless there is close control over the mud properties and analysis. The mud engineer has equipment available to make routine chloride ion tests, but the routine test is not adequate to show these subtle changes. Additional tests will have to be run."

Today, service companies are equipped to measure the *gain or loss* of chlorides between inlet and flow line of the mud stream (Fig.4.23) and attempt to relate apparent variations to drilling and pressure conditions. Also many of the modern drilling rig data units present the possibility of recording both inlet and flow-line mud resistivity.

Several other novel concepts, for example, specific ion variations or redox and pH measurements on the ingoing and outgoing mud stream (corrected for lag time), which have been studied in detail on drill cuttings (Fertl and Timko, 1973), may offer additional possibilities as continuous indicators. However, several factors, such as drilling mud additives, downhole chemical reactions, etc., would probably greatly affect interpretation of such indicators.

Pit level and total pit volume

Variations in the total mud volume can be monitored by pit level indicators. Such devices register any large mud-volume reduction, as caused by lost circulation, and monitor large amounts of fluid entry into the borehole, e.g. from unexpected high formation pressures.

First indication of a kick while going into the hole is the observation that the pit level increases in excess of the mud displacement by the pipe run into the hole.

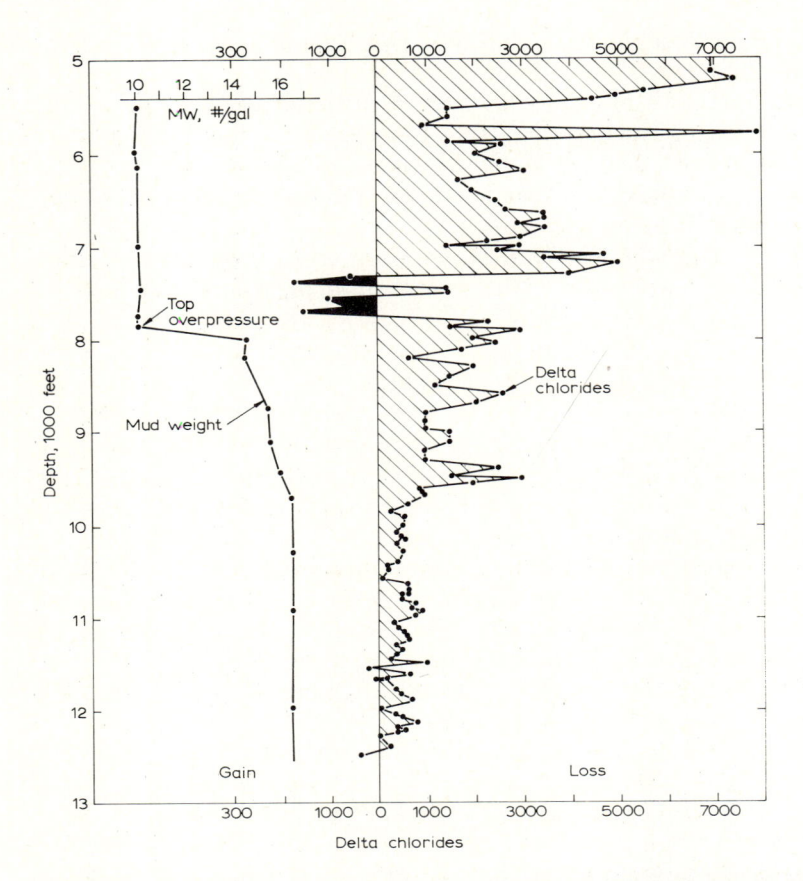

Fig.4.23. Differential (delta) chloride plot on inlet and mud flow line plus mud weight requirements in the subject well. Delta chlorides expressed in ppm.

Recently, Dupin de Saint Cyr (1973) discussed the development of ultrasonic equipment which accurately measures drilling fluid levels in mud tanks. These measurements are made without any contact with the mud, which makes this method particularly effective on floating, offshore drilling rigs.

Hole fill-up

If the drill string is pulled, the mud volume needed to fill the hole should equal the pipe volume. Table 4.V shows bottom-hole pressure reductions under given operating conditions (Goins, 1968). Keeping the hole full is even more critical at the time drill collars are pulled, since if the same length of collars as that of drill pipe is pulled, the level of the drilling mud in the hole will fall four to five times as fast. Furthermore, there may be a temporary

TABLE 4.V

Drill pipe displacement (after Goins, 1968)

Size (in. OD)	Designation	Nominal weight (lb/ft)	Average actual weight (lb/ft)	Displacement (barrels)			
				bbl/ft	1—90' stands	5—90' stands	10—90' stands
$2\frac{3}{8}$	tubing T/C	4.70	4.59	0.00167	0.15	0.75	1.50
	tubing TJ	4.70	5.13	0.00187	0.17	0.84	1.68
$2\frac{7}{8}$	tubing T/C	6.50	6.34	0.00233	0.21	1.05	2.11
	tubing TJ	6.50	7.22	0.00263	0.24	1.18	2.37
	LWDP	6.85	7.33	0.00267	0.24	1.20	2.40
	drill pipe	10.40	10.90	0.00397	0.36	1.79	3.57
$3\frac{1}{2}$	tubing T/C	9.50	9.20	0.00335	0.30	1.51	3.02
	tubing TJ	9.50	10.13	0.00369	0.33	1.66	3.32
	LWDP	9.50	10.27	0.00374	0.34	1.68	3.37
	drill pipe	9.50	10.27	0.00374	0.34	1.68	3.37
	drill pipe	13.30	13.80	0.00503	0.45	2.26	4.53
		15.50	16.1	0.00587	0.53	2.64	5.28
$4\frac{1}{2}$	tubing T/C	12.75	12.83	0.00467	0.42	2.10	4.20
	tubing TJ	12.75	14.58	0.00531	0.48	2.39	4.78
	LWDP	13.75	15.08	0.00549	0.49	2.47	4.94
	drill pipe	16.60	17.80	0.00648	0.58	2.92	5.83
	drill pipe	20.00	21.60	0.00776	0.70	3.49	6.98
5	LWDP	15.00	17.69	0.00644	0.58	2.90	5.80
	drill pipe	19.50	20.60	0.00750	0.68	3.38	6.75

Nomenclature: tubing T/C = tubing with threads and couplings; tubing TJ = tubing with tool joints; LWDP = light-weight drill pipe.

Ten stands of pipe represent a small volume of steel and accurate measurements are required for checking. This means that flow back must be prevented with the hole completely filled. Mud in the drill pipe should be slugged with barite to prevent flow back. Rig crews frequently and deliberately fill the hole only partly to make certain the pipe pulls dry, but this practice will hide partial swabbing and can lead to trouble.

pressure reduction while the string is pulled due to bit balling, high mud viscosity, thick mud cake, back pressure valve on string, etc.

If salt water, oil, or gas from the formation has entered the well, the mud volume required to fill the hole will be less than the volume of the pipe pulled and gives the first kick indication. Measurements of mud volume used to fill the hole can be checked from changes in the pit mud level.

Mud flow rate

Any abnormal rise in pit level caused by mud flow from the annulus will also be reflected in an increasing flow rate, which can be measured by a standard flowmeter.

Actually, flow rate measurements are superior to pit level checks since small flows can be detected before they become sufficiently large to show on any pit level device. If such small flows are noticed immediately, they are — as yet — not as critical and more time is still available to take proper control measures.

Shale cuttings parameters

Shale bulk density

Bulk density in normally compacted shales increases with depth. It can be measured by the *high-pressure mercury pump technique* (Boatman, 1967), the *fluid density gradient column*, the *mud balance method* (Matthews, 1969), etc.

The *density gradient column method* uses a mixture of two fluids of different densities in a graduated column such that the density of the mixture varies with column height.

Shale cuttings density is determined every 5, 10, or 30 ft [1.5, 3.05, or 9.15 m] and even larger increments in soft, fast-drilling intervals by measuring the distance the selected drill cuttings settle in the density column. This value is then entered in proper calibration curves to obtain the corresponding shale density value.

Naturally, care must be taken in selecting and preparing the cuttings for analysis, which includes (1) washing, (2) screening to discard large cavings and/or smooth recirculated cuttings, and (3) air or spin drying until cutting surface has dull appearance. Multiple cuttings have to be tested due to variance in sample data. The average density value, for a given time, is then plotted versus depth.

Mud balance measurements to determine shale bulk density, in our opinion, is a faster and, due to the larger sample size involved, probably more representative method. The procedure includes the following steps:

(1) Place shale cuttings into cup of the mud balance so that the balance indicates 8.3 lb/gal [1.0 kg/dm^3] with the cap on.

(2) Finish filling the cup with water and weigh again (W_2).

(3) Shale bulk density then equals $[8.3/(16.7 - W_2)]$; or use the graphical solution in Fig.4.25.

Shale bulk density values are then plotted versus depth and normal compaction trend lines are established. These data can be used to forecast densities at greater depth. Since shale porosity increases in pressure transition and overpressure zones, any decrease in bulk density values may reflect the presence of such abnormal-pressure environments (Fig.4.24).

In principle, the cuttings density method is similar to bulk density measurements by well logs (see Chapter 5). However, the cuttings method has the advantage of a much shorter lag time, since the only delay is the bottom-up time during mud circulation.

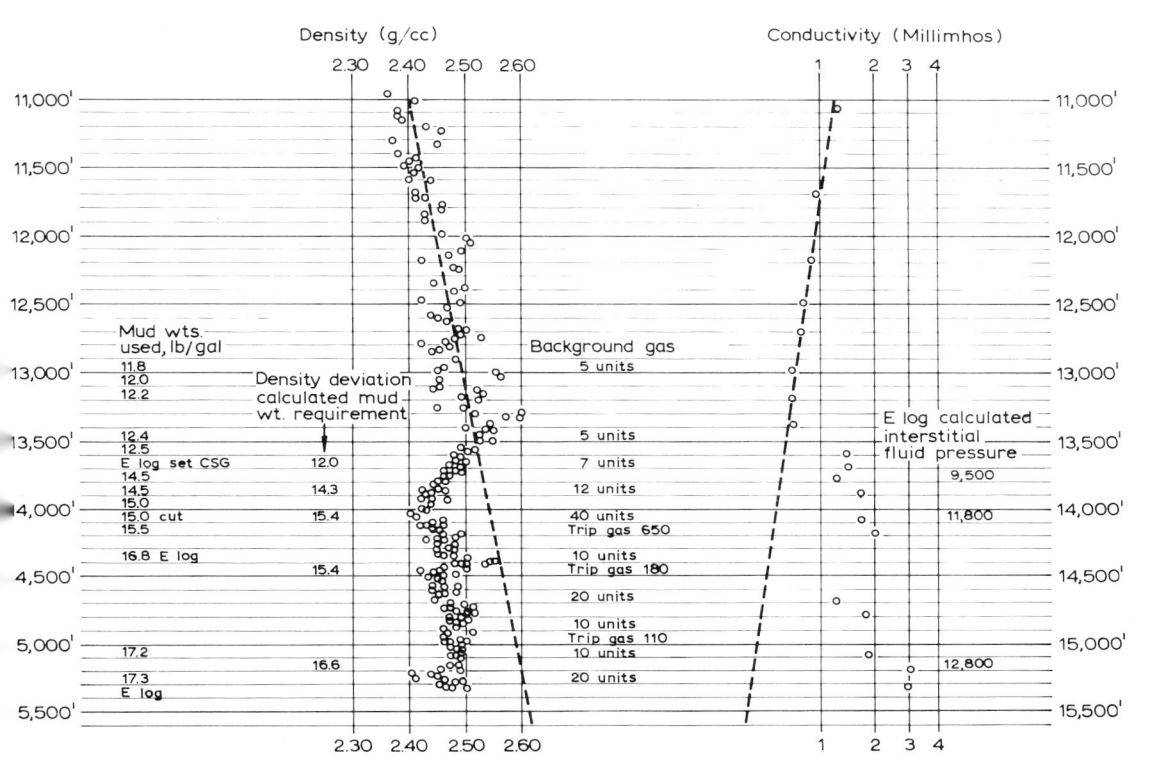

Fig.4.24. Shale bulk density variations in normal and overpressured zones plus well-log conductivity curve and mud weight requirements (after Boatman, 1967). (Courtesy of *World Oil*.)

Quantitative pressure evaluation from shale density data is possible by the *equivalent depth method* (same principle as shown for the *d*-exponent previously) or from *empirical* curves established for a given area (Fig.4.26). Proposed variations of these basic concepts will yield minor variations in the quantitative analysis (Griffin and Bazer, 1968). The classic methods are quite satisfactory, however, if one considers the accuracy and reliability of the basic well-site measurement involved.

Occasionally or, depending on the area, frequently several limitations may be encountered, such as:

(1) Presence of shale gas in the cuttings decreases apparent bulk density values.

(2) Cavings and/or recirculated cuttings are part of the sample investigated.

(3) Reliability of density data depends on the care taken by personnel to catch samples at shale shaker and in running the tests.

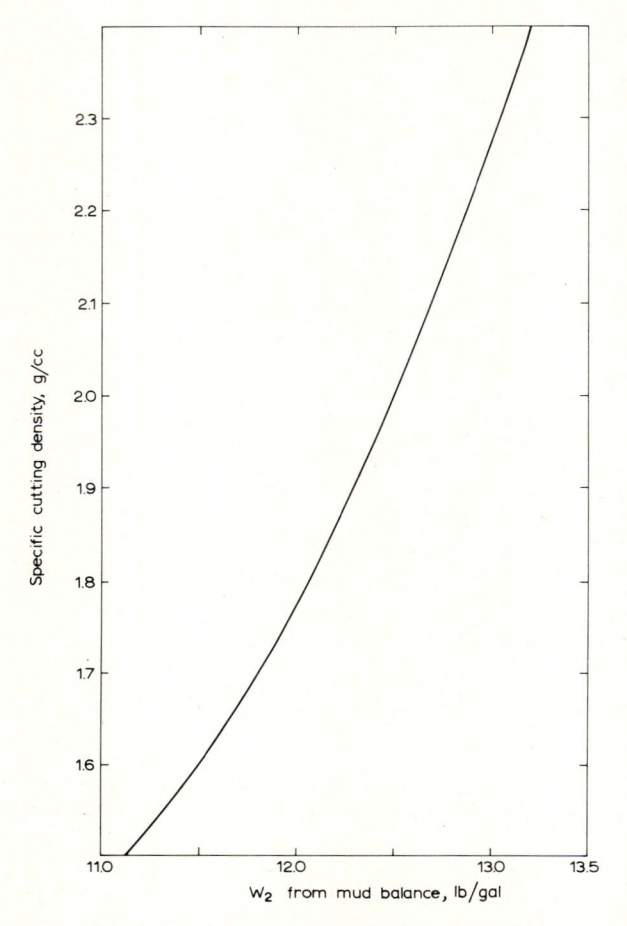

Fig.4.25. Graphical shale bulk density determination for mud balance technique.

Fig.4.27 illustrates possible complications due to human failure. Shale density measurements were carried out in a well located offshore of Louisiana by two contract services — a mud-logging contractor and a drilling engineering consulting firm. The data of both checked closely but failed to detect the transition zone at 13,000 ft [3963 m] and the well kicked when a saltwater sand was penetrated 600 ft [183 m] deeper.

The well was a rank wildcat in an area where little was known about geopressures and a concentrated effort was made to detect pressure through the shale density data. The failure prompted further investigation to determine the cause of failure, if possible. Shale density measurements were rerun on retained cutting samples through the interval of concern. The results of the rerun shale densities indicate that in this instance inaccuracy in density

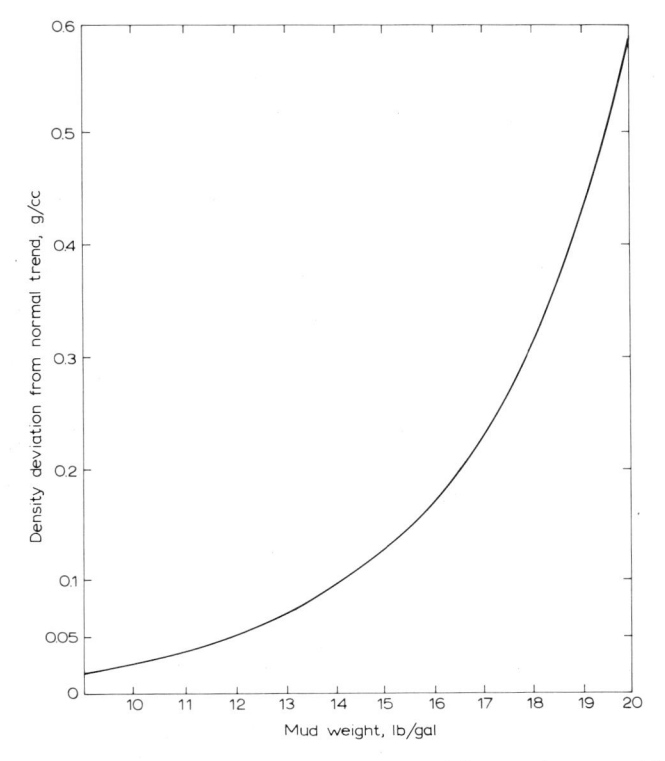

Fig.4.26. Empirical chart for mud weight requirements derived from shale bulk density variations, for local U.S. Gulf Coast area (after Boatman, 1967). (Courtesy of *World Oil*.)

measurement had occurred in the initial field work. Although run from the same samples, considerable difference is apparent between the density plots for the field work and the rerun. The rerun densities are higher than the original, which is normal for dry samples and thus is insignificant. The difference in density variations is most significant, however. A break from the normal compaction trend is apparent on the rerun plot, while none can be seen on the original.

It is also of interest to note that neither penetration rate nor the gas content of the mud indicated the presence of the transition zone. However, the break on the rerun density plot corresponded precisely to the resistivity decrease of the resistivity (IES) log.

(4) Age boundaries, unconformities, etc., may also greatly influence normal compaction trend lines. Fig.4.28 illustrates a sharp shift across an unconformity in a Venezuelan well.

(5) Lithological variations (high carbonate content, silty and/or sandy shales, mudstones, marls, etc.) can result in erroneous pressure evaluations.

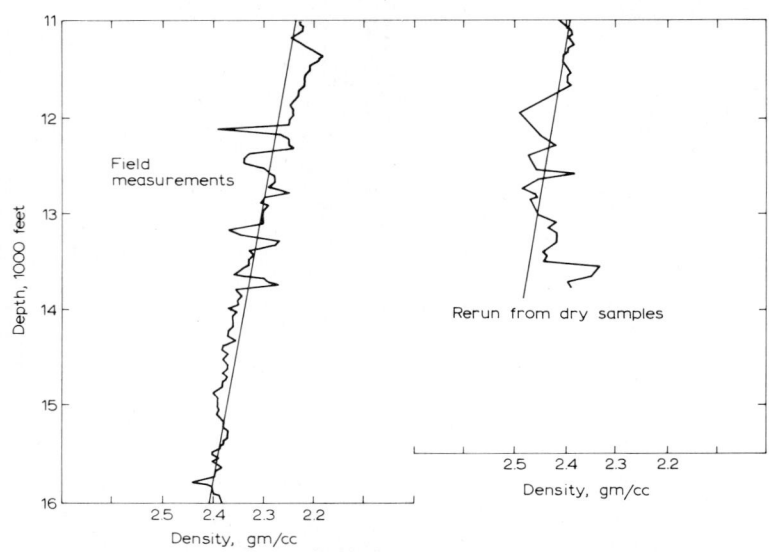

Fig.4.27. Comparison of shale density measurements in a wildcat well, offshore Louisiana, U.S.A. (after Fertl and Timko, 1971). (Courtesy of the Society of Petroleum Engineers of A.I.M.E.)

For example, substantial carbonate content may cause large variations in shale density determinations. With the exception of pinpointing pressure caps, i.e. limey shale barriers on top of pressure transition zones (Boatman, 1967; Rochon, 1967; Fertl and Timko, 1970), large and varying $CaCO_3$ content in shales throughout the geologic section is detrimental to straightforward application of density data as pressure indicators. Two field examples illustrate this effect (Fig.4.29).

(6) The presence of heavy minerals, such as pyrite (Permian Basin, U.S.A.; offshore Cameroon, Africa; South China Sea area etc.), siderite (McKenzie Delta, Canada; South China Sea) etc., will increase shale bulk density and as such may mask the onset (top) of overpressures.

Despite shortcomings, as listed above, shale bulk density plots are often successful in pressure work (Fig.4.30). This statement applies to worldwide exploratory drilling operations, provided limitations and shortcomings are recognized immediately.

Fig.4.31 shows plots of bulk density versus depth by two methods for an offshore Louisiana well. Bulk density measurements can be carried with wireline-logging methods or on shale cuttings. The bulk density determined from cuttings is less than that from the log, a quite frequent observation. The exception is in the cap rock from 10,500 to 11,000 ft [3201 to 3354 m], where cutting alteration has not been severe because of lower cap rock permeability. The magnitude of bulk density data here is similar. However, the pressure top at 11,000 ft [3354 m] is defined clearly by both methods.

SHALE CUTTINGS PARAMETERS 155

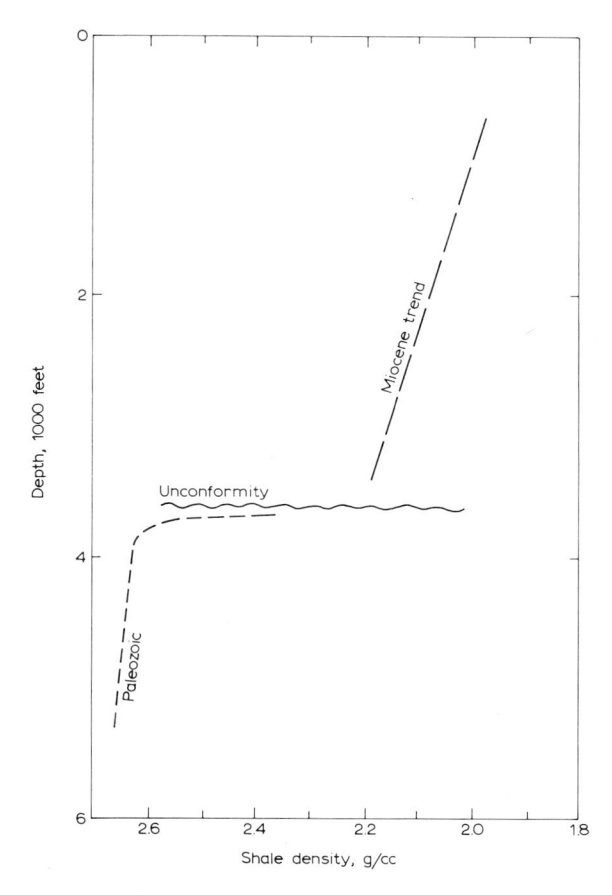

Fig.4.28. Shale bulk density change across an unconformity in a Venezuelan well, South America.

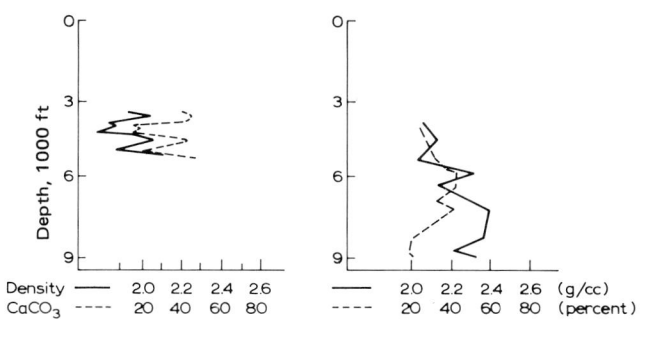

Fig.4.29. Effect of the carbonate content on the density of shales.

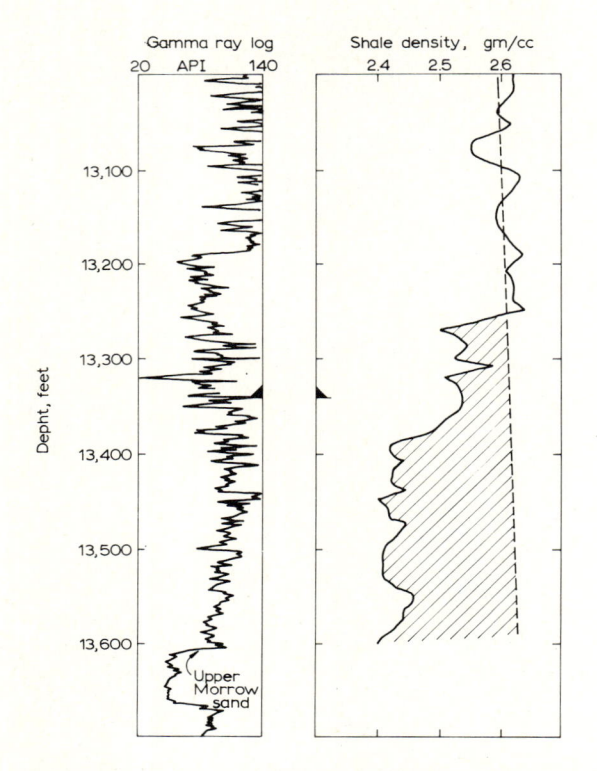

Fig.4.30. Shale bulk density plot in a gas discovery well, Texas Panhandle area. The $10\frac{3}{4}$-inch casing seat is at 13,342 ft [4068 m].

After a normal trend is established, the top of an overpressured zone shows up as a decrease in bulk density.

As observed on all bulk density figures shown, the values are plotted on a linear scale. Obviously, cuttings bulk density data can also be plotted on a logarithmic scale. This will accentuate minor density variations.

Combined with the so-called *shale factor*, shale bulk density measurements are useful in detecting the tight barrier or cap rock, which is often found on top of overpressured zones.

Shale factor

The application of elaborate X-ray diffraction analyses for the investigation of the type and amount of mineral components in reservoir rocks is restricted to the laboratory. However, some simplified rig-site approaches have been developed.

One method that has proved successful is the methylene blue test (Nevins and Weintritt, 1967; Hang and Brindley, 1970), better known to geologists

SHALE CUTTINGS PARAMETERS

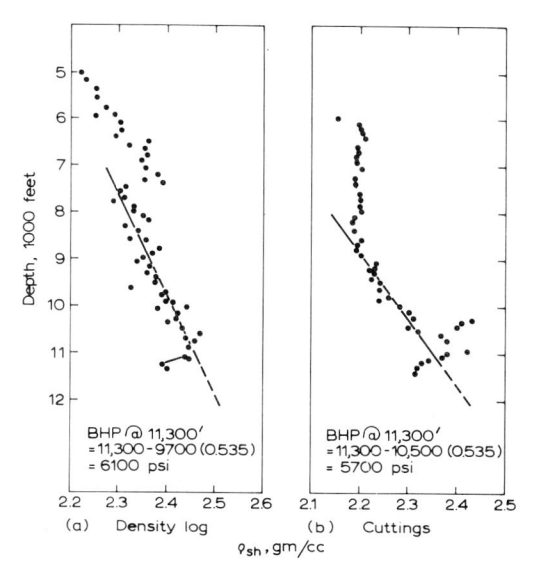

Fig.4.31. In an offshore Louisiana well, shale bulk densities measured by the density log (a) are higher than corresponding values determined from cuttings analysis (b). However, both plots accurately show the top of an overpressured zone at 11,000 ft [3354 m] (after Fertl and Timko, 1970). (Courtesy of *Oil and Gas Journal*.) 0.535 = gradient in psi/ft.

TABLE 4.VI

General shale classification proposed for drilling operations (after Mondshine, 1969)

Class	Texture	Methylene blue capacity (me/100 g)	Water content	Water (wt.%)	Clay type	Clay (wt.%)	Density (g/cm^3)
A	soft	20—40	free and bound	25—70	montmorillonite and illite	20—30	1.2—1.5
B	firm	10—20	bound	15—25	illite and mixed-layer montmorillonite—illite	20—30	1.5—2.2
C	hard	3—10	bound	5—15	trace of montmorillonite, high in illite	20—30	2.2—2.5
D	brittle	0—3	bound	2—5	illite, kaolin, chlorite	5—30	2.5—2.7
E	firm-hard	10—20	bound	2—10	illite and mixed-layer montmorillonite—illite	20—30	2.3—2.7

and engineers as the *shale formation factor method* (Gill, 1968; Mondshine, 1969; Gill and Weintritt, 1970; Gill, 1972). Generally, this shale factor may be equated with the cation exchange capacity of solids carried out of the borehole, which, in turn, may be related to the montmorillonite content, and, thus, the water-holding capacity of cuttings. The shale factor not only identifies clay composition (Table 4.VI), but it is also representative of the clay/sand abundance ratio in shale cuttings or, in general, the clay content of the section. Knowledge of the latter is of extreme importance when drilling thick carbonate sections with few shale streaks, where density data may easily give erroneous interpretations. The same is true in silty, sandy shales.

More recently, shale factor logs have been used to locate formations that cause severe borehole instability.

The shale factor also appears to be a supplementary and useful method for the detection of the impermeable cap rock often found on top of high-pressure sections (Fig.4.15).

Volume, shape, and size of shale cuttings

Experience indicates entry into overpressured environments to be characterized by an increase in penetration rate, which results in an increased volume of cuttings over the shale shaker. Additionally, the shape and size of the cuttings will change. In the transition zones, the shape of cuttings is angular and sharp, rather than rounded as in normal-pressure environments. In addition, cuttings from overpressured environments can become unusually large and generally splintery in appearance.

Novel, miscellaneous detection methods

Recently, several unconventional methods for detecting and possibly even predicting the presence of overpressures while drilling have been investigated (Fertl and Timko, 1973a,b). These include shale cuttings resistivity, filtration rate of shale cuttings slurry, filtrate (shale water) color index, shale cuttings moisture index, litho-function plot, bicarbonate—overpressure relationship, redox and pH potentials of cuttings slurry and slurry filtrate, and specific ion distribution in slurry filtrate. Several interesting observations have been made. However, additional research is still necessary to study each one of these indicators in more detail.

Shale slurry resistivity

Analysis of electrical properties of shales from well logs is one of the best methods for pressure detection and quantitative formation pressure evaluation. In overpressured shales, the short normal curve from the induction-electrical (IES) log, for example, shows a departure from the "normal" trend to lower than normal shale resistivity (see Chapter 5).

Taking this abnormal pressure detection concept one step further, similar electrical measurements may be made on representative shale samples (cuttings) at the well site. Tests can be made on shale cuttings, shale cuttings slurry, slurry filtrate, or by measuring minute changes in mud stream resistivity (salinity) at both in- and outgoing flow lines.

Borel and Lewis (1969) measured surface resistivity on shale cuttings, but measurements were difficult and tedious to perform. At the same time, Conoco* was testing new concepts for predicting abnormal pressures. For proprietary reasons, the encouraging results were not released at the time.

More recently, Overton (1970) presented a resistivity log from measurements made on clay and shale cuttings collected at the shale shaker. The log includes resistivity of a slurry of shale fines plus distilled water, resistivity of slurry filtrate, and sodium ion potential of the filtrate. Since then, several mud-logging companies started providing similar services, including comparative resistivity (salinity) measurements on in- and outgoing mud stream.

Weaver and Beck (1971) made shale resistivity measurements on a series of sidewall samples from a deep well in the South Pass area, Louisiana. As in previous investigations, shale resistivity values decreased in overpressure zones.

The complexity of interaction between clay minerals and water has long been studied. The resistivity of a shale formation is a function of in-situ pressure and temperature, composition of shales and associated water, porosity, lithologic variations, etc. Water adsorbed on clay-mineral surfaces has physical and thermodynamic properties that differ from those of normal water. Most importantly, it must be recognized that the resistivity of shale and that of shale slurry will differ significantly.

Proposed test procedure. Shale cuttings collected at the shaker are cleaned to remove mud and screened to eliminate sand, large cavings, and small recirculated shale particles. Drained, wet cuttings or spin-dried samples having a constant weight are blended in distilled water for a given length of time. We have also successfully used the method on wells drilled with oil-base mud. However, the cuttings had to be cleaned by special washing techniques.

Localized major lithology variations, such as limey shales, in the abnormal-pressure environment must be recognized as such and not as apparent formation pressure decrease. This holds true for log-derived shale resistivity values and also for shale slurry resistivity values. For the latter, empirical corrections can be developed easily.

A battery-operated conductivity bridge has been used to measure shale cuttings slurry conductivity (Fig.4.32). Shale slurry resistivity values then were computed for a constant temperature or converted to estimated downhole conditions. Resulting slurry resistivity values plotted versus depth were

---

*Continental Oil Company.

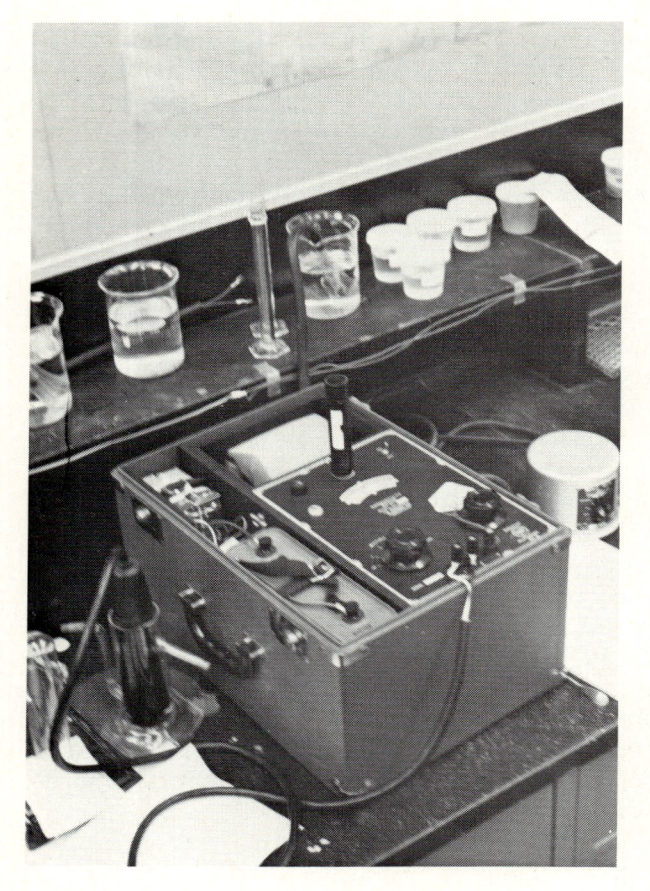

Fig.4.32. Battery-operated conductivity bridge can be used to measure conductivity of shale cuttings slurry. These data then are used to compute and plot the shale slurry resistivity (after Fertl and Timko, 1973b). (Courtesy of *World Oil*.)

compared with conventional log-derived resistivity plots. Results of two field tests are reported next.

Onshore Texas field test. Well-site results from Aransas County, Texas, are shown in Fig.4.33. Shale slurry resistivities are given at a constant temperature of 80°F [26.7°C] and at downhole conditions using a constant temperature gradient of 1.4°F/100 ft [2.55°C/100 m]. A comparison of these latter values with well-log data, such as shale resistivity from the short normal curve, is shown in Fig.4.34.

Offshore Louisiana field test. Shale slurry resistivities versus depth in Fig.4.35 are compared to (1) shale resistivity from the short normal curve,

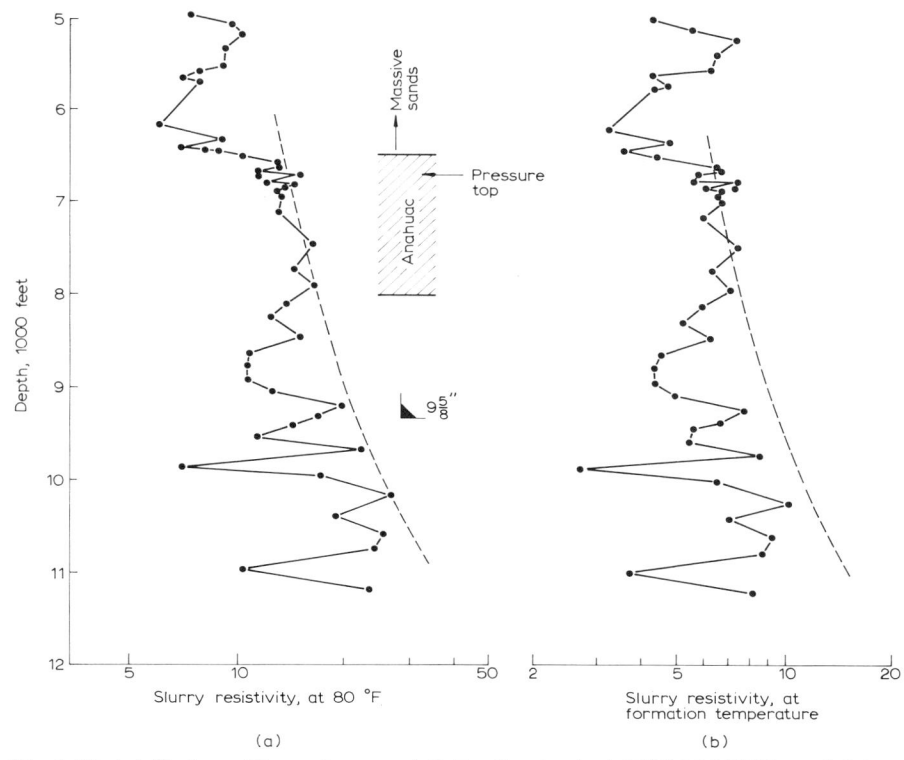

Fig.4.33. (a) Shale cuttings slurry resistivity (in Ωm) at 80°F [26.7°C], and (b) at postulated formation temperature in an onshore Texas well (after Fertl, 1973b). (Courtesy of *World Oil*.)

(2) bulk densities of shale cuttings as measured by a service company at the well site, and (3) mud weight requirements for drilling the well.

Average trend of shale slurry resistivity in the normally pressured section is similar to the trend line of normal compaction established on the short normal plot. Slope of both trend lines is similar. However, because of different measuring techniques, resistivity values are different. Of greatest significance is the clear delineation of the high-pressure top by the drastic decrease in shale slurry resistivity at 13,130 ft [4003 m] and the required mud weight increase.

Filtration rate of shale slurry

When overpressure environments are approached during drilling, several geochemical changes are encountered in subsurface formations. Similarly, a change will occur in filtration rate from shale cuttings slurry under constant pressure. Factors affecting filtration rate include the shale/sand ratio of cutting samples; particle size, shape and type of clay minerals present; pH and electrolyte concentration (salinity) of interstitial fluids in shales; etc.

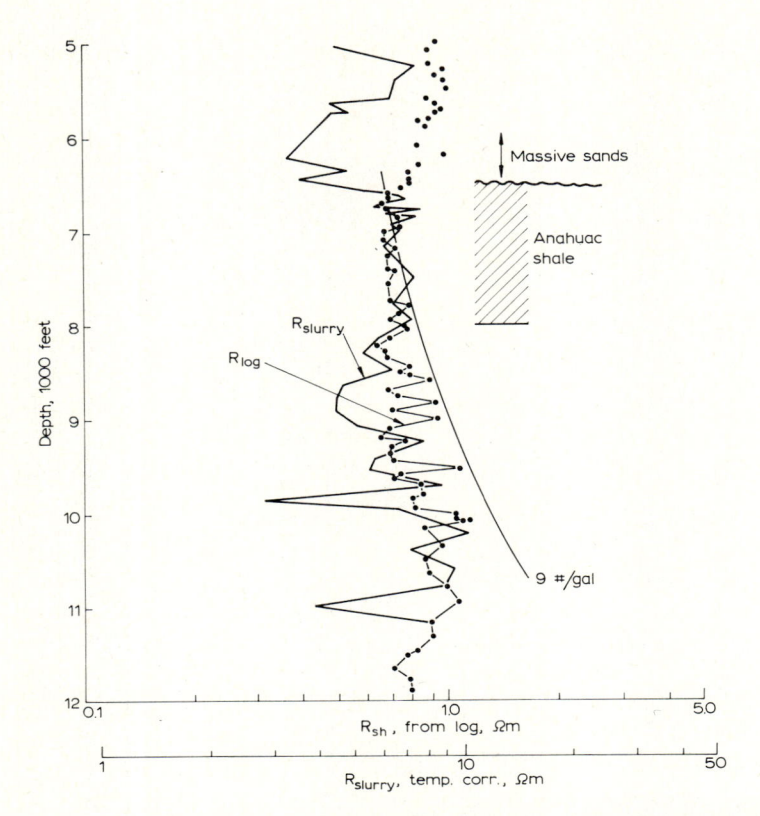

Fig. 4.34. Comparison of temperature-corrected resistivity of shale cuttings slurry and log-derived shale resistivity for the same onshore Texas well as shown in Fig. 4.33 (after Fertl and Timko, 1973b). 9 lb/gal is the mud weight used in normal-pressure environment. (Courtesy of *World Oil*.)

The arrangement assumed by clay particles during sedimentation in water depends on the pH and electrolytes of the aqueous environment. Particles form an edge-to-face structure in environments of low pH, a face-to-face flocculated structure in environments of high salt content, and a dispersed structure in environments of low electrolyte content and high pH (Van Olphen, 1963).

Experimental data on montmorillonite, kaolinite, and natural clays indicate that at the same pressure, orientation of clay particles is better with pore solutions of low electrolyte concentration. For clay compressed at a certain pressure, flow rate diminishes and orientation increases with decreasing electrolyte concentration.

Filtration times were measured on shale cuttings slurries from an offshore Louisiana well. Filtration time was the time required to obtain 50 ml of

SHALE CUTTINGS PARAMETERS 163

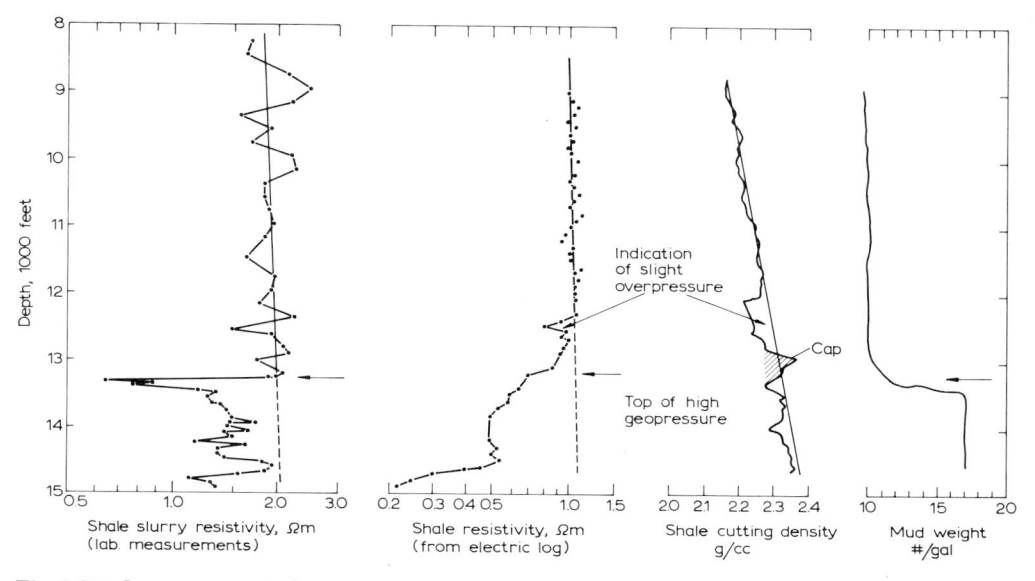

Fig.4.35. Overpressure indicated by a shale slurry resistivity plot, which compares closely with those of other more commonly used pressure indicators, such as bulk density and log-derived shale resistivity (after Fertl and Timko, 1973b). (Courtesy of *World Oil*.)

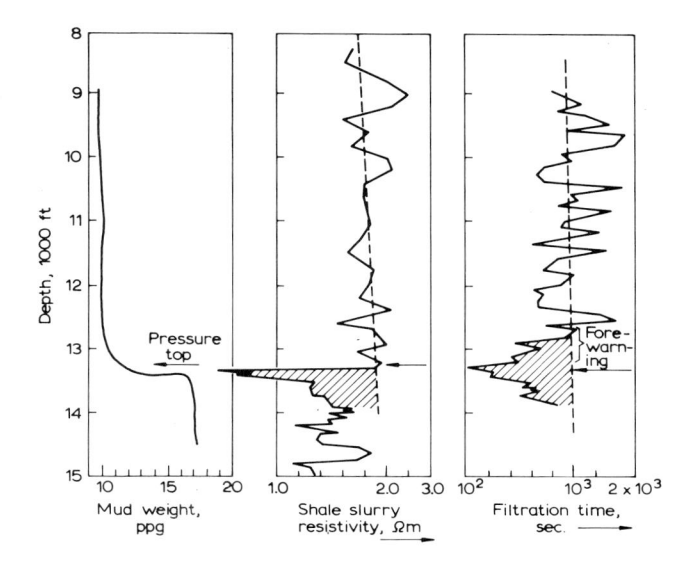

Fig.4.36. Filtration rate of shale cuttings slurry under constant pressure changes as overpressure is approached (after Fertl and Timko, 1973b). This concept may give a forewarning of impending overpressure before it is actually penetrated by the bit. (Courtesy of *World Oil*.)

filtrate with filtration pressure of 35 psi [2.46 kg/cm^2]. The amount of filtrate could also be measured for a given time.

The time of filtration stays fairly constant over the normally pressured interval (Fig.4.36). However, prior to entering the high-pressure environment, filtration time markedly decreases, the lowest value coinciding with the onset of high pressure as pinpointed by other pressure detection techniques and mud weight requirements.

Similar to the often-observed decrease in formation water salinities in sandstones overlying cap rocks of overpressured environments (see Chapter 5, *The salinity principle*), filtration tests may be one way to predict the proximity of overpressured zones prior to entering them. Filtration measurements are simple and can be used simultaneously with the color-index indicator.

Shale water color

Water squeezed from shale samples in abnormal-pressure environments has frequently been found to be amber to brown in color resulting from organic compounds. In several instances, an apparent correlation exists between the color of shale slurry filtrate and in-situ formation pressure variations.

Use of filtrate color changes as an additional pressure indicator (see Fig.4.40) might be promising but can be greatly impaired by mud additives, drilling through lignite zones, etc.

Moisture index

Use of bulk density measurements of shale cuttings for on-site delineation of cap rocks and abnormal pressures is well established. Knowing shale bulk density, grain matrix density (e.g. from composition), and fluid density in pore space allows calculation of shale porosity. Going one step further, one can measure the shale moisture content (MC) index, which is related directly to shale porosity. Abnormal pressures are reflected in higher-than-normal shale porosity and moisture content at a given depth.

The MC index of shale cuttings can be determined easily at the well site by measuring weight loss between the wet and dried sample (which requires only a hot plate and balance) or by commercially available moisture analyzers. Prior to MC-index measurements, shale cuttings must be cleaned and surface water must be removed.

The MC-index method is independent of the mud system used and works well in oil-base, oil-emulsion, or water-base systems.

Litho-function plotting

As discussed previously, major lithological variations complicate straightforward interpretation of bulk density data for pressure detection.

It has been our experience that for limey shales, plotting the litho-function (the product of shale bulk density and MC index) may greatly minimize deviations from the normal compaction trend. This can be done since bulk density and MC index are obtained by independent measurements.

Bicarbonate—overpressure relationship

Geochemical studies indicate that virtually no environment exists on or near the earth's surface where pH and oxidation-reduction reactions are incompatible with organic life. Because CO_2 is the main byproduct of organic oxidation and is the building material of plants and numerous bacteria, it is important.

CO_2 dissolves in water, producing the bicarbonate ion and a free hydrogen ion. Equilibrium conditions of carbon dioxide, carbonic acid, and bicarbonate ions can be expressed as:

$$H_2O + CO_2 = H_2CO_3 = HCO_3^- + H^+ = 2H^+ + CO_3^{2-} \qquad (4.9)$$

In a closed system, this reaction moves to the right with increasing temperature or in the presence of overpressure.

Both organic and inorganic theories have been developed to explain the origin of CO_2 in the subsurface. Organic theories include decay of organic matter and oxidation of hydrocarbons (Zobell, 1946; Kartsev, 1959; Kuznetsov, 1967; Picard, 1962). Inorganic theories include metamorphism of carbonates, solution of limestone by groundwater and igneous emanation (Germann, 1941; Lang, 1959; Zartman, 1961; Kissin, 1967).

Gulf Coast formation brines. Data from the Louisiana Gulf Coast indicate that bicarbonate content of formation waters is 2—20 times that in the Gulf of Mexico water (Jones, 1968).

Laboratory study of shale samples from a well in Bayou Long, Louisiana, showed a high concentration of HCO_3^-; it was higher in the overpressured section than in overlying normally compacted sediments. This agrees with Weaver and Beck (1971) who studied a series of sidewall cores of shales and silty shales from a deep well in the South Pass area. HCO_3^- was found to be the dominant anion in interstitial waters, with SO_4^{2-} being next in abundance, followed by Cl^-.

Concentration of HCO_3^- generally decreases with depth in normally compacted Gulf Coast sediments for two possible reasons: (1) CO_2 escapes through clays readily, perhaps as uncharged H_2CO_3 molecules; (2) in-situ CO_2 regeneration at greater depths cannot maintain a supply of HCO_3^-.

Presence of a low-permeability cap rock could be responsible for two effects: (1) reduction in the amount of bicarbonate escaping over geologic time; and (2) in the presence of some carbonate or calcareous rocks (limey shales), a supply of HCO_3^- could be regenerated to more than compensate for any diffusion, which may occur, on a very reduced scale, through the pressure seal.

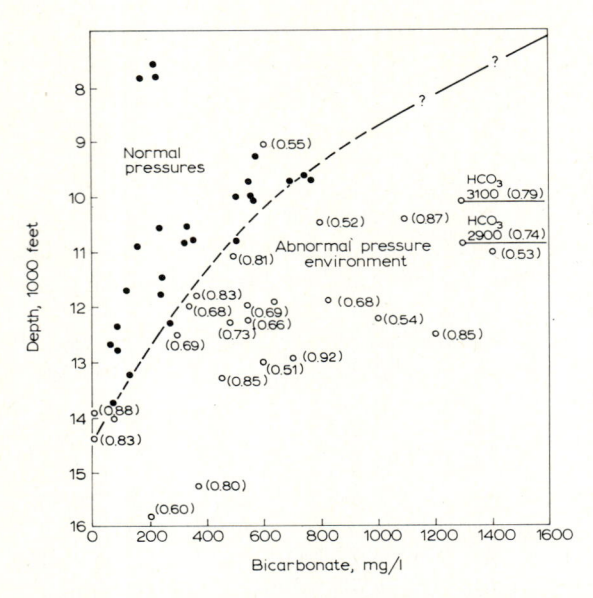

Fig.4.37. Bicarbonate concentration in formation waters from permeable abnormally pressured U.S. Gulf Coast intervals is higher than in normally compacted intervals (after Fertl and Timko, 1973a). Figures in parentheses represent the hydrostatic pressure gradients in psi/ft. (Courtesy of *World Oil*.)

Fig.4.37 plots HCO_3^- concentration versus depth based on data by Collins (1970) and Jones (1968), plus pressure data in the wells studied. Again, formation waters from abnormally pressured zones exhibit HCO_3^- concentration higher than normal for a specific depth.

Gulf Coast experience shows formation brines to be fresher near and in overpressure environments, and bicarbonate content in formation brines to increase with decreasing salinity. High pressure and temperature prevailing in overpressure environments will enhance this effect.

Field observations. Fig.4.38 shows log-derived shale parameters used to indicate downhole formation pressure environments in an onshore Louisiana well. A sharp pressure break is evident at about 13,300 ft [4055 m]. Analysis of shale slurry (shale cuttings mixed with distilled water) filtrate from this well is also shown in Table 4.VII.

The chloride ion concentration decreases and the HCO_3^- content increases under high pressure. Chloride probably migrates or diffuses easier, while the HCO_3^- content is related to diffusion ability, decomposition of organic matter, and calcite content.

In an offshore Louisiana well (Fig.4.39), the pressure top is at 13,130 ft [4003 m]. Na_2CO_3 increases at the overpressure top and decreases drastically

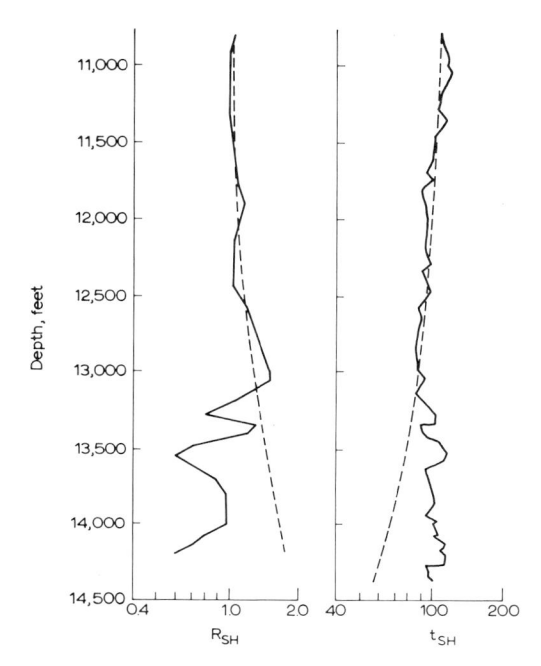

Fig.4.38. Log-derived shale parameters indicate overpressures below 13,300 ft [4055 m] in this Louisiana well (after Fertl and Timko, 1973c). Shale cuttings slurry filtrate analysis (Table 4.VII) shows a correspondingly higher bicarbonate concentration as compared to the normally pressured section. R_{SH} = shale resistivity in Ωm; t_{SH} = shale transit travel time in μsec/ft. (Courtesy of *World Oil.*)

TABLE 4.VII

Shale slurry analysis from cuttings in the Louisiana well shown in Fig.4.38 (after Fertl and Timko, 1973a)

Depth (ft)	Formation pressure	Ion concentration (mg/l)			
		Cl^-	HCO_3^-	Ca^{2+}	Mg^{2+}
9,064	normal	21	183	2	4
13,440	abnormal	10	357	0	3

in the high-pressure section. $NaHCO_3$ is constant through the normally pressured interval and increases in the cap rock. The apparent pressure top coincides with the lowest $NaHCO_3$ values. Deeper in the overpressured zone, the $NaHCO_3$ values are similar to those in the normally pressured section. Similar results occurred in an onshore Texas well.

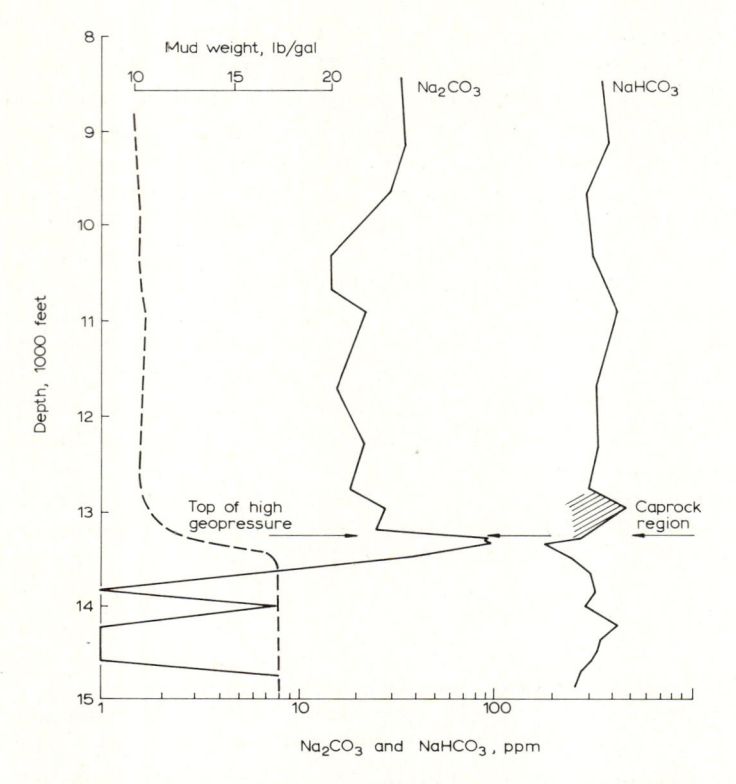

Fig.4.39. How overpressures relate to filtrate analysis of shale cuttings from an offshore Louisiana well (after Fertl and Timko, 1973a). (Courtesy of *World Oil*.)

Of course, such trends may be masked by drastic lithological variations or while drilling in massive carbonate sections.

Redox and pH potentials

Recently, the development of a method that will allow prediction of overpressures *ahead* of the bit has been emphasized (Fertl and Timko, 1973a).

In several wells, oxidation-reduction potential and pH variations were measured on shale cuttings during drilling. Variation of oxidation potential values detected the overcompacted zone before its presence was indicated by other physical means, such as shale density measurements. Despite some shortcomings, application of redox potential measurements warrants continued investigation and consideration as a well-site pressure indicator.

Redox potential determines the mobility of an element of variable valence and fixes its dispersion or localization in the earth's crust. Early attempts at measuring redox potentials in sedimentary rocks were unsuccessful, but today, realistic measurements can be made if certain precautions are taken.

Definitions. The local environment of sediments is characterized by physical and chemical norms. Principal chemical factors are oxidation-reduction (redox) potential, pH (acidity or alkalinity), ion concentration or salinity, and temperature of the depositional medium. The type of mineral precipitation and microorganisms are related closely to Eh and pH.

The pH value is the negative logarithm to the base 10 of the hydrogen ion activity of a solution. The pH value characterizes the acid-alkaline equilibrium, with a value of 7 being neutral, values above 7 indicating an alkaline solution, and values below 7 indicating an acid solution.

The redox potential, also called oxidation-reduction or Eh potential, measures the tendency of a chemical species to change its oxidation or reduction state in an ionic solution. The reactions can be expressed as "reduced state $\rightleftharpoons$ oxidized state + n electrons". Redox potential is thus a quantitative measure of the electron-escaping tendency of a reversible reduction-oxidation system.

Most reactions in the neutral environment involve both electrons and protons. Thus, pH and Eh ultimately reflect concentrations of protons and electrons. Since protons neutralize electrons (and vice versa), an abundance of one means a deficit of the other. A detailed theoretical discussion on geochemical sedimentary environments has been given by Garrels and Christ (1965).

Field example. Continuous on-site measurements of redox and pH values were made on cuttings from a Louisiana well. Comparative data were obtained at the well site on untreated shale slurries and later on buffered slurries in the laboratory.

Fig.4.40 shows mud weight, penetration rate, redox index, and pH values of the shale slurry and color of the shale slurry filtrate. Generally, the trend of redox index decreases slightly with depth in the normally pressured section to about 12,700 ft [3872 m], then drastically increases over the next 400 ft [122 m]. Starting at 12,850 ft [3918 m], bulk density of shale cuttings increased, indicating presence of a cap rock, which also correlates with a lower penetration rate. Redox values thus indicated the pressure boundary (cap rock) some 150 ft [45.7 m] earlier than did the densities of the cuttings.

In the overpressured section, several major redox deflections occur. Also, pay zones coincide with lower redox values for several wells investigated. This indication may have application as an exploratory tool.

Based on experience in shale/sand sequences and massive carbonate section, properly measured redox values are related to subsurface pressure variations, hydrocarbon environments, and lithology. However, effects are sometimes superimposed, and interpretation may be complicated. Other indicators (drilling parameters, well logs, etc.) thus will still be required.

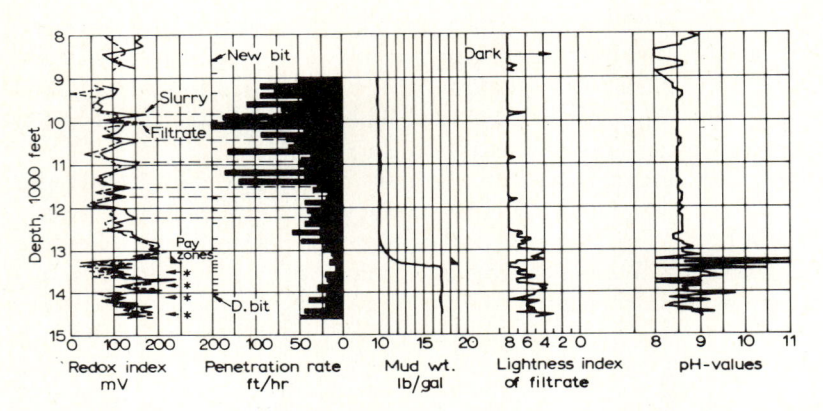

Fig.4.40. Redox potential measurements showed the pressure top approximately 150 ft [45.7 m] earlier than the currently used methods. In effect, pressure was detected ahead of the bit. In addition, redox measurements also indicated pay zones (after Fertl and Timko, 1973a). (Courtesy of *World Oil*.)

Specific ion distribution

Filtrates of shale cuttings slurries were investigated for anions, such as chloride (Cl^-) and sulfate (SO_4^{2-}); for monovalent cations, such as sodium (Na^+) and potassium (K^+); and for divalent cations, such as calcium (Ca^{2+}) and magnesium (Mg^{2+}).

Shale cuttings, cleaned from drilling mud by washing with distilled water, were collected at the well site on a 12-mesh screen. Shale samples of 10 g were then dispersed by blending in a microblender for five minutes in 100 ml of distilled water and the slurries filtered under a 35-psi [2.46 kg/cm^2] pressure, using a mud filter press. The contents of the various ions in the filtrate were then determined in the laboratory by atomic absorption techniques and titration analysis.

Fig.4.41 compares these ionic concentrations with well-log and mud data. Cl^- values (measured by two independent methods) increased in the overpressured section. Concentration of SO_4^{2-} was less dominant than Cl^-. Generally, SO_4^{2-} was constant down to 12,900 ft [3933 m] (top of cap rock) but decreased in the cap rock itself. In the high-pressure region below 13,130 ft [4003 m], a marked SO_4^{2-} increase occurs, and largest increases correlate with known pay zones.

Since SO_4^{2-} concentration may be related to organic content of the shales, an attempt was made to correlate concentration values with the color of shale slurry filtrate. However, this attempt was unsuccessful.

The concentration of SO_4^{2-} in the overpressured region may be an in-situ result of complex geochemical reactions and/or due to selective filtration effects. Also SO_4^{2-} is not very mobile.

The well-log-derived presence of a physical barrier (and possible geo-

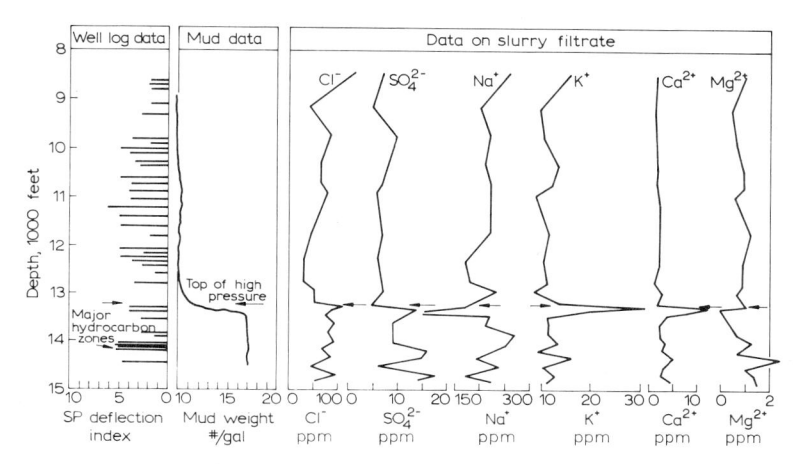

Fig.4.41. Apparent concentration variations of specific ions in shale cuttings slurries in normal and overpressured zones (after Fertl and Timko, 1973a). SP deflection index gives the maximum value of spontaneous potential (SP) curve for sandstone reservoirs, in mV. (Courtesy of *World Oil*.)

chemical barrier) might explain retention of higher SO_4^{2-} concentration in the overpressured interval.

Na^+ is fairly consistent with depth down to 11,900 ft [3628 m], markedly decreases from 11,900 to 12,900 ft [3628 to 3933 m], but again shows an increase above the overpressured zone. At the top of the overpressure zone, Na^+ decreases drastically but deeper in the overpressured section again reaches values equal to or higher than the average concentration in the normally compacted section.

K^+ concentration is much less than Na^+ but drastically increases at the top of the overpressured zone. In the deeper part of the overpressured section, K^+ values again are close to or only slightly higher than in the normally compacted section.

Ca^{2+} concentration in all slurry filtrates is generally low. An exception is the drastic increase at the pressure top and near the limey cap region, which usually has a fairly high calcite content.

No drastic variation of Mg^{2+} occurred in samples from the normally pressured section. This was followed by a decrease at the pressure top and a general increase deeper in the overpressured section.

Limitations

Indicators may be affected by the type of drilling mud (water or oil base), downhole environmental changes during drilling, and exposure to the mud stream while lifting cuttings to the surface. Ion exchange will take place between cuttings, drilling mud, and its additives.

Unfortunately, it is at present impossible to predict accurately the order of cation replaceability in a mixed cationic system, particularly in a complex system such as a drilling mud containing cuttings.

Pressure and temperature changes occurring while cuttings are carried to the surface also are important. For example, the magnitude and order of cation exchange selectivity changes when determinations are made with a dispersed or compacted clay.

Shale cuttings investigated in the laboratory were washed at the shale shaker and again cleaned in the laboratory. During washing, some ions will be lost.

References

Angelopulo, O.K., Paramonov, V.S. and Anderson, B.A., 1965. Interpretation of gas-log data during drilling of deep high-pressure wells. *Razved. Okhr. Nedr*, 11: 48—50.

Birch, A.F. and Clark, H., 1940. The thermal conductivity of rocks and its dependence upon temperature and composition. *Am. J. Sci.*, 238: 529—558; 613—635.

Boatman, W.A., 1967. Shale density key to safer, faster drilling. *World Oil*, 165(2): 69—74; also *J. Pet. Technol.*, 19: 1423—1431, 1967.

Bolt, D.B., 1972. How to detect overpressure in drilling worldwide. *World Oil*, 174(6): 107—111.

Boone, D.E., 1972. Porosity and pressure log performs well in the North Sea. *Petrol. Petrochem. Int.*, 12: 36—38.

Borel, W.J. and Lewis, R.L., 1969. Ways to detect abnormal formation pressures. *Pet. Eng.*, 41(7): 49—63; 41(10): 101—109; 41(11): 82—94; 41(12): 80—101.

Bourgoyne, A.T., 1971. A graphic approach to overpressure detection while drilling. *Pet. Eng.*, 43(9): 76—78.

Bourgoyne, A.T., Rizer, J.A. and Myers, G.M., 1971. Porosity and pore pressure logs. *Drill. Contract.*, 36—45, May—June.

Bridgman, P.W., 1924. The thermal conductivity and compressibility of several rocks under high pressures. *Am. J. Sci.*, 7: 81—102.

Bullard, E.C., Maxwell, A.E. and Revelle, R., 1956. Heat flow through the deep sea floor. In: H.E. Landsberg (Editor), *Advances in Geophysics*, Vol. 3. Academic Press, New York, N.Y., pp.153—181.

Clark, H., 1941. The effect of simple compression and wetting on the thermal conductivity of rocks. *Trans. Am. Geophys. Union*, 22: 543—544.

Collins, A.G., 1970. Geochemistry of some petroleum-associated waters from Louisiana. *U.S. Bur. Min., Rep. Invest.*, No. 7326, 38 pp.

Combs, G.F., 1968. Prediction of pore pressure from penetration rate. *SPE 2162, 43rd AIME Fall Meet.*, Houston, Texas, September.

Dupin de Saint Cyr, P., 1973. Ultrasonic device monitors mud pit levels on floaters. *World Oil*, 177(7): 57—58.

El-Hadidi, S., 1970. *Use of well logging data for predicting of rock drillability*. M.S. Thesis, Univ. of California, Berkeley, Calif., 90 pp.

Eucken, A., 1911. Ueber die Temperaturabhaengigkeit der Waermeleitfaehigkeit fester Nichtmetalle. *Ann. Phys.*, 39(2): 185—211.

Fertl, W.H., 1973. What to remember when interpreting mud-gas cutting. *World Oil*, 177(4): 67—72.

Fertl, W.H. and Timko, D.J., 1970. How abnormal pressure detection techniques are applied. *Oil Gas J.*, 68(2): 62—71.

REFERENCES

Fertl, W.H. and Timko, D.J., 1971. Parameters for identification of overpressure formations. *SPE 3223, 5th Conf. on Drilling and Rock Mechanics*, Univ. of Texas, Austin, Texas, January.
Fertl, W.H. and Timko, D.J., 1972. How downhole temperatures, pressures affect drilling, 2. Detecting and evaluating formation pressures. *World Oil*, 175(1): 45—50.
Fertl, W.H. and Timko, D.J., 1973a,b. How downhole temperatures, pressures affect drilling, 9 (1973a), and 10 (1973b). *World Oil*, 176(2): 47—51; 176(4): 62—66.
Forgotson, J.M., 1969. Indication of proximity of high pressure fluid reservoir, Louisiana and Texas Gulf Coast. *Bull. Am. Assoc. Pet. Geol.*, 53: 171—173.
Galle, E.M. and Woods, H.B., 1963. Best constant bit weight and rotary speed for minimum drilling cost. *Oil Gas J.*, 61(41): 147—166.
Garnier, A.J. and Van Lingen, N.H., 1959. Phenomena affecting drilling rates at depth. *Trans. AIME*, 216: 232—239.
Garrels, R.M. and Christ, C.K., 1965. *Solutions, Minerals and Equilibria.* Harper and Row, New York, N.Y., 390 pp.
Germann, F.E., 1941. The origin of underground carbon dioxide. *J. Chem. Phys.*, 46: 61—68.
Gill, J.A., 1968. Applied drilling technology, an engineered package for pressure detection and control. *Drill. Contract.*, 128—140, March—April.
Gill, J.A., 1972. Shale mineralogy and overpressure: some case histories of pressure detection worldwide utilizing consistent shale mineralogy parameters. *SPE 3890, 3rd Symp. on Abnormal Subsurface Pore Pressure*, Louisiana State Univ., Baton Rouge, La., May.
Gill, J.A. and Weintritt, D.J., 1970. Shale factor: a diagnostic tool for formation logging. *11th Prof. Well Log Analysts Symp.*, Los Angeles, Calif.
Gnirk, P.F. and Cheatham, J.B., 1969. A theoretical description of rotary drilling for idealized downhole bit/rock conditions. *Trans. AIME*, 246: 443—450.
Goddard, R.D., Guest, R.J. and Anderson, T.O., 1973. High resolution fluid measurements improve drilling. *Drill. Contract.*, 55—64, March—April.
Goins, W.C., 1968. Guidelines for blowout prevention. *World Oil*, 167(4): 88—106.
Goins, W.C., 1969. *Blowout Prevention.* Gulf Publishing Co., Houston, Texas, 260 pp.
Goins, W.C. and O'Brien, T.B., 1962. How to detect and control threatened blowouts. *Oil Gas J.*, 60(42): 143—151.
Goldsmith, R.C., 1972. Why gas-cut mud is not always a serious problem. *World Oil*, 175(5): 51—54.
Griffin, D.G. and Bazer, D.A., 1968. A comparison of methods for calculating pore pressures and fracture gradients from shale density measurements using the computer. *SPE 2166, 43rd AIME Fall Meet.*, Houston, Texas, September.
Gstalder, S. and Raynal, J., 1966. Measurement of some mechanical properties of rocks and their relationship to rock drillability. *Trans. AIME*, 237: 991—996.
Guyod, H., 1946. Temperature well logging. *Oil Weekly*, October 28.
Hang, P.T. and Brindley, G.W., 1970. Methylene blue adsorption by clay minerals. Determination of surface areas and cation-exchange capacities. *Clays Clay Miner.*, 18: 203—212.
Herbert, W.E. and Young, F.S., 1972. Estimation of formation pressure with regression models of drilling rate. *J. Pet. Technol.*, 24: 9—15.
Ioffe, V.I., 1972. Effect of chemical reagents introduced in clay suspension on gas-logging data. *Razved. Geofiz.*, 50: 152—154.
Issenmann, O. and Lucon, C., 1971. Present state of gas-logging techniques with the use of a continuous mud weight recorder. *J. Can. Pet. Technol.*, 12: 9—11.
Jones, P.H., 1968. Hydrodynamics of geopressure in the northern Gulf of Mexico Basin. *SPE 2207, 43rd AIME Fall Meet.*, Houston, Texas, September.
Jorden, J.R. and Shirley, O.J., 1966. Application of drilling performance data to overpressure detection. *J. Pet. Technol.*, 18: 1387—1394.

Kartsev, A.A., 1959. *Geochemical Methods of Prospecting and Exploration for Petroleum and Natural Gas* (English translation). Univ. of California Press, Los Angeles, Calif., 349 pp.

Kennedy, J.L., 1970. Drilling porosity log proves accurate. *Oil Gas J.*, 68(34): 53—55.

Kennedy, J.L., 1973. Drilling issue. *Oil Gas J.*, 71(39): 119—154.

Kissin, I.G., 1967. The possibility of carbon dioxide generation at depth at moderately high temperatures. *Dokl. Akad. Nauk S.S.S.R.*, 174(2): 451—454.

Kuznetsov, S.I., 1962. *Introduction of Geologic Microbiology*. Izd. Akad. Nauk S.S.S.R., Moscow, 315 pp.

Kvet, R., 1970. Genesis of the principal components of natural gas from Hluk. *Gas. Mineral. Geol.*, 4: 357—367.

Lang, W.B., 1959. The origin of some natural carbon dioxide gases. *J. Geophys. Res.*, 64: 127—131.

Langseth, M.G., 1965. Techniques of measuring heat flow through the ocean floor. In: *Terrestrial Heat Flow — Am. Geophys. Union Man.*, 8: 58—77.

Lewis, C.R. and Rose, S.C., 1970. A theory relating high temperatures and overpressures. *J. Pet. Technol.*, 22: 11—16.

Lutz, J., Raynaud, M., Gstalder, S., Quichaud, C., Raynal, J. and Muckleroy, J.A., 1972. Instantaneous logging based on a dynamic theory of drilling. *J. Pet. Technol.*, 24: 750—758.

Matthews, W.R., 1969. Drilling variables show transition zone. *Oil Gas J.*, 67(44): 69—76.

Maurer, W.C., 1962. The perfect-cleaning theory of rotary drilling. *Trans. AIME*, 225: 1270—1274.

Maurer, W.C., 1966. How bottom-hole pressure affects penetration rate. *Oil Gas J.*, 64(2): 61—66.

Mondshine, T.C., 1969. New technique determines oil-mud salinity needs in shale drilling. *Oil Gas J.*, 67(28): 70—75.

Nevins, M.J. and Weintritt, D.J., 1967. Determination of cation-exchange capacity by methylene blue adsorption. *Bull. Am. Ceram. Soc.*, 46: 587—592.

Nukiyama, S. and Yoshizawa, Y., 1934. The thermal conductivity of seawater and some water solutions. *J. Soc. Mech. Eng. (Japan)*, 37: 347—351.

O'Brien, T.B. and Goins, W.C., 1960. The mechanics of blowouts and how to control them. *API Drill. Prod. Pract.*, 27: 41—55.

Overton, H.L., 1970. Resistivity logging from shale slurries. *11th Prof. Well Log Analysts Symp.*, Los Angeles, Calif., Paper C.

Overton, H.L. and Timko, D.J., 1969. The salinity principle — a tectonic stress indicator in marine sands. *Log Analyst*, 10(6): 34—43.

Pechorin, O.M., 1971. Thermocatalytic processes and generation of hydrocarbons in a well. *Geol. Nefti Gaza*, 1: 22—23.

Picard, M.D., 1962. Occurrence and origin of Mississippian gas in the Four Corners region. *Bull. Am. Assoc. Pet. Geol.*, 46: 1681—1700.

Pixler, B.O., 1945. Some recent developments in mud analysis logging. *SPE 2026, AIME Fall Meet.*, Houston, Texas.

Ratcliffe, E.W., 1960. The thermal conductivities of ocean sediments. *J. Geophys. Res.*, 65: 1535—1541.

Rehm, W.A., 1969. Pressure control in drilling. *Oil Gas J.*, 67(31) through 68(7).

Rehm, W.A. and McClendon, R., 1971. Measurement of formation pressure from drilling data. *SPE 3601, 46th AIME Fall Meet.*, New Orleans, La., October.

Rochon, R.W., 1967. Relationship of mineral composition of shales to density. *Bull. South Texas Geol. Soc.*, 7, July 14.

Rochon, R.W., 1968. The effect of mud weight in mud logging gas anomalies. Monarch Logging Co., San Antonio, Texas, 12 pp.

REFERENCES

Rodionov, V.A., 1965. The effect of drilling technique on the results of gas logging. *Prikl. Geofiz.*, 41: 181—188.

Skelly, W.G. and Kjellstrand, J.A., 1966. The thermal degradation of modified lignosulfonates in drilling muds. *API Div. Prod., Southern District Spring Meeting.*

Somerton, W.H., 1959. A laboratory study of rock breakage by rotary drilling. *Trans. AIME*, 216: 92—97.

Van Olphen, H., 1963. *Clay Colloid Chemistry.* Interscience, New York, N.Y., 301 pp.

Vries, D.A., 1952. Het warmtegeleidings-vermogen van grond. *Mededel. Landbouwhogesch. Wageningen*, 52: 73—80.

Wardlaw, H.W.R., 1969. Drilling performance optimization and identification of overpressure formations. *SPE 2388, 4th Conf. on Drilling and Rock Mechanics*, Univ. of Texas, Austin, Texas.

Weaver, C.E. and Beck, K.C., 1971. Clay water diagenesis during burial: how clay becomes gneiss. *Geol. Soc. Am., Spec. Paper*, 134, 93 pp.

Wilson, G.J. and Bush, R.E., 1973. Pressure prediction with flow line temperature gradient. *J. Pet. Technol.*, 25: 135—142.

Young, F.S., 1968. Computerized drilling control. *SPE 2241, 43rd AIME Fall Meet.*, Houston, Texas, September.

Zamora, M., 1972. Slide rule correlation aids d-exponent use. *Oil Gas J.*, 70(51): 68—72.

Zartman, R.E., 1961. Helium, argon, and carbon dioxide in some natural gases. *J. Geophys. Res.*, 66: 227—306.

Zierfuss, H., 1969. Heat conductivity of some carbonate rocks and clayey sandstones. *Bull. Am. Assoc. Pet. Geol.*, 53: 251—260.

ZoBell, C.E., 1946. Studies on redox potential of marine sediments. *Bull. Am. Assoc. Pet. Geol.*, 30: 447—514.

Zoeller, W.A., 1970. The drilling porosity log. *SPE 3066, 45th AIME Fall Meet.*, Houston, Texas, October.

Chapter 5. GEOPHYSICAL WELL-LOGGING TECHNIQUES DETECT AND EVALUATE ABNORMAL FORMATION PRESSURES

Introduction

Abnormal formation pressures can be detected and evaluated by several methods. One of the best methods is well logging — any operation in which characteristic formation properties are recorded in terms of depth.

Wireline-logging methods and their evaluation are "after-the-fact" techniques — after penetration of the formation by the bit. Nevertheless, these methods significantly aid engineering, even though very short zones sometimes have to be logged to monitor pore pressure variations continually. Logs that can detect abnormally high formation pressures include:

(1) *Electrical* surveys, such as the *short normal* curve, *conductivity* curve of induction log, and *spontaneous potential* (SP) curve; the *shale formation factor* method is included here.

(2) *Acoustic (sonic)* measurements.

(3) *Bulk density* measurements.

(4) *Neutron* measurements.

(5) *Pulsed neutron* measurements.

(6) *Nuclear magnetic resonance* techniques.

(7) *Downhole gravity data* measurements.

Except for the SP curve data, all parameters recorded in shale formations, rather than in sands, are plotted versus depth. Trend lines are then established for normal compaction. The shape and slope of a particular trend line may be characteristic for a formation over an entire geologic region or be representative for one field and, sometimes, for a single fault block only.

Interpretation of such plots depends on the departure from the normal trend. The divergence depends on the degree of abnormality in the formation pressure from the normal hydrostatic pressure at a specific depth. The interpretation methods are often straightforward, but complications occasionally arise. These may include drastic lithology changes, age changes, large hole washouts, steep and thin formations, proximity to massive salt bodies (salt domes, etc.), logging-tool malfunctions, etc. Such limitations and possible pitfalls are discussed at the end of this chapter.

Electrical surveys

The short normal curve

Basic principles

In conventional non-focused electrical logging, two electrode circuits are utilized, the current circuit and the measuring circuit. When current is caused to flow in the current circuit, an electromagnetic field is set up between the two current electrodes, the shape of which is modified by the relative resistivities of the substances existing within the field. The electrodes of the measuring circuit are placed in this electromagnetic field at preselected distances apart from the current electrodes. The difference of electrical potential between the two measuring electrodes is then recorded with depth (Pirson, 1963; Wyllie, 1963). This implies that such logging devices do not work in nonconductive muds (such as oil base), air, or gas. Generally, the longer the spacing, the deeper the logging device investigates into the formation. Thus, the short normal curve (usually, a 16-inch spacing) is a rather shallow investigation tool.

In formation pressure evaluation work, this short normal curve, or preferably its electronically five-fold amplification, i.e. the *amplified short normal* curve, has become one of the most reliable tools.

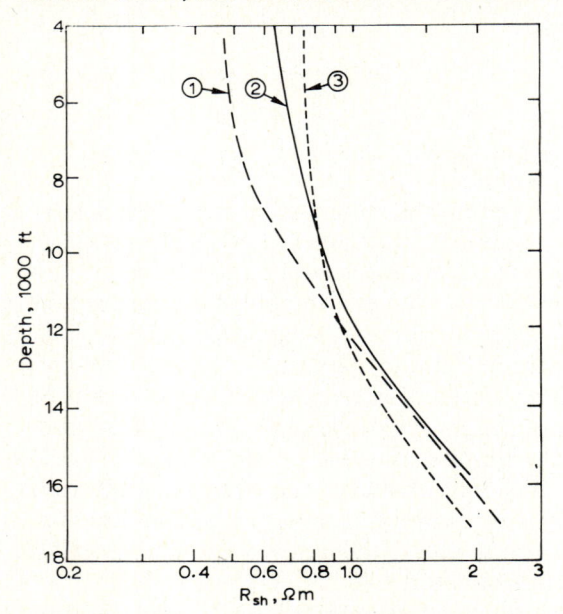

Fig.5.1. Shale resistivity versus depth of burial (after Hottman and Johnson, 1965). *1* = average trend for Oligocene/Miocene shales, southwestern Louisiana; *2* = Miocene, Jefferson Parish, Louisiana; *3* = Miocene, Iberia Parish, Louisiana. (Courtesy of *Journal of Petroleum Technology*.)

ELECTRICAL SURVEYS

Proposed by Hottmann and Johnson (1965), the method relies upon the observation that under normal compaction the shale resistivity (R_{sh}) increases with depth, since porosity decreases. The exact shape and slope of the *normal* trend line will vary with the age and type of shales present (Fig.5.1). In other words, normal compaction trends will not be identical along the U.S. Gulf Coast, North Sea region, the South China Sea area, or elsewhere.

R_{sh}-values are plotted on a logarithmic scale versus linear depth (semilog graph paper). All data points in normal-pressure environments will group along a definite trend line, R_{sh} increasing with depth (Fig.5.2). In overpressured shales, the short normal curve shows a departure from this normal trend to lower than normal resistivity. This is due to increasing porosity and resulting higher water content of these shales (Hottmann and Johnson, 1965; Ham, 1966). It seems that the relative salinities of the well-compacted versus undercompacted shales should also be taken into consideration (G. V. Chilingar, personal communication, 1975).

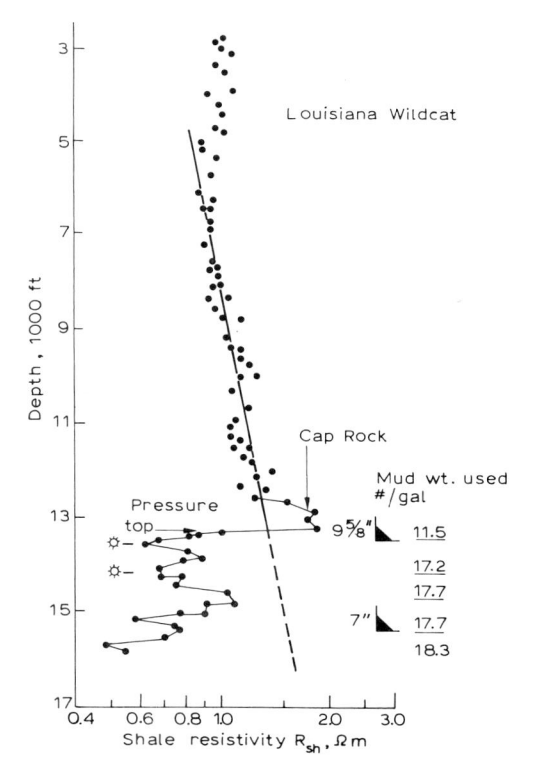

Fig.5.2. Shale resistivity plot in a U.S. Gulf Coast well. Note limey cap rock (resistivity increase) immediately above the $9\frac{5}{8}$-inch casing point and the sharp pressure transition zone below.

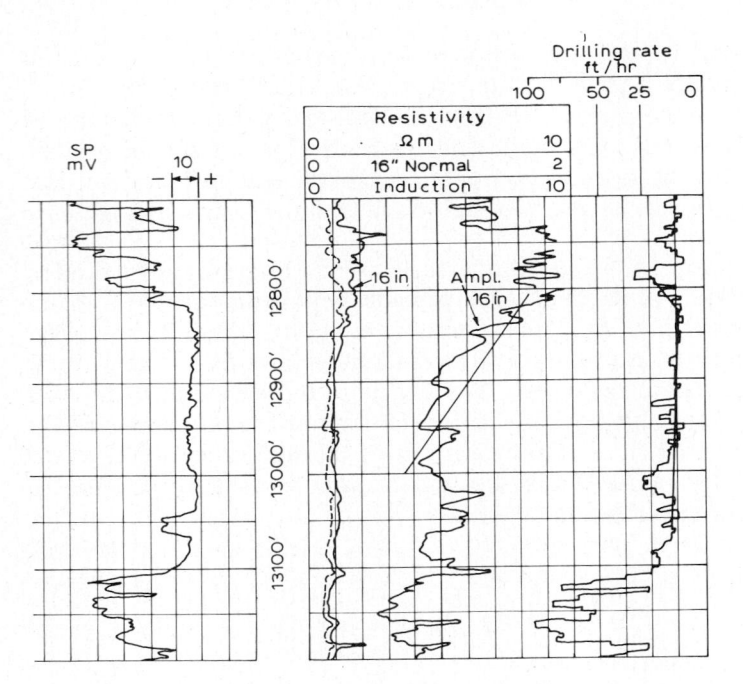

Fig.5.3. Induction-electric surveys over transition zone in a well located in Jefferson County, Louisiana. Note decreasing resistivity gradient and increase in penetration in the pressure transition zone (i.e. shale interval). The dashed curve is the induction curve.

Fig.5.3 shows an electric log from Jefferson County, Louisiana. Casing was set at 13,005 ft [3965 m] in a 16-lb/gal [1.92 kg/dm^3] environment. Besides decreasing shale resistivity, the penetration rate also increases with depth in the shale section while penetrating overpressured zones.

Quantitative pressure evaluation

Quantitative formation pressure evaluation can be carried out in several ways.

Approach A.

(1) Plot shale resistivity data and establish the normal compaction trend line.

(2) Pressure top is at the depth at which the plotted values start to diverge from the normal trend.

(3) Determine the formation pressure at a specific depth as follows: (a) divergence of observed shale resistivity value, $R_{sh(ob)}$, from the extrapolated (normal trend line) value, $R_{sh(n)}$, determines the shale resistivity ratio (normal R_{sh}/observed R_{sh}); (b) from Fig.5.4 the fluid pressure gradient (FPG) corresponding to the shale resistivity ratio is found.

ELECTRICAL SURVEYS

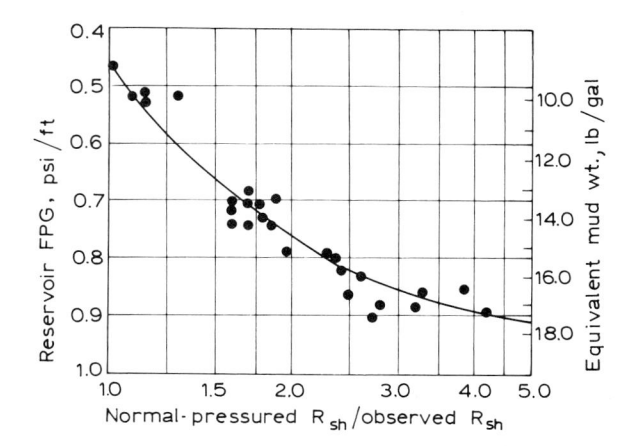

Fig. 5.4. Relationship between shale resistivity ratio ($R_{sh,normal}/R_{sh,observed}$) and reservoir fluid pressure gradient (FPG) in overpressured Miocene/Oligocene formations, U.S. Gulf Coast area (after Hottman and Johnson, 1965). (Courtesy of *Journal of Petroleum Technology*.)

(4) The FPG value is multiplied by the subject depth to obtain the formation pressure.

A pressure gradient profile can be constructed for the subject well by this procedure.

The relationship between the shale resistivity ratio and formation pressure, as proposed by Hottmann and Johnson (1965) for the Miocene and Oligocene formations along the U.S. Gulf Coast area, is shown in Fig. 5.4. Basic input data used to establish this *empirical* relationship are listed in Table 5.I. The standard deviation from the average line representing these data in Fig. 5.4 is 0.020 psi/ft [0.046 kg cm^{-2} m^{-1}].

Similar calibration charts must be developed for other areas with different geologic settings. Furthermore, one must keep in mind that all these methods are restricted to areas in which overpressures have resulted primarily from compaction processes.

Approach B is the *equivalent depth method*. This method is based on the following mathematical relationship, which is valid for any logging parameter, including shale resistivity, travel time, conductivity, density, etc.

The technique includes the following steps:
(1) Plot the R_{sh}-values and establish the normal trend line.
(2) Determine the formation pressure from:

$$P_f = G_o \cdot D_A - D_E(G_o - G_H) \tag{5.1}$$

$$P_f = D_A - 0.535 \cdot D_E \tag{5.2}$$

where P_f = formation pore pressure in psi; D_A = depth of interest in

TABLE 5.I

Formation pressures and shale resistivity ratios in overpressured Miocene/Oligocene formations, U.S. Gulf Coast area (after Hottman and Johnson, 1965)

Parish or County and State	Well	Depth (ft)	Pressure (psi)	FPG* (psi/ft)	Shale resistivity ratio** (Ω m)
St. Martin, La.	A	12,400	10,240	0.83	2.60
Cameron, La.	B	10,070	7,500	0.74	1.70
Cameron, La.	B	10,150	8,000	0.79	1.95
	C	13,100	11,600	0.89	4.20
	D	9,370	5,000	0.53	1.15
Offshore	E	12,300	6,350	0.52	1.15
St. Mary, La.	F	12,500	6,440	0.52	1.30
		14,000	11,500	0.82	2.40
Jefferson Davis,	G	10,948	7,970	0.73	1.78
La.	H	10,800	7,600	0.70	1.92
		10,750	7,600	0.71	1.77
Cameron, La.	I	12,900	11,000	0.85	3.30
Iberia, La.	J	13,844	7,200	0.52	1.10
		15,353	12,100	0.79	2.30
Lafayette, La.	K	12,600	9,000	0.71	1.60
		12,900	9,000	0.70	1.70
	L	11,750	8,700	0.74	1.60
	M	14,550	10,800	0.74	1.85
Cameron, La.	N	11,070	9,400	0.85	3.90
Terrebonne, La.	O	11,900	8,100	0.68	1.70
		13,600	10,900	0.80	2.35
Jefferson, Tex.	P	10,000	8,750	0.88	3.20
St. Martin, La.	Q	10,800	7,680	0.71	1.60
Cameron, La.	R	12,700	11,150	0.88	2.80
		13,500	11,600	0.86	2.50
		13,950	12,500	0.90	2.75

*Formation fluid pressure gradient.
**Ratio of resistivity of normally pressured shale to observed resistivity of overpressured shale: $R_{sh(n)}/R_{sh(ob)}$.

abnormal-pressure environment in feet; D_E = normal, *equivalent* depth in feet, corresponding to D_A (Fig.5.5); G_H = hydrostatic pressure gradient in psi/ft (equal to 0.433 psi/ft [1.0 kg cm^{-2} m^{-1}] for fresh and brackish formation waters, e.g. Mid-Continent, Rocky Mountains, etc., or equal to 0.465 psi/ft [1.07 kg cm^{-2} m^{-1}] for salt water, e.g. Gulf Coast area, etc.); and G_o = overburden pressure gradient in psi/ft. The concept of variable overburden gradient as proposed by Eaton (1969) requires area data. Often a constant overburden gradient of 1.0 psi/ft (corresponding to an average bulk density of 2.31 g/cm^3) is a satisfactory assumption.

ELECTRICAL SURVEYS

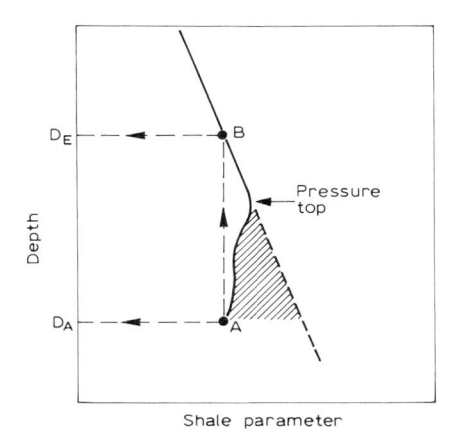

Fig.5.5. Schematic representation of D_A and D_E used in the equivalent depth method.

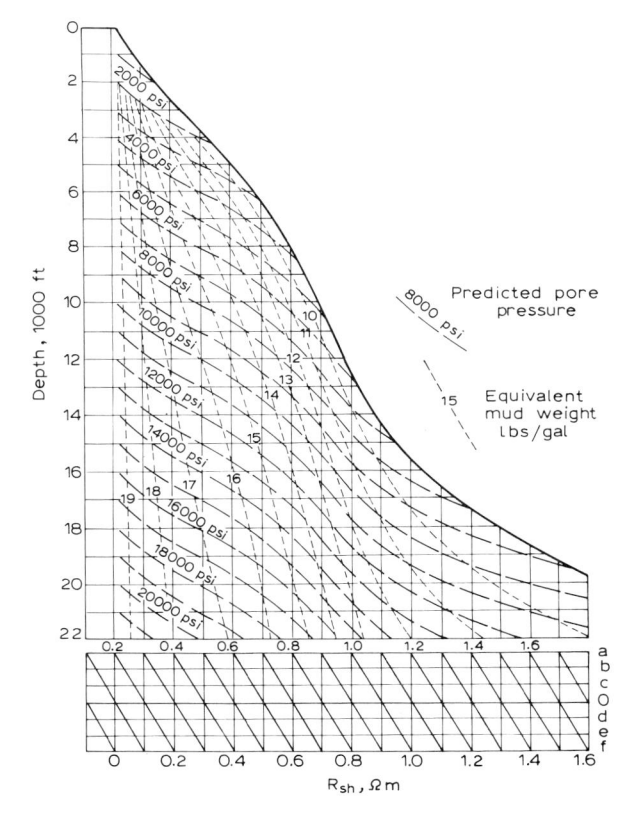

Fig.5.6. Nomograph for estimation of formation pressure from Louisiana Gulf Coast shale resistivities (after Ham, 1966).

Eq. 5.1 above also holds true for normal, hydrostatic pressure environments, since in this case $D = D_A = D_E$. This simplifies eq.5.1 to:

$$P_f(\text{psi}) = G_H \cdot D \tag{5.3}$$

Once the normal compaction trend has been established for a given area, then it is quite easy to construct a pressure evaluation chart for these specific conditions. This holds true for any Tertiary basin in the world. Fig.5.6 shows such a nomograph for South Louisiana conditions (Ham, 1966). The grid at the bottom of the chart allows R_{sh}-scale shifts to take local conditions and/or geologic age into account. Plot several shale points over the normal-pressure section of the well on this nomograph in order to determine which baseline (a, b, c, 0, d, e, and f) should be used for the pressure evaluation.

For example, at 14,350 ft [4375 m] the shale resistivity from the short normal curve is 0.8 Ωm. For local conditions the "0"-scale has been determined to be representative. Enter 14,350 ft [4375 m] on the ordinate and 0.8 Ωm on the "0"-abscissa. At the point of intersection of both lines, one determines (1) an estimated pore fluid pressure of 9000 psi [633 kg/cm^2], and (2) an equivalent mud weight of 12.0 lb/gal [1.44 kg/dm^3] necessary to balance that pressure.

Consequently, it is obvious that for any given compaction trend line (using any shale parameter, not just resistivity), which corresponds to normal, hydrostatic pressure, a transparent chart of parallel lines for equivalent mud weight requirements can be established. With such a chart, superimposed on plotted shale parameters, mud weight requirements and/or formation pressures can be read directly. This method is illustrated later on for conductivity plots in areas such as South Texas and Italy (see Fig.5.13).

Possible pitfalls

Formation resistivity measurements are influenced by several factors, including porosity, temperature, formation water salinity, mineral composition of shale, etc. Possible limitations and/or related pitfalls will be discussed in detail at the end of this chapter.

The conductivity curve

Basic considerations

Focused induction-logging devices record the formation conductivity (mmhos) and its reciprocal, resistivity (Ωm). The logging sonde transmits high-frequency alternating current into the formation. This creates a magnetic field, which causes electromotive forces (eddy currents) to flow in a circular path coaxial to the sonde. These currents, in turn, create magnetic fields which induce signals in the logging-tool receiver. These recorded signals are essentially proportional to the formation conductivity (Doll, 1949; Tixier et al., 1960). The tool can be run in any type of drilling mud.

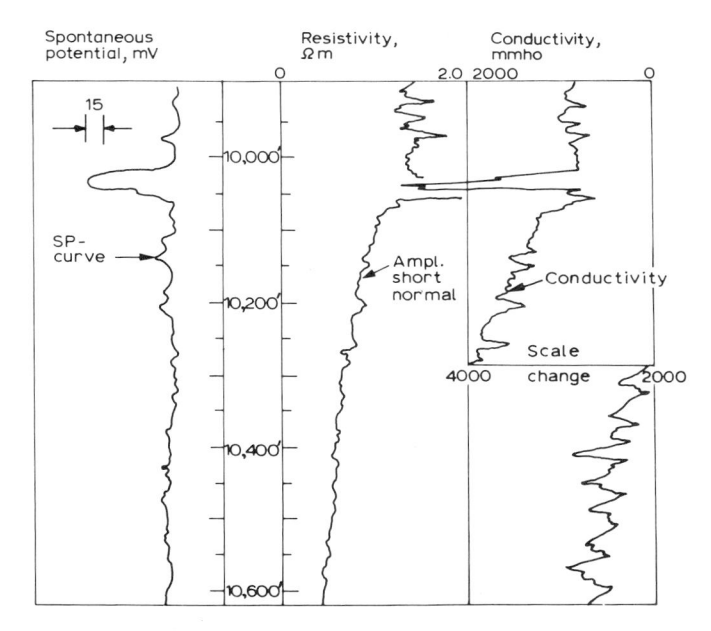

Fig.5.7. Conductivity curve of induction log and short normal curve clearly define increase of formation pressure in the shale section from 10,060 to 11,000 ft [3067 to 3354 m].

Conductivity values in normal-compacted shales decrease with depth. However, transition to overpressured shales is indicated by greater-than-normal conductivity resulting from higher-than-normal water content and porosity (Wallace, 1964, 1965a,b; MacGregor, 1965).

Fig.5.7 shows an actual field log with a marked conductivity gradient as caused by a pressure transition. Fig.5.8 illustrates conductivity plots in normal and overpressured wells. Pressure tops are defined by the divergence from the normal trend line.

Quantitative pressure evaluation

Again, several methods are applicable.

Method A. One of the first applications of the conductivity curve for quantitative pressure evaluation (Wallace, 1964) is shown in Figs.5.9 and 5.10. In a South Louisiana well a departure of 1200 mmhos at a depth of 12,200 ft [3720 m] was observed. Using an empirical calibration chart for well-log conductivity and formation pressure gradient (Fig.5.10) developed for this specific area, a formation pressure of 9882 psi [695 kg/cm^2] is determined. Formation tests in the sand immediately below this shale zone measured a pressure of 9900 psi [696 kg/cm^2].

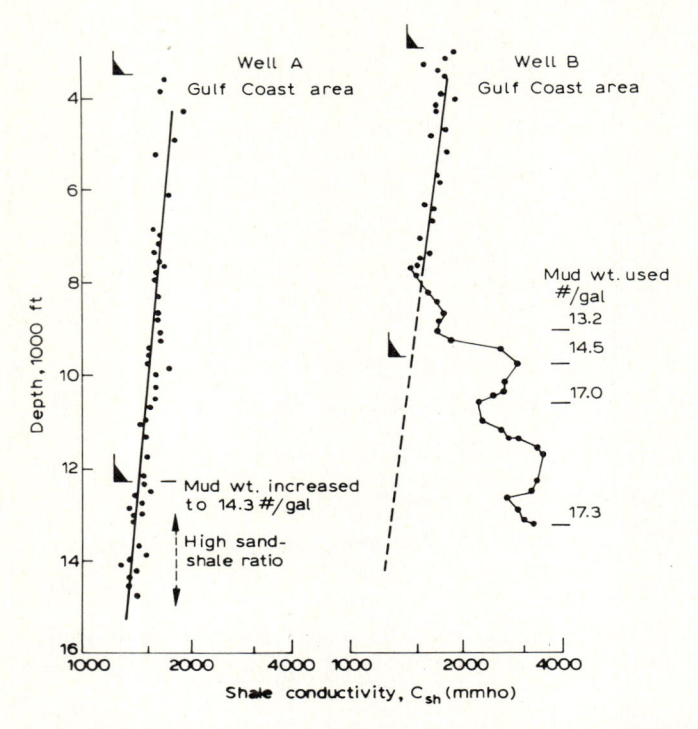

Fig.5.8. Relationship between formation pressure and shale conductivity (induction log) (after Fertl and Timko, 1971b). Well A has normal pressures to total depth (TD); thus casing point at 12,200 ft [3720 m] was unnecessary. Shale conductivity values sharply increase below 7800 ft [2378 m] in well B, indicating transition from normal pressures to overpressures.

Fig.5.11 shows a conductivity ratio calibration for Gulf Coast Pleistocene formations as proposed by Matthews (1971).

In summary, Method A is based on the empirical correlation of actual formation pressure data (tests, well kicks, etc.) and shale conductivity values for a given, specific area. Wallace (1965b) developed such a nomograph for quantitative pressure evaluation in Louisiana, claiming an accuracy within 5—10%.

Method B. Similar to other logging parameters, the *equivalent depth method* can again be used (see eq. 5.1). This approach was used in an offshore Louisiana well to illustrate the presence of a drastic formation pressure increase (Fig.5.12). Note that the plot in Fig.5.12 is on semilog paper, whereas in Fig.5.8 the values are plotted on a linear scale. Both options are equally well applicable.

ELECTRICAL SURVEYS 187

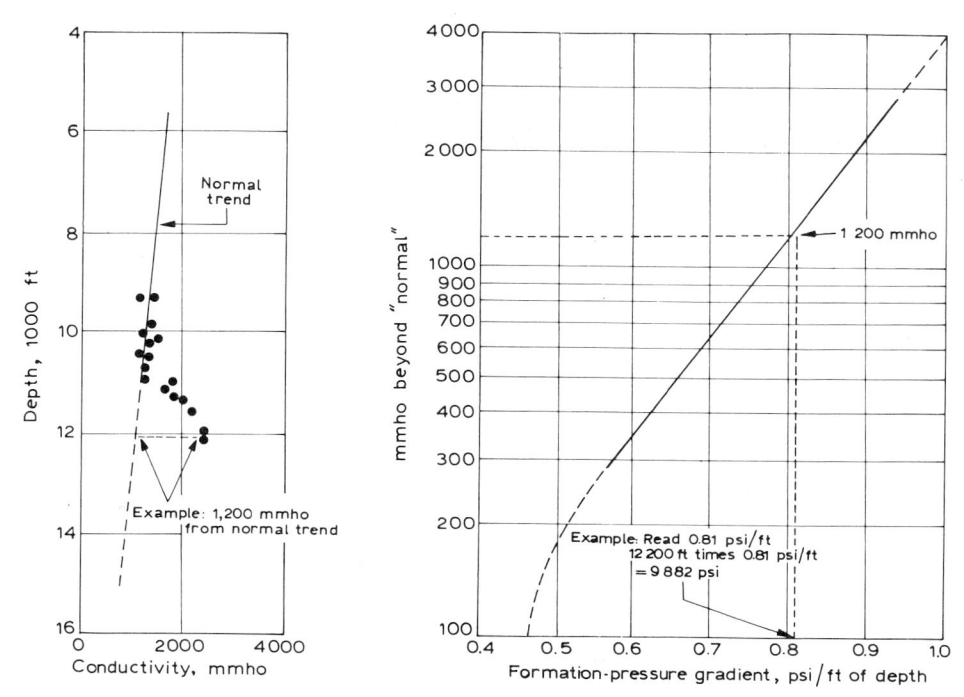

Fig.5.9. Field case of quantitative formation pressure evaluation in a South Louisiana wildcat (after Wallace, 1965a). There is a departure of 1200 mmhos at a depth of 12,200 ft.

Fig.5.10. Relationship between the shale conductivity and formation pressure gradients in South Louisiana (after Wallace, 1965a).

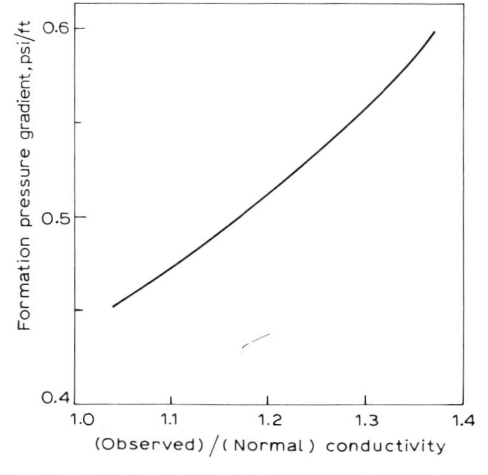

Fig.5.11. Relationship between the shale conductivity ratio ($C_{sh(ob)}/C_{sh(n)}$) and formation fluid pressure gradient (FPG) in Pleistocene U.S. Gulf Coast formations (after W.R. Matthews, personal communication, 1972).

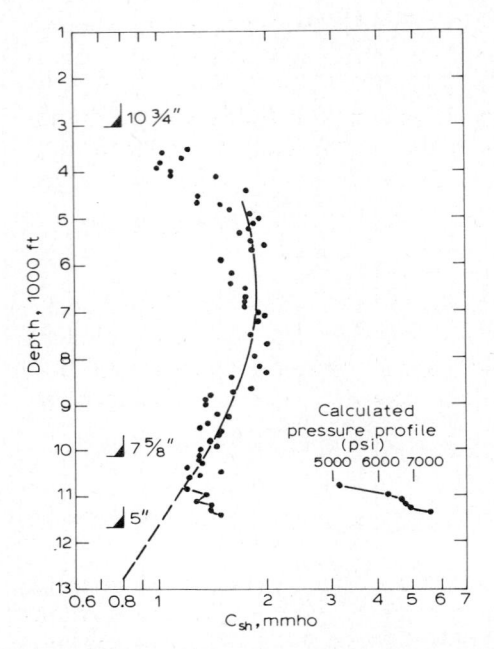

Fig.5.12. Conductivity plot and formation pressure profile calculated by the equivalent depth method in an offshore Louisiana well.

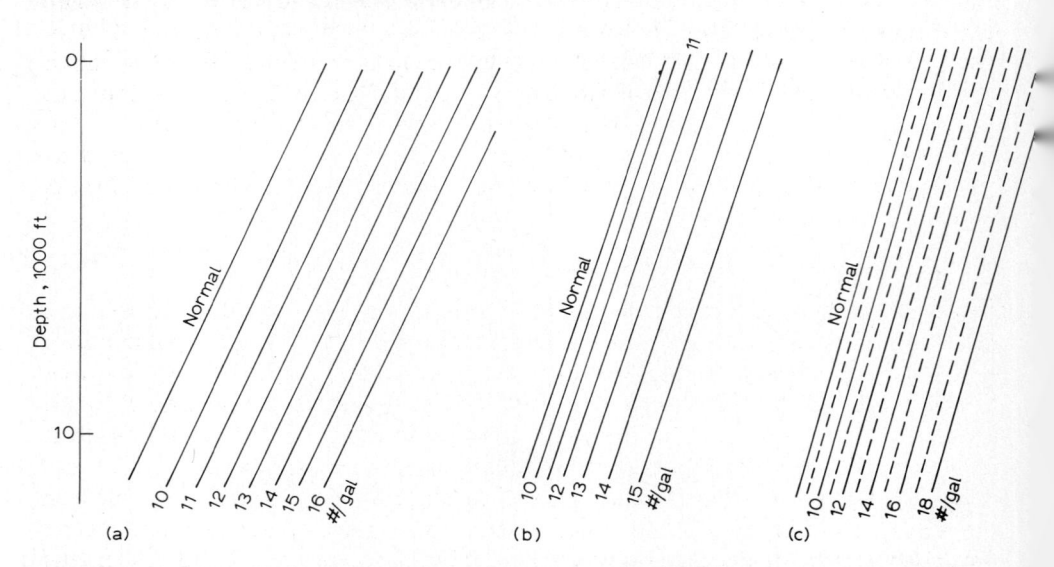

Fig.5.13. Equivalent mud weight requirements (overlays) for use with conductivity plots in (a) Frio Formation, South Texas; (b) Wilcox Formation, South Texas; and (c) Pleistocene and Pliocene formations, Italy (after W.R. Matthews, personal communication, 1972).

ELECTRICAL SURVEYS

Another version is the development of a transparent overlay of equivalent mud weight lines parallel to the normal compaction trend, which corresponds to hydrostatic conditions. Overlaying such calibration charts on the conductivity plot allows one to read mud weights and/or formation pressures directly from the conductivity plot. Fig.5.13 shows such charts for South Texan and Italian formations. As mentioned previously, such an overlay approach is valid for any shale parameter in pressure evaluation.

Possible pitfalls

Again, shale porosity, temperature, formation water salinity, mineral composition of shales, etc., affect conductivity measurements. Limey shales, proximity to salt domes, etc., will be discussed later in this chapter. Particular tool problems, however, may be encountered in very high-conductivity shales (2000 mmhos and higher).

The spontaneous potential (SP) curve

Basic principles

The SP curve is a millivolt recording (with depth) of naturally occurring potential differences between the fixed potential of a surface electrode and a movable electrode, i.e. the logging sonde, in a borehole filled with conductive fluid (Schlumberger and Leonardon, 1934; Doll, 1948). The basic components of the SP are: (1) electrochemical potentials, such as the shale membrane potential (Mounce and Rust, 1944) and the liquid-junction (diffusion) potential (Wyllie, 1949), and (2) electrokinetic potentials, also called streaming potential or electrofiltration potential (Wyllie, 1951; Gondouin et al., 1957; Wyllie et al., 1958a; Hill and Anderson, 1959; Althaus, 1967).

Since in clean, essentially shale-free, permeable formations the total electrochemical potential (E_c) is the sum of the shale membrane and liquid-junction potential, it equals:

$$E_c = -K \log (a_w / a_{mf}) \tag{5.4}$$

where a_w and a_{mf} are the chemical activities of the formation water and mud filtrate at formation temperature. K, the electrochemical coefficient, is directly proportional to the absolute temperature and, for sodium chloride, is equal to 71 at 77°F [25°C]. Furthermore, for pure sodium chloride solutions that are not too concentrated, resistivity values are inversely proportional to activities. However, this inverse proportionality does not hold exactly at high concentrations or for all fresh waters with divalent ions prevailing. Empirical K-values and correction charts for formation waters have been developed (Gondouin et al., 1957; Moore et al., 1966; Desai and Moore, 1969).

Since under normal circumstances the SP curve mainly depends on the salinity (resistivity) contrast of drilling mud filtrate and formation water, the use of freshwater muds, in general, gives the best results. However, if the salinity contrast is small or nonexistent, such as in wells drilled with salt mud, little if any SP development will occur, and the SP curve becomes rather featureless. Furthermore, SP cannot be recorded in boreholes filled with nonconductive muds, such as oil base, air and gas, or in cased holes. SP also loses character in high-resistivity beds and is affected by invasion of mud filtrate into permeable formations, presence of divalent ions, saltwater flows (i.e. influx of formation water into borehole if underbalanced drilling conditions prevail), stray currents causing spurious potentials, streaming potentials under overbalanced drilling conditions, etc.

The principal uses of the SP curve include: (1) detection and correlation of permeable formations, such as sands in a shale/sand sequence; (2) determination of formation water resistivity (salinity); (3) estimation of the clay content in potential reservoir rocks; and (4) indication of tectonic stress for use in abnormal formation pressure detection. Implications of the latter concept will be discussed in more detail later.

The salinity principle

Soviet investigators have cited field observations which illustrate a marked freshening or decrease of formation water salinities with increasing depth (Lindtrop, 1927; Sukharev and Krumbol'dt, 1962; Pol'ster et al., 1967; Simkhaev and Samedov, 1969). Similar observations are well documented for the U.S. Gulf Coast area for formation waters in both sands and shales (Timm and Maricelli, 1953; Myers and Van Siclen, 1964; Foster and Whalen, 1965; Fowler, 1968; Jones, 1968; Overton and Timko, 1969; Fertl and Timko, 1970a,b,c; Weaver and Beck, 1971; Schmidt, 1973), for the U.S. Mid-Continent region (Breeze, 1970; Ferran, 1973), and many other Tertiary basins around the world.

For illustration, some data on the formation water salinity and associated formation pressures for the U.S. Gulf Coast area are given in Table 5.II. Low salinity values were observed in many of these abnormally high-pressured wells.

Based upon such observations, it has been proposed by Overton and Timko (1969) to plot salinity variations as calculated from the SP curve for abnormal formation pressure detection work. The authors suggested a remarkably simple relationship, the so-called salinity principle, between the salinity of clean sands, C_W, and the porosity of adjacent shales, ϕ_{sh}:

$$C_W \cdot \phi_{sh} = \text{constant} \tag{5.5}$$

That is, formation water salinity (with assumed equilibrium between sands and shales) will vary inversely with the porosity in adjacent shales.* Under

---

*This assumption has been questioned by G. V. Chilingar (personal communication, 1975; see also Chapter 2).

TABLE 5.II

Geostatic ratios and formation water composition* in wells located in the Texas and Louisiana Gulf Coast area (after Jones, 1968)

Depth (ft)	Original pressure (psi)	Geostatic ratio**	Dissolved solids (mg/l)	Sodium (mg/l)	Potassium (mg/l)	Calcium (mg/l)	Magnesium (mg/l)	Bicarbonate (mg/l)	Sulfate (mg/l)	Halides (as chloride) (mg/l)
12,992	11,960	0.92	92,000	38,000	520	4,300	580	700	260	52,300
10,400	10,400	0.87	98,000	38,000	200	1,200	200	1,100	26	56,700
12,500	10,500	0.85	99,000	36,000	200	1,100	200	1,200	27	56,600
11,106	9,030	0.81	111,000	40,000	320	2,700	360	500	—	—
10,082	8,000	0.79	55,000	19,000	130	1,100	200	3,100	82	30,200
10,875	8,110	0.74	41,000	15,000	100	340	60	2,900	700	21,600
10,401	7,700	0.74	146,000	46,000	320	6,700	600	—	100	76,000
10,870	7,900	0.73	136,000	43,000	500	5,400	700	400	53	73,000
12,552	8,690	0.69	175,000	63,000	540	6,100	700	300	—	98,000
16,064	9,600	0.60	93,000	30,000	480	4,200	300	200	10	53,500
9,051	5,000	0.55	14,000	4,800	30	290	60	600	60	7,700
11,000	5,950	0.53	66,000	24,000	100	1,300	300	1,400	25	38,200
12,200	6,500	0.54	97,000	36,000	220	1,300	300	1,000	—	55,500
11,200	5,830	0.52	72,000	26,000	210	1,100	200	100	12	40,900
10,500	5,440	0.52	100,000	35,000	210	1,600	400	800	—	57,000
13,000	6,600	0.51	45,000	16,000	100	560	100	600	—	26,000
Gulf of Mexico water			35,800	10,970	429	423	1,324	147	2,750	19,770

*In sands.
**Ratio of aquifer fluid pressure to overburden pressure.

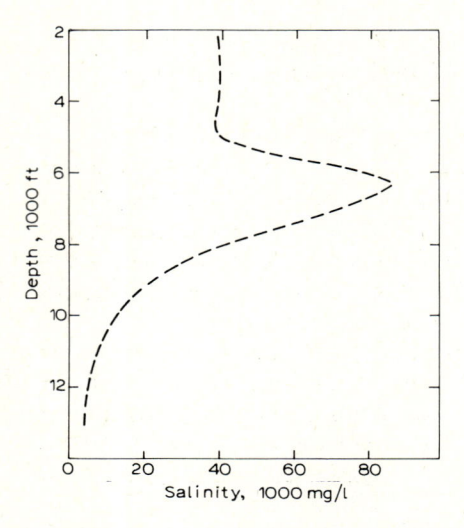

Fig.5.14. Decrease in the formation water salinity, as indicated by the SP curve, can indicate approach or proximity to abnormal-pressure environments. Salinity trend in Matagorda County, Texas (after Myers and Van Siclen, 1964).

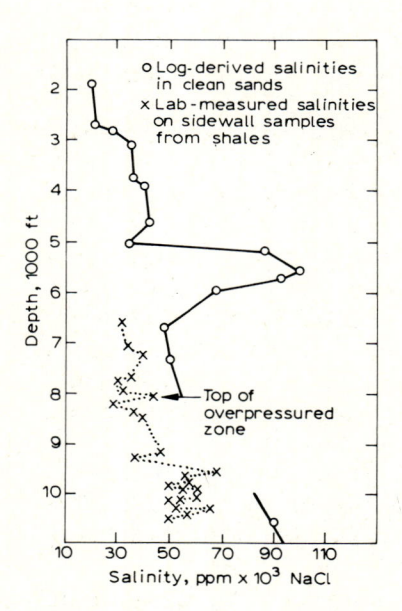

Fig.5.15. Salinity trend in an offshore Louisiana well (after Fertl and Timko, 1970a). Comparison of log-derived salinities in clean sands and laboratory-measured salinities of sidewall samples from shale zones. (Courtesy of *Oil and Gas Journal*.)

normal compaction, shale porosity decreases with increasing depth (Athy, 1930a,b; Hedberg, 1936; Von Engelhardt, 1961), whereas formation water salinity tends to increase (Dickey, 1969).

Abnormal formation pressure environments cause divergence from such normal trends. Apparent shale porosity increases in overpressured zones which in turn suggests a decrease in formation water salinity provided the salinity principle (eq. 5.5) is applicable. Decrease in the formation water salinity, as indicated by the SP curve, then would indicate proximity of abnormal formation pressures. Field cases from Matagorda County, Texas (Fig.5.14), and offshore Louisiana (Fig.5.15) illustrate these observations.

The plot of salinity versus depth in Fig.5.15 combines salinities in the shale zones as determined on sidewall samples in the laboratory and those derived from the SP curve in clean sands. The normal salinity trend in the sands shows a consistent increase from the surface to about 5500 ft [1677 m]. As shown in Fig.5.15 there is an apparent salinity reduction (freshening) from about 2000 ft [610 m] above the overpressured zones and in the latter. Also note that throughout the section pore waters are less saline in the shales than in associated sandstones. This observation is in agreement with other investigations (Sukharev and Krumbol'dt, 1962; Hedberg, 1967; Schmidt, 1973).

The concept of using the SP curve as a tectonic stress indicator in shale/sand sequences has in many instances proven quite successful. Often freshening of formation water has been observed prior to the drill bit actually penetrating overpressures. This gives a forewarning of impending downhole pressure changes.

Comparative plots of shale resistivity and variation of pore water salinities in associated sandstones are shown in Fig.5.16 for a U.S. Gulf Coast well, drilled in Iberia Parish, Louisiana. Changes in shale resistivity and salinity values suggest that the well has cut a fault. The presence of this fault was confirmed by additional geological data. The presence of a limey sealing barrier (cap rock) on top of the overpressured interval is also readily observed by the increase in shale resistivity at a depth of about 11,800 ft [3598 m], whereas the salinities in the sand show a freshening at about 11,400 ft [3476 m] giving about 400 ft [122 m] forewarning of the presence of overpressures. In addition to these data, mud weight requirements to drill the well are shown.

Another field example from the U.S. Gulf Coast area (Fig.5.17) shows a drastic pressure increase below a well-developed cap rock around 13,000 ft [3963 m]. In the subject well the transition from the normal to abnormally high pressure gradient is rather abrupt.

Based upon worldwide experience, such transition zones may exhibit pore fluid pressure changes of 5000 psi [352 kg/cm^2] over shale sections as short as 300—400 ft [91—122 m] making closely controlled drilling operations essential. On the other hand, transition zones extending over several thousand

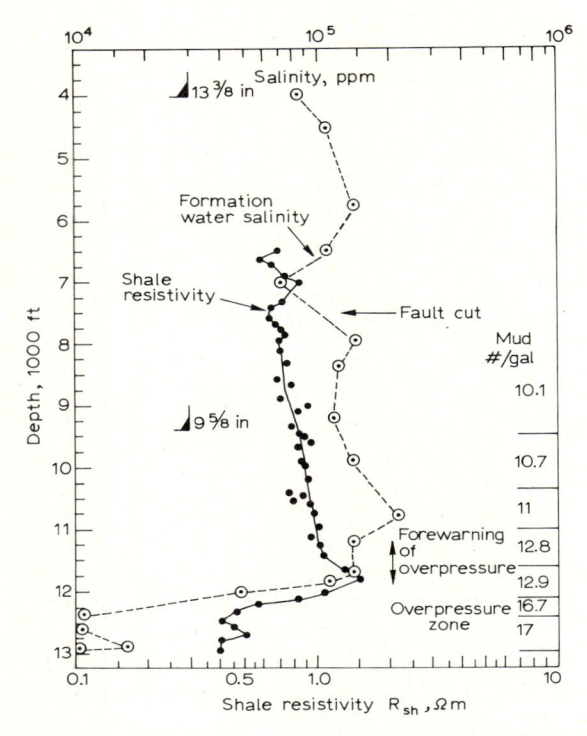

Fig.5.16. Comparison of shale resistivity and formation water salinity plots in a Gulf Coast well, Iberia County, Louisiana (after Fertl and Timko, 1970c).

feet have been drilled without any problem due to a gradual pressure increase.

In the subject well, the onset of a drastic salinity decrease occurred about 800 ft [244 m] prior to drilling into overpressured formations. There was an excellent agreement of the formation water salinity (40,000 ppm NaCl equivalent) as calculated from the SP curve in sands and that measured on water samples produced during formation tests.

In some United States areas, such as the Mid-Continent region (Oklahoma, etc.), in the Himalayan foothill belt of India (Sahay, 1972), the South China Sea area, and elsewhere, salinity reductions in overpressure environments at greater depth become so drastic that these salinity variations of the formation water even cause SP reversals, i.e. the recorded SP potential changes polarity and SP deflections in overpressured sands become positive.

Straightforward application of the salinity principle is not always possible, and correlations of formation pressures with salinity variations may be limited locally. This may be the result of (1) problems inherent to SP measurements as discussed previously, and/or (2) the fact that formation water salinities can vary quite markedly with faulting, salt dome uplift,

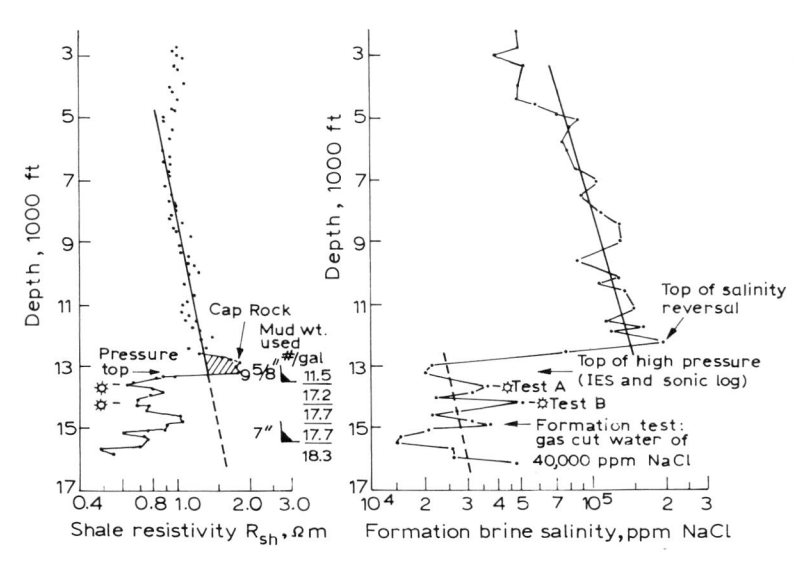

Fig.5.17. Louisiana wildcat exhibits a sudden salinity divergence from the normal trend about 800 ft [244 m] above overpressure top (after Fertl and Timko, 1971b). Overpressure top is defined from logs (such as shale resistivity plot on left) and drilling data. Casing points, mud weights, and formation tests are indicated.

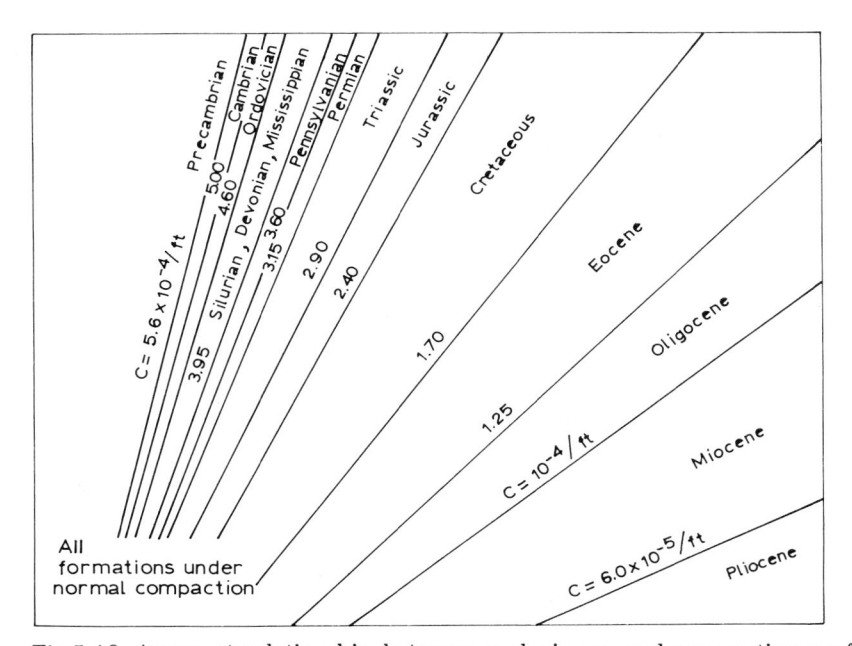

Fig.5.18. Apparent relationship between geologic age and compaction coefficient (after Zanier, 1969). Ordinate in semilogarithmic plot represents depth and abscissa represents the log-derived salinity in the formations. C = slope of trend lines for different geologic ages. (Courtesy of *Oil and Gas Journal.*)

unconformities and other local or regional geological factors (Fertl and Timko, 1970b). In the absence of any of these geological features being superimposed on the overburden compaction stress, the formation water salinities may increase with depth at a predictable gradient, depending on the geologic age of the interval (Fig.5.18).

For example, drastic salinity variations across unconformities have been reported in many areas, such as Kansas (Case, 1945; Levorsen, 1967) and Israel (Bentor, 1969). Similar salinity variations are found in the Morrow Formation in Oklahoma. Using formation water salinity plots, derived from the SP curve, a well-known unconformity could be correlated over a large lateral extent (Zanier and Timko, 1970).

The shale formation factor method

Basic concepts

As pointed out by Hubbert and Rubey (1959), for a given clay there exists for each porosity value, ϕ, a maximum value of effective compressive stress, σ, which the clay can support without further compaction. This state of abnormal pressure condition can be expressed as an exponential function as follows:

$$\phi = \phi_i \cdot e^{-K\sigma} \tag{5.6}$$

where ϕ = shale porosity, ϕ_i = initial sediment porosity (zero overburden), K = constant, σ = net overburden pressure or vertical component of rock frame stress (see Chapter 1, eq. 1.4).

Empirical observations relate porosity to the degree of consolidation (cementation factor m) and the formation factor, F, of the rock:

$$F = \phi^{-m} \tag{5.7}$$

Basically, the formation factor, F, is a dimensionless quantity by which the formation water resistivity, R_w, is to be multiplied in order to obtain the resistivity of the reservoir rock, R_o, when 100% saturated with formation water:

$$F = R_o/R_w \tag{5.8}$$

Combining eqs. 5.6 and 5.7 and taking natural logarithms of both sides, one obtains:

$$\ln F = m(K \cdot \sigma - \ln \phi_i) \tag{5.9}$$

or:

$$\log F = m(0.434 K \cdot \sigma - \log \phi_i) \tag{5.10}$$

Thus, log F can be plotted against σ as a straight line. However, it is simpler to plot log F versus depth, D, since for normal compaction the vertical rock frame stress equals $0.535D$ (Foster and Whalen, 1965).

Quantitative pressure evaluation

Formation pressure estimates by the Foster and Whalen method include the following steps:

(1) Construct a formation water resistivity (R_w) profile by computing R_w-values from the SP curve in clean, shale-free water sands (Pirson, 1963).

(2) Plot shale resistivity, R_{sh}, from the short normal or conductivity logs in pure, lime-free shales.

(3) Compute formation factor F_{sh} from R_{sh} and R_w at the corresponding depth ($F_{sh} = R_{sh}/R_w$).

(4) Plot F_{sh} versus depth on semilog paper.

In the normal-pressure environment, all F_{sh}-values fall along a straight line, the normal compaction trend. Onset (top) of overpressured formations is characterized by departure from this normal trend, i.e. the F_{sh}-values become less than under normal compaction conditions.

Fig.5.19 shows the F_{sh} plot in an offshore Louisiana well. Note the pressure top at about 11,750 ft [3581 m].

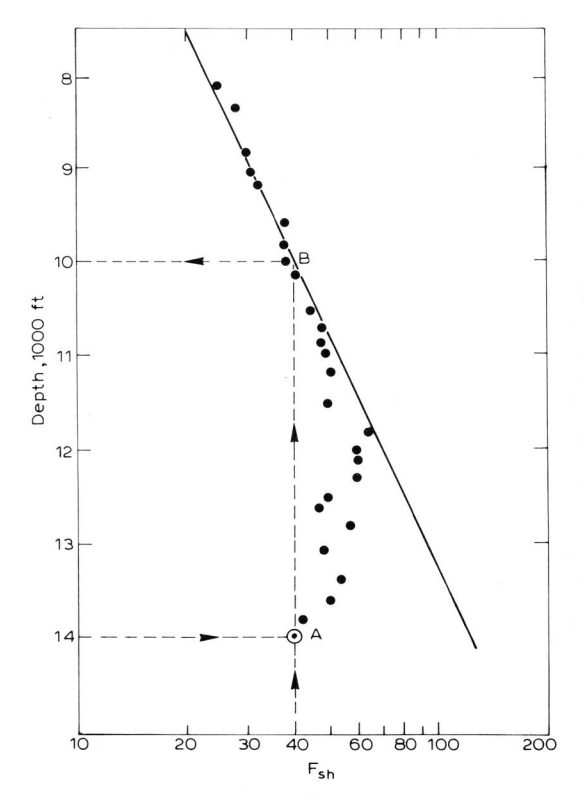

Fig.5.19. Shale formation factor plot in a U.S. Gulf Coast well.

Quantitative pressure data is easily obtained using the *equivalent depth method*. For example, formation pressures in the subject well (Fig.5.19) at 14,000 ft (point A) are calculated as follows:

(1) At point A, F_{sh} = 40.
(2) From point A move vertically up to intersect the normal trend line. Intersection establishes point B at an equivalent depth (D_E) of 10,000 ft.
(3) $\sigma = 0.535 D_E$ = 5350 psi at point A (14,000 ft).
(4) $P_f = P_o - \sigma$
 = 1.0(depth at A) $-$ 0.535(D_E at B)
 = 14,000 $-$ 5350 = 8650 psi.

(P_o = total overburden stress and σ = grain-to-grain or effective stress).

That is, a pore fluid pressure of 8650 psi [608 kg/cm^2] exists at the depth of 14,000 ft [4268 m] in the subject well.

Possible limitations

The method assumes that formation water salinities are the same in sands and adjacent shales. However, this is not usually the case (Sukharev and Krumbol'dt 1962; Von Engelhardt and Gaida, 1963; Hedberg, 1967; Fertl and Timko, 1970a; Schmidt, 1973).

Lithology variations, shale gas, large washouts, etc., will affect the R_{sh}-values and thus the computed F_{sh}. Furthermore, pressures in shales and sands are not necessarily the same; field cases of this observation are shown in Figs.5.42 and 5.43 later on.

The method is rather time-consuming, since both the R_w- *and* R_{sh}-values have to be determined first in order to compute F_{sh}. Nevertheless, application of this method has been helpful and successful in many instances. Foster and Whalen (1965) have claimed an accuracy within 5% for pressure predictions in over 80 wells.

Acoustic (sonic) surveys

Basic principles

This logging device records the transit time of elastic waves through formations over a predetermined distance. The log is used for estimating porosity and for measuring sound energy transmission characteristics for use in differentiating fluid content (Hicks and Berry, 1956; Wyllie et al., 1958b; Tixier et al., 1960; Pickett, 1963). In cased holes, elastic wave logs are used to indicate cement-bond conditions (Fertl et al., 1973).

A transducer creates elastic wave pulses which travel through the formation and are picked up by receivers in the logging sonde. The time required for the signal to travel the predetermined distance in the formation is recorded as the travel time. The total wave train may be impressed on an oscilloscope screen and photographed for later detailed study. Signal amplitudes may also be recorded.

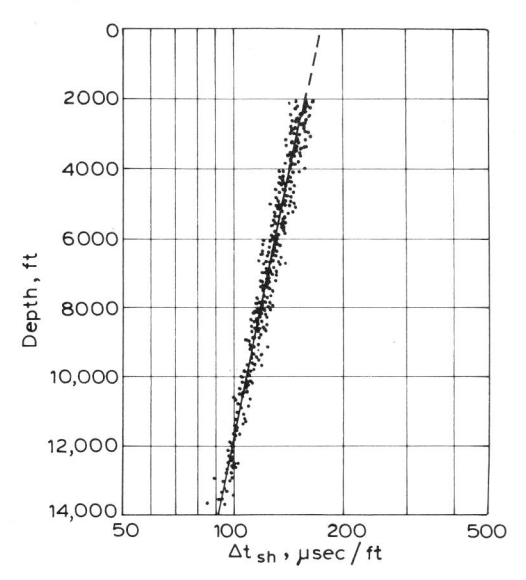

Fig.5.20. Shale sonic travel time versus depth of burial for Miocene and Oligocene shales, Upper Texas and Southern Louisiana Gulf Coast area (after Hottman and Johnson, 1965). (Courtesy of *Journal of Petroleum Technology*.)

Interval transit travel time (the reciprocal of wave velocity) is recorded in microseconds per foot (Δt in μsec/ft). In a given formation, Δt depends upon lithology, degree of compaction (consolidation), porosity, and the fluid content in the pore space. The sonic log can be run in all drilling muds, with the exception of air- or gas-filled holes. It is relatively unaffected by changes in hole size, temperature, and formation water salinity. Gas-cut mud may cause cycle skipping.

In *formation pressure evaluation* work the transit travel time, Δt, of shales is plotted versus depth on semilog paper. In normal-pressure environments all data points fall along the normal compaction trend line, characteristic for the subject well (Fig.5.20). Pressure anomalies, such as overpressures, can be detected by an increase in transit time above normal Δt-values. This deflection is the result of the significant increase in porosity in overpressured shales (Hottman and Johnson, 1965). Fig.5.21 shows such a Δt_{sh} increase in a pressure transition zone.

Quantitative pressure evaluation

As for any other logging parameter, several methods are applicable for the Δt-values:

Method A.

(1) Plot shale transit time, Δt, and establish the normal compaction trend line.

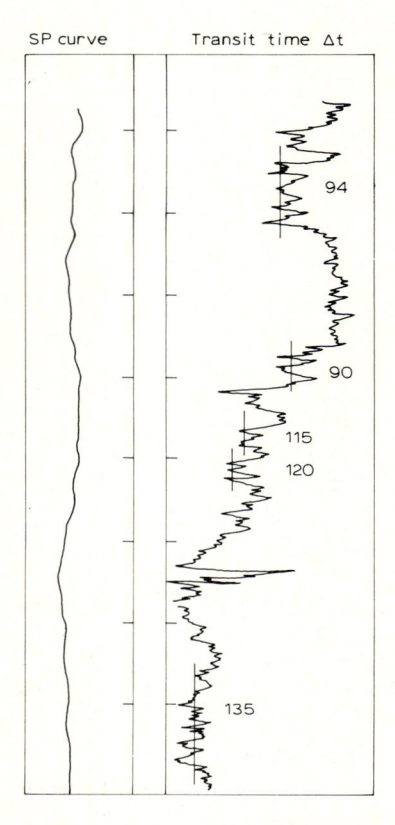

Fig.5.21. Acoustic (sonic) log response in an overpressured shale zone (modified after Matthews, 1971). The numbers represent the transit time values in μsec/ft.

(2) Pressure top is located at the depth at which plotted data points diverge from normal trend.
(3) Determine the formation pressure at a specific depth as follows: (a) divergence of actual (observed) shale value, Δt_{ob}, at a given depth from the value on the extrapolated normal trend line, Δt_n (see Fig.5.22); (b) from Fig.5.23 the fluid pressure gradient (FPG) corresponding to the difference in Δt-values ($\Delta t_{ob} - \Delta t_n$) is found.
(4) The FPG is multiplied by the subject depth to obtain the formation pressure.

A pressure gradient profile then can be constructed for the well. The relationship between the shale transit time values ($\Delta t_{ob} - \Delta t_n$) and FPG for Miocene and Oligocene formations along the U.S. Gulf Coast area is shown in Fig.5.23. Basic input data used to establish this empirical relationship are listed in Table 5.III. The standard deviation from the average trend line representing these data in Fig.5.23 is 0.020 psi/ft [0.046 kg cm^{-2} m^{-1}].

ACOUSTIC (SONIC) SURVEYS

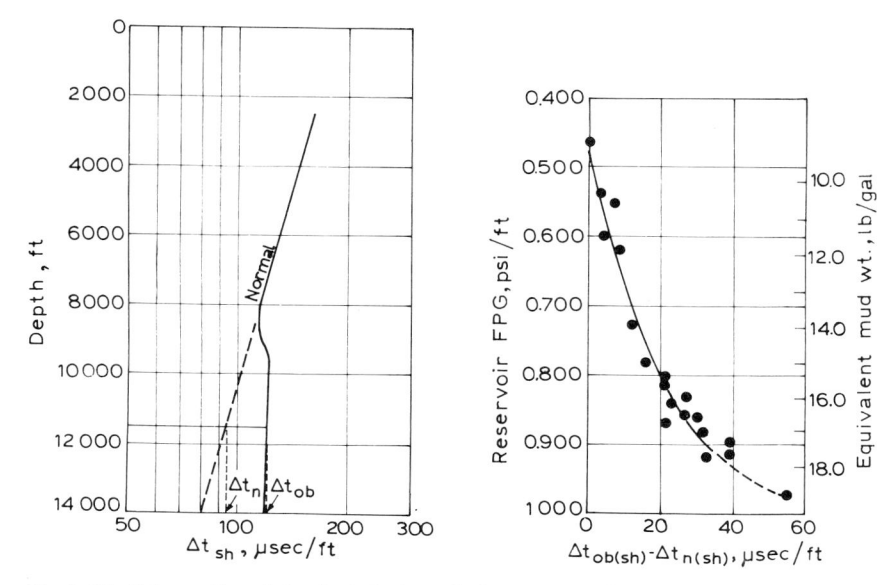

Fig.5.22. Schematic plot of shale travel time versus depth of burial (after Hottman and Johnson, 1965). (Courtesy of *Journal of Petroleum Technology*.)

Fig.5.23. Relationship between the shale acoustic parameter difference ($\Delta t_{ob(sh)} - \Delta t_{n(sh)}$) and reservoir fluid pressure gradient (FPG) (after Hottman and Johnson, 1965). (Courtesy of *Journal of Petroleum Technology*.)

TABLE 5.III

Formation pressure and shale acoustic log data in overpressured Miocene/Oligocene formations, U.S. Gulf Coast area (after Hottman and Johnson, 1965)

Parish or County and State	Well	Depth (ft)	Pressure (psi)	FPG* (psi/ft)	$\Delta t_{ob(sh)} - \Delta t_{n(sh)}$ (μsec/ft)
Terrebonne, La.	1	13,387	11,647	0.87	22
Offshore Lafourche, La.	2	11,000	6,820	0.62	9
Assumption, La.	3	10,820	8,872	0.82	21
Offshore Vermilion, La.	4	11,900	9,996	0.84	27
Offshore Terrebonne, La.	5	13,118	11,281	0.86	27
East Baton Rouge, La.	6	10,980	8,015	0.73	13
St. Martin, La.	7	11,500	6,210	0.54	4
Offshore St. Mary, La.	8	13,350	11,481	0.86	30
Calcasieu, La.	9	11,800	6,608	0.56	7
Offshore St. Mary, La.	10	13,010	10,928	0.84	23
Offshore St. Mary, La.	11	13,825	12,719	0.92	33
Offshore Plaquemines, La.	12	8,874	5,324	0.60	5
Cameron, La.	13	11,115	9,781	0.88	32
Cameron, La.	14	11,435	11,292	0.90	38
Jefferson, Tex.	15	10,890	9,910	0.91	39
Terrebonne, La.	16	11,050	8,951	0.81	21
Offshore Galveston, Tex.	17	11,750	11,398	0.97	56
Chambers, Tex.	18	12,080	9,422	0.78	18

*Formation fluid pressure gradient.

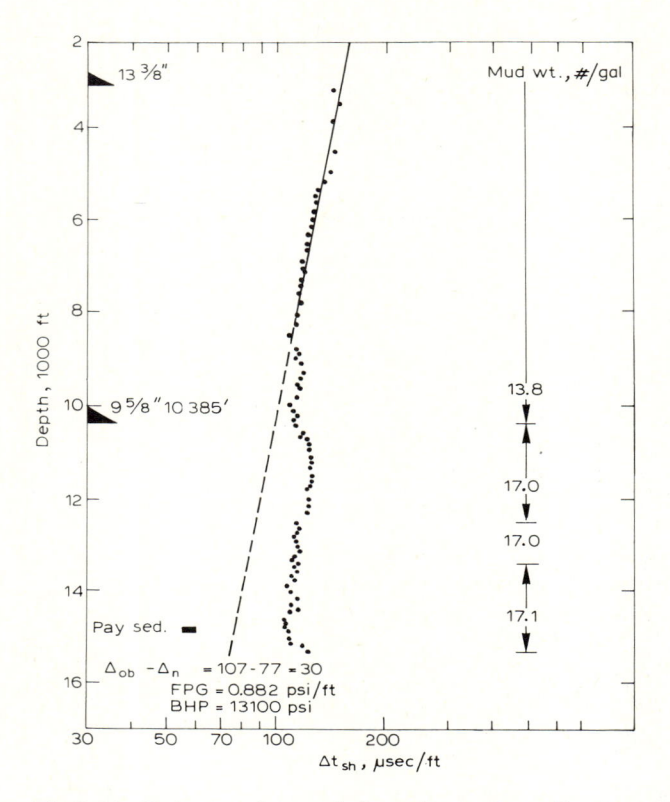

Fig.5.24. Shale transit travel-time plot and application of equivalent depth method for quantitative pressure evaluation. Casing points and mud weight requirements are also given.

As for any other logging parameter used in pressure evaluation work, similar calibration charts have to be developed for areas with different geologic setting, such as the North Sea area and the South China Sea region.

Fig.5.24 applies this method in a U.S. Gulf Coast well. The observed Δt difference is 30 μsec/ft which results in an FPG of 0.882 psi/ft [2.04 kg cm^{-2} m^{-1}]. At the subject depth this gives a pressure of 13,100 psi [921 kg/cm^2]. The maximum mud weight used in this well is 17.1 lb/gal [2.06 kg/dm^3].

Method B. Another possibility again is using the *equivalent depth technique*. The stepwise outline for this method has been discussed previously in eqs. 5.1 and 5.2.

Once the normal compaction trend has been established for a particular field, area, and/or region, then proper nomographs for pressure evaluation can be constructed for these specific conditions. This is illustrated for the

ACOUSTIC (SONIC) SURVEYS 203

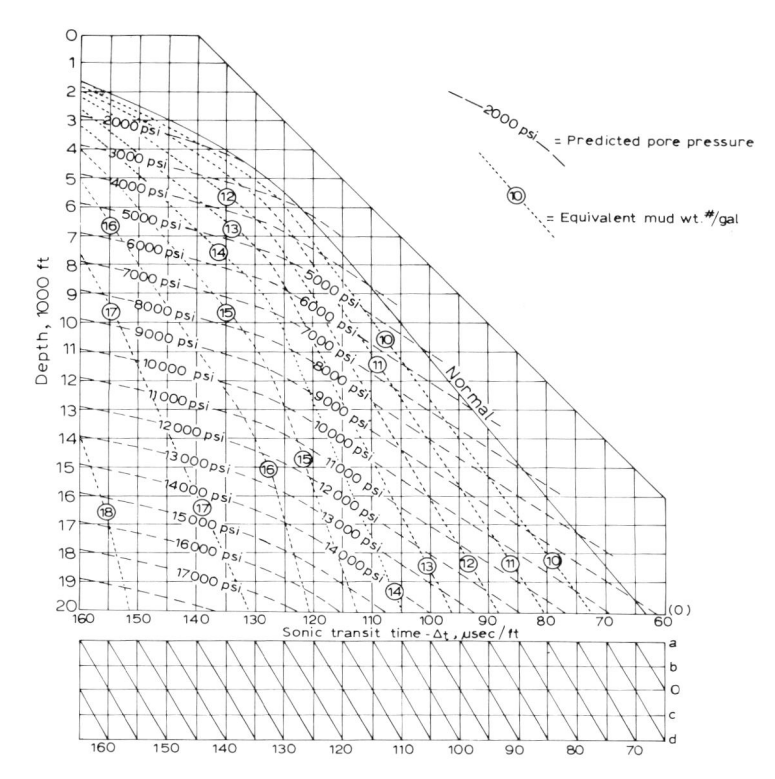

Fig.5.25. Nomograph for estimation of formation pressures from sonic logs, U.S. Gulf Coast area (after Ham, 1966).

South Louisiana area in Fig.5.25. Similar to the shale resistivity nomograph (see Fig.5.6), the grid at the bottom of the acoustic chart allows shifts in the Δt-scale to take local conditions and/or geologic age into account. Plot several shale points over the normal-pressure section of the well on this chart in order to determine which baseline (*a*, *b*, 0, *c*, or *d*) should be used for the formation pressure determination.

For example, if a shale travel transit time of 132 μsec/ft is recorded in a shale at a depth of 12,500 ft [3811 m], locate 132 μsec/ft on the chart's (Fig.5.25) horizontal "0"-scale, and move upward to 12,500 ft [3811 m] on the vertical scale. The resulting formation pressure is 10,200 psi [717 kg/cm^2] and mud weight needed to balance the formation pressure as read from the chart is 15.8 lb/gal [1.89 kg/dm^3].

Possible limitations and pitfalls

It seems that use of Δt-values, in general, results in the best and most reliable quantitative pressure evaluation from well logs. Nevertheless, as with any other well-logging measurement, there are certain limitations inherent to Δt measurements. This, in turn, reflects upon the validity of pressure evaluation using this logging parameter. A detailed discussion of possible pitfalls is given at the end of this chapter.

Formation density measurements

Basic principles

The density-logging device consists of a radioactivity source which irradiates the formation with medium-energy gamma rays. Collision with electrons in sedimentary rocks causes a gamma ray to lose energy. This is called Compton scattering. The number of these Compton-scattering collisions is directly related to electron density (i.e. number of electrons) surrounding the borehole (Alger et al., 1963; Tittman and Wahl, 1965). Electron density is related to the true bulk density, ρ_b, in g/cm³, which in turn depends on the density of the rock-matrix material, formation porosity, and density of fluids in the pore space.

A detector records intensity of back-scattered rays, which varies with the bulk density of the formation of interest. Generally, a marked departure (i.e. decrease) from the normal compaction trend indicates the presence of overpressures. However, in cases of severe hole washouts in shale sections, as is frequently observed, the reliability of bulk density plots decreases drastically. Washouts cause the density log to read too low a bulk density, which is reflected in too high a computed formation pressure. Further complications may arise from the high barite content in weighted drilling fluids.

Quantitative pressure evaluation

The *equivalent depth method* probably represents the most versatile approach. The density data in clean, not washed-out shale sections can be plotted on either a linear or logarithmic scale. For illustration, results of both types of plotting are shown for the same well in Fig.5.26.

However, experience indicates that unless borehole conditions are ideal, these devices will not be as accurate or reliable for pressure evaluation as other techniques based on acoustic and/or shale resistivity parameters.

Neutron measurements

Basic considerations

Neutron logs are used mainly for the delineation of porous reservoir rocks and determination of their porosity. The logging sonde responds primarily to the amount of hydrogen present in the formation. High-energy neutrons are emitted continuously from a radioactive source in the tool. Collisions of these neutrons with nuclei in the reservoir rock cause the neutrons to lose some of their energy. Since neutrons have a mass almost identical to the mass

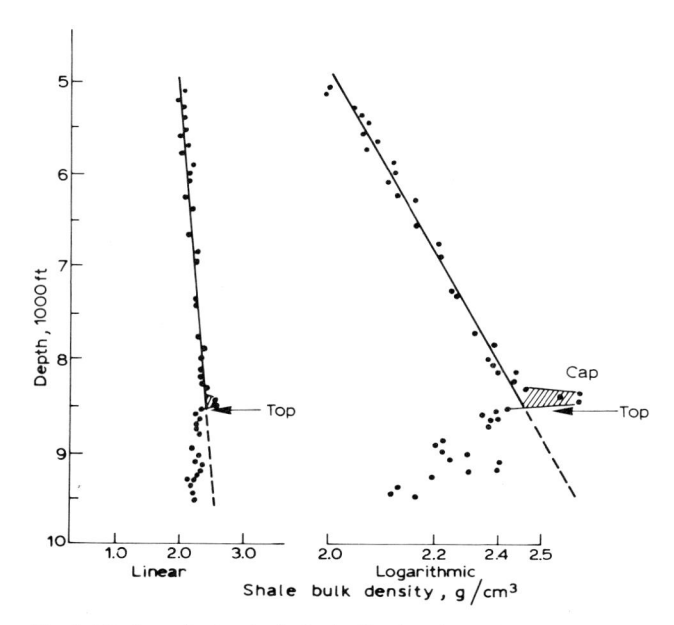

Fig.5.26. Log-derived shale bulk density plots on a linear and logarithmic scale for the same well.

of a hydrogen atom, and the greatest energy loss occurs when colliding with a nucleus of equal mass, the slowing down of neutrons depends largely on the amount of hydrogen present in the formation. Depending on the tool used, either the high-energy gamma ray of capture or neutrons themselves are counted by a detector (Dewan and Allaud, 1953; Tittle, 1961; Tittman et al., 1966).

Increase in porosity and corresponding water content of shales in overpressure environments allows the use of these tools for formation pressure evaluation. However, experience shows that unless borehole conditions are ideal, these devices are not as accurate and reliable as other logging tools.

Pulsed neutron measurements

Basic principles

The thermal neutron lifetime indicates the rate at which thermal neutrons are captured or absorbed by a material. This lifetime is actually controlled by the microscopic thermal neutron capture cross section (Σ) of the material. The larger this cross section, the more quickly the neutrons are captured. The lifetime is the time, in microseconds, required for the neutron population to be reduced to one-half.

The chlorine ion is the strongest absorber of all the commonly occurring elements in sedimentary rocks. The log response is related to the same basic parameters exploited by the chlorine log. The neutron generator for this device is an electromechanical source which emits 14-MeV neutrons in precisely controlled repetitive bursts. After each burst there is a period of quiescence during which the neutron population dies away. This neutron population is sampled by a scintillation counter which detects the gamma rays produced when the thermal neutrons are captured. This gamma-ray intensity is proportional to the neutron population.

The early part of the neutron die-away is very complex, being a combination of the instrument environment (tool housing, casing, cement, borehole, fluid) and formation characteristics. After some period of time, these borehole perturbations are gone and the rate of change of the neutron population is controlled by the formation. The lifetime measured at this time is that of the formation. The counting of capture gamma rays at two separate and known times during the latter phase of the die-away supplies sufficient data to calculate the formation lifetime or its reciprocal, the capture cross section (Youmans et al., 1964, 1966; Wichmann et al., 1967).

The advantage of the pulsed neutron logging devices over conventional neutron measurement is that casing and cement have little effect on the log response. Its general features resemble open-hole electrical resistivity logs, allowing easy correlation. The potential for applying these special logging techniques, such as *neutron lifetime* and *thermal decay time logs* (Clavier, 1969) as monitoring devices for the detection of overpressures and/or pressure depletion behind pipe has been discussed recently (Fertl and Timko, 1970b). It has been shown that the Σ-values in shales decrease in a regular fashion with depth in normally compacted intervals. Overpressures are flagged by divergence from this normal trend to higher Σ-values.

Quantitative pressure evaluation

The *equivalent depth method* has proven quite successful. Fig.5.27 shows the application of a neutron lifetime log to determine abnormal formation pressures in a cased well located in Louisiana. Computations indicate that a mud weight of 10.2 lb/gal [1.22 kg/dm^3] should be sufficient. Drilling of a sidetrack with a 10.4-lb/gal [1.25 kg/dm^3] mud was performed without any complications.

In other words, the pulsed neutron log accomplishes several things:

(1) The log gives a quantitative indication of saturation changes in a pay zone behind the pipe.

(2) The log can also be applied in pressure evaluation work. A log can be run through the drill pipe or in already cased holes to detect and evaluate pressure variations.

Pulsed neutron logging devices also help in investigating special reservoir problems, such as communication with, and depletion of pay zones by, wells in adjacent fault blocks, and recharge of aquifers with oil, and/or their repressuring from a nearby overpressured pay zone. Such observations are

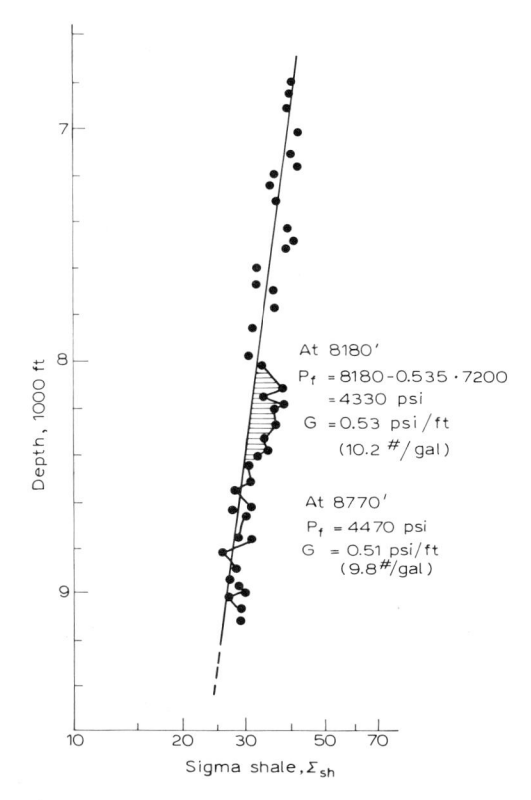

Fig.5.27. Use of pulsed neutron logging for quantitative formation pressure evaluation in cased holes (after Fertl and Timko, 1970b). Σ_{sh} = neutron capture cross section of shale in 10^{-3} cm^{-1}. (Courtesy of *Oil and Gas Journal*.)

At 8180'
$P_f = 8180 - 0.535 \cdot 7200$
$= 4330$ psi
$G = 0.53$ psi/ft
(10.2 #/gal)

At 8770'
$P_f = 4470$ psi
$G = 0.51$ psi/ft
(9.8 #/gal)

important in reservoir engineering and when drilling in partially pressure-depleted areas.

Nuclear magnetic resonance techniques

Nuclear magnetic resonance is not a new phenomenon; the basic concepts were investigated as early as 1946 (Bloch, 1946; Purcell, 1946). Basically, "atomic nuclei possess magnetic moments and spins; that is, they are similar in some respect to bar-magnet and gyroscope combinations. Molecules and their nuclei are subject to thermal motion, which has a scrambling effect, tending to leave as many nuclear spins oriented in any one direction as in any other. However, if a magnetic field is applied, the magnetic nuclei tend to align in the direction of the field. The scrambling and aligning forces compete with each other, with the result that a few more spins are oriented parallel to the field than in other directions. This gives a net magnetization, or polarization, which is directly proportional to the strength of the applied magnetic field (aligning influence) and inversely proportional to the absolute temperature scrambling influence" (Brown, 1960).

The basic operating principle of the nuclear magnetism log, then, is a magnetic field being set up within the reservoir rock by means of an energy source and a polarizing coil. The coil is then de-energized, and the voltage induced in the coil by precessing hydrogen nuclei is measured. The magnitude of the voltage induced is a measure of the number and geometry of the hydrogen nuclei. The time of realignment of the hydrogen nuclei with the earth's gravitational field is related to the type of fluid containing these nuclei. Thus, the use of nuclear magnetism in well logging is of special interest because it offers a way of making direct measurements on the hydrogen (proton) in the formation fluids and not on the rock matrix. Theories of nuclear magnetic resonance of protons of a hydrogenous liquid in the pore spaces of a porous solid have been presented by Korringa (1962) and Senturia and Robinson (1970). Applications of these theories to the estimation of the permeability of sandstones from nuclear magnetism resonance (NMR) measurements have been discussed by Loren and Robinson (1970). Detailed evaluation of pulsed NMR data for estimating porosity, permeability, and irreducible water saturation of sandstones was reported by Timur (1969). More recently, the possibility of determining residual oil saturation with the nuclear magnetism log (NML) has been explored.

Seevers (1972) discussed the use of pulsed NMR measurements on drill cuttings (shale) as a method for determining the permeability to fluid flow of the shales in an overpressured section. These permeabilities can be combined with pressure data on adjacent sands and yield the present-day flux of water through the shales. A knowledge of the variation of water flux with depth will aid in understanding the mode of pressure origin in the particular area studied.

Two very striking features are shown in the vertical water flux versus depth plot (Fig.5.28). One is that the flux out of the sand at 16,500 ft [5030 m] is much less than the influx, and the other is the very rapid buildup of flux between 14,000 and 15,000 ft [4268 and 4573 m]. Supplementary data include the mud weight used and the formation temperature in the subject well.

The discontinuity in the vertical flux implies that the sand at 16,500 ft [5030 m] is leaky. In engineering units the influx is 6 barrels/acre-year and the vertical outflux is about 1 barrel/acre-year. Thus, about 5 barrels/acre-year flush through the sand horizontally.

An estimate for the rate of increase in flux due to normal compaction is about 10^{-10} ml cm^{-3} yr^{-1}, whereas the apparent rate of increase between 14,000 and 15,000 ft [4268 and 4573 m] is 6.5×10^{-7} ml cm^{-3} yr^{-1}. This large rate implies some source of water other than that due to compaction. One possible source that has been suggested is the diagenetic alteration of the

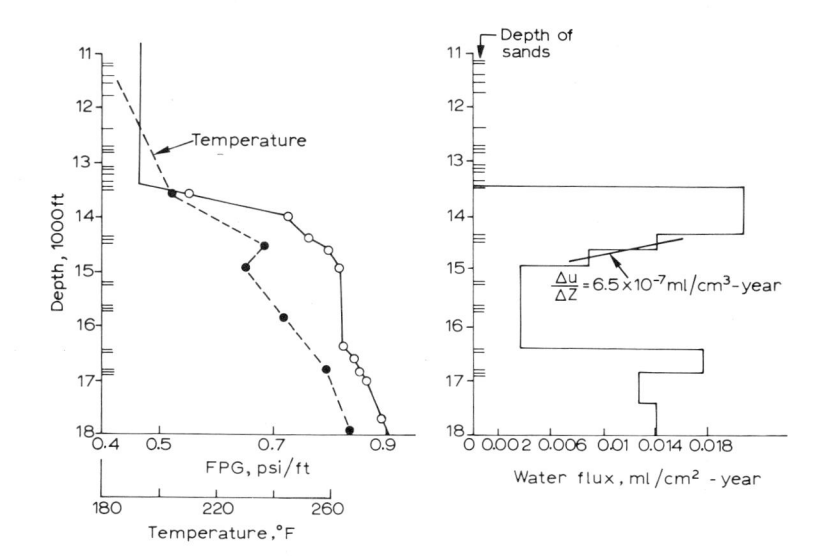

Fig.5.28. Comparison of fluid pressure gradient and formation temperature gradient with the water flux computed from NMR-logging devices (after Seevers, 1972). $\Delta u/\Delta Z$ is the rate of flux due to compaction, where u = water flux in ml cm^{-2} yr^{-1} and Z = elevation in cm. (Courtesy of the Society of Petroleum Engineers of A.I.M.E.)

clay mineral montmorillonite to illite (Powers, 1967; Burst, 1969; Perry and Hower, 1970).

Downhole gravity data measurements

The downhole gravity meter is a high-precision, large-volume, bulk density tool (Smith, 1950; McCulloh, 1968; Jones et al., 1971, 1972a,b). The difference in gravitational attraction between any two points below the earth's surface is a function of the free-air gradient and the mass of material between the two points. This is summarized in the equation:

$$G_1 - G_2 = \Delta G = F\Delta Z - \beta \Delta Z \tag{5.11}$$

where G_1, G_2 = gravitational attraction at stations 1 and 2, respectively; F = free-air correction, 0.09406; $\beta = 4\pi K\rho_b = 0.02551\rho_b$; ρ_b = bulk density in g/cm^3; K = gravitation constant (6.67 × 10^{-8} in c.g.s. units); and ΔZ = difference in elevation. From this:

$$\rho_b \text{ (g/cm}^3\text{)} = 3.687 - 39.18\Delta G/\Delta Z \tag{5.12}$$

Since ΔG is measured with the downhole meter and ΔZ with the wireline, downhole gravity data directly measures bulk density and no calibration is required. It is also apparent that bulk density measurement involves a large volume of rock.

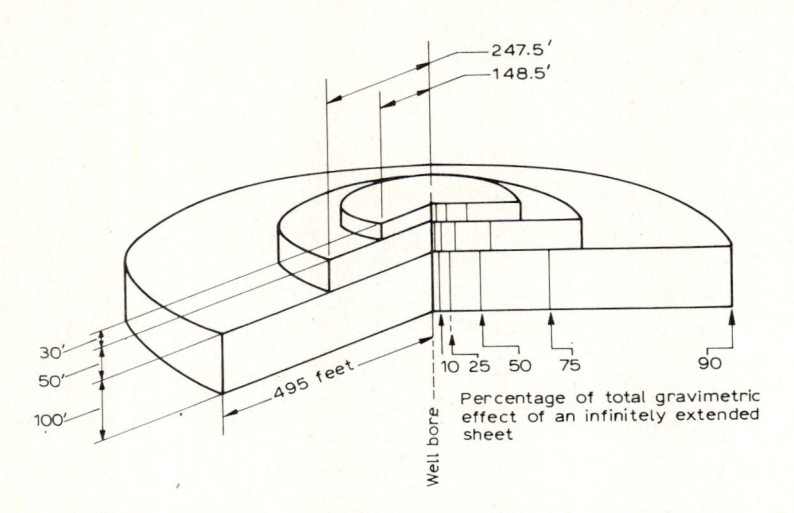

Fig.5.29. Difference in gravity effect on downhole sonde at two elevations indicates density of rock in the interval (after Jones, 1972b). 90% of the effect comes from the rock volume within a radius of about five times interval thickness, as illustrated here for 30- [9.1 m], 50- [15.2 m], and 100-ft [30.5 m] intervals. (Courtesy of *World Oil*.)

Fig.5.29 illustrates rock volume involved in density measurement. 95% of the effect comes from a rock cylinder ΔZ ft thick and approximately $5\Delta Z$ ft in radius. For example, ΔZ is 100 ft [30.48 m], the cylinder is 100 ft [30.48 m] thick, with a radius of 495 ft [150.9 m] or nearly with a 500-ft [152.4 m] depth of investigation (Jones, 1972b).

This means that measurements are practically unaffected by mud cake, infiltrated zones, washouts or casing. The borehole gravity log, then, is basically a density-logging tool (Fig.5.30). Its radius of investigation is tens of feet rather than tens of inches.

Of particular interest is Jones' (1972b, p.58) statement that "bulk densities measured with the borehole gravity log ... do not show a density change in the overpressured shale zones." These observations are limited to a localized area and may be explained by frequent lithological variations in the subject wells, due to the presence of limey shales in several areas of the U.S. Gulf Coast.

Well logs recommended for conventional pressure evaluation

The *acoustic (sonic) log*, in our opinion, is usually the best log for quantitative pressure evaluation since the log is relatively unaffected by changes in hole size, formation temperature, and formation water salinity. However, the recorded interval transit time in a given formation still depends on several parameters, including the degree of compaction, porosity, fluid content, and lithology.

CONVENTIONAL PRESSURE EVALUATION

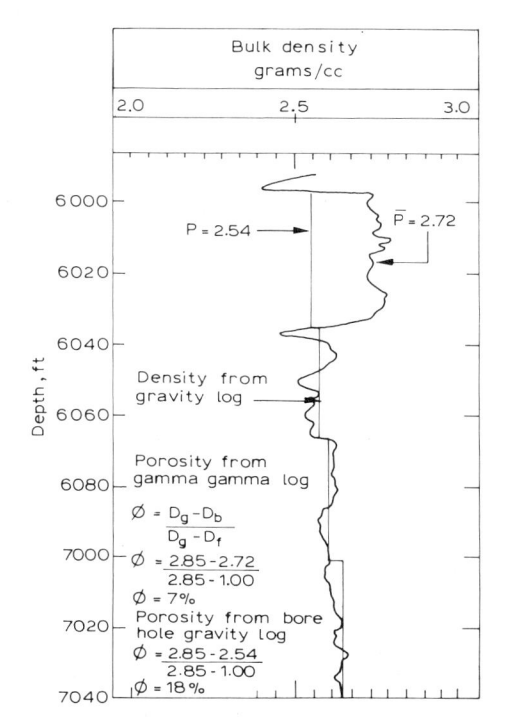

Fig.5.30. Bulk density values from gravity log in the three lower intervals agrees closely with bulk density values recorded by the density (gamma-gamma) log in this southwestern United States well (after Jones, 1972b). Difference of 0.18 g/cm³ in interval 5990—6037 ft [1826.2—1840.5 m] could mean that the large-volume gravity—density log possibly includes productive porosity away from the well-bore. P = bulk density values from gravity log in g/cm³, $\bar{P}$ = bulk density values from density log in g/cm³. (Courtesy of *World Oil*.)

Another frequently used logging curve is the *short normal* curve or, preferentially, its electronically five-fold amplification, the *amplified short normal* curve. The latter allows a more accurate selection of proper shale resistivity values for pressure plots. This logging curve is part of the induction-electrical log (IES) or the electrical log (ES), the latter usually being available in older wells. As in any other resistivity measurement, the response of this log is affected by large variations of formation temperature and salinity. Lithology variations, proximity to massive salt deposits, presence of steep and rather thin formations, etc., also largely affect recorded shale resistivity values. Recently, a focused spherical short normal curve has become available. Comparative field tests have shown that the newer generation sondes perform satisfactorily in quantitative pressure evaluations.

From this discussion, it is obvious that preference is given to these two logging devices. However, it does not mean that other well-logging devices are considered obsolete. Frequently, they too give very valuable information.

However, it appears to be a prudent practice not to rely on them alone or completely.

Limitations and pitfalls of wireline methods

Despite an enormous volume of technical contributions to the subject of formation pressure detection and evaluation using wireline methods, the practical "how to do" approach for drilling personnel and geologists in the field has only been touched on or often completely neglected. Seldom have questions been raised and explicitly answered, such as:

(1) Which well logs are superior for quantitative pressure evaluation?

(2) How should one quickly and efficiently recognize, select, plot, and evaluate logging parameters for in-situ formation pressure changes?

(3) What limitations and possible pitfalls are inherent?

To highlight these basic but important questions is the ultimate goal of the following discussion.

With the exception of data on formation water salinity derived from the SP curve of permeable zones, all other logging parameters are recorded in shale zones. These parameters are plotted versus a preselected depth scale, usually 1000 ft [304.8 m] per 1 inch [2.54 cm] or, for detailed studies of formation pressure variations, on a more expanded scale.

Formation pressure plots basically rely on four main considerations: (1) the logging sonde was working properly; (2) major detrimental effects on log response, such as washouts, bit size changes, lithology variations, etc., are either not present, can be avoided in plotting data, or at least are recognized; (3) relatively pure and thick shale zones are present; and (4) realistic trend lines for normal compaction have been established.

Possible pitfalls

Proper calibration of logs is a prerequisite for any realistic formation pressure evaluation. Nevertheless, there is always the possibility of malfunction, especially in ultra-deep and very hot wells or, generally speaking in boreholes where the bottom-hole temperature approaches or exceeds the temperature rating of the specific logging device used. Dimensions, temperature and pressure ratings of conventional logging sondes are found in the appropriate manuals available from all well-logging service companies.

How can one avoid or at least recognize such malfunction? First of all, the rating of the tools should be compared with the expected downhole conditions (pressure and temperature) in the subject well. If logs from offset wells are available, it is a good idea to have log copies and the corresponding pressure plots at the rig site. However, this is not always the best solution to one's problem either, since adjacent wells may test different fault blocks, and faults are known to act either as pressure continuities or discontinuities. In other words, pressure profiles encountered in adjacent wells, but separated by major faults, could be identical, similar, or completely different.

Fig.5.31 shows two conductivity logs in two offshore wells, Vermilion area, Louisiana. The wells are only 2000 ft [610 m] apart but separated by a well-known fault. Whereas well A was drilled in essentially normal, hydrostatic pressure environment to below the casing point at 12,500 ft depth [3811 m], well B encountered abnormally high pressures below 11,000 ft [3354 m]. Mud weight requirements over the next 1000 ft [305 m] of drill-

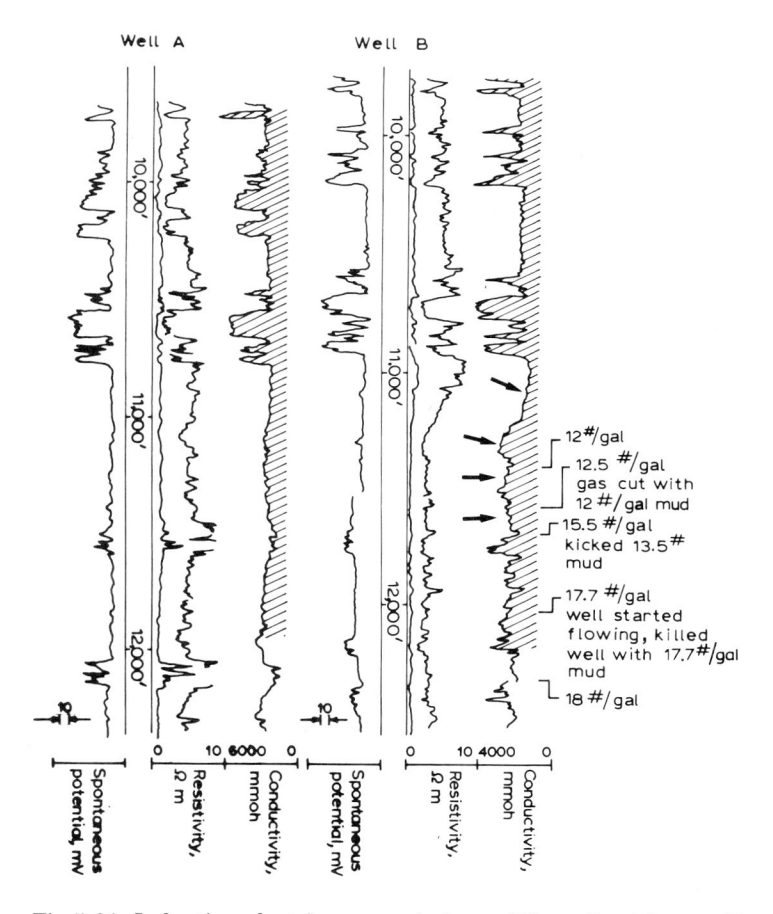

Fig.5.31. Induction-electric surveys in two offshore Louisiana wells only 2000 ft [610 m] apart, but located in different fault blocks (after Wallace, 1965b). The well on the left is normally pressured, whereas the second well on the right is overpressured. Note the abrupt increase in the conductivity curve in the second well. From left to right the curves represent the spontaneous potential (SP) curve in mV, induction resistivity curve in Ωm, short normal resistivity curve in Ωm, and induction conductivity curve in mmhos. The arrows show the conductivity increase in the pressure transition zone.

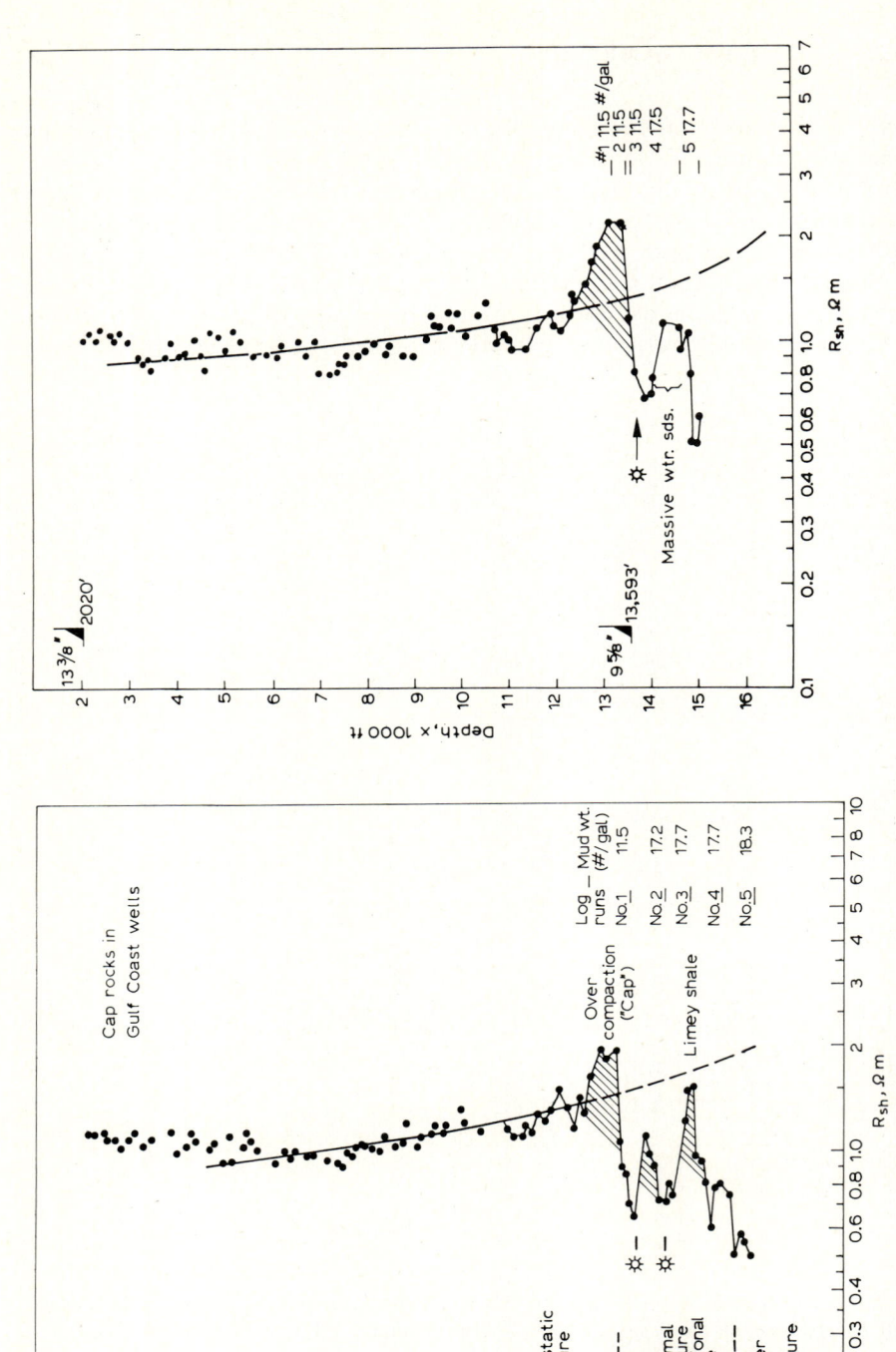

Fig.5.32. Shale resistivity plots show massive pressure cap rock at about 13,000 ft [3963 m] (after Fertl and Timko, 1970d). Note correlative features of plots in two adjacent offshore wells. (Courtesy of the Society of Petroleum Engineers of A.I.M.E.)

ing called for up to 18 lb/gal [2.16 kg/dm^3] mud due to gas cutting and several well kicks.

Fig.5.32 illustrates the correlative shale resistivity features in a new well with those in the well drilled earlier in the same fault block. We have successfully used similar formation pressure correlations, using several conventional methods of measuring shale properties, not only on a local but also on a regional scale. The importance of such correlative features becomes apparent when drilling a new well in a development area. Correlating drilling data from an active well (under drilling) with logging and drilling data of previously drilled wells helps the driller to know where the drill bit is with respect to the formation pressure variations. This "know where you are" approach has proven useful in many drilling operations on a worldwide basis.

Obviously a good practice is to have engineers and geologists with the proper background and experience on location. This improves the odds that well logs are run properly and the measured shale parameters are plotted and evaluated correctly. If there is still some doubt as to the log quality, one should consider relogging the interval in question with a backup logging tool of the same type and design. One could also switch over to another type of log, provided this option is feasible in the subject well.

This latter possibility is illustrated by a field case in a Texas well (Fig.5.33). Both the shale resistivity from the short normal curve and the shale travel time from the acoustic log were available for this well. Note the considerably different characteristics of the two plots over the interval from about 8000 ft [2439 m] to 10,000 ft [3048 m]. Subsequent examination and comparison of all three log runs with the IES log clearly indicated that the logging sonde during run No. 2 was malfunctioning, recording very low resistivity values. Thus, formation pressure computations based upon the IES log would have been erroneous.

Occasionally, due to certain reasons, different service companies may run the same type of log over different intervals in the same well. Fig.5.34 shows comparative pressure plots using short normal curves run by two different service companies over the same interval in one and the same wildcat well. The shape of the resistivity profiles are similar, but obviously the plotted shale resistivity values are quite different. Assuming the hypothetical case that the log of service company B only would have been run over the lower part of the hole, i.e. interval Z, then the resulting lower shale resistivity would have been interpreted as an increase in formation pressure, calling for much higher mud weight than was actually needed to safely drill this well.

Unfortunately, it is not always easy and sometimes it is even unlikely to recognize such pitfalls. This is especially true if only one log is available for in-situ pressure evaluation and additional pressure indicators, such as drilling parameters, etc., are not used properly or are simply ignored. *It cannot be stressed often enough that data from every source available should be considered and studied carefully.*

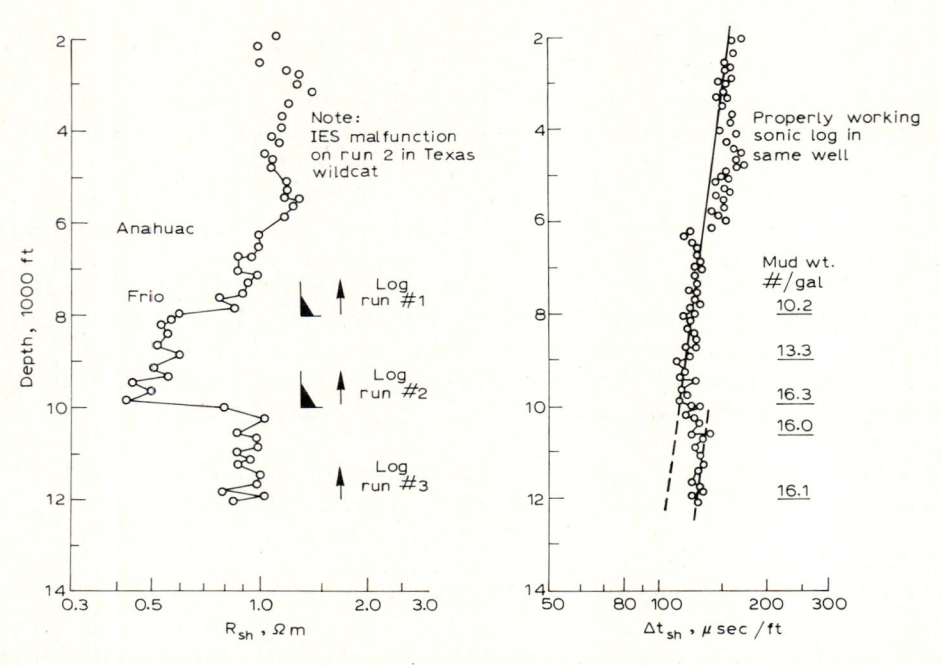

Fig.5.33. Effect of logging-sonde malfunction on the R_{sh} plot (log run No. 2) (after Fertl and Timko, 1971b).

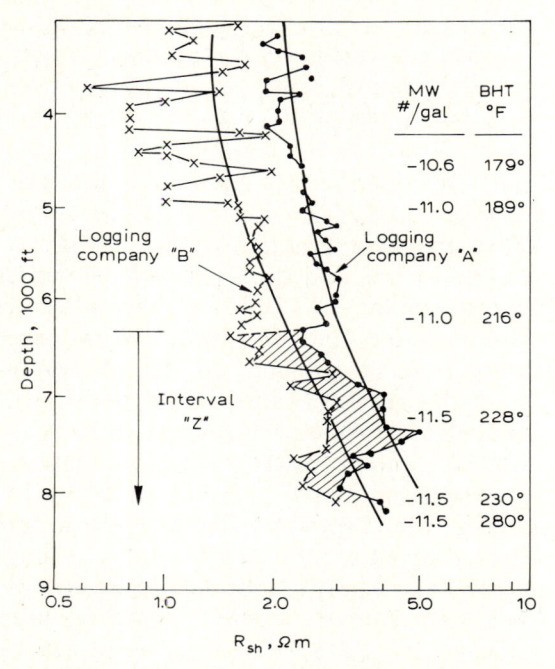

Fig.5.34. Shale resistivity plots from short normal logs recorded by different service companies in the same borehole, offshore Far East.

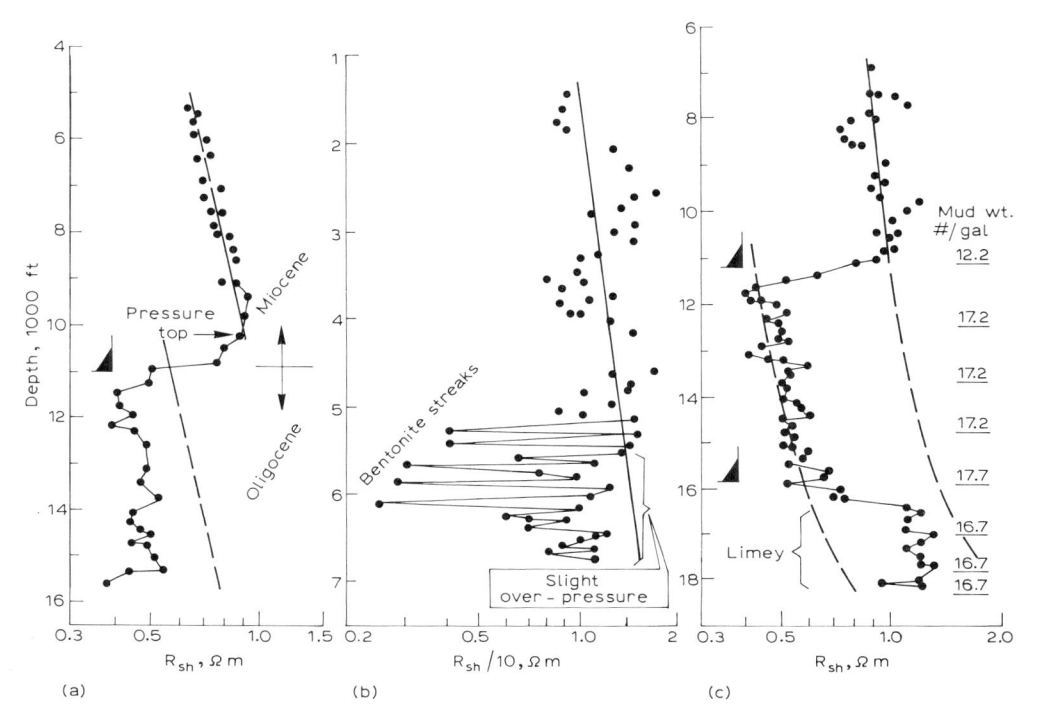

Fig.5.35. Possible effects on resistivity (shale parameter) plots, such as (a) age boundary effect in a Louisiana Gulf Coast well; (b) fluid-sensitive shales in a Mid-Continent, U.S.A., well; and (c) major lithology effects in a Louisiana Gulf Coast well (after Fertl and Timko, 1971b).

In addition, major lithology or age changes, washouts, etc., or even drastic but unexpected formation pressure variations can also result in "unusual" pressure profiles (Figs.5.35 and 5.36).

It is always a good idea to have an overlap section of subsequent logging runs. This allows a "tie-in" and check with shale values of the previous logging run. However, prolonged exposure of shale zones to the drilling mud may alter some of the shale properties (Louden and Woods, 1970). Significant variations in shale resistivities are sometimes observed from one logging run to the next (Fertl and Timko, 1972).

These effects appear to be related to the type of drilling mud — its composition, pH value and weight; the time the open hole is exposed to mud; the magnitude of formation pressures; the downhole temperatures; the presence or absence of shale gas; and shale mineralogy.

Recently, Reynolds et al. (1973) have discussed possible shale hydration effects as observed on acoustic (sonic) logs. One of their field cases is shown in Fig.5.37. The conventional normal compaction trend suggests overpressures

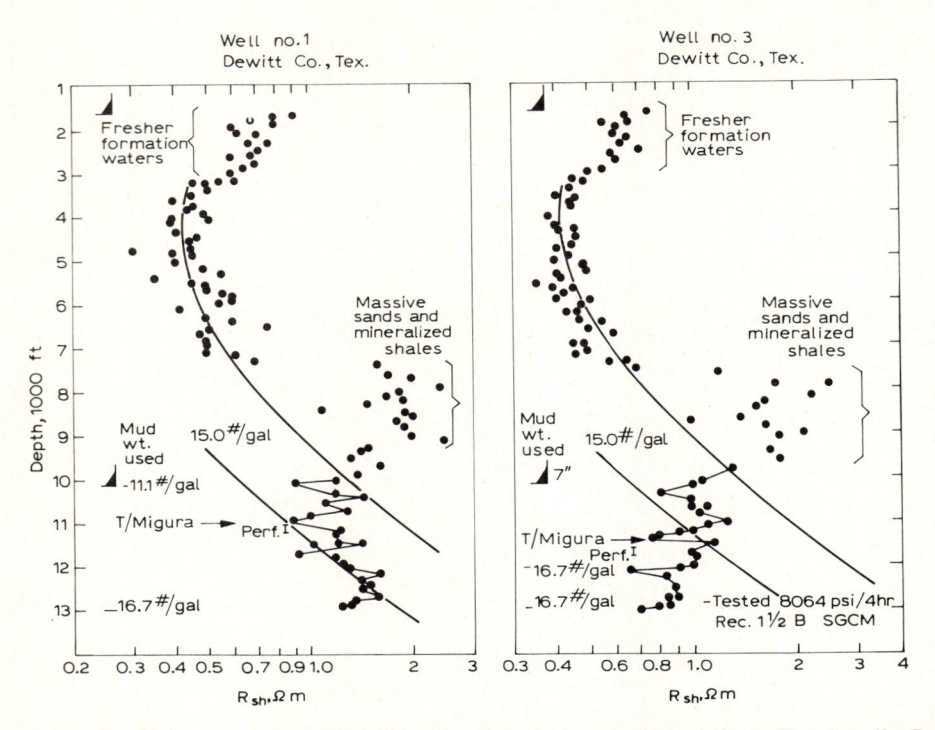

Fig.5.36. Effects on resistivity (shale parameter) plots in two onshore Texas wells, DeWitt County. Note presence of fresher shale water (higher resistivity) down to 3500 ft [1067 m], limey shales between 7000 and 9000 ft [2134 and 2744 m], and presence of high overpressures in lower part of both boreholes. T/Migura indicates the top of the Migura Formation. Rec. 1½ B SGCM means 1½ barrels of slightly gas-cut mud recovered during the 4-hour test.

starting at 7000 ft [2134 m] and extending to total depth. However, the well was drilled with a 10.5-lb/gal [1.26 kg/dm^3] mud and no significant problems were encountered, which suggests the presence of essentially normal formation pressures. The authors commented, "What, then, is causing the shale transit times to depart from the normal? We have a clue at just below 5200 ft [1585 m] where part of the hole has been logged twice. The first log was run within a few hours after the interval was drilled. The second log was run after the well had been deepened another 8000 ft [2439 m] which took almost 5 weeks.

To study in greater detail the interval from 5200 to 5500 ft [1585 to 1677 m] examine the actual sonic logs. At first glance, it appears that there is fair to good agreement between the two runs. However, closer scrutiny shows that the better shales (those with higher gamma-ray counts) are associated with the larger separations between the sonic curves. It is felt that these separations result when the shales hydrate, increasing the sonic transit time.

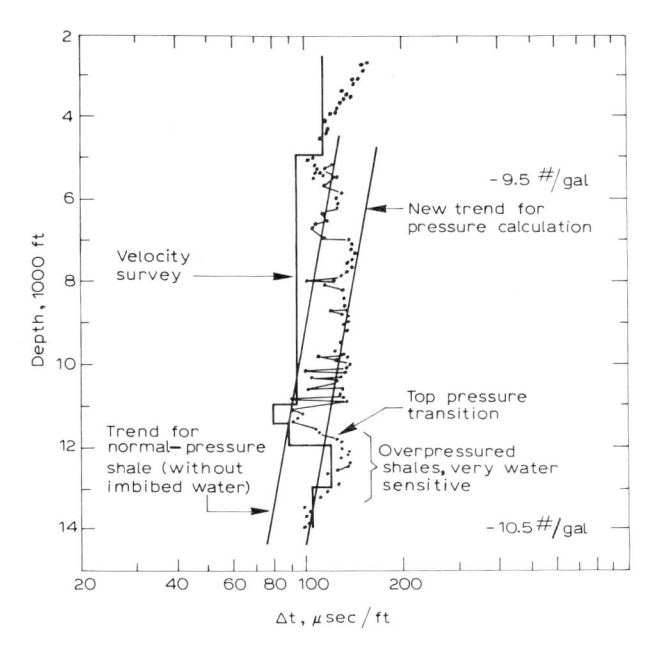

Fig.5.37. Acoustic log-derived shale pressure plot of onshore well with water-sensitive shales (after Reynolds et al., 1973). Velocity survey is the seismic data interpretation, in μsec/ft. (Courtesy of *Journal of Petroleum Technology*.)

Supporting evidence for shale alteration near the borehole over the interval from 5200 to about 10,500 ft [1585 to 3201 m] came from the velocity survey. As is shown in Fig.5.37, the overall velocity of the predominantly shale section is considerably higher according to the velocity survey than according to the sonic log. Since the velocity survey records velocities through the virgin formation and the sonic log records velocities in disturbed formations, Fig.5.37 conveniently shows that the shales have been hydrated enough by the drilling fluid to alter their true transit times by as much as 40 μsec/ft [122 μsec/m].

In an attempt to correct for the hydrational effects, a new 'normal trend' was established through the normally pressured, maximum transit time shales and parallel to the original 'normal trend'. Now it appears from the plot that only the interval from 11,000 to 13,000 ft [3354 to 3963 m] is overpressured."

Sometimes both oil and service company personnel are confused when IES logs (induction-electrical surveys) are run and the short normal and induction-log curves indicate quite different resistivity values in the shale sections. As a matter of fact, some field engineers consider such logs miscalibrated. Others use this apparent separation or difference in recorded shale resistivity values as criteria to determine if a well is being drilled overbalanced, balanced, or underbalanced. Both concepts are erroneous. They neglect to consider (1) differences of basic well-logging principles as inherent to conventional electrical measurements, such as the short normal curve, and induction-type measurements, as recorded by the induction curve; and (2) the fact that the horizontal layering of plate-like particles, such as clay minerals in shales, results in both physical and electrical anisotropy. Frequently, shale resistivity values measured by the short normal curve exceed values recorded by the induction curve. These concepts are confirmed by laboratory data of the electrical anisotropy coefficient λ, which, by definition, equals $\sqrt{R_V/R_H}$, where R_H is the longitudinal resistivity (horizontal, parallel to bedding planes) and R_V is the transversal resistivity (vertical to bedding planes). Data by Kunz and Moran (1958) show λ-values in shales ranging from 1.0 to about 2.5.

Considerations in plotting shale parameters

— Use intervals at least 10—30 ft [3.05—9.15 m] thick.
— Use data for relatively pure and clean shales and, if possible, avoid too

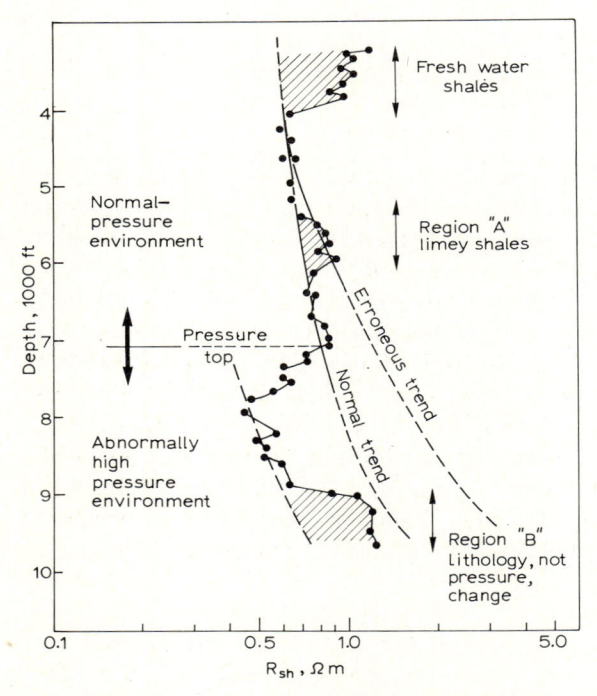

Fig.5.38. Erroneous normal compaction trend lines and several other possible pitfalls of pressure evaluation using shale parameters.

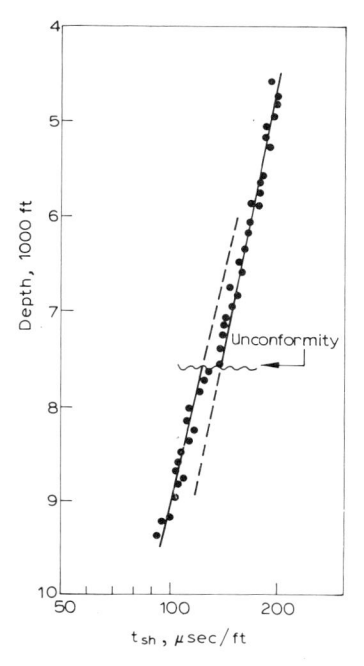

Fig.5.39. Unconformity causes shift in the normal compaction trend line.

limey and/or too silty (sandy) shale streaks. Use drill cutting analysis reports as additional check.

— Try to avoid plotting shale resistivity values down to about 2500—3500 ft [762—1067 m] (critical compaction depth). Usually these shales contain formation water fresher than seawater, resulting in high resistivity values (Fig.5.38). However, locally empirical correlations can be worked out.

— Geologic age changes, major unconformities, etc., are often reflected in drastic changes in shale properties. Miocene and Oligocene shales in the U.S. Gulf Coast area have different electrical properties, causing a marked shift in the normal trend lines. The effect of an unconformity on acoustic (sonic) log data is illustrated in Fig.5.39, with a shift in the normal compaction trend line across the unconformity. Similar observations with resistivity, acoustic, and density data were made worldwide in many exploratory areas.

— Avoid, if possible, plotting data for steep, thin, and tectonically overturned shales.

— Shales proximal to large salt masses, such as salt domes, etc., have very low resistivities due to increase of salinity in pore water. This falsely indicates higher-than-actual formation pressures. The associated salinity variation in adjacent sands is illustrated in Fig.5.40.

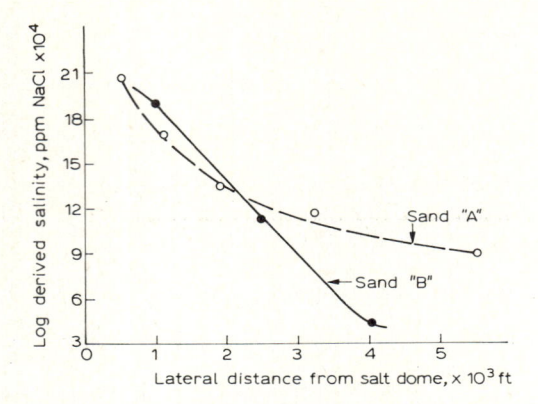

Fig.5.40. Salinity variations in sands as a function of the proximity to a salt dome, U.S. Gulf Coast area (after Fertl and Timko, 1971c).

— Extremely fresh formation waters, i.e. 5000 ppm NaCl (equivalent) or less, at great depths increase shale resistivity values which then indicate lower-than-actual formation pressures.
— Presence of "shale gas" in the shales can drastically affect both sonic and resistivity measurements. Observed variations have no direct relationship with formation pressure variations.
— Considerable scattering of acoustic energy occurs in gas-cut muds and drilling fluids containing lost circulation additives (mica, etc.). Sometimes

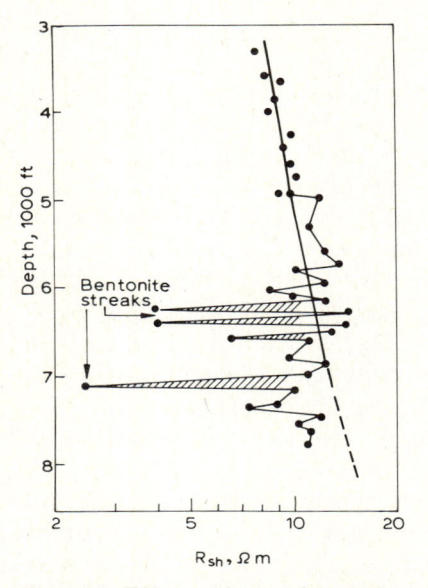

Fig.5.41. Effect of bentonitic shale streaks on shale resistivity plots.

attenuation of these muds is so great and variable that no realistic measurements can be obtained.

— Bentonitic and fluid-sensitive shales exhibit very low resistivity values, suggesting very high formation pressures which are usually nonexistent. Fig.5.41 shows a field example of such a case.

— Avoid — if at all possible — shale zones plagued by severe washouts. All logging-tool responses are affected by them.

— Major changes in bit size (hole size) also affect resistivity parameters.

— Recognize major lithology changes, i.e. the presence of limey shales. They may occur in the normal and/or abnormally high formation pressure environment. Apparent shale resistivities increase, whereas acoustic shale travel times decrease. Fig.5.38 indicates the possibility of an erroneous pressure evaluation due to a false normal trend line, resulting in computed mud weight requirements which are much too high. Furthermore, it has been our experience in the U.S. Gulf Coast area and elsewhere that in a specific temperature range there usually occurs a major change in shale properties. The resulting shift to higher shale resistivity has no relation to formation pressure variations (region B, Fig.5.38).

Recent correction methods attempt to take such lithology effects into account (Matthews, 1971). These methods are helpful locally but caution should be exercised when applied universally.

— In thick, massive shale sections, it is often observed that the pressure gradient in the shales is a function of proximity to the permeable sands.

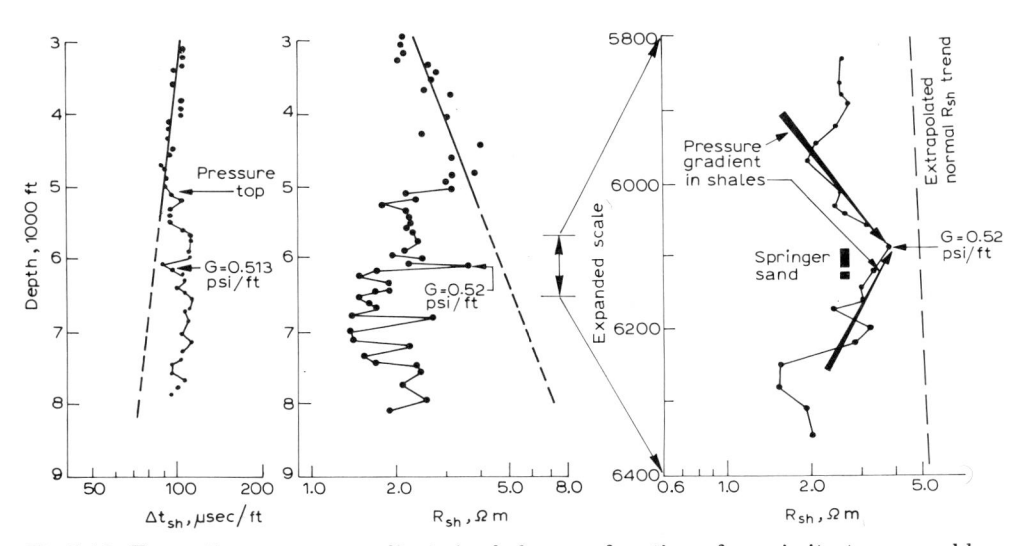

Fig.5.42. Formation pressure gradients in shales as a function of proximity to permeable sands. Note pressure gradients on expanded shale resistivity plot in the vicinity of Springer Sand, Oklahoma.

Usually, pressure gradients somewhat decrease when approaching sands. This has clear implications for drilling, testing, and completion techniques. Higher mud weights are necessary to prevent the shales from heaving and sloughing, but at the same time, circulation loss occurs in the adjacent lower-pressured sands. In such cases, one should replot the shale resistivity values on an expanded depth scale and plot all shale data as close as possible to the sands. Fig.5.42 clearly illustrates this observation for the Springer Sand in a well drilled in Oklahoma.

Fig.5.43 shows similar observations in a well located offshore in Louisiana. Shale resistivity and travel time were plotted conventionally on a depth scale of ½ inch [1.27 cm] = 1000 ft [304.8 m]. Gas zones occur at 7820—7890 ft [2384.15—2405.49 m] and 10,160—10,210 ft [3097.6—3112.8 m]. Both plots confirmed normal hydrostatic pressure in the upper zone and a gradient of about 0.75 psi/ft [1.73 kg cm^{-2} m^{-1}] in the lower zone.

In drilling the interval 9600—10,500 ft [2927—3201 m], mud weights exceeded 15 lb/gal [1.80 kg/dm^3]. This on-site data further confirmed log-calculated high pressures. Because of the expected high pressure in the lower zone, it was decided not to dual-complete the well but to complete only in the lower zone through a tie-back string all the way to the surface. Testing of the lower zone showed that the high pressure does not exist — actual pressure gradient was 0.55 psi/ft [1.27 kg cm^{-2} m^{-1}] instead of 0.75 psi/ft [1.73 kg cm^{-2} m^{-1}].

This well was restudied using an expanded-scale plot of ½ inch [1.27 cm] = 100 ft [30.48 m] for the interval 9600—10,500 ft [2927—3201 m]. The shales exhibited a decreasing pressure gradient from 9925 ft [3026 m] to the sand and an increasing gradient below the sand. Thus, the sand is not at the same high pressure as the shale 100—150 ft [30.48—45.7 m] above and below.

— Make sure the normal trend line used is realistic, since it is the basis for picking the pressure top and any quantitative pressure interpretation. Proper normal compaction trend lines are of utmost importance in ultra-deep wells, since with increasing depth any deviation from the correct trend line becomes enhanced, thus aggravating any erroneous quantitative pressure evaluation. Proper trend lines are also important in boreholes plagued by major washouts, presence of abrupt age and/or lithology changes, thick and massive pressure caps, etc. There are no fixed and universal rules on how to establish proper trend lines — and there is no substitute for experience. However, one should always use all data available, not just the well logs.

— Finally, even if it turns out that the interpretation was wrong, one should not be prudent but rather go back and restudy the pressure plot and all other available data. Establish the exact cause for the erroneous interpretation. Experience gained in such a way may prevent many of the technical and financial problems when drilling the next well.

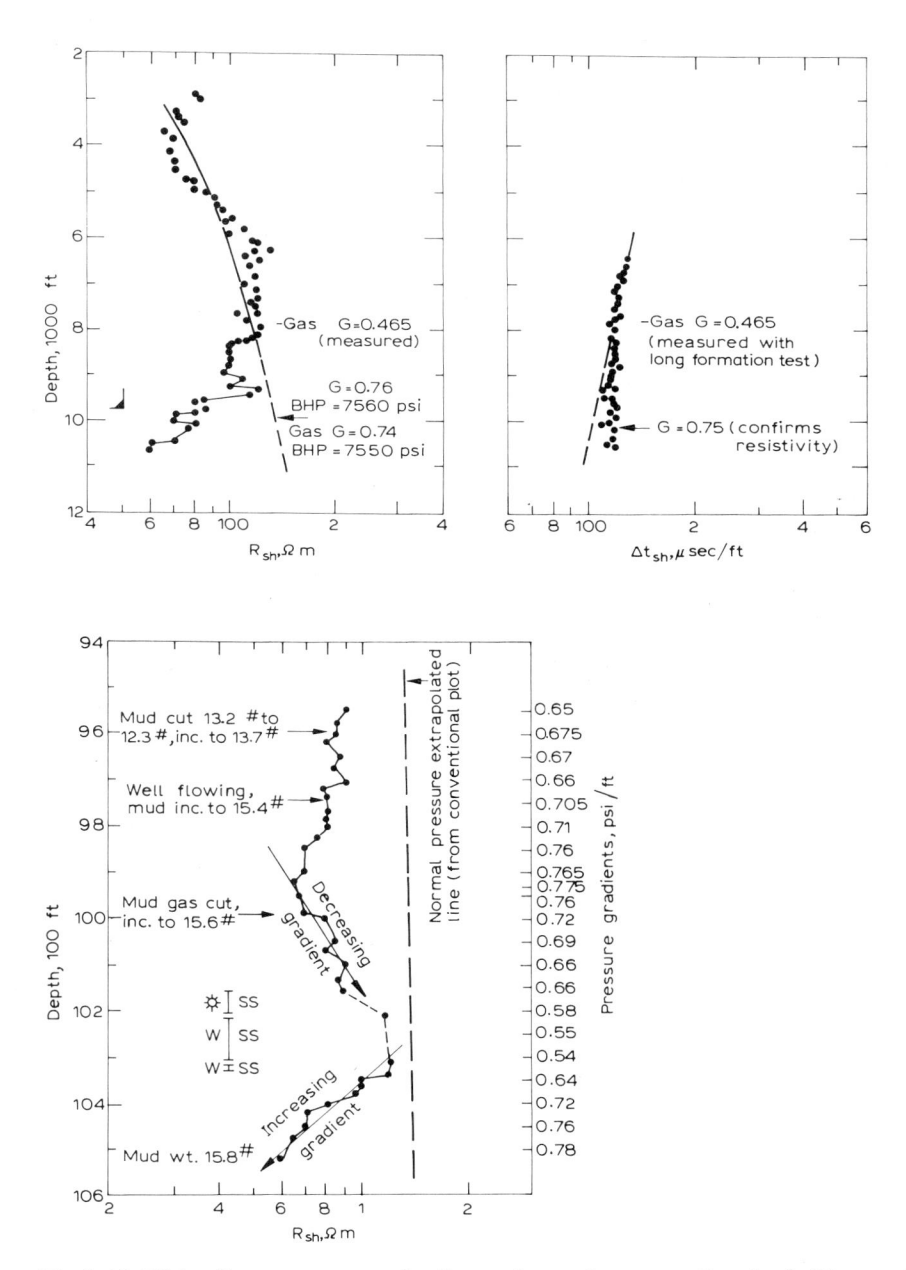

Fig.5.43. Misleading pressure evaluation using only conventional plotting methods of shale resistivity and travel time (after Fertl and Timko, 1970b). Note gradient difference of 0.2 psi/ft [0.46 kg cm^{-2} m^{-1}] between conventional and detailed plotting technique. In the figures, G denotes the pressure gradient in psi/ft, W stands for water and SS for sandstone reservoir. All mud weights are in lb/gal. (Courtesy of *Oil and Gas Journal.*)

Conclusions

Both surface and subsurface techniques are available to the oil industry for detection and evaluation of abnormally pressured environments. These include geophysical methods (Chapter 3), drilling parameters, and measurements on the drilling mud and drilling cuttings (Chapter 4), well-logging methods, and direct downhole pressure measuring devices. These techniques are numerous, with a multitude of technical publications describing theoretical considerations and practical implications.

Some of the best qualitative and quantitative abnormal-pressure detection and evaluation techniques are based on wireline-logging methods, even though these methods are "after-the-fact" techniques, i.e. the well bore has to be drilled prior to logging. Sometimes several rather short logging runs are necessary.

Since the classic paper by Hottman and Johnson (1965) discussing methods to estimate formation pressures from log-derived shale properties using resistivity and acoustic logs, numerous investigators have shown the potential of several other logging devices for qualitative and quantitative formation pressure evaluation in open hole and — under special circumstances — through casing or drill pipe. Fig.5.44 summarizes how several of these well-logging parameters respond to hydrostatic, normal pressures and overpressures. The causes of these specific log responses have been reviewed in this chapter.

The acoustic and short normal logs are the tools most valuable for qualitative and quantitative in-situ formation pressure evaluations. Such techniques have been successful in the past and certainly will be so in the

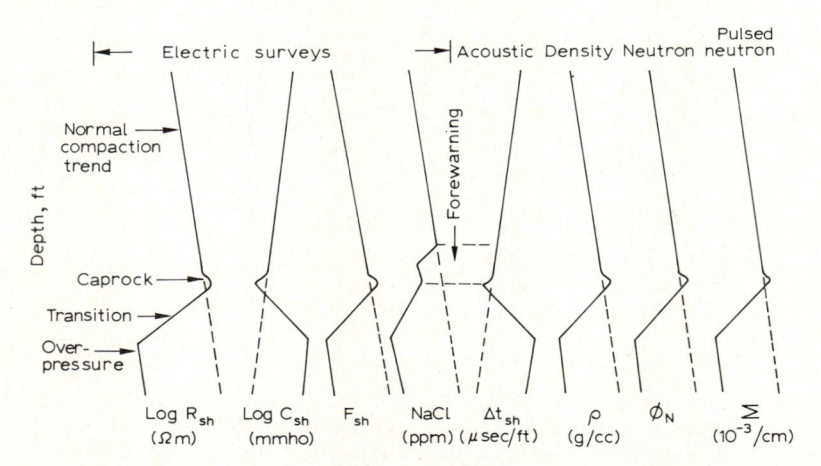

Fig.5.44. Schematic response of well-logging parameters to normal- and overpressured environments (after Fertl and Timko, 1971a). (Courtesy of the Society of Petroleum Engineers of A.I.M.E.)

future. However, application of these methods is not always simple and straightforward; more often than not complications may arise due to unrecognized logging-tool problems, severe borehole conditions, and unusual formation characteristics. One has to be aware of such possible pitfalls in order to utilize fully the potential of using well logs as one of our industry's best methods to detect and evaluate abnormal formation pressures.

References

Alger, R.P., Raymer, L.L., Hoyle, W.R. and Tixier, M.P., 1963. Formation density log applications in liquid-filled holes. *J. Pet. Technol.*, 15: 321—333.
Althaus, V.E., 1967. Electrokinetic potential in South Louisiana sediments. *Log Analyst*, 8(1): 29—34.
Athy, L.F., 1930a. Density, porosity, and compaction of sedimentary rocks. *Bull. Am. Assoc. Pet. Geol.*, 14: 1—24.
Athy, L.F., 1930b. Compaction and oil migration. *Bull. Am. Assoc. Pet. Geol.*, 14: 25—35.
Bentor, Y.K., 1969. On the evaluation of subsurface brines in Israel. *Chem. Geol.*, 4: 83—100.
Bloch, F., 1946. Nuclear induction. *Phys. Rev.*, 70: 460—472.
Breeze, A.F., 1970. *Abnormal—subnormal relationships in the Morrow sands of northwestern Oklahoma*. M.S. Thesis, Univ. of Oklahoma, Tulsa, Okla., 122 pp.
Brown, R.J.S., 1960. Nuclear magnetism logging. *Trans. AIME*, 219: 199—207.
Burst, J.F., 1969. Diagenesis of Gulf Coast clayey sediments and its possible relation to petroleum migration. *Bull. Am. Assoc. Pet. Geol.*, 53: 73—93.
Case, L.C., 1945. Analysis of water sands, Russell County, Kansas. *Oil Weekly*, p.54, October 29.
Clavier, C., 1969. Quantitative interpretation of TDT logs. *SPE 2658, 44th AIME Fall Meet.*, Denver, Colo., September.
Desai, K.P. and Moore, E.J., 1969. Equivalent NaCl solutions from ionic concentrations. *Log Analyst*, 10(2): 23—29.
Dewan, J.T. and Allaud, L.A., 1953. Environmental basin for neutron logging interpretation. *Pet. Eng.*, 25(9): B49—B54.
Dickey, P.E., 1969. Increasing concentration of subsurface brines with depth. *Chem. Geol.*, 4: 361—370.
Doll, H.G., 1948. The SP log: theoretical analysis and principles of interpretation. *Trans. AIME*, 179: 146—186.
Doll, H.G., 1949. Introduction to induction logging and application to logging of wells drilled with oil-base mud. *Trans. AIME*, 186: 148—162.
Ferran, L.H., 1973. *Evaluation of abnormally high and low pressured Morrow sands in northwestern Oklahoma using well logs and water sample data*. M.S. Thesis, Univ. of Tulsa, Tulsa, Okla., 110 pp.
Fertl, W.H. and Timko, D.J., 1970a. Occurrence and significance of abnormal pressure formations, 1. *Oil Gas J.*, 68(1): 97—108.
Fertl, W.H. and Timko, D.J., 1970b. How abnormal pressure detection techniques are applied, 2. *Oil Gas J.*, 68(2): 62—71.
Fertl, W.H. and Timko, D.J., 1970c. Association of salinity variations and geopressures in soft and hard rocks. *11th Prof. Well Log Analysts Symp.*, Los Angeles, Calif.
Fertl, W.H. and Timko, D.J., 1970d. Occurrence of cemented reservoir roof rock and geopressure cap rock and its implication in petroleum geology and geohydrology. *SPE 3085, 45th AIME Fall Meet.*, Houston, Texas, October.

Fertl, W.H. and Timko, D.J., 1971a. Parameters for identification of overpressure formations. *SPE 3223, 5th Conf. on Drilling and Rock Mechanics*, Univ. of Texas, Austin, Texas, January.

Fertl, W.H. and Timko, D.J., 1971b. Application of well logs to geopressure problems in the search, drilling, and production of hydrocarbons. *Colloq. A.R.T.E.P.*, French Petroleum Institute, Rueil, June, Paper No. 4.

Fertl, W.H. and Timko, D.J., 1971c. Salinity plots evaluate subsurface formations. *World Oil*, 173(1): 93—100.

Fertl, W.H. and Timko, D.J., 1972. How downhole temperatures, pressures affect drilling, 4. Pitfalls in overpressure prediction. *World Oil*, 175(4): 45—50.

Fertl, W.H., Pilkington, P.E. and Scott, J.B., 1973. A look at cement bond logs. *SPE 4512, 48th AIME Fall Meet.*, Las Vegas, Nev., October.

Foster, J.B. and Whalen, H.E., 1965. Estimation of formation pressure from electrical surveys, offshore Louisiana. *SPE 1200, 40th AIME Fall Meet.*, Denver, Colo., October; also *J. Pet. Technol.*, 18: 165—171, 1966.

Fowler, Jr., W.A., 1968. Pressure, hydrocarbon accumulation and salinities — Chocolate Bayou field, Brazoria County, Texas. *SPE 2226, 43rd AIME Fall Meet.*, Houston, Texas, September; also *J. Pet. Technol.*, 22: 411—432, 1970.

Gondouin, M., Tixier, M.P. and Simard, G.L., 1957. An experimental study of the influence of the chemical composition of electrolytes on the SP curve. *Trans. AIME*, 210: 58—72.

Ham, H.H., 1966. A method of estimating formation pressures from Gulf Coast well logs. *Trans. Gulf Coast Assoc. Geol. Soc.*, 16: 185—197.

Hedberg, H.D., 1936. Gravitational compaction of clays and shales. *Am. J. Sci.*, 31: 241—287.

Hedberg, W.H., 1967. *Pore-water chlorinities of subsurface shales.* Ph.D. Thesis. Univ. of Wisconsin, Madison, Wisc., 121 pp.

Hicks, W.G. and Berry, J.E., 1956. Application of continuous velocity logs to determination of fluid saturation of reservoir rocks. *Geophysics*, 21: 739—754.

Hilchie, D.W., 1969. Some aspects of pulsed neutron logging. *Log Analyst*, 10(2): 7—18.

Hill, H.J. and Anderson, A.E., 1959. Streaming potential phenomena in SP log interpretation. *Trans. AIME*, 216: 203—208.

Hottman, C.E. and Johnson, R.K., 1965. Estimation of formation pressures from log-derived shale properties. *J. Pet. Technol.*, 17: 717—723.

Hubbert, M.K. and Rubey, W.W., 1959. Role of fluid pressure in mechanics of overthrust faulting, 1. Mechanics of fluid-filled porous solids and its application to overthrust faulting. *Bull. Geol. Soc. Am.*, 70: 115—166.

Jones, P.H., 1968. Hydrodynamics of geopressures in the northern Gulf of Mexico Basin. *SPE 2207, 43rd AIME Fall Meet.*, Houston, Texas, September; also *J. Pet. Technol.*, 21: 802—810, 1969.

Jones, B.R., 1972a. The use of downhole gravity data in formation evaluation. *13th Prof. Well Log Analysts Symp.*, Tulsa, Okla.

Jones, B.R., 1972b. Downhole gravity tool spots distant porosity. *World Oil*, 175(2): 56—59.

Jones, B.R., Jageler, A. and Nettleton, L.L., 1971. Applications of borehole gravity data. *41st Annu. SEG Meet.*, Houston, Texas, November.

Korringa, J., 1962. Theory of spin pumping and relaxation in systems with a low concentration of electron spin resonance centers. *Phys. Rev.*, 127: 1143—1149.

Kunz, K.S. and Moran, J.H., 1958. Some effects of formation anisotropy on resistivity measurements in boreholes. *Geophysics*, 23: 770—794.

Levorsen, A.I., 1967. *Geology of Petroleum*. W.H. Freeman and Co., San Francisco, 2nd ed., 724 pp.

Lindtrop, N.T., 1927. Outline of water problems in New Grozny oil field, Russia. *Bull. Am. Assoc. Pet. Geol.*, 11: 1037—1043.
Loren, J.D., 1970. Relations between pore size, fluid and matrix properties and NML measurements. *Trans. AIME*, 249: 268—278.
Loren, J.D. and Robinson, J.D., 1970. Relations between pore size, fluid and matrix properties, and NML measurements. *Soc. Pet. Eng. J.*, 10: 268—278.
Louden, L.R. and Woods, E.W., 1970. Is shale remineralization a cause for formation damage? *World Oil*, 170(2): 55—58.
MacGregor, J.R., 1965. Quantitative determination of reservoir pressures from conductivity log. *Bull. Am. Assoc. Pet. Geol.*, 49: 1502—1511.
Matthews, W.R., 1971—1972. Well logs: a basic drilling tool. 7-part series in *Oil Gas J.*, November 15, 1971, to January 24, 1972.
McCulloh, T.H., 1968. Application of gravity measurements in wells to problem of reservoir evaluation. *9th Prof. Well Log Analysts Symp.*, New Orleans, La.
Moore, E.J., Szasz, S.E. and Whitney, B.F., 1966. Determining formation water resistivity from chemical analysis. *J. Pet. Technol.*, 18: 373—377.
Mounce, D.W. and Rust, W.M., 1944. Natural potentials in well logging. *Trans. AIME*, 155: 49—57.
Myers, R.L. and Van Siclen, D.C., 1964. Dynamic phenomena of sediment compaction in Matagorda County, Texas. *Trans. Gulf Coast Assoc. Geol. Soc.*, 14: 241—252.
Overton, H.L. and Timko, D.J., 1969. The salinity principles — a tectonic stress indicator in marine sands. *10th Prof. Well Log Analysts Symp.*, Houston, Texas.
Perry, E. and Hower, J., 1970. Burial diagenesis in Gulf Coast pelitic sediments. *Clays Clay Miner.*, 18: 165—177.
Pickett, G.R., 1963. Acoustic character logs and their applications in formation evaluations. *J. Pet. Technol.*, 15: 659—668.
Pirson, S.J., 1963. *Handbook of Well Log Analysis.* Prentice-Hall, Englewood Cliffs, N.J., 326 pp.
Pol'ster, L.A., Viskovskiy, Y.A., Guseva, A.N., Parnov, E.I. and Plaskova, A.G., 1967. *Physicochemical Characteristics and Hydrogeological Factors of Migration of Natural Solutions in Relation to the Study of Oil and Gas-Bearing Basins.* Izd. Nedra, Leningrad, 172 pp.
Powers, M.C., 1967. Fluid-release mechanism in compacting marine mud-rocks and their importance in oil exploration. *Bull. Am. Assoc. Pet. Geol.*, 51: 1240—1245.
Purcell, E.M., 1946. Resonance absorption by nuclear magnetic moments in a solid. *Phys. Rev.*, 69: 37—45.
Reynolds, E.B., Timko, D.J. and Zanier, A.M., 1973. Potential hazards of acoustic-log shale pressure plots. *J. Pet. Technol.*, 25: 1039—1045.
Sahay, B., 1972. Abnormal subsurface pressures: their origin and methods employed for prediction in India. *SPE 3900, 3rd Symp. on Abnormal Subsurface Pore Pressure.* Louisiana State Univ., Baton Rouge, La., May.
Schlumberger, C. and Leonardon, E.G., 1934. A new contribution to subsurface studies by means of electrical measurements in drill holes. *Trans. AIME*, 110: 237—289.
Schmidt, G.W., 1973. Interstitial water composition and geochemistry of deep Gulf Coast shales and sandstones. *Bull. Am. Assoc. Pet. Geol.*, 57: 321—337.
Seevers, D.O., 1972. A method of determining permeability by means of pulsed NMR and an application of the method to the study of a Gulf Coast supernormally pressured well. *SPE 3847, 3rd Symp. on Abnormal Subsurface Pore Pressure*, Louisiana State Univ., Baton Rouge, La., May.
Senturia, S.D. and Robinson, J.D., 1970. Nuclear spin — lattice relaxation of liquids confined in porous solids. *Soc. Pet. Eng. J.*, 10: 237—244.
Simkhaev, V.Z. and Samedov, S.S., 1969. Hydrological criteria for oil and gas-condensate exploration in northern Dagestan. *Dokl. Akad. Nauk. Az. S.S.R.*, 4: 18—31.

Smith, N.J., 1950. The case for gravity data from boreholes. *Geophysics*, 15: 605—635.
Sukharev, G.M. and Krumbol'dt, T.S., 1962. New data on the genesis of ground waters in the productive beds of Azerbaidzhan. *Dokl. Akad. Nauk S.S.S.R.*, 145(5): 23—27.
Timm, B.C. and Maricelli, J.J., 1953. Formation waters in southwest Louisiana. *Bull. Am. Assoc. Pet. Geol.*, 37: 394—409.
Timur, A., 1969. Pulsed neutron magnetic resonance studies of porosity, movable fluid, and permeability of sandstones. *Trans. AIME*, 246: 775—786.
Tittle, C.W., 1961. Theory of neutron logging, I. *Geophysics*, 26: 27—39.
Tittman, J. and Wahl, J.S., 1965. The physical foundations of formation density logging (gamma-gamma). *Geophysics*, 30: 284—294.
Tittman, J., Sherman, H., Nagel, W.A. and Alger, R.P., 1966. The sidewall epithermal neutron porosity log. *J. Pet. Technol.*, 18: 1351—1363.
Tixier, M.P., Alger, R.P. and Tanguy, D.R., 1960. New developments in induction and sonic logging. *J. Pet. Technol.*, 12: 79—88.
Von Engelhardt, W., 1961. Neuere Ergebnisse der Tonmineralienforschung. *Geol. Rundsch.*, 51: 457—477.
Von Engelhardt, W. and Gaida, K.H., 1963. Concentration changes of pore solutions during the compaction of clay sediments. *J. Sediment. Petrol.*, 33: 919—930.
Wallace, W.E., 1964. Will induction log yield pressure data? *Oil Gas J.*, 62(37): 124—126.
Wallace, W.E., 1965a. Abnormal subsurface pressures measured from conductivity or resistivity logs. *6th Prof. Well Log Analysts Symp.*, Dallas, Texas.
Wallace, W.E., 1965b. Application of electric-log measured pressures to drilling problems and a new simplified chart for well site pressure computation. *Log Analyst*, 6(6): 26—38.
Weaver, C.E. and Beck, K.C., 1971. Clay water diagenesis during burial: how clay becomes gneiss. *Geol. Soc. Am., Spec. Paper*, 134, 93 pp.
Wichmann, P.A., Youmans, A.H. and Hopkinson, E.C., 1967. Advances in neutron lifetime logging. *J. Pet. Technol.*, 19: 479—487.
Wyllie, M.R.J., 1949. A quantitative analysis of the electro-chemical component of the SP curve. *J. Pet. Technol.*, 1: 17—26.
Wyllie, M.R.J., 1951. An investigation of the electrokinetic component of the self-potential curve. *Trans. AIME*, 192: 1—15.
Wyllie, M.R.J., 1963. *The Fundamentals of Well Log Interpretation*. Academic Press, New York, N.Y., 3rd ed., 238 pp.
Wyllie, M.R.J., De Witte, A.J. and Warren, J.E., 1958a. On the streaming potential problem. *Trans. AIME*, 213: 409—417.
Wyllie, M.R.J., Gregory, A.R. and Gardner, G.H.F., 1958b. An experimental investigation of factors affecting elastic wave velocities in porous media. *Geophysics*, 23: 459—494.
Youmans, A.H., Hopkinson, E.C., Bergan, R.A. and Oshry, H.I., 1964. Neutron lifetime, a new nuclear log. *J. Pet. Technol.*, 16(3): 319—328.
Youmans, A.H., Hopkinson, E.C. and Wichmann, P.A., 1966. Neutron lifetime logging in theory and practice. *7th Prof. Well Log Analysts Symp.*, Tulsa, Okla.
Zanier, A.M., 1969. Use of the salinity principle to find formation geologic age. *Oil Gas J.*, 67(50): 84—85.
Zanier, A.M. and Timko, D.J., 1970. Prediction of Morrow sand performance and geological environment by well log salinity. *API Mid-Continent Spring Meet.*, Wichita, Kansas, Paper 851-44-B.

Chapter 6. DRILLING AND COMPLETION CONCEPTS IN OVER-PRESSURED FORMATIONS

Introduction

In present-day drilling and completion practices, both maximum well control (i.e. engineered drillable condition) and minimum cost are key factors.

The aim of good well planning (such as tentative drilling mud and casing programs) and actual drilling operations is to avoid or at least minimize the dangers of well kicks, stuck pipe, lost circulation, etc. Hence, a basic understanding of two key formation parameters is a prerequisite, namely, *formation pore pressure* and *fracture pressure* (see also Chapter 1). Both parameters are equally important in cementing, hydraulic fracturing, fluid injection, sand consolidation, etc. Today seismic, drilling, and well-logging data greatly assist in the effort of predicting both pore and fracture pressures prior to spudding a well.

Pertinent literature related to the theory of consolidation and the state of stress in compacting sediments has been recently reviewed by Rieke and Chilingarian (1974). Several methods of fracture pressure gradient prediction available to the industry are discussed in this chapter.

Regardless of the prediction method used, pore and fracture pressure gradient curves versus true vertical depth are the two basic parameters for any well drilling and completion plan. However, one has to keep in mind that problems may still occur although everything was supposedly well planned, as a result of, for instance, incorrect information and human negligence.

Basic concepts of the pore pressure—fracture pressure relationship

In petroleum engineering operations, one often seeks to deduce the natural state of stress in rocks by locally imposing a supplementary stress system and observing displacements or load variations that result from this artificial change. This concept applies to hydraulic fracturing in well stimulation, well drilling, casing-shoe tests, etc. For example, Hubbert and Willis (1957) reasoned that the general condition of subsurface stress is characterized by three unequal principal stresses and that hydraulic injection pressures must be nearly equivalent to the least-principal compressive stress. In the schematic presentation of Fig.6.1, the *breakdown pressure* of a formation is also

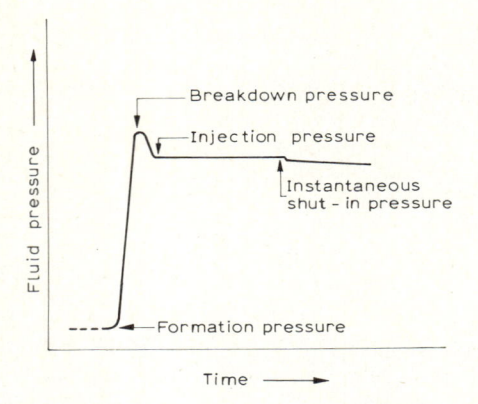

Fig.6.1. Schematic presentation of subsurface pressure relationships.

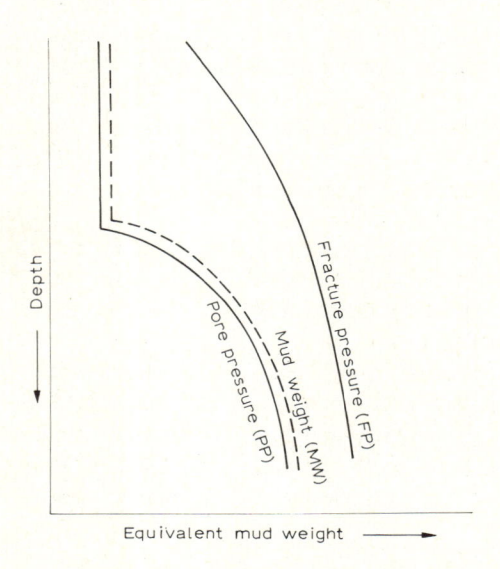

Fig.6.2. Generalized trends of the three key parameters, namely, formation pore pressure, ideal mud weight requirements, and fracture pressure in a high-pressure well.

known as the *fracture initiation pressure*, whereas the *injection pressure* corresponds to the *fracture propagation pressure*.

Generally speaking, formation fracture pressure gradients are related to formation pore pressure, lithology, age and depth of the formation, and to in-situ rock stress environments.

Fig.6.2 illustrates a generalized trend of the three key parameters, namely, pore pressure (*PP*), mud weight (*MW*), and fracture pressure (*FP*) versus depth. Trends shown are somewhat typical within many Tertiary basins (i.e.

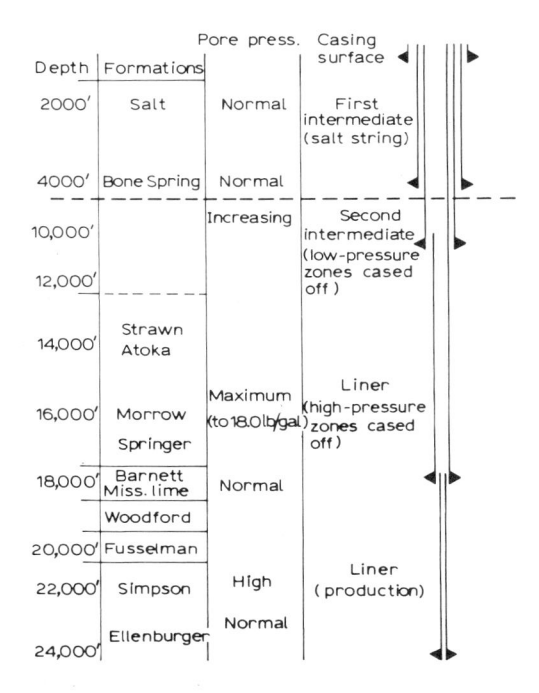

Depth	Formations	Pore press.	Casing
			surface
2000'	Salt	Normal	First intermediate (salt string)
4000'	Bone Spring	Normal	
10,000'		Increasing	Second intermediate (low-pressure zones cased off)
12,000'			
14,000'	Strawn Atoka		
16,000'	Morrow	Maximum (to 18.0 lb/gal)	Liner (high-pressure zones cased off)
	Springer		
18,000'	Barnett Miss. lime	Normal	
	Woodford		
20,000'	Fusselman		
22,000'	Simpson	High	Liner (production)
	Ellenburger	Normal	
24,000'			

Fig.6.3. Geologic section, pore pressure variations, and standard casing program in the Delaware Basin, West Texas (after Reid, 1970). (Courtesy of *Oil and Gas Journal*.)

shale/sand sequences). However, in many areas around the world, particularly in so-called hard-rock formations as encountered in West Texas, Oklahoma, and elsewhere, the trends of observed pore and fracture pressure gradients are quite different (Fig.6.3). This will be illustrated in this chapter in more detail later on.

Fig.6.2 also indicates that in both normal and overpressured zones, the recommended mud weight should exceed formation pressures by about 0.2—0.4 ppg [0.024—0.048 kg/dm^3]. This practice allows the drill string to be pulled without swabbing, results in reasonable drilling mud costs and good penetration rates, reduces possibility of differential pipe sticking, etc. It is obvious that such close tolerances between pore pressure and mud weight require accurate calculations.

Casing point selection

Casing in the borehole protects both penetrated formations and the wellbore from contamination and drilling hazards. For simplicity, we assume that a perfectly balanced condition prevails, i.e. the pore pressure equals the hydrostatic pressure of the mud column. Fig.6.4 shows the casing point (*CP*) at depth D_1. Under the given conditions in this well, drilling can be continued until depth D_2 is reached. *Region A* corresponds to the depth interval in which the well can be circulated and, if necessary, can be killed (*FP* > *PP*). Drilling operations in *Region B* (i.e. below depth D_2) would result in lost circulation somewhere below the casing shoe (*CP*), and well kicks could not be dealt with successfully (*PP* > *FP*).

An indispensable, basic requirement is a good cement job at the casing shoe. Otherwise, channeling behind the casing uphole will result in much lower fracture pressure gradients than one would otherwise anticipate. Furthermore, there is always the possibility of the presence of weak zones somewhere along the open hole, including coal (lignite) beds, lime streaks (cap rocks, shell beds), small faults or joint systems intersecting the wellbore, etc.

A few simple calculations show the practical importance of the basic concepts shown in Fig.6.4. For example, a surface casing of $13\frac{3}{8}$ inches has been set at 5000 ft [1524 m] where the fracture pressure is estimated to be 15.0 ppg (equivalent mud weight) [1.80 kg/dm³]. Hence, the maximum pressure at the casing shoe without loss of borehole fluid into the formation is:

15.0·0.052·5000 = 3900 psi

To carry out such a test at the casing shoe, the maximum surface pressure

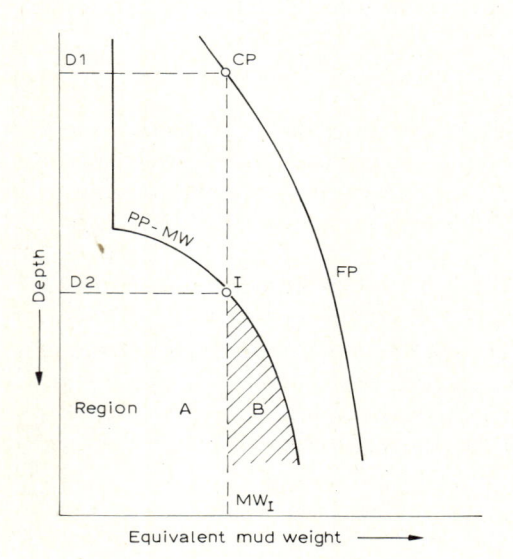

Fig.6.4. Casing point selection under perfectly balanced conditions. Note "safe" drilling and well killing conditions in Region A, whereas in Region B well kicks cannot be handled and circulation is lost. *CP* = casing point; *FP* = fracture pressure; *PP* = pore pressure; *MW* = mud weight.

PORE PRESSURE—FRACTURE PRESSURE RELATIONSHIP

which can be tolerated corresponds to the pressure at the casing shoe minus the hydrostatic mud column head ($MW = 10$ lb/gal):

$3900 - (10.0 \cdot 0.052 \cdot 5000) = 3900 - 2600 = 1300$ psi

Fig.6.5 gives a quick graphical solution that can be used with any mud weight and any applied surface pressure. This chart also indicates that even small surface pressures have a major effect on the effective mud weight with shallow-surface casing. For example, at a depth of 5000 ft if the surface pressure is increased from 2600 to 3900 psi, the effective mud weight increase equals 5 ppg.

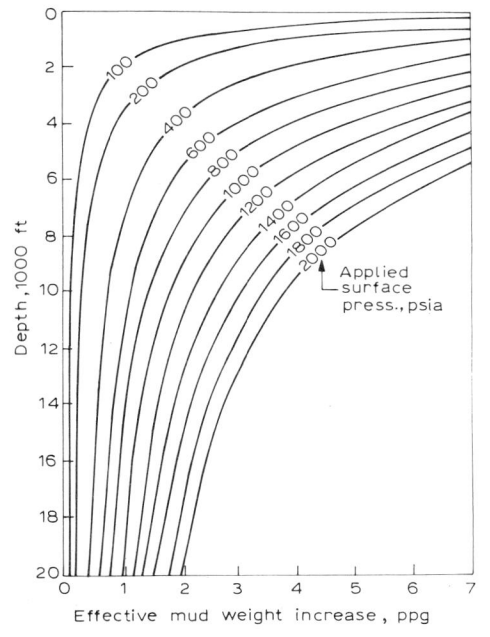

Fig.6.5. Effect of surface pressure (in pounds per square inch absolute) on the effective mud weight increase as a function of depth.

Additional but more sophisticated kick-tolerance computations with and without gas influx into the borehole, etc., have been discussed in the excellent literature on this subject (see references listed in Chapter 4). Generally speaking, however, boreholes tend to "kick" when the actual mud weight in the well is lower than the weight needed to control the excessive formation pore pressure of any given permeable, uncased interval in the section penetrated by the drill bit.

Drill pipe sticking

Actual pipe-sticking forces depend on (1) how long the pipe remains motionless against the formation, (2) the slickness of filter cake, (3) permea-

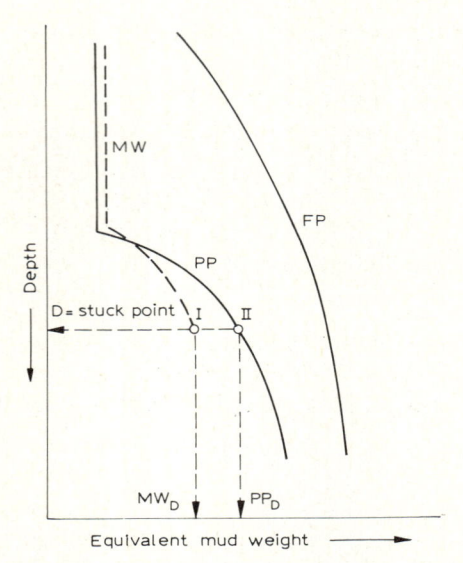

Fig.6.6. Depth versus equivalent mud weight: schematic diagram showing reasons for drill pipe sticking due to an underbalanced mud system.

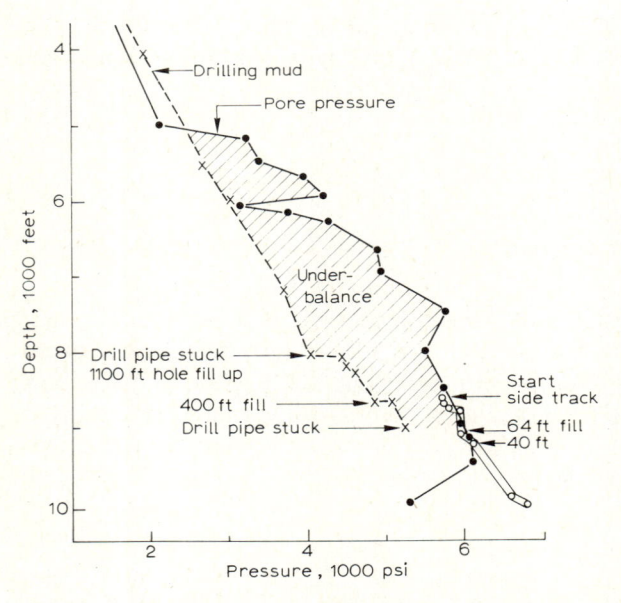

Fig.6.7. Relationship of depth to pore pressure and drilling mud pressure resulting in stuck pipe, hole fill-up, and sidetracked hole due to underbalance in a U.S. Mid-Continent well, located in Oklahoma.

bility and (4) thickness of zone, and, most important, (5) the differential pressure between drilling mud and pore pressure.

Differential pipe sticking may occur as a result of large underbalance as illustrated schematically in Fig.6.6. In the overpressured section, the equivalent mud weight exceeds the pore pressure down to a depth D, where the drill pipe will get stuck. While the drill bit approaches this stuck-point, the borehole starts caving in and the torque increases simultaneously. Fig.6.7 shows a typical field case in the U.S. Mid-Continent area. Note that several times drilling problems occurred: the pipe stuck and the hole filled up. Consequently, a sidetracked borehole became necessary.

Differential pipe sticking may be due to such underbalance (*negative* differential sticking) but may be also caused by a large overbalance (i.e. mud weight greatly exceeds pore pressure — *positive* differential sticking) or may occur during a kick.

Lost circulation

Lost circulation may occur in several ways, including loss into vuggy and/or fissured formations, into formations due to excessively weighted drilling muds, and into weak zones (lignites, fault and joint systems). For our present discussion, the prevention of mud losses (lost circulation) depends on the practical utilization of proper estimates of fracture pressure gradients and pore pressures.

Fig.6.8 illustrates a well being drilled with too heavy a mud weight (i.e. large overbalance). The shaded area shows the region of unnecessary and

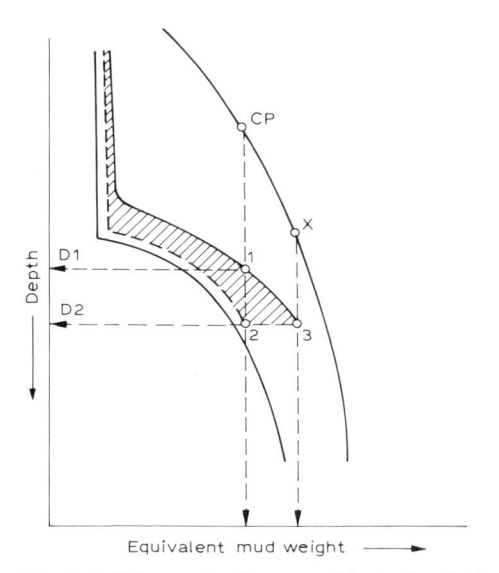

Fig.6.8. Schematic diagram of lost circulation due to an overbalanced mud system.

dangerous overbalance. With this heavy mud in the hole, drilling has to be stopped at depth D_1 (point 1), whereas a properly engineered mud weight would have allowed drilling down to depth D_2 (point 2) without an additional protection pipe. Excessive mud weight at depth D_2 (point 3) would cause pipe sticking and lost circulation.

Frequently, circulation of the drilling mud can be reestablished by cutting back the mud weight. However, a large overbalance often causes the protection pipe (casing) to be set far too high to cure lost circulation.

Prediction methods for fracture pressure gradients

In this section, a review is given of the various methods developed and employed by the oil industry to calculate hydraulic fracture pressure gradients. The latter are equally important in drilling, well completion and stimulation operations, secondary recovery, etc.

The Hubbert and Willis approach

In a theoretical and experimental examination of the mechanics of hydraulic rock fracturing, Hubbert and Willis (1957) reasoned that the general condition of subsurface stress is characterized by three unequal principal stresses and that hydraulic injection pressures must be nearly equal to the least-principal compressive stress. The authors also suggested that "in geologic regions where normal faulting is taking place, the greatest stress should be approximately vertical and equal to the effective pressure of the overburden, while the least stress should be horizontal and most probably between one-half and one-third the effective pressure of the overburden."

The overburden pressure, P_o, equals the sum of formation pressure, P_f, and the vertical matrix stress, σ (see also Chapter 1, eq. 1.4):

$$P_o = P_f + \sigma \tag{6.1}$$

or:

$$\sigma = P_o - P_f \tag{6.2}$$

The injection or fracture propagation pressure, FP, has to overcome the pore pressure, P_f, and the horizontal rock matrix stress, σ_H:

$$FP = P_f + \sigma_H \tag{6.3}$$

where, according to Hubbert and Willis (1957):

$$\sigma_H = (\tfrac{1}{3} \text{ to } \tfrac{1}{2})\sigma = (\tfrac{1}{3} \text{ to } \tfrac{1}{2})(P_o - P_f) \tag{6.4}$$

Substitution of eq. 6.4 into eq. 6.3 and dividing by the depth D, the fracture pressure gradient is expressed by:

$$FP/D = P_f/D + (\tfrac{1}{3} \text{ to } \tfrac{1}{2})(P_o - P_f)/D \tag{6.5}$$

Hence, under normal hydrostatic pressure conditions of 0.46 psi/ft [0.106 kg cm^{-2} m^{-1}] and an overburden gradient (P_o/D) of 1.0 psi/ft [0.231 kg cm^{-2} m^{-1}], such as for the U.S. Gulf Coast area, the limits of the fracture pressure gradients (FPG) can be computed as follows:

$$\text{FPG}_{\text{min}} = P_o/3D + 2P_f/3D = 0.64 \text{ psi/ft} \quad (\text{equivalent fluid weight} = 12.3 \text{ ppg}) \tag{6.6}$$

$$\text{FPG}_{\text{max}} = 0.5(P_o/D + P_f/D) = 0.73 \text{ psi/ft} \quad (\text{equivalent fluid weight} = 14.1 \text{ ppg}) \tag{6.7}$$

From the above, it can be seen that fracture pressure gradients are dependent on the overburden and pore pressure gradients and the rock frame stress. It should also be recalled that the relationship of $\sigma_H = \sigma/3$ corresponds to a Poisson's ratio of 0.25.

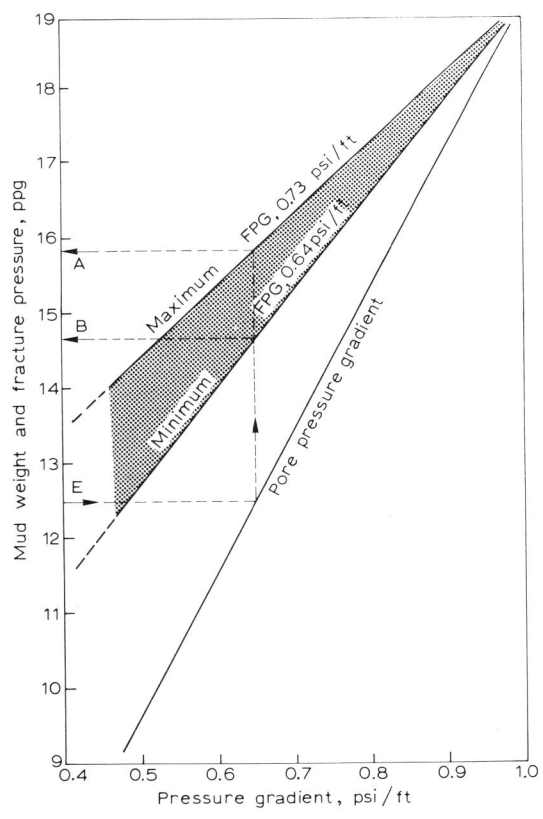

Fig.6.9. Graphical presentation of maximum and minimum fracture pressure gradients as proposed by Hubbert and Willis (1957).

As observed from the above mathematical relationships, for all normally pressured formations, the fracture pressure gradients remain *constant* with increasing depth. Fig.6.9 allows a quick graphical solution for U.S. Gulf Coast conditions:

(1) Determine mud weight requirements to balance formation pressures (using well logs, offset well data, etc.).

(2) Enter this mud weight at the ordinate (for example, point E) and intersect horizontally with the pore-pressure gradient line.

(3) Intersect vertically with fracture-pressure gradient lines.

(4) Read corresponding fracture pressure (in ppg equivalent) on the ordinate of the graph.

As noted in Fig.6.9, the difference between maximum and minimum fracture pressure gradients decreases with increasing pore pressure gradient. Hence, surge pressures (which occur while going in and out of the borehole, starting mud pumps, etc.) become more and more important with higher pore pressure gradients.

Field experience indicates that in many areas a constant fracture pressure gradient with increasing depth will not be realistic. For example, in the U.S. Gulf Coast area, fracture-pressure gradient predictions from eq. 6.5 are usually on the low side compared with actual field data.

Empirical data correlations

Least-square curve fitting techniques and empirical field data have been used in many drilling areas around the world which successfully correlate the fracture pressure gradient with pore pressure gradient and formation depth data.

Method of Matthews and Kelly

In 1967, Matthews and Kelly published a fracture-pressure gradient relationship which differs from the Hubbert-Willis model in that a *variable matrix stress coefficient*, K_i (i.e. variable horizontal to vertical stress ratio) was introduced:

$$FP/D = P_f/D + K_i \sigma/D \tag{6.8}$$

where FP = fracture pressure at point of interest in psi; P_f = pore fluid pressure at point of interest in psi; D = depth at point of interest in feet; σ = matrix stress at point of interest in psi, with $\sigma = P_o - P_f$ (see eq. 6.2); and K_i = matrix stress coefficient for the depth at which the value of σ would be the *normal* matrix stress, dimensionless.

Values for K_i, the most significant variable, were obtained by substituting empirical field data of fracture initiation (breakdown) pressures into eq. 6.8 and solving for K_i. The variable matrix stress ratio K_i is shown in Fig.6.10 as a function of depth for both the South Texas Gulf Coast and offshore

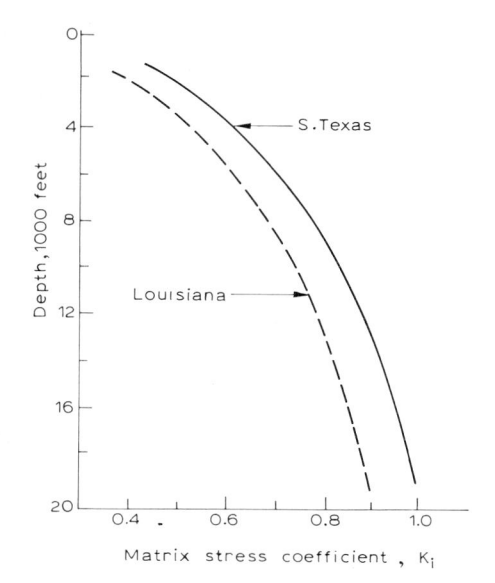

Fig.6.10. Concept of the variable matrix stress coefficient, K_i, in the U.S. Gulf Coast area (after Matthews and Kelly, 1967). (Courtesy of *Oil and Gas Journal*.)

Louisiana. Note the significantly higher stress coefficient for the South Texas sands, which are generally much shalier than sands in offshore Louisiana.

In the normally pressured formations, where P_o/D = 1.0 psi/ft [0.231 kg cm^{-2} m^{-1}] and P_f/D = 0.465 psi/ft [0.106 kg cm^{-2} m^{-1}], the fracture-pressure gradient expression simplifies to:

$$FP/D = 0.465 + 0.535 K_i \quad (6.9)$$

In eq. 6.9, the value for K_i is now the value of the matrix stress coefficient at the depth of interest.

The stepwise procedure of calculating the fracture pressure gradient by the Matthews and Kelly method includes:

(1) Determine the pore pressure (from well logs, offset well data, etc.).

(2) Calculate the effective stress, σ, where (assuming a constant overburden load of 1.0 psi/ft) the value of σ equals $(1.0 - P_f/D)D$.

(3) Determine depth D_i, in feet, for which the σ-value would be the *normal* value [$\sigma = (1.0 - 0.465)D_i$]:

$$D_i = \sigma/0.535 \quad (6.10)$$

(4) Calculate the fracture pressure gradient from eq. 6.8.

For convenient, quick interpretations at the well site, a graphical solution is given in Fig.6.11. Enter the graph at the depth of interest and horizontally intersect with the mud weight equivalent of the pore pressure (as determined

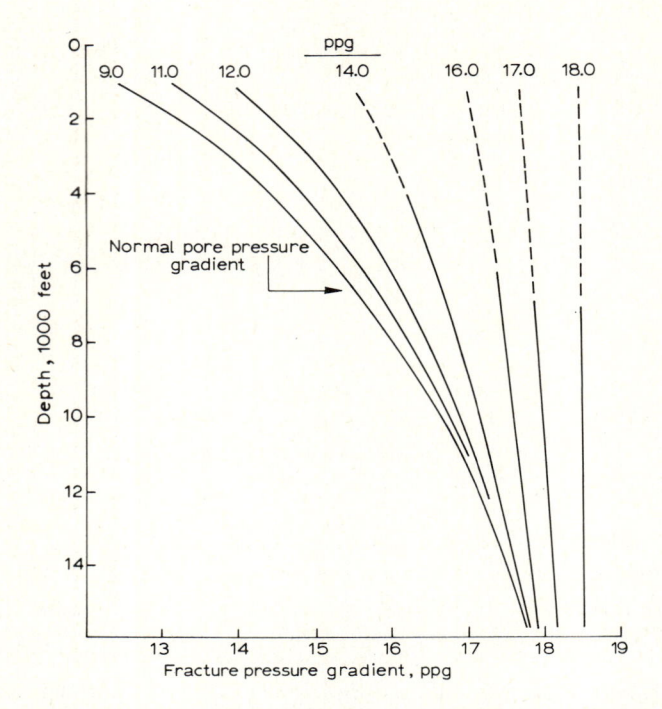

Fig.6.11. Graphical fracture gradient prediction for the U.S. Gulf Coast area using the Matthews-Kelly (1967) concept. Figures on top represent mud weight equivalents of the pore pressure in ppg.

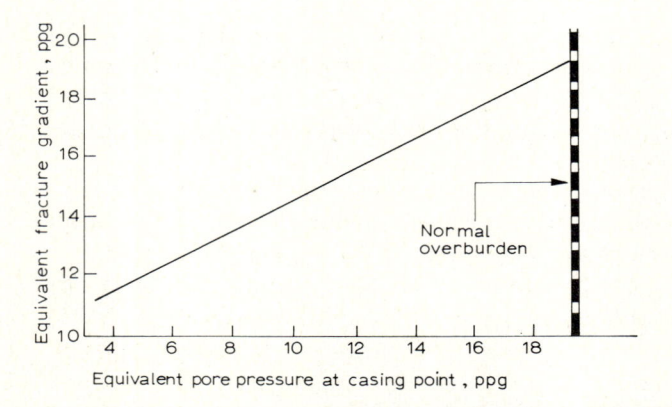

Fig.6.12. Fracture pressure gradient estimates in the Delaware Basin, West Texas (after Mead and Reid, 1969). Normal overburden is the equivalent pore pressure of 19.24 ppg, which corresponds to a constant overburden gradient of 1.0 psi/ft. (Courtesy of *Oil and Gas Journal.*)

above in step 1); then go vertically to the abscissa to read the fracture pressure gradient value in ppg equivalent.

Application of this fracture-pressure gradient prediction method in areas other than the U.S. Gulf Coast area requires the knowledge of the local K_i variation with depth. Obviously, K_i can be easily determined from eq. 6.8 if all other input parameters are known. For example, based on the Matthews-Kelly concept, the fracture pressure gradient in the Delaware Basin in West Texas has been found to vary directly with the pore pressure (Mead and Reid, 1969):

$$FP = 9.6 + K \cdot P_f \tag{6.11}$$

where K is a constant value of 0.5. Fig.6.12 illustrates this fracture pressure gradient and pore pressure relationship, expressed in ppg equivalent.

Method of Eaton

Assuming earth formations to be elastic, Eaton (1969) related the horizontal stress, σ_H, and the vertical stress, σ_V, by Poisson's ratio, ν, as expressed in Hooke's law:

$$\sigma_H = \frac{\nu \sigma_V}{(1-\nu)} \tag{6.12}$$

Eaton then extended the concept proposed earlier by Matthews and Kelly (1967) by introducing Poisson's ratio into the expression for the fracture pressure gradient:

$$FPG = P_f/D + \left(\frac{\nu}{1-\nu}\right)\sigma/D \tag{6.13}$$

where σ = matrix stress which equals $(P_o - P_f)$.

Basically, the author assumed all of the independent variables to be functions of depth, i.e. overburden load and Poisson's ratio of rocks. Pressure gradients may vary with depth.

Generally speaking, the overburden pressure at a given depth D equals the cumulative weight of the rocks above that depth:

$$P_o = 0.4335 \int_0^D \rho_b(h)dh \tag{6.14}$$

where $\rho_b(h)$ = formation bulk density as a function of depth h, and 0.4335 = constant for converting g/cm^3 to psi/ft.

Hence, local or regional overburden pressures can be determined from eq. 6.14 by using a simple plot of shale bulk density versus depth (derived from density logs or other shale density measurements) and converting them into an overburden load gradient curve. Figs.6.13 and 6.14 illustrate this procedure for typical U.S. Gulf Coast conditions as derived by Eaton (1969). Additional overburden load gradients for other areas have been previously shown in Fig.1.1.

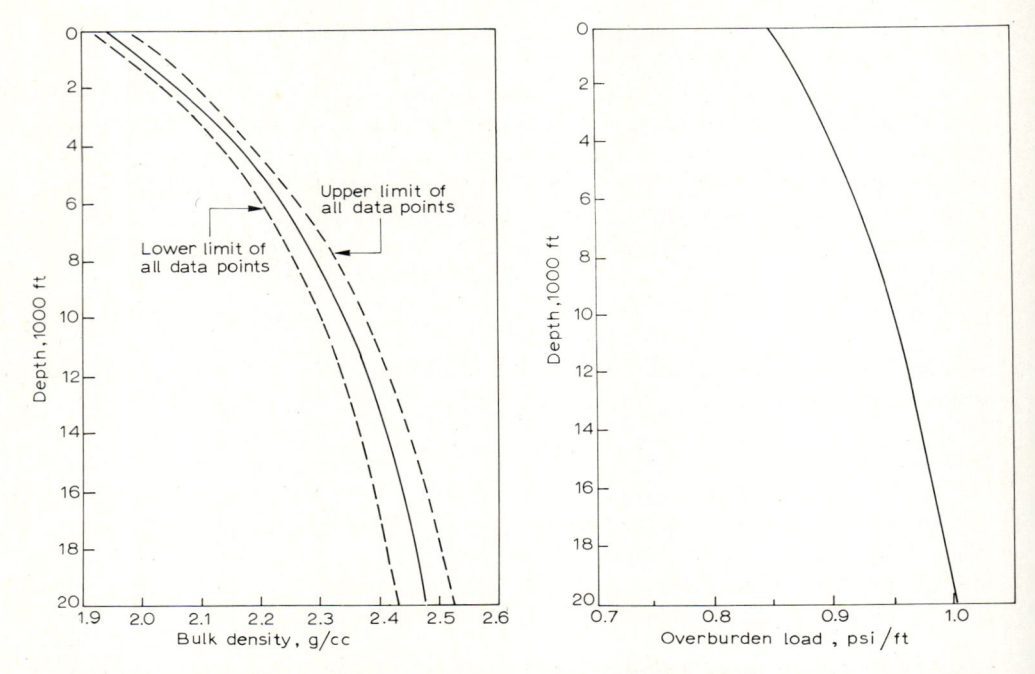

Fig.6.13. Bulk density variation with depth in U.S. Gulf Coast shales (after Eaton, 1969). (Courtesy of *Journal of Petroleum Technology*.)

Fig.6.14. Relationship between overburden load gradient and depth for U.S. Gulf Coast shales (see Fig.6.13 — after Eaton, 1969). (Courtesy of *Journal of Petroleum Technology*.)

Besides overburden-load gradient and fracture-pressure gradient data at several depths (such as actual fracturing data, cement squeeze, and lost circulation values), formation pore pressures at corresponding depths are required.

Using these data and rewriting eq. 6.13:

$$\frac{\nu}{1-\nu} = \frac{\text{FPG} - P_f/D}{(\sigma/D)} \tag{6.15}$$

Poisson's ratios in a given area can be back-calculated and plotted versus depth. Fig.6.15 shows a constant Poisson's ratio with depth for West Texas (where $\nu = 0.25$, $P_o/D = 1.0$), based on hydraulic fracturing treatment data (Crittendon, 1959), and increasing Poisson's ratios with depth for the U.S. Gulf Coast area as developed by Eaton (1969). Similar trends can be established for other areas of interest. Simplified local or regional fracture-pressure gradient prediction charts then can be developed which assist in any engineered well planning. Fig.6.16 shows such a chart for U.S. Gulf Coast area.

Eaton's fracture-pressure gradient prediction method is one of the best

PREDICTION METHODS FOR FRACTURE PRESSURE GRADIENTS 245

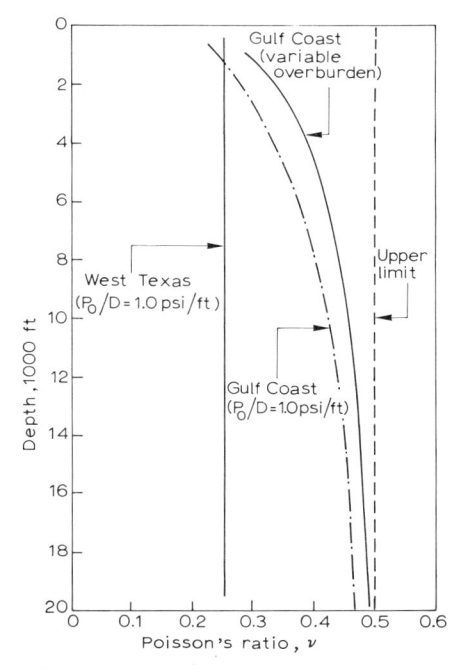

Fig.6.15. Magnitude and variation of Poisson's ratio with depth in West Texas and the U.S. Gulf Coast area (after Eaton, 1969). (Courtesy of *Journal of Petroleum Technology*.)

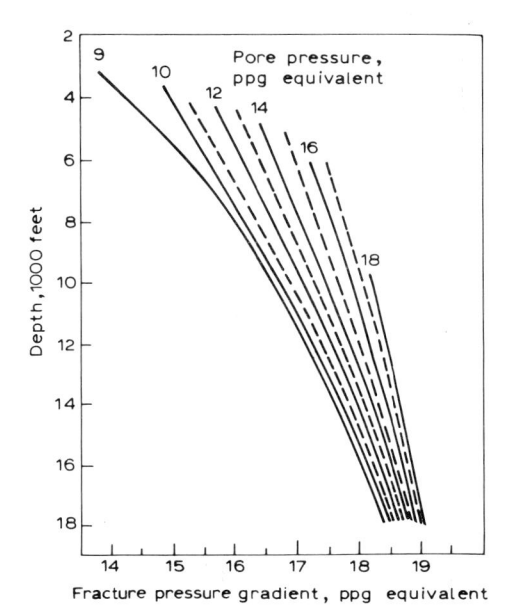

Fig.6.16. Graphical fracture-pressure gradient estimates in the U.S. Gulf Coast area based on Eaton's approach (1969).

TABLE 6.I

Formation fracture pressure gradients (psi/ft) of Delaware Basin fields (after Gibbs, 1966)

Formation	Coyanosa	Gomez	Waha (west)	Mendel	Hamon
Alluvium	0.80*	0.80*	0.80*	0.80*	0.80*
Rustler	0.90	0.90*	0.90*	0.88	0.85
Salt	0.70*	0.70*	0.70*	0.70*	0.70*
Anhydrite	0.66*	0.66*	0.66*	0.66*	0.66
Tansil	np	0.78	np	np	np
Yates	np	0.70	np	np	np
Capitan Reef	np	0.55	np	np	np
Delaware sand	0.58	0.59	0.61	0.59	0.57
Cherry Canyon	0.56	0.57	0.60*	0.55	0.56
Brushy Canyon	0.58	0.59	0.60	0.59*	0.58*
Bone Spring	0.63	0.62	0.63	0.62	0.64
Wolfcamp, top	0.72	0.70	0.68	0.70	0.66
Wolfcamp, bottom	0.88	0.90	0.86*	0.90	0.80
Pennsylvanian		0.90	0.88*	0.90*	0.90*
Mississippian	0.65*	0.66	0.65*	0.65*	0.65*
Devonian	0.63	0.63	0.63	0.63	0.63*
Silurian	0.65*	0.65	0.65*	0.65*	0.65*
Simpson	0.65*	0.65	0.65	0.73	0.65*
Ellenburger	0.63	0.64	0.63	0.64	0.64

*Gradients these formations have consistently supported (not fracturing gradients).
np = no pressure gradient data available.

TABLE 6.II

Several formation fracture gradients in Wyoming (after Rehm, 1969)

Basin	Formation	Formation fracture gradient (psi/ft)
Powder River	Sussex	0.75
	Shannon	0.64
	Muddy	0.75
	Dakota	0.81
	Minnelusa	0.75
Wind River	Fort Union	0.76
	Cody	0.61
	Fronteer	0.76
	Muddy	0.84
	Dakota	0.90
	Phosphoria	0.78
	Tensleep	0.83
Green River (south)	Mesa Verde	0.63
	Fronteer	0.63
	Tensleep	0.85
	Madison	0.75
	Caprian	0.75

techniques available to the oil industry today and is used worldwide both on- and offshore.

Table 6.I shows fracture pressure gradients in several Delaware Basin fields in West Texas, whereas Table 6.II lists such data for formations in the Powder River, Wind River, and Green River Basins, all located in Wyoming, U.S.A. All these data fall within the range of the theoretical predictions.

Field data study by Taylor and Smith

Taylor and Smith (1970) studied injection tests under controlled conditions while abandoning wells to obtain new information and knowledge of how to improve fracture-pressure gradient estimates in offshore drilling (Table 6.III) (see pages 248, 249). Only data from wells without stress disturbances, such as salt diapirism or major faults, were considered. Wells are located offshore of Louisiana near the Mississippi River delta.

Taylor and Smith (1970) also stated that "acceptable drilling operations with a low risk of lost circulations due to formation fracturing cannot be designed with only theoretical fracture pressure predictions." Fracture-pressure gradient predictions from theoretical concepts may become speculative due to the required assumptions for input parameters. However, the theoretical concepts may provide the necessary guidance in properly engineered field operations.

In their study, these authors only considered fracturing in normally pressured formations. Fig.6.17 shows measured fracture propagation pres-

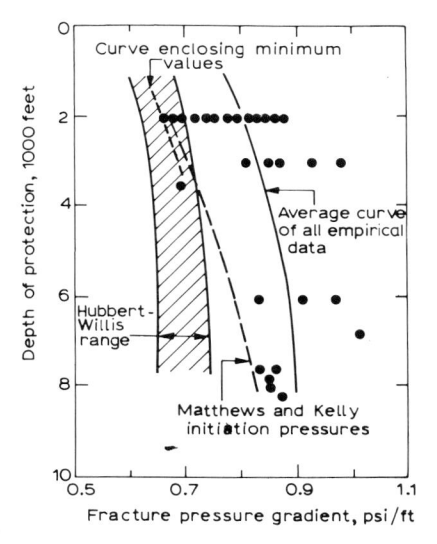

Fig.6.17. Comparison of predicted and empirical fracture pressure gradients in the U.S. Gulf Coast area (after Taylor and Smith, 1970). (Courtesy of *Oil and Gas Journal.*)

TABLE 6.III

Fracture propagation gradient data (after Taylor and Smith, 1970)

Depth of casing protection (ft)	Surface injection pressure (psi)	Measurement datum (ft, a.s.l.)	Water depth (ft)	Mud weight in well (ppg)	Propagation pressure at depth of protection (psi)	Propagation pressure gradient* at depth of protection (psi/ft)
1,999	250	92	370	12.5	1,549	0.7750
2,000	300	92	375	12.5	1,600	0.8000
2,000	550	93	325	11.6	1,756	0.8782
2,002	90	92	380	12.2	1,360	0.6793
2,002	600	93	263	10.2	1,661	0.8301
2,004	340	88	185	10.5	1,434	0.7156
2,006	250	93	248	10.2	1,313	0.6550
2,009	400	88	230	13.1	1,768	0.8803
2,012	600	100	180	11.9	1,845	0.9170
2,014	350	93	283	10.0	1,397	0.6937
2,018	400	93	340	10.5	1,501	0.7442
2,018	400	92	250	12.0	1,659	0.8222
2,020	380	88	140	13.0	1,745	0.8641
2,024	600	93	360	12.0	1,862	0.9204
2,025	400	93	350	11.5	1,610	0.7955
2,029	250	93	400	12.0	1,516	0.7472
2,038	680	88	340	10.4	1,782	0.8744
2,047	500	88	185	12.9	1,873	0.9150
2,050	550	93	310	11.0	1,722	0.8402
2,055	350	93	310	11.9	1,621	0.7891
2,923	600	101	155	11.3	2,317	0.7928
3,000	450	90	268	12.4	2,384	0.7948
3,002	600	93	310	12.8	2,598	0.8654
3,003	200	93	390	13.0	2,230	0.7426
3,014	450	93	225	13.4	2,550	0.8461
3,015	550	90	240	12.0	2,431	0.8064
3,024	780	90	295	13.0	2,824	0.9339
3,024	950	93	183	12.9	2,978	0.9849

3,500	620	86	240	10.0	2,440	0.6971
5,914	875	93	160	15.8	5,733	0.9695
5,998	650	96	155	15.4	5,453	0.9091
6,000	1,835	86	240	10.0	4,955	0.8258
6,790	1,050	88	140	16.4	6,840	1.007
7,600	2,350	86	240	10.0	6,302	0.8292
7,700	2,625	86	240	10.0	6,629	0.8609
7,800	2,580	86	240	10.0	6,636	0.8507
8,100	2,680	86	240	10.0	6,892	0.8508
8,200	2,875	86	240	10.0	7,139	0.8706

*Calculated from measurement datum.

a.s.l. = above sea level.

sure gradients versus the depth of casing protection. A regression-type curve, fitting all data, and the curve enclosing the minimum fracture pressure gradients are presented.

Variation of measured values at a given depth is apparent. Similar observations have been made by Anderson et al. (1973) as will be shown later in this chapter. Also note in Fig.6.17 the similarity of the fracture initiation pressure gradient given by Matthews and Kelly (1967) with the minimum fracture propagation values (after Taylor and Smith, 1970) at shallow depths in normally pressured formations. However, a much lower range of fracture propagation pressure gradients was theoretically predicted by Hubbert and Willis (1957).

Method of MacPherson and Berry

MacPherson and Berry (1972) investigated the possibility of predicting fracture pressure gradients from well logs, such as correlations of the wellbore elastic moduli (determined in situ from both the sonic and density log) with fracture pressure gradient data in ten oil fields, located offshore of West Louisiana. Data presented in this study allow the following conclusions:

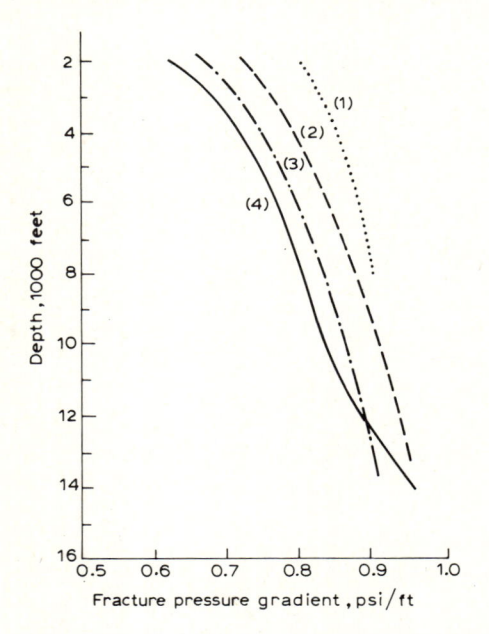

Fig.6.18. Comparison of several fracture-pressure gradient predictions in U.S. Gulf Coast area: *1* = East Louisiana (after Taylor and Smith, 1970); *2* = South Texas (after Matthews and Kelly, 1967); *3* = Louisiana Gulf Coast (after Matthews and Kelly, 1967); *4* = average empirical trend for ten offshore fields in West Louisiana (after MacPherson and Berry, 1972).

PREDICTION METHODS FOR FRACTURE PRESSURE GRADIENTS

(1) Fracture pressure gradients versus depth are different in all ten offshore fields but appear to be related to the sand/shale ratio, depth of overpressures, geologic age, and the location in the basin. An average trend of all ten fields studied is plotted in Fig.6.18. For comparison purposes, the results of other investigations are also presented.

(2) Fracture pressure gradients in shales appear to be significantly higher than in sands.

(3) At equivalent depth, the heavier the mud weight used, the higher the fracture pressure gradient. This observation is in accordance with the previous findings of Taylor and Smith (1970).

(4) Breakdowns in overpressured intervals are only slightly different from normal pressure trends in the direction of higher fracture pressure gradient with increasing overpressures.

(5) Most important, a quantitative relationship exists between the ratio of sand elastic modulus to overburden pressure and the fracture pressure (Fig.6.19). This should generally provide a prediction within 700 psi [49.3 kg/cm^2] of true breakdown (fracture initiation) pressure.

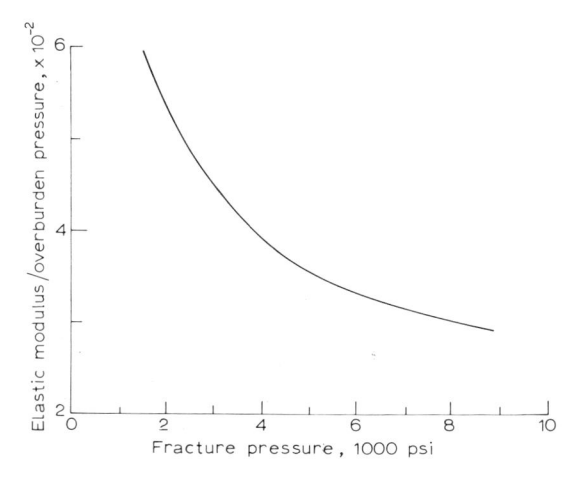

Fig.6.19. Quantitative correlation of the ratio of sand elastic modulus to overburden pressure with fracture pressure in offshore Louisiana (after MacPherson and Berry, 1972).

Method of Christman

Reliable fracture-pressure gradient prediction is fundamental in present-day and particularly future deepwater drilling operations. Drilling in 500 ft [152.4 m] of water has become a daily routine. Currently a well can be spudded in over 2000 ft [610 m] of water, and oil companies have started acquiring concessions in water depths of 4000 ft [1220 m] and deeper.

Recently, Christman (1973) investigated fracture-pressure gradient concepts that would assess the impact of water depth on deepwater drilling operations in the Santa Barbara Channel, offshore California. Having developed a *typical* onshore area gradient based on Eaton's (1969) concepts, offshore fracture pressure gradients were then calculated essentially in the same manner.

The overburden load gradient at depth D corresponds to the cumulative weight of all formations above depth D:

$$P_o/D = (0.4335/D) \int_0^D \rho_b(h)dh \tag{6.16}$$

where the constant 0.4335 converts g/cm³ to psi/ft. For practical use, however, a simple mathematical average of the bulk density, ρ_b, in g/cm³ is adequate:

$$P_o/D = 0.4335\rho_b \tag{6.17}$$

Since in deepwater drilling the uppermost interval is water, which is considerably less dense than rock, this has to be taken into account by modifying eq. 6.17:

$$P_o/D = (0.4335/D)(\rho_b D_1 + \rho_w D_w) \tag{6.18}$$

where D = depth below datum (ocean surface) in feet, D_1 = depth below mud line in feet, and D_w = depth of water in feet.

Since seawater has a density of 1.02 g/cm³, eq. 6.18 simplifies to:

$$P_o/D = (0.44 D_w + \rho_b D_1)/D \tag{6.19}$$

Fig.6.20 illustrates the significant decrease in the fracture pressure gradient with increasing water depth, and Table 6.IV lists field data from the Santa

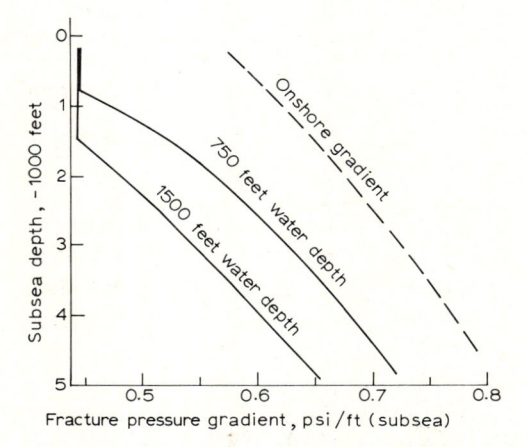

Fig.6.20. Offshore fracture pressure gradients as a function of water depth, Santa Barbara Channel, offshore California (after Christman, 1973). Subsea indicates depth below ocean floor. (Courtesy of *Journal of Petroleum Technology*.)

Barbara Channel at depths up to 5000 ft [1524 m]. Fracturing is assumed to occur at the depth of highest stress ratio and lowest density. Based on the data in Table 6.IV, a nearly perfect correlation results when plotting these stress ratios versus corresponding rock densities.

TABLE 6.IV

Stress ratio versus rock density correlation, Santa Barbara Channel (after Christman, 1973)

Data depth,* subsea (ft)	Water depth (ft)	Overburden gradient (psi/ft)	Bulk density (g/cm^3)	Applied pressure gradient (psi/ft)	Equivalent stress ratio**	Results
414	141	0.73	2.05	0.62	0.62	lost circulation
725	285	0.69	2.00	0.53	0.36	satisfactory
1,157	640	0.67	2.20	0.52	0.34	satisfactory
1,205	719	0.62	1.95	0.53	0.52	lost circulation
1,792	1,299	0.55	1.93	0.49	0.40	lost circulation
1,982	1,497	0.54	1.90	0.48	0.40	satisfactory
2,020	994	0.63	1.90	0.55	0.61	lost circulation
2,258	768	0.74	2.08	0.61	0.56	fracture test
2,319	719	0.75	2.00	0.61	0.53	fracture test
2,605	1,395	0.63	1.97	0.53	0.45	satisfactory
2,797	1,497	0.63	1.95	0.50	0.30	satisfactory
3,117	561	0.77	2.10	0.58	0.41	satisfactory
3,167	310	0.88	2.20	0.66	0.49	satisfactory
3,219	719	0.78	1.63	0.59	0.45	lost circulation
4,114	1,299	0.73	2.11	0.59	0.46	satisfactory
4,154	762	0.79	2.16	0.66	0.64	fracture test
4,322	1,032	0.74	1.94	0.58	0.47	fracture test
4,509	941	0.82	2.20	0.56	0.32	satisfactory
4,695	1,437	0.73	2.20	0.66	0.74	fracture test
4,846	1,046	0.74	1.60	0.58	0.45	lost circulation

*Data depth = depth of formations for which calculations and fracture tests were carried out.
**Formation pressure gradient = 0.44 psi/ft, and "equivalent stress ratio" equals $(FPG - P_f/D)/(P_o/D - P_f/D)$.

In the Santa Barbara Channel area, application of a stress ratio (i.e. horizontal/vertical stress) versus bulk density correlation yields better results than techniques proposed previously. Applicability of this concept in other deepwater areas is suggested, since fracture pressure gradients are substantially lower in deepwater and particularly shallow formations, as has previously been shown in Fig.6.20.

Method of Anderson, Ingram, and Zanier

It has been observed previously that fracture pressure gradients may vary widely at a given depth in the same geologic area (Taylor and Smith, 1970;

MacPherson and Berry, 1972). Frequently, this variation is not accounted for in predicting fracture pressure gradients. However, the possible effect of the shaliness (shale content) of U.S. Gulf Coast sands has been suggested previously by the different matrix stress coefficient K_i for South Texas and offshore Louisiana (Fig.6.10) formations, since higher K_i-values are found in the shalier South Texas sands.

Anderson et al. (1973) studied 29 fracture-pressure gradient measurements in twelve Gulf Coast wells during fluid injection for sand consolidation. Twelve of these measurements were made in normally pressured formations, and — similar to previous data by Taylor and Smith (1970) — show variation at a given depth (Fig.6.21).

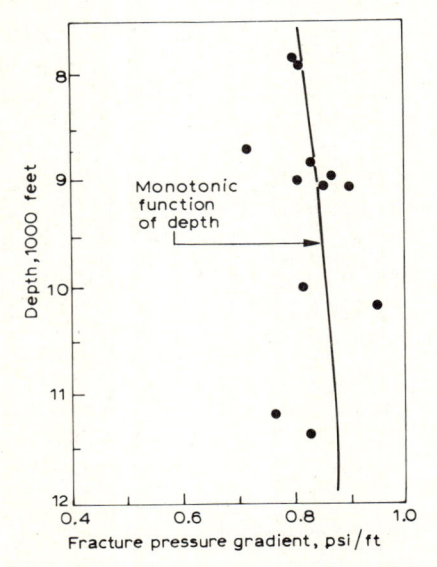

Fig.6.21. Field measurements of the fracture pressure gradient in the U.S. Gulf Coast area (after Anderson et al., 1973). (Courtesy of *Journal of Petroleum Technology*.)

Using Terzaghi's (1923) concept of effective stress, the fracture pressure gradient FPG can be expressed as:

$$\text{FPG} = \frac{2\nu}{1-\nu}(P_o/D) + \frac{1-3\nu}{1-\nu}(P_f/D) \qquad (6.20)$$

On the other hand, using Biot's (1955) stress-strain relationship, one derives the following fracture-pressure gradient expression:

$$\text{FPG} = \frac{2\nu}{1-\nu}(P_o/D) + \alpha \cdot \frac{1-3\nu}{1-\nu}(P_f/D) \qquad (6.21)$$

where ν = Poisson's ratio, and α = a rock compressibility factor, in terms of values derived from logs (e.g. porosity); $\alpha = 1 - C_r/C_b$ where C_r is the intrinsic compressibility of the solid rock material, and C_b the compressibility of the porous rock skeleton.

Hence, eqs. 6.20 and 6.21 differ only by the presence or absence of the coefficient α. According to Anderson et al. (1973), use of eq. 6.21 "is more pleasing in that the pore pressure term disappears at zero porosity, since in that case C_r equals C_b and α becomes zero." Presence of the coefficient α in eq. 6.21 shows a porosity dependence that attenuates the importance of the pore pressure term.

Eq. 6.21 also clearly shows Poisson's ratio to be a sensitive formation property governing fracture pressures. Thus, Anderson et al. (1973) established an important, empirical relationship between Poisson's ratio and the shaliness (shale content) of Gulf Coast sands. Shale content was determined from the sonic and density log, respectively. As previously discussed in Chapter 5, these two logs also provide a means of calculating pore pressure and overburden load gradient, which are both required to compute the fracture pressure gradient.

Table 6.V shows a comparison of measured field data and the computed values in several Gulf Coast wells. The computed fracture pressure gradient is expressed as a function of overburden and pore pressures, porosity, Poisson's ratio, and depth. Digital data handling of sonic and density log data can be used to give a continuous log of the fracture pressure gradient versus depth. Fig.6.22 shows a comparison of such computed values with measured data in a U.S. Gulf Coast well.

TABLE 6.V

Comparison of calculated fracture pressure gradient with field-measured values (after Anderson et al., 1973)

(1) Depth (ft)	(2) Measured (psi/ft)	(3) Biot* (psi/ft)	(4) Terzaghi** (psi/ft)	Col. 2 minus col. 3	Col. 2 minus col. 4
8,360	0.830	0.834	0.837	−0.004	−0.006
8,430	0.830	0.846	0.861	−0.016	−0.031
8,450	0.830	0.780	0.762	0.050	0.067
8,690	0.720	0.800	0.880	−0.080	−0.162
8,710	0.720	0.800	0.888	−0.080	−0.170
9,280	0.860	0.814	0.802	0.046	0.057
9,560	0.830	0.800	0.768	0.030	0.062
9,800	0.855	0.840	0.938	0.015	−0.084
9,850	0.945	0.900	0.942	0.045	0.002
10,240	0.843	0.836	0.838	0.007	0.005
11,670	0.880	0.878	0.860	0.002	0.020
12,260	0.856	0.856	0.879	0.000	−0.023

*Standard deviation using Biot's model = 0.045 psi/ft.
**Standard deviation using Terzaghi's model = 0.083 psi/ft.

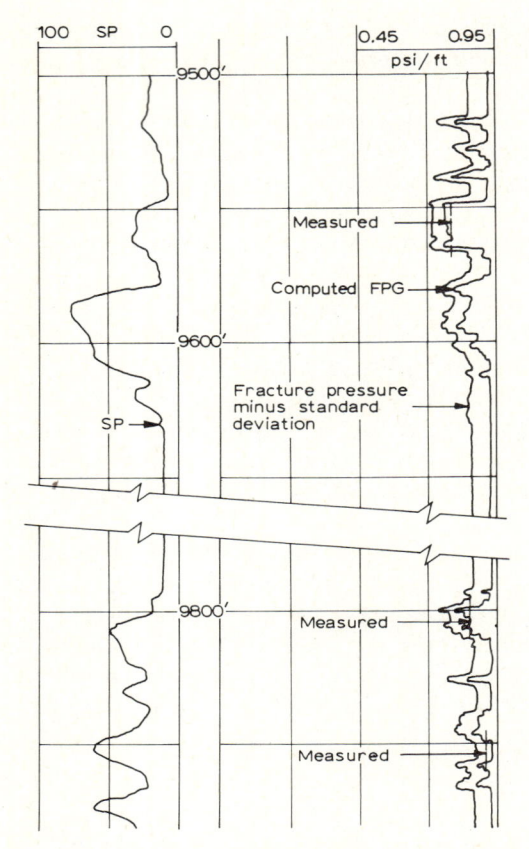

Fig.6.22. Computed and measured fracture-pressure gradient log in a U.S. Gulf Coast well (after Anderson et al., 1973). Self potential (SP) curve is presented for lithology identification. (Courtesy of *Journal of Petroleum Technology*.)

Typical casing programs in overpressured areas

General comments

There is no single, unique casing program. It has to be engineered, tentative, flexible, and continuously reappraised for each well under consideration. Hence, *optimum* casing programs are the result of continuously evaluating and incorporating additional data and new experience, and are not the outgrowth of a one-time, single decision.

Fig.6.23 shows the planned and actual final casing programs in a West Texas well (Cobb, 1967). Sometimes an engineered compromise of underbalanced drilling versus the hazards of pipe sticking and possible kicks allows

TYPICAL CASING PROGRAMS IN OVERPRESSURED AREAS 257

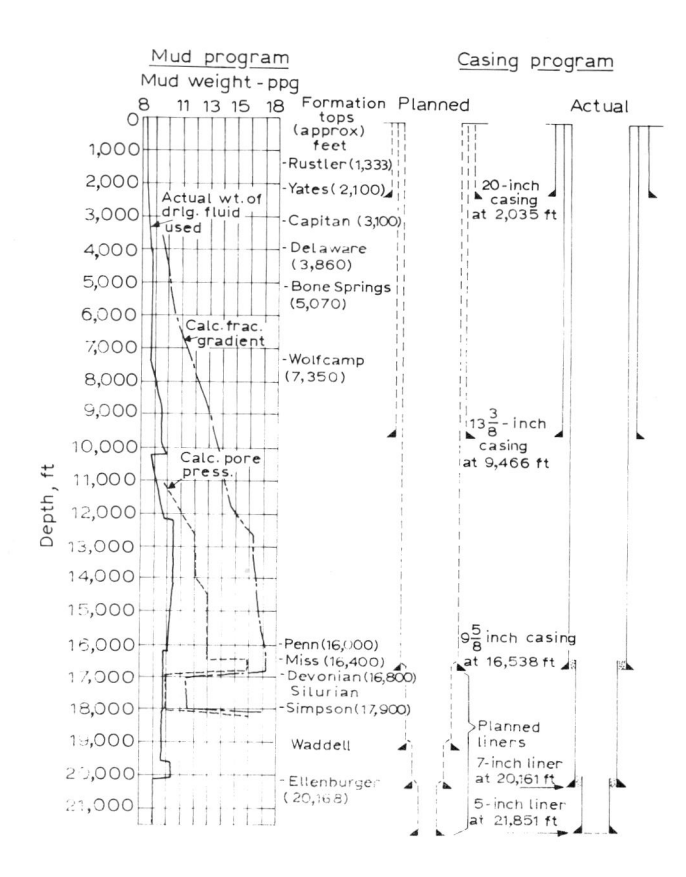

Fig.6.23. Anticipated and actual mud and casing requirements for a Delaware Basin wildcat (after Cobb, 1967). Spud mud was used in the surface hole; brine mud (saltwater mud) was then used to total depth. Underbalanced drilling permitted production of high-pressure low-volume gas from the Lower Wolfcamp and Simpson formations. The low-weight brine mud prevented circulation losses into Devonian and Ellenburger formations. (Courtesy of *World Oil*.)

the intermediate pipe to be set deeper. Fig.6.24 shows such a field case in a deep Wilcox well, located in South Texas (Hanger, 1968).

Properly engineered well plans must be effective, safe, and soundly applied for optimum results. Whereas drilling progresses from the surface to bottom, engineered well planning is a reverse process (i.e. it starts at the bottom of the hole). After both the pore-pressure and fracture-pressure gradient curves have been established versus vertical depth, the proposed ideal mud weight curve can be drawn in, which frequently exceeds the pore pressure by about 0.2—0.4 ppg [0.024—0.048 kg/dm^3].

Fig.6.25 shows an engineered well plan for a Louisiana offshore well. The pore-pressure gradient curve was derived from shale resistivity data and the

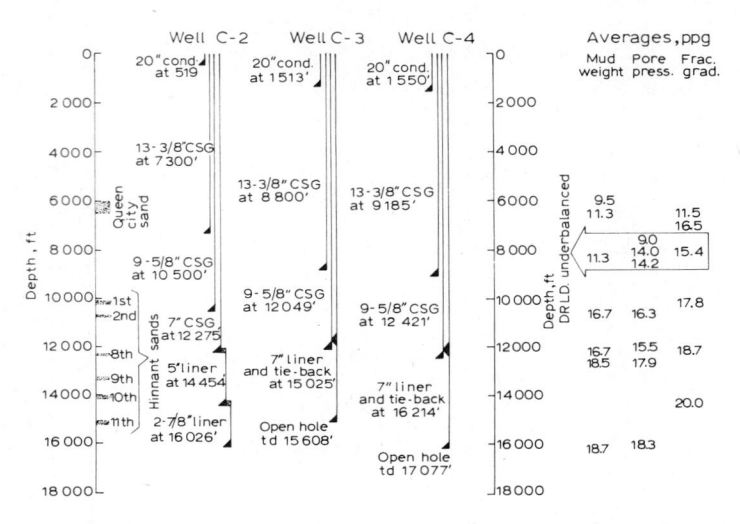

Fig.6.24. Development of casing programs used in deep South Texas wells (after Hanger, 1968). Underbalanced drilling of an overpressured shale at 7500 ft [2286.6 m] and lowering intermediate casing depth to about 9000 ft [2743.9 m] (wells C-3, C-4) made it possible to eliminate one protection string previously set at 10,500 ft [3201 m] (well C-2). The hookup used in well C-4 has proven satisfactory. (Courtesy of *World Oil.*)

fracture-pressure gradient curve by the method proposed by Eaton (1969). Proper casing setting depths were then determined as follows (Eaton, 1970):

(1) A vertical line (*a*) is drawn upwards to the fracture-pressure gradient curve from the mud weight curve at the point of highest mud weight (18 ppg in this case). This line represents the static head of the mud column which extends to the surface. The intersection of the vertical line and fracture-pressure gradient curve indicates that 18-ppg mud standing in the open hole will break down the formation at about 10,100 ft [3078 m] or above that point.

(2) The liner-casing point is selected at a depth just below the point where line (*a*) intersects the fracture-pressure gradient curve. Some safety factor must be considered since the mud weight gradient does not allow for the true circulating density.

(3) A horizontal line (*b*) is then drawn at the selected depth of the casing shoe to the required mud weight curve.

(4) A vertical line (*c*) is extended upwards to the fracture-pressure gradient curve from where line (*b*) intersects the mud weight curve.

(5) The intermediate casing point is selected just below the depth of the latter intersection, again allowing an appropriate safety margin.

(6) The same procedure using horizontal line (*d*) and vertical line (*e*) is used to select the surface casing point.

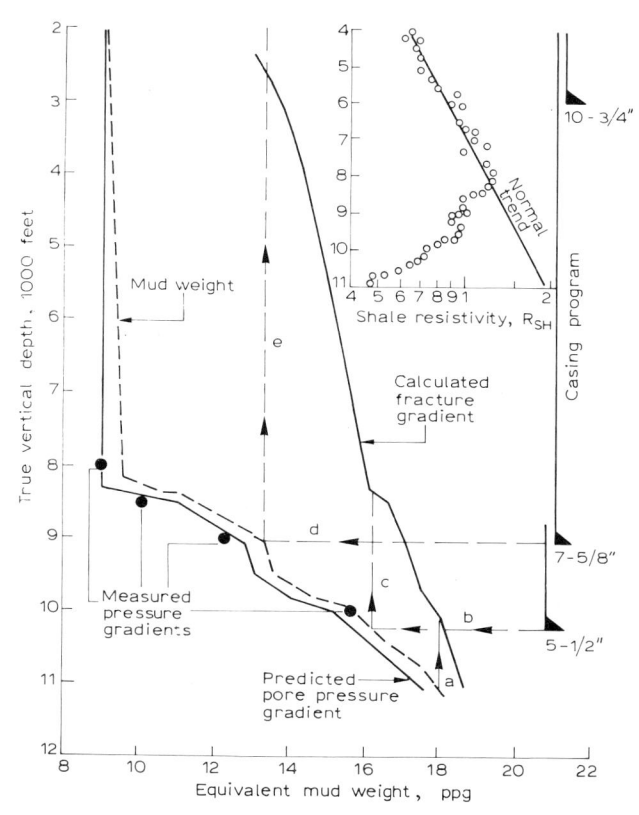

Fig.6.25. Well-planning chart for a Louisiana offshore location providing guidelines for safe and economic well-drilling operations (after Eaton, 1970). Pore-pressure gradient curve was computed from shale resistivity data in a nearby well. (Courtesy of *World Oil*.)

Formation pressure measurements in this well confirmed that the pore pressure values were predicted fairly accurately. Such plots also clearly show that excessive mud weight (large overbalance) will result in too high casing seats, thereby requiring additional casing strings.

Naturally, we have touched on some of the basic considerations only. As expected, fracturing occurs at the weakest formation that is uncased; thus the zone most likely to fracture first may or may not be the shallowest formation exposed below the casing shoe.

Casing-shoe pressure tests give information on the quality of the cement job and fracture pressure gradient. Whereas Eaton (1970) recommended that at least 0.5 ppg [0.06 kg/dm^3] below the fracture pressure gradient should be allowed as insurance that breakdown will not occur when testing, one may want to set the maximum pressure (i.e. leak-off or rupture pressure) at a preselected value (Moore, 1973) depending on the operator's further object-

TABLE 6.VI

Generalized typical casing programs in several abnormally pressured U.S. basins

Region*	Typical casing sizes (inches)	Approximate average casing point depth (ft)		
Delaware Basin (West Texas)	(1) 20″ conductor (2) $13\frac{3}{8}″$ (3) $9\frac{5}{8}″$ or $10\frac{3}{4}″$ (4) $7\frac{7}{8}″$ liner (6) liner tie-back** to surface, $7″(7\frac{5}{8}″) \times 5″(5\frac{1}{2}″)$ or $7\frac{7}{8}″$ only	*Ward Co.* 5,000 11,000 14,500 20,000	*Reeves Co.* 5,000 11,000 16,000 21,000	*Pecos Co.* 1,700 10,500 15,500 22,000	*Winkler Co.* 5,000 13,000 17,000 22,500
U.S. Gulf Coast (Texas, Louisiana)	(1) 20″ (locally plus 30″) (2) $13\frac{3}{8}″$ or $10\frac{3}{4}″$ (3) $9\frac{5}{8}″$ or $7\frac{5}{8}″$ (4) 7″ liner and/or $5\frac{1}{2}″$ (or 5″)	*South Texas* — 9,000 12,500 15,000	*Upper Texas* — 3,000 12,000 15,500	*Onshore Louisiana* — 3,000—4,000 13,000—15,000 15,000—18,000 (plus)	*Offshore Louisiana* 4,000 13,000 20,000 (plus)
Anadarko Basin (Texas, Panhandle, western Oklahoma)	For 20,000-ft wells: (1) 24″ conductor (2) 20″ (or 16″) conductor (3) $10\frac{3}{4}″$ (or $10\frac{3}{4}″$) (4) $7\frac{5}{8}″$ (or $9\frac{5}{8}″$) liner (5) $5\frac{5}{8}″$ (or 7″) liner	*Oklahoma* no 2,000 12,000 19,000 to TD	*Texas* yes 4,000 13,500 16,700 to TD		
Rocky Mountain area (Utah, Colorado)	(1) 20″ conductor (2) $13\frac{3}{8}″$ (3) $9\frac{5}{8}″$ (4) 7″ liner	*Uinta Basin (Utah)* 1,500 9,500—12,500 to TD			
	(1) $13\frac{3}{8}″$ conductor (2) $9\frac{5}{8}″$ (3) 7″ (4) 5″ liner	*Piceance Basin (Colorado)* 5,000 14,500 to TD			

*Schematics shown in Fig. 6.26.

TYPICAL CASING PROGRAMS IN OVERPRESSURED AREAS

ives. However, whatever one's further objectives, if casing joints are not selected properly, drilling problems are invited.

Delaware Basin, West Texas

Drilling operations in the Delaware Basin (see Fig.6.3) located in West Texas, frequently encounter several problems. These require casing off: shallow freshwater zones to prevent their contamination; shallow gas zones (for example in Yates Sand); shallow lost-return zones; massive evaporite zones; incompetent, sloughing shales; zones of low fracture pressure gradients above and/or below overpressured intervals; etc. Hence, the properly engineered casing program has to consider all these problem zones. Typical, generalized casing programs using liner tie-back strings in deep Ellenburger wells are listed for several West Texas counties in Table 6.VI and shown in Fig.6.26.

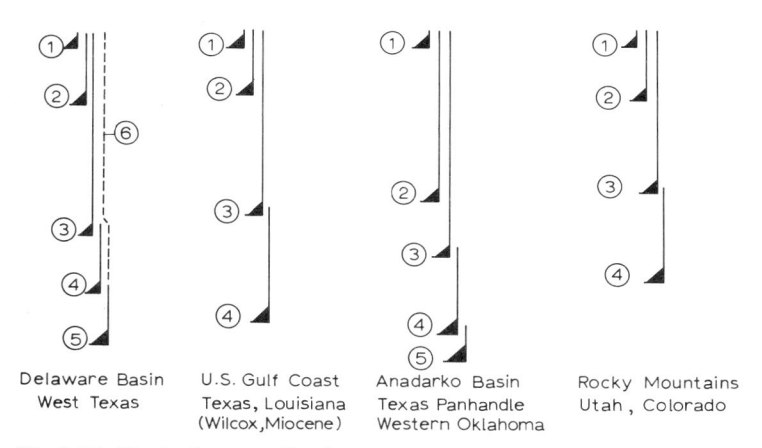

Fig.6.26. Typical, generalized casing programs in overpressured areas in the United States. The encircled numbers refer to Table 6.VI.

U.S. Gulf Coast area

In the U.S. Gulf Coast area, the presence of abnormally high formation pressures in shale/sand sequences of both on- and offshore Texas and Louisiana is the key factor in selecting optimum casing programs. Overpressures can cause high-pressure gas and saltwater flows, borehole instability, pipe sticking due to differential pressure, lost circulation, high formation temperature, etc. For the drilling operator to handle these problems adequately, particularly in the light of *day-rate* versus *footage-rate* drilling contracts, most of today's well-established pore- and fracture-pressure gradient prediction techniques have been initially developed and field-tested in this area. Table 6.VI lists typical casing programs for Wilcox wells in Texas and Miocene wells in Louisiana, as schematically shown in Fig.6.26.

Anadarko Basin, U.S. Mid-Continent area

In the U.S. Mid-Continent area, such as the Anadarko Basin in the Texas Panhandle and western Oklahoma, conditions are somewhat similar to the Delaware Basin discussed previously. Typical casing programs (Table 6.VI, Fig.6.26) for deep Anadarko wells have to consider the following: protection of freshwater and casing off of lost circulation zones, evaporites (salts, gypsum, etc.), drastically different lithologies, and pore pressure reversals due to the presence of normal, subnormal, and/or abnormally high formation pressures.

Rocky Mountain area

Most of the problems and hazards that are inherent to drilling operations throughout the world can be found to some degree in the Rocky Mountain area including Wyoming, Utah, and Colorado. Drastic variations in terrain and well-site elevation; lost circulation; presence of sloughing or heaving shales; flows of fresh, brackish, salt, or trona waters from Green River Formation; and wide variation of pore pressures have large impact on how and where to set casing seats (Table 6.VI, Fig.6.26). For example, in the Uinta Basin, located in Utah, the Duchesne River, Uinta, and Upper Green River Formations exhibit pore pressures ranging from subnormal to normal, whereas the Lower Green River and Wasatch Formations are normally or abnormally high pressured.

Table 6.VI lists generalized casing programs typical for the Uinta Basin in Utah and the Piceance Basin in Colorado.

Comments on cementing operations

Well-cementing technology in both relatively straight and high-angle directional holes has advanced dramatically since the first casing was cemented in 1903. Besides the everyday cementing needs in "problem-free" boreholes, recent engineered improvements successfully deal with cementing of arctic wells, ultra-deep and hot holes, and water-sensitive, incompetent, fractured, or highly permeable formations.

The basic requirements for obtaining a successful primary cement job have been known for years. Good design characteristics are based on a knowledge of formation, cement, and pipe properties and controlled placement techniques that consider fracture pressure gradients. Also important are (1) an understanding of minimum practical mud density and viscosity, (2) cement type, (3) turbulent flow conditions, (4) the optimum size of preflushes,

(5) centralizing of casing, the use of scratchers, and the handling of pipe, and (6) the proper choice of casing.

Present-day planning and laboratory-testing of cement slurries and additives begin well in advance of the actual job on ultra-deep and overpressured wells in order to meet specific well conditions. For example, numerous additives are available which retard or accelerate thickening times, vary the slurry density and flow properties, and aid in filtration and lost circulation control.

Since — besides several other factors — slurry placement pressures are controlled by fracture pressure gradients, a detailed engineering, hydraulic evaluation should be carried out to determine the maximum pressures to be encountered during any deep cementing operations. Both pressure and temperature criteria become very critical if very long cement columns are used. Today, single- and multistage placement techniques are available to cope with specific downhole pressure environments. Variation in both formation fracture pressure gradients and the hydrostatic gradient of cement columns for each placement stage can be calculated and then plotted versus depth to decide upon the optimum cement placement method in subject well. For example, in well Green No. 1, located in Beckham County, western Oklahoma, a cement column was successfully placed from 24,000 to 10,000 ft [7315 to 3048 m] in a single-stage process (Wheeler and Moriarty, 1969).

In deep, overpressured sour-gas wells in Mississippi, U.S.A., experience has shown that the best way to properly cement the $9\frac{5}{8}$-inch casing is in stages (Kirk, 1972).

In the prolific Delaware Basin in western Texas, cementing operations face several critical problems, including (1) cementing long sections across incompetent, weak formations; (2) high-pressure gas zones that can be drilled with relatively light muds due to rather small permeability but which are not under complete control during cementing; (3) large temperature differences between top and bottom of very long cement columns; and (4) extremely long liners set in small holes across weak formations (Gibbs, 1966). Where a well is being drilled, as in the case of casing programs, the cementing procedure should be evaluated continuously and subsequently upgraded as more experience becomes available in a given area.

In the Delaware and Anadarko Basins of West Texas, and under similar conditions elsewhere, most difficulties occur when it becomes necessary to set drilling liners across high-pressure gas zones. Major problems include low tolerances between hole and pipe size, gas-cut cement columns, high collapse loading on the liner, liner contact with corrosive sour gas and fluids, etc. Hence, experience indicates that no single cementing procedure is universally applicable in all areas.

The most difficult of all liner cementing problems occurs across high-pressure gas-bearing intervals, and several techniques have been developed recently (Van Hook and Feemster, 1969; Lindsey and Bateman, 1973; Mahony and Barrios, 1974). This current technology offers several promising and widely accepted liner cementing methods, including:

(1) *Single-stage* cementing, with the liner being cemented in a single, continuous operation from bottom to top. A variation, the controlled reversing method, allows the operator to hold pressure against the gas-producing interval.

(2) *Two-stage* cementing, with about two-thirds of the liner cemented in a conventional one-stage operation; then the liner top is subjected to a planned squeeze job.

(3) Refined versions of the above basic methods have been discussed in detailed outlines and were successfully applied in the field (Lindsey and Bateman, 1973; Mahony and Barrios, 1974).

Utilizing such present-day technology and experience, a 10,925 ft [3330 m], $7\frac{5}{8}$-inch liner was successfully cemented in the United States recently.

Specific completion and production considerations

Geological, technological, safety, and economic considerations influence size, setting depth, and design factors (burst, collapse, tension) in the various sections of the production string in overpressure environments. However, present-day technology has progressed to the point that if overpressures were the only factor, the completion requirements would be rather straightforward. Frequently, numerous additional related problems are being encountered, particularly in high-pressure sour-gas environments; in some areas gas may consist of up to 75% H_2S and/or CO_2. Some of the specific problems, such as permafrost-freezeback pressures, hydrate formation, salt precipitation, packer fluid requirements, casing collapse, and tubing leaks are briefly discussed next.

Abnormally high permafrost-freezeback pressures around Arctic wells

Unique abnormal formation pressure phenomena, such as talik and permafrost freezeback, are found in the Arctic (see Chapter 1). Arctic oil-well drilling and production are known to result in extensive thawing of permafrost (Couch and Keller, 1970). If a well is allowed to refreeze, i.e. the fluids confined within the pipe freeze, then the pressure rises and casing damage may occur (Bleakley, 1971). A second mechanism involves freezeback of fluids or thawed permafrost outside the casing. Extreme external pressures, which may result in collapse loading of casing, are generated by the phase, and hence volume, change of water to ice (Shalavin and Klyushin, 1972; Goodman and Wood, 1973).

Large-scale field testing of permafrost freezeback was carried out recently on the North Slope at Prudhoe Bay in Alaska (Perkins et al., 1973). In experimental wells, pressure transducers and thermistors were used in situ to record temperature variations and external pressures generated during freezeback. Measured data and digital model studies results showed good agreement of depth- and time-dependent freezeback pressure gradients. Freezeback pressures are of such magnitude that they have to be considered in casing design for permafrost areas. Freezeback pressure gradients range from 0.66 psi/ft [0.152 kg cm^{-2} m^{-1}] to as high as 1.44 psi/ft [0.333 kg cm^{-2} m^{-1}]. Nevertheless, experience in the Prudhoe Bay, Alaska, has shown that the $13\frac{3}{8}$-inch, 72-lb, N-80 casing will safely withstand the maximum pressure generated by permafrost freezeback in the subject area.

Hydrates

In many producing high-pressure gas wells with some heavier hydrocarbons or sour-gas production, the possibility of hydrate formation exists. Such hydrate formation can often be eliminated by pressure reduction (if feasible) using subsurface chokes or regulators, heaters, hot oil circulation, alcohol injection, and use of sweet gas in sour-gas wells. However, presence and increasing content of H_2S and CO_2 increase the temperature for hydrate formation in sour gases appreciably above the data published for sweet gas.

Another, however completely different, hydrate problem relates to the well-established presence of gas hydrates in the Arctic. For example, in the Mackenzie Delta, Canada, wells penetrated what appeared to be exceptionally prolific high-pressure gas zones, but it was discovered later that some of the high pressures were caused by decomposing gas hydrate deposits.

Precipitation of salts, sulfur, etc.

Precipitation of salts from highly concentrated, i.e. very salty, formation waters often is experienced in high-pressure gas wells. This can be eliminated by injection of fresh water below the point of precipitation to dilute the saturated brine.

Precipitation of elemental sulfur in deep, high-pressure sour-gas wells has been experienced in Alberta, Canada, the Zechstein carbonates in North Germany, in the United States in Mississippi and Oklahoma, and elsewhere. Often tubings are plugged completely after short test or production periods, and snubbing operations become necessary.

Phase-behavior studies of sulfur basically indicate that it can exist in the vapor phase, dependent on the composition of the gas, temperature, and pressure. A reduction of pressure and temperature, such as may occur during tests or production, causes precipitation of sulfur in the formation and/or tubing. Similar problems were encountered recently in the record borehole Bertha Rogers No. 1, drilled to 31,441 ft [9585.7 m] in Oklahoma, where sulfur from the Arbuckle zone precipitated in the well causing mechanical problems and subsequently preventing extensive tests.

Plugging of tubing and surface chokes in high-pressure wells may occur as a result of thermal dehydration of mud solids, weighting material, and occasional plugs of shales, silt, sand, or cement chips. High-pressure snubbing units using small diameter clean-out strings and high-pressure floor manifolds, which allow change of chokes without shutting-in the well, are often used to combat these problems.

Packer fluid requirements

Generally speaking, completion or packer fluids should be (a) physically and chemically stable over long periods, (b) able to prevent corrosion, (c) sufficiently heavy to control abnormally high formation pressures, (d) easy to handle, and (e) relatively cheap. Such systems include (1) *weighted*, heavy drilling muds, and (2) *unweighted*, light fluid systems.

Weighted mud systems are available at no extra cost; they may be sufficient to kill a well in the event of a tubing leak, or to seal it off in the event of a casing leak. However, several disadvantages significantly affect the well later on: these include barite settling, tubular stress failures due to lignosulfonate degradation, and workover operations due to high-temperature gellation of lime base fluids.

Unweighted, light fluid systems have been investigated and classified by Speel (1967) and recently by Mayell and Stein (1973). The latter authors summarized their evaluation of the more common packer fluids as follows:

(1) *Inhibited saltwater systems* have inhibitive but no sealing properties.

(2) *Oil-base muds* are inhibitive, but are expensive, difficult to handle offshore, and have barite-settling problems.

(3) *Polymer-carbonate systems* have inhibitive and sealing properties, possible long-term stability, and can be handled without any special problems.

(4) *Saltwater gel systems* are inhibitive, have long-term stability, probable sealing properties, and are handled with ease.

Based on both laboratory test data and actual field experiences, the polymer-carbonate systems, and especially saltwater gels, offer the best overall qualifications for high-pressure wells (Mayell and Stein, 1973).

General comments on casing and tubing requirements

Completion considerations related to casing and tubing requirements are numerous, and solutions are not unique. Often specific, and sometimes rather unusual, mechanical completion hookups are used in a given area for a given problem. Since it is virtually impossible to even review all of these concepts, an attempt is made to highlight some of the factors involved.

Casing design factors, such as *burst*, *collapse*, and *tension*, as well as the availability of choices for heavier and higher-grade material, require special attention in areas where abnormally high formation pressure environments are known to exist. Selection of absolutely gas-tight, metal-to-metal casing joints, application of thread-sealing components, properly cleaned and adequately torqued joints, etc., are additional important considerations. Completion requirements in sour-gas environments have resulted in the development of special tubular goods. For example, C-90 grade material has been successfully used in deep, high-pressure Smackover wells in Mississippi.

Both API* and NACE** specifications are available for these special conditions. Furthermore, careful material inspection and appropriate testing at the plant site and at the field location are a prerequisite.

Planning decisions are often complicated due to requirements for material of extremely high strength and the possibility of hydrogen embrittlement.

These and similar problems are encountered worldwide: examples include the Smackover Formation in northeastern Texas (Mottley, 1968) and Mississippi (Kirk, 1972), the area along the deep Wilcox trend of South Texas (Hanger, 1968), deep drilling areas of West Texas and Oklahoma, the Zechstein evaporites and overpressured sandstones in North Germany (Speel, 1967; Werner and Becker, 1968; Petersen et al., 1973), and the deep gas fields in the foothills of the Pyrenees in southwest France (Jean, 1973).

A decision sometimes has to be made between running a full string or a liner. Cementing a liner inside the casing is known to increase the combined collapse strength by 25—70% (Cheatham and McEver, 1964).

"Flowing" salt and shales not only cause drilling problems but can also deform the casing, causing minor deformations or a complete casing collapse (Clegg, 1971). Dual completions are often not feasible because of a large pressure differential between two prolific pay sections, such as is found in Oklahoma between the high-pressure Morrow/Springer section and the deeper, but lower-pressured, Hunton interval.

Tubing strings may cause problems and need special considerations similar to the ones mentioned for casing (Epperson, 1958). Tubing strings in the wellbore are continuously subjected to constantly changing alternate forces, thereby causing tension, bending, buckling, and direct and reverse ballooning. Several mathematical models are available which attempt to estimate tubing movement caused by temperature and pressure variations, such as during production (high flow rates) and stimulation treatments (high rate/high pressure).

Basically, tubing strings have to be robust, strong, resistant to chemical attack by H_2S and CO_2 (to avoid stress-corrosion cracking of high-strength tubular strings) and must provide absolutely gas-tight joints even under

---

*American Petroleum Institute.
**National Association of Corrosion Engineers.

extreme differential pressures. For the latter situation, special threaded connections, thread-sealing lubricants, and synthetic seal rings are available. Nevertheless, field experience indicates (West and Lindsey, 1968) that often, for no apparent reason, tubing leaks develop in the connections despite adequately torqued pipe. Investigations of deep gas wells in South Louisiana have shown that at high flow rates, the tubing string tends to develop *reverse torque* (Goeken, 1966).

Many deep wells show a tubing-casing annulus pressure buildup shortly after installing the Christmas tree. This may be caused by a leak in the casing, tubing, or subsurface equipment. According to West and Lindsey (1968), probably more than 75% (!) of the deep wells in West Texas have a problem with pressure buildup on one or more casing strings. Whereas remedial work depends on the type of leak, the best solution is preventive action.

Contingency planning for drilling and producing high-pressure sour-gas wells

Particularly during the last two decades, the necessary technology has been developed to guarantee safe and efficient operations in hostile overpressure environments. This includes careful selection, quality control, and proper inspection of materials used, properly planned and designed drilling and production operations, personnel training in both technological and safety concepts, development of appropriate contingency plans for protecting operating personnel, nearby residents and their properties, and the environment (on- and offshore) in general. Needless to say, the oil industry is successfully putting large efforts into all these areas. Both academic, industrial, and governmental guidelines are available in the attempt to manage safely operations in hazardous environments, such as deep, high-pressure, sour-gas reservoirs.

Recently, Hamby and Smith (1972) discussed in detail such a contingency plan which was successfully used in Mississippi, where the high-pressure sour-gas well Cox No. 1 blew out of control near Piney Woods in 1970. To determine the level of response required for a particular situation, three operating conditions have been discussed (Table 6.VII). In addition, a detailed two-phase evacuation plan was developed (Table 6.VIII). Phase I entails the action that must be taken by the operations personnel who will be closest to the emergency and it involves the immediate evacuation of all personnel within the danger zone. Phase II involves the action to be taken after the situation has been assessed — namely, the removal of persons other than those in the immediate vicinity of the well.

TABLE 6.VII

Operating conditions with H_2S present (after Hamby and Smith, 1972)

Condition I: potential danger to life	Condition II: moderate danger to life	Condition III: extreme danger to life
Warning sign: Yellow with numeral 1 in center	*Warning sign:* Orange with numeral 2 in center	*Warning sign:* Red with numeral 3 in center
Characterized by: Minor gas leaks from mud or equipment. Poisonous gases present at threshold concentrations that may or may not be detectable by odor	*Characterized by:* Moderate gas leaks from mud and equipment. Poisonous gases present above threshold concentrations	*Characterized by:* Full scale blowout. Poisonous gas concentrations above hazardous limits
Probable occurrence Drilling: as a drill show or trip gas (i.e. when circulating bottoms up). Producing operations: minor leaks from valves, flanges, etc.	*Probable occurrence* Drilling: as trip gas or well kick. Producing operations: severe leaks from valves, flanges, flowline ruptures, etc.	*Probable occurrence* Drilling and production operations: complete loss of well control
General action (all personnel): Check safety equipment for proper functioning. Keep it with you. Be alert for a condition change. Follow instructions of supervisor	*General action* (all personnel): Stay in the "safe briefing area" if not working to correct the situation. Follow instructions of supervisor	*General action:* Stay in the "safe briefing area" if not working to correct or control the situation. Follow instructions of supervisor. Supervisor will direct ignition of the well and any other necessary operations with an absolute minimum of personnel. All persons working in the immediate hazard area will wear a self-contained breathing apparatus. All other personnel will restrict their movements to those directed by supervisor. If the well is ignited, H_2S will be converted to sulfur dioxide (SO_2) which is also poisonous. *Do not* assume that the area is safe after the gas is ignited. Continue to observe emergency procedures and follow the instructions of the supervisor

TABLE 6.VIII

Outline of evacuation plan (after Hamby and Smith, 1972)

Area	Action	
	Phase I (field)	Phase II (staff organization)
Danger zone (½ mile radial area around well)		
Lethal concentration of H_2S/SO_2 may be present	Foreman will evacuate	Monitor H_2S/SO_2 Limited access until emergency is controlled
Health hazard sector (45° sector up to 3 miles)		
Concentration of H_2S/SO_2 may exceed TLV*	Foreman will monitor H_2S/SO_2	Monitor H_2S/SO_2
Concentrations not considered lethal	Foreman may evacuate	Emergency support team may evacuate Emergency support team may allow residents to return
Polluted sector (many miles)		
All concentrations of H_2S/SO_2 are less than 2.0 ppm	No action	Monitor H_2S/SO_2 Evacuation highly unlikely

*Threshold limit value: concentration to which it is believed that a normal healthy worker may be repeatedly exposed 8 hours/day without adverse effect.

Relief-well planning

Blowout control basically attempts to block off escaping high-pressure reservoir fluids or gas either in the borehole or in the formation itself. It involves recapping of a wild well or drilling of one or several relief wells through which the wild well is brought under control. In recent years, several novel concepts in planning and drilling relief wells have been developed and successfully applied in the field. A discussion of some of these methods follows:

(1) Relief wells may be drilled to total depth, i.e. to the *problem source*, of the wild well to establish communication and kill the blowing well by pumping heavy mud or cement slurry at high rates into it. However, the time factor, which is directly proportional to the depth of the formation, directional drilling control problems, drillability of overlying geologic section, etc., is critical.

(2) Relief well(s) is (are) again drilled to total depth, and water is injected through them into the formation, but direct communication between all wells is avoided. According to Lehner and Williamson (1974), "this restricts the escape of reservoir fluid by a pressure buildup resulting from the flow of water through the formation, and by the continuous narrowing of the passageways open to the escaping reservoir fluid between spreading water-saturated volumes. When all passageways are closed off by the water, the wild well is under *dynamic* control and may produce a large fraction of the water that is being continuously injected. A final plugging operation is still necessary to gain permanent control over the well. The termination of relief-well water injection must then be timed carefully, particularly when dealing with an overpressured gas reservoir." For this, the authors also presented a very elaborate mathematical analysis of the basic considerations and factors involved.

(3) An intermediate-depth relief well can be drilled to intersect and parallel the wild wellbore over a short distance far *above* the in-situ *blowout source*. Then the two strings are cemented together, perforated to establish an interwell communication, and heavy mud or cement slurry is pumped into the wild well to choke it off. Basically, such a method was successfully used to kill a recent blowout in the high-pressure sour-gas well Cox No. 1, located in Mississippi, U.S.A. (Bruist, 1972).

(4) A modification of Method 3 consists of a relief well converging to within a short distance of the wild well, washing a large cavity at the bottom of the hole, pumping conventional explosives into the cavity, and detonating on command. Feasible explosive termination concepts were analyzed by McLamore and Suman (1972) recently. The detonation would create a larger cavity, crush and rubble the rock, collapse and shutter the casing in the wild well, and — as the ultimate aim — stem the flow of gas at that depth. The authors concluded that to be most effective, a high-energy shot should be detonated in a relatively weak rock, such as plastic shale. Furthermore, the center-to-center distance between the wild and the relief well should be less than about six times the size of the explosive charge if the shot is to be effective.

(5) Explosive termination of a blowing well using nuclear devices has been successfully demonstrated in the U.S.S.R. in two wild, high-pressure gas wells (Anonymous, 1970). Lack of precise directional data on the blowing wells, pressure gradients as high as 0.9 psi/ft [0.208 kg cm^{-2} m^{-1}], and severe gas breakthrough into the nearby boreholes and water wells prevented application of conventional killing methods. Hence, a 30-kiloton nuclear explosive was placed in the relief well less than 230 ft [70.1 m] away from the wild well and the shot was detonated in a shale section. The wild well was plugged over a 200-ft [61 m] interval, and the uncontrolled gas flow (up to 55 million cubic feet of gas per day) was completely shut off.

Besides the old multi-shot directional surveys, recently two independent *well-logging methods* were utilized to determine (a) the *distance* from a relief well to the wild well, and (b) *distance* and *direction* from an open hole to a cased wellbore.

Mitchell et al. (1972) have investigated the potential of both the *standard electric logging* sondes and the newer *ULSEL arrays* (Runge et al., 1969) to determine distances between wells. It was found that resistivity measurements can be applied to estimate quantitatively distances between open and cased holes in the subsurface at distances up to 50 ft [15.24 m], whereas qualitative indications of proximity to casing can be obtained at distances as far as 70—80 ft [21.34—24.38 m]. Applications of these concepts include relief-well drilling, sidetracking operations, and critical passing situations in multi-well platform drilling.

Remanent magnetization in casing, caused by certain casing inspection procedures, can be detected in adjacent wells by wireline surveys. Such data then can be used to calculate both distance and direction from a relief well to a wild well (Robinson and Vogiatzis, 1972). For example, in well Cox No. 1 in Mississippi, this method assisted to intersect the wild well at a depth of 10,000 ft [3048 m] within 9 inches (!) [0.23 m] of the point predicted. Two magnetostatic methods are available, both of which use measurements in the relief well of the total magnetic field, the apparent azimuth from a gyrocompass, and the apparent azimuth from a magnetic compass. According to Robinson and Vogiatzis (1972), "the first method uses the regular periodicity of the field from the casing to estimate distance and direction from points of observation in the relief well. It is applicable if the magnetization of the casing has not been altered by large mechanical stresses. The second method comprises a least-square fit to the total magnetic field observed over a few hundred feet of the relief well and a comparison of calculated and observed magnetic declination over the same interval. It is applicable even when the casing has been subjected to large positive and negative stresses that alter the regular periodicity of the field outside the casing."

It is obvious that regardless of the method used, the capability of determining both distance *and* direction from a wild well to the relief hole is one of the most important factors in properly engineered relief-well drilling.

References

Anderson, R.A., Ingram, D.S. and Zanier, A.M., 1973. Determining fracture pressure gradients from well logs. *J. Pet. Technol.*, 25: 1259—1268.
Anonymous, 1970. Survey of possible uses of nuclear explosions for peaceful purposes within the national economy of the Soviet Union. Paper presented at *General Conf. Int. At. Energy Agency*, Vienna, September.
Biot, M.A., 1955. Theory of elasticity and consolidation for a porous anisotropic solid. *J. Appl. Phys.*, 26(2): 115—135.

REFERENCES

Bleakley, W.B., 1971. North Slope operators tackle production problems. *Oil Gas J.*, 69(43): 89—92.
Bruist, E.H., 1972. A new approach in relief well drilling. *J. Pet. Technol.*, 24: 713—722.
Burley, J.D. and Drouin, A.H., 1971. A solution to ground subsidence problems in casing strings and well heads. *J. Pet. Technol.*, 23: 654—660.
Cheatham, J.B. and McEver, J.W., 1964. Behavior of casing subject to salt loading. *J. Pet. Technol.*, 16: 1069—1075.
Christman, S.A., 1973. Offshore fracture gradients. *J. Pet. Technol.*, 25: 910—914.
Clegg, J.D., 1971. Casing failure study — Cedar Creek anticline. *J. Pet. Technol.*, 23: 676—684.
Cobb, J.B., 1967. Underbalanced drilling with brine cuts deep well costs. *World Oil*, 165(5): 85—90.
Couch, E.J. and Keller, H.H., 1970. Permafrost thawing around producing oil wells. *J. Can. Pet. Technol.*, 9: 107—116.
Crittendon, R.D., 1959. The mechanics of design and interpretation of hydraulic fracture treatments. *J. Pet. Technol.*, 11: 21—29.
Eaton, B.A., 1969. Fracture gradient prediction and its application in oil field operations. *J. Pet. Technol.*, 21: 1353—1360.
Eaton, B.A., 1970. How to drill offshore with maximum control. *World Oil*, 171(5): 73—77.
Epperson, M.J., 1958. Completing high-pressure wells. *Pet. Eng.*, 30(3): B53—B57.
Gibbs, M.A., 1966. Delaware Basin cementing — problems and solutions. *J. Pet. Technol.*, 19: 1281—1285.
Goeken, R.J., 1966. Equipping and operating a high-capacity gas field. *World Oil*, 162(7): 96—100.
Goodman, M.A. and Wood, D.B., 1973. A mechanical model for permafrost freezeback pressure behavior. *SPE 4589, 48th AIME Fall Meet.*, Las Vegas, Nev., October.
Hanger, D.M., 1968. Why deep Wilcox wells are difficult to drill, complete. *World Oil*, 166(7): 60—63.
Hubbert, M.K. and Willis, D.G., 1957. Mechanics of hydraulic fracturing. *Trans. AIME*, 210: 153—168.
Jean, L.A., 1973. Problems and techniques in producing gas wells in southwest France. *Erdoel Erdgas Z.*, 89(3): 107—111.
Kirk, W.L., 1972. Deep drilling practices in Mississippi. *J. Pet. Technol.*, 24: 633—641.
Lehner, F. and Williamson, A.S., 1974. Gas-blowout control by water injection through relief wells — a theoretical analysis. *Soc. Pet. Eng. J.*, 14(4): 321—329.
Lindsey, H.E. and Bateman, S.J., 1973. Improve cementing of drilling liners in deep wells. *World Oil*, 177(5): 65—68.
MacPherson, L.A. and Berry, L.N., 1972. Prediction of fracture gradients from log-derived elastic moduli. *Log Analyst*, 13(5): 12—19.
Mahony, B.J. and Barrios, J.R., 1974. Cementing liners through deep high pressure zones. *Pet. Eng.*, 46(3): 61—67.
Matthews, W.R. and Kelly, J., 1967. How to predict formation pressure and fracture gradient. *Oil Gas J.*, 65(8): 92—106.
Mayell, M.J. and Stein, F.C., 1973. Saltwater gel packer fluid for high pressure completions. *SPE 4611, 48th AIME Fall Meet.*, Las Vegas, Nev., October.
McLamore, R.T. and Suman, G.O., 1972. Explosive termination of a wild well — evaluation of a concept. *J. Pet. Technol.*, 24: 1084—1094.
Mead, J.L. and Reid, C.A., 1969. Instrumentation and analysis of on-site drilling data to achieve an optimized drilling program. *Southwest District API Meet.*, March, Paper 906-14-B.
Miller, R.T. and Clements, R.L., 1972. Reservoir engineering techniques used to predict blowout control during the Bay Marchand fire. *J. Pet. Technol.*, 24: 234—240.

Mitchell, F.R., Robinson, J.D., Vogiatzis, J.P., Pehoushek, F., Moran, J.H., Ausburn, B.E. and Berry, L.N., 1972. Using resistivity measurements to determine distance between wells. *J. Pet. Technol.*, 24: 723—740.

Moore, P.L., 1973. Casing-seat-testing — why and how. *Oil Gas J.*, 71(32): 72—74.

Mottley, J.R., 1968. Completing and producing hot corrosive gas wells. *World Oil*, 166(7): 69—72.

Perkins, T.K., Rochon, J.A. and Knowles, C.R., 1973. Studies of pressures generated upon refreezing of thawed permafrost around a wellbore. *SPE 4588, 48th AIME Fall Meet.*, Las Vegas, Nev., October.

Peterson, H., Ott, H. and Kloeckner, A., 1973. Probleme beim Bohren auf Hochdruckgaslagerstatten in Norddeutschland. *Erdoel Erdgas Z.*, 89(7): 240—248.

Record, L., 1968. Engineered concept optimizes deep drilling. *Pet. Eng.*, 40(3): 66—74.

Rehm, B., 1969. Pressure control in drilling, 5. Pressures and pressure gradients. *Oil Gas J.*, 67(39): 78—82.

Reid, C.A., 1970. Here are drilling fluids being used in Permian basin's pressured formations. *Oil Gas J.*, 68(17): 78—83.

Rieke, III, H.H. and Chilingarian, G.V., 1974. *Compaction of Argillaceous Sediments*. Elsevier, Amsterdam, 424 pp.

Robinson, J.D. and Vogiatzis, J.P., 1972. Magnetostatic methods for estimating distance and direction from a relief well to a cased wellbore. *J. Pet. Technol.*, 24: 741—749.

Runge, R.J., Worthington, A.E. and Lucas, D.R., 1969. Ultra-long spaced electric log (ULSEL). *Trans. Soc. Prof. Well Log Analysts*, H-1-22.

Shalavin, A.M. and Klyushin, G.P., 1972. Method for determining the pressure on the casing pipes with freezing of the flushing fluid in the well. *Burenie*, 9: 24—26.

Shishov, V.A., Kuksov, A.K., Sukhanov, V.B., Ponomarev, V.A. and Burak, L.A., 1970. Collapse of casing during sectional cementing of deep wells. *Gazov. Prom.*, 9: 3—6.

Speel, L., 1967. High-pressure gas well completions. *Proc. 7th World Pet. Congr.*, Mexico City, 2: 431—444.

Taylor, D.B. and Smith, R.K., 1970. Improving fracture gradient estimates in offshore drilling. *Oil Gas J.*, 68(15): 67—72.

Terzaghi, K.V., 1923. Die Berechnung der Durchlaessigkeitsziffer des Tones aus dem Verlauf der Hydrodynamischen Spannungserscheinungen. *Sitzungsber. Akad. Wiss. Wien, Math. Naturwiss. Kl. Abts.*, 2A: 105—132.

Van Hook, W.A. and Feemster, W.E., 1969. Permanent completion techniques used in the high-pressure portion of the Anadarko Basin. *API Prod. Div., Mid-Continent District Meet.*, Hot Springs, Arkansas, April, Preprint 851-43-H.

West, E.R. and Lindsey, H.E., 1968. Equipping and completing deep West Texas gas wells. *World Oil*, 166(6): 126—132.

Werner, K. and Becker, J., 1968. Zweizonenfoerderung in Hochdruck-Gasbohrungen. *Erdoel Kohle*, 21(12): 757—763.

Wheeler, R. and Moriarty, D.G., 1969. World's largest/strongest casing set. *Pet. Eng.*, 41(5): 105—111.

Chapter 7. RESERVOIR ENGINEERING CONCEPTS IN ABNORMAL FORMATION PRESSURE ENVIRONMENTS

Introduction

Today, reservoir engineering has emerged as a highly technical and specialized field of petroleum engineering with the objective to describe reservoir and production behavior based upon geologic and petrophysical data, reservoir fluid properties, in-situ thermodynamic (pressure, temperature) regimes, etc. (Muskat, 1937; Pirson, 1958; Craft and Hawkins, 1959).

Over the last decade, increased attention has been focused on the analysis of reservoir reserves, behavior, and possible mechanisms important to the production from abnormally high-pressured reservoir rocks. Frequently, these overpressured reservoirs in shale/sand sequences, containing mainly gas, are of limited areal extent and/or have limited aquifers, and often do not behave as volumetric gas reservoirs. This complicates gas-in-place estimates and usually results in much too optimistic reserve figures.

To explain these apparent limitations with conventional reservoir evaluation and prediction methods, several possible reservoir mechanisms have been proposed in the literature to describe more reliably overpressured hydrocarbon deposits such as those found in geologic environments similar to the U.S. Gulf Coast area.

Suggested mechanism for overpressure reservoirs include (1) water influx from shale zones adjacent to potential pay sands (*shale water influx*), (2) *rock compressibility* and *rock failure*, and (3) water influx into the reservoir from *limited aquifers*, etc.

However, published studies show that essentially any of these energy sources can be used to match mathematically observed field performance from a single field. This is illustrated in this chapter by the NS2B reservoir in the Ossum gas field, Lafayette Parish, Louisiana.

The following discussion presents several of these concepts and models and, whenever possible, illustrates these mechanisms by actual field observations using both production and well-logging data. However, it is left to the reader to evaluate in his own mind the possible merits of each of the proposed mechanisms.

Reservoir mechanisms

Shale water influx

Gas-in-place reserves are frequently predicted using a graphical solution for the gas material balance equation. A special case of this material balance equation, under specific restrictive assumptions, is the *linear (straight-line) relationship between P/Z and the cumulative gas production* (P = reservoir pressure, Z = gas deviation factor). Based on the equation of state and corrected for compressibility, the reservoir is considered to be closed and without any water encroachment.

For years it has been observed that in many gas reservoirs the production curves show a slow decline in the early-life history after which they become steeper. Fig. 7.1 shows such a typical trend for an overpressured gas reservoir in the U.S. Gulf Coast area.

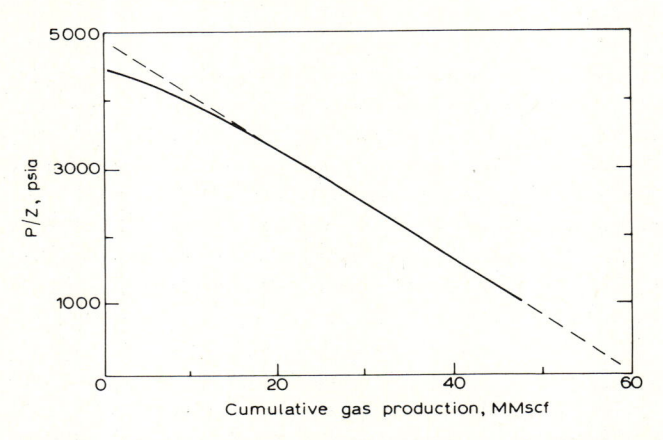

Fig.7.1. Typical *P/Z* versus cumulative production behavior for an overpressured gas reservoir, U.S. Gulf Coast area. Cumulative gas production is in millions of standard cubic feet (MMscf; standard conditions: 60°F and 14.7 psi).

A detailed study of the effect of water influx on plots of *P/Z* versus cumulative gas production has been made by Bruns et al. (1965). Various modes of gas production and water encroachment, such as proposed by Schilthuis (1936), Hurst (1943), and Van Everdingen and Hurst (1949), were investigated. Results of this study definitely illustrate that not only errors in the field measurements but also water influx can cause the typical curved portion during the early-life production history of a gas reservoir.

Therefore, the authors correctly concluded that it is erroneous to extrapolate early production data on a straight line without considering possible water influx. The study by Bruns et al. (1965) suggested the possibility of overestimating gas-in-place reserves by as much as 100% (Fig.7.2).

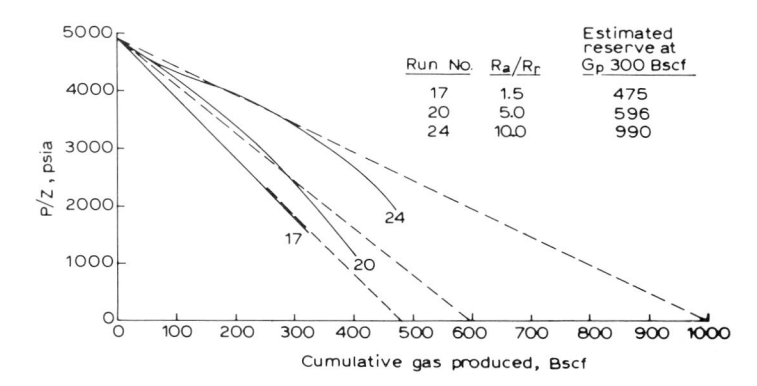

Fig.7.2. Comparison of P/Z curves for increasing aquifer size (R_a = radius of aquifer, R_r = radius of reservoir), Van Everdingen — Hurst's radial finite method (after Bruns et al., 1965). Cumulative gas production is in billions of standard cubic feet (Bscf). (Courtesy of *Journal of Petroleum Technology*.)

Wallace (1969) has studied reservoir performance data of several overpressured gas reservoirs located in South Louisiana. He also suggested that: "Shale compaction may be resumed by withdrawal of gas or fluids from the reservoir, thus creating a pressure differential across the shale/sand interface. This initiates a type of abnormally pressured water influx drive that we will call *shale water drive*. This shale water drive is controlled by the factors involved in Darcy's law, i.e.: (1) the permeability of the shale, (2) the pressure differential across the shale/sand interface, (3) the cross sectional area of the sand interface, and (4) the viscosity of the water."

Thus, Wallace (1969) proposed shale water influx as a potential reservoir energy to support numerous field observations, a concept challenged by Hurst (1969). However, field observations, such as reservoir performance data and well-log responses, appear to lend some support to the shale water influx concept. Published reservoir performance data include the Anderson "L" reservoir, Nueces County, Texas (Duggan, 1972), the Chocolate Bayou field, Brazoria County, Texas (Fowler, 1968), and several studies in gas fields in Louisiana (Wickenhauser, 1966; Lavaquial, 1971).

Recently, Bourgoyne et al. (1972) gave a detailed review of shale properties as related to the shale water influx theory in overpressured reservoirs. The authors noted in their model and field case studies that the pressure distribution in the shales should be very steep and lower in the vicinity of permeable sands. This concept is supported by independent mathematical model studies carried out by Bredehoeft and Hanshaw (1968) and Smith (1973), and well-log-derived field observations (Fertl and Timko, 1970a). The latter are illustrated and discussed later in this chapter.

Bourgoyne et al. (1972) also proposed an appropriate method for extrapolating early P/Z versus cumulative gas behavior in overpressured gas reservoirs and illustrated this technique on the NS2B reservoir in the Ossum field, Lafayette Parish, Louisiana. Basically, the model computes unsteady-state water influx with permeability and compressibility being a function of pressure. In the two field cases cited, a shale permeability of the order of 10^{-5} millidarcy (md) and shale bulk compressibility of 40×10^{-6} psi^{-1}* would account for the pressure support mechanism. Furthermore, shale water influx appears to be approximately proportional to the square root of the permeability-compressibility product.

Rock compressibility and rock failure

In 1969 Harville and Hawkins discussed the performance study of the NS2B gas reservoir in the North Ossum field located in Lafayette Parish, Louisiana. Pertinent reservoir data are given in Table 7.I. Furthermore, good geologic data on the reservoir and faulting in this area suggest a closed reservoir with a limited aquifer. Production data of the reservoir are shown in Fig.7.3 and P/Z versus cumulative production is presented in Fig.7.4. Note that extrapolation of early-life production data (Fig.7.4) indicates initial gas reserves as high as 220 billions of standard cubic feet (Bscf; standard conditions: 60°F and 14.7 psi), approximately twice the volumetric gas-in-place estimate given in Table 7.I.

TABLE 7.I

Reservoir data of NS2B reservoir, North Ossum field, Lafayette Parish, Louisiana, U.S.A. (after Harville and Hawkins, 1969)

Depth, ft	12,500
Pressure, psia (pound per square inch absolute)	8,921
Gradient, psi/ft	0.725
Temperature, °F	248
Gas/water contact, ft	12,580
Average gross sand, ft	100
Porosity (33 ss*; quite uniform)	0.235
(electric logs)	0.24
Connate water (electric logs)	0.34
Permeability, md (33 ss*)	200
Producing wells	4
Geologic control wells	18
Dew-point pressure, psia	6,920
Initial GOR, bbl/MMscf	160
Condensate gravity, °API	47
Net bulk gas volume, MMscf	2,480
Initial Z factor	1.472
Initial gas in place, Bscf (volumetric)	114
Initial gas compressibility	30×10^{-6} psi^{-1} at 8,921 psia

*ss = sandstone.

*1 psi^{-1} = 14.223 cm^2/kg.

RESERVOIR MECHANISMS 279

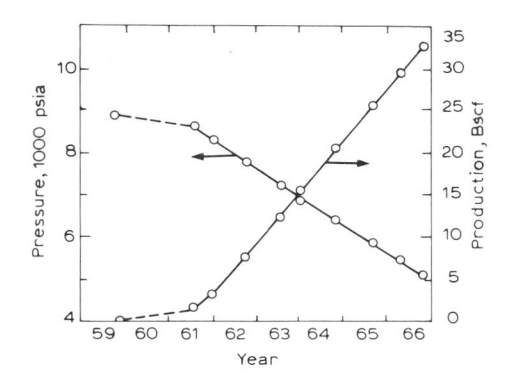

Fig.7.3. Pressure—production history, NS2B reservoir, North Ossum field, Lafayette Parish, Louisiana (after Harville and Hawkins, 1969). (Courtesy of *Journal of Petroleum Technology*.)

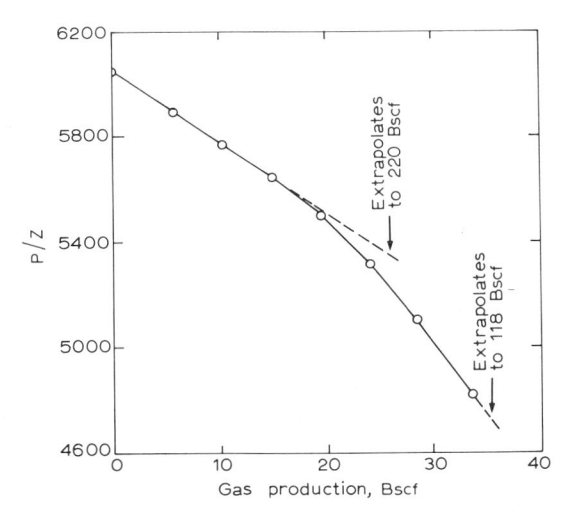

Fig.7.4. *P/Z* versus production for the NS2B reservoir, North Ossum field, Lafayette Parish, Louisiana (after Harville and Hawkins, 1969). (Courtesy of *Journal of Petroleum Technology*.)

The authors suggest that as reservoir pressure is depleted, the increase in net overburden pressure causes inelastic rock compaction or rock failure, and as the failure continues with decreasing pore pressure, rock compressibility decreases until it eventually reaches normal values in the range of 6 microsip (i.e. microsip = 10^{-6} psi^{-1}). Fig.7.5 illustrates rock compressibility for the subject reservoir as a derivative of the pore volume computation (using eq. 7.1) as a function of pore fluid pressure:

$$V_p = \frac{GB_{gi}S_w}{1-S_w}\underbrace{[1 + c_w(P_i - P)]}_{\text{water}} + \underbrace{(G - G_p)B_g}_{\text{gas}} \qquad (7.1)$$

where B_g = gas volume factor at pressure P in cubic feet/standard cubic feet (ft³/scf), B_{gi} = gas volume factor at initial pressure in ft³/scf, c_w = connate water compressibility in psi⁻¹ (3×10^{-6} psi⁻¹ was used), G = initial gas in place in scf, G_p = gas produced when pressure is P in scf, P = average pore pressure at later time in psia, P_i = initial reservoir pore pressure in psia, S_w = connate water saturation expressed as a fraction of pore volume, and V_p = total pore volume in ft³.

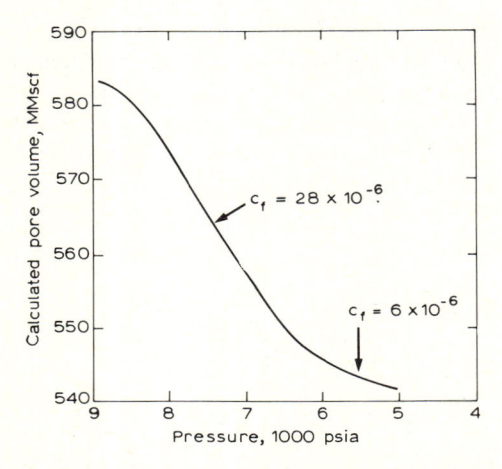

Fig.7.5. Calculated pore volume of the NS2B reservoir, North Ossum field, Lafayette Parish, Louisiana, assuming no water influx and gas-in-place of 114 Bcf (after Harville and Hawkins, 1969). c_f = formation compressibility (see Chapter 2) in psi⁻¹. (Courtesy of *Journal of Petroleum Technology*.)

As shown in Fig.7.5, a maximum rock compressibility of 28 microsip occurs during early reservoir depletion when rock failure is presumed to occur. At about 6000 psi [422 kg/cm²], the rock compressibility again "normalizes" (6 microsip). Low rock compressibility plus higher gas compressibility (due to lower pressure) causes the reservoir performance to match closely that of a constant volume system, which should result in reasonably accurate *P/Z* extrapolation and thus more realistic reserve estimates.

This is confirmed by field observations on the subject reservoir since after production of some 20 Bscf, which caused the pressure decline to 6500 psi [457 kg/cm²], the *P/Z* extrapolation results in a 118-Bscf gas-in-place estimate (Fig.7.4). This is in excellent agreement with the volumetric reserve estimate of 114 Bscf (Table 7.I).

The authors, however, wisely cautioned that these results do not prove, of course, that rock failure is the major source of pressure and thus energy in the reservoir studied, despite some independent supporting rock mechanics data on the Sespe Sandstone (Fatt, 1954). However, this study does suggest that a mechanism of rock compressibility and associated rock failure is a possibility in overpressured reservoir rocks.

Peripheral water influx

Numerically generated and actual field performance data of abnormally pressured gas reservoirs were analyzed recently by Bass (1972). Starting with the concept of a combination of rock compressibility, shale water influx, and peripherical water influx as superimposed mechanisms, a generalized mathematical volume balance expression was developed. Gas reservoir performance data were studied to investigate the effect and magnitude of each of these three possible energy sources in overpressured reservoirs. Bass concluded that peripherical water influx makes one of the greatest long-term energy contributions and, based on this concept, presented a method for gas-in-place estimates. Neglecting any shale water influx, the initial gas-in-place is described by the following equation:

$$G = \frac{G_{p2}\left[-(P/Z)^{-2}\frac{d(P/Z)}{dG_p} + (P/Z)^{-1}\right]_2 + G_{p1}\left[-(P/Z)^{-2}\frac{d(P/Z)}{dG_p} + (P/Z)^{-1}\right]_1}{\left[\left(-(P/Z)^{-2}\frac{d(P/Z)}{dG_p}\right)_2 + \left((P/Z)^{-2}\frac{d(P/Z)}{dG_p}\right)_1\right]} \tag{7.2}$$

where G = original gas in place in scf, G_p = cumulative gas produced in scf, P = reservoir pressure in psi, and Z = gas deviation factor.

Attempts by the author to apply this method using conventional field data available from several overpressured gas reservoirs have met with inconsistent success. Based upon this study, however, Bass concluded that:

(1) If data is taken with care, the original gas-in-place can be determined early in the productive life of a gas reservoir.

(2) Special pressure data is required to determine the performance of gas reservoirs.

(3) It is recommended that pressures be taken after 4, 6, and 8 months during a period of high gas production rate, and these pressures be used to determine gas-in-place by substitution into eq. 7.2.

Hammerlindl's prediction methods in overpressured depletion-type gas reservoirs

According to Hammerlindl (1971), who supported observations made previously, two distinct slopes are evident on P/Z plots in overpressured reservoirs with the final slope being steeper than the initial one. Consequently, gas-in-place estimates based on early production history are erroneously high. Hammerlindl attributed the initial slope to gas expansion, reservoir rock compaction, crystal expansion, and water expansion. As soon as pressure approaches normal pressure (depletion), the reservoir will behave as a normal gas expansion reservoir. At this stage, effective compressibility of the system is about equal to gas compressibility, which is the cause of the second, steeper slope. In other words, reserve estimates using early-life bottom-hole shut-in pressure and cumulative production should be adjusted to allow for pore-volume change and water expansion.

Hammerlindl developed two prediction methods for initial gas-in-place estimates from early production history using reservoir rock compressibility, gas production, and shut-in bottom-hole pressure.

Method A utilizes the ratio of average values for effective compressibility (Craft and Hawkins, 1959) and gas compressibility, while Method B includes compressibility in the expansion term of the material balance equation.

Method A

P/Z correction to obtain a realistic gas-in-place (GIP) estimate comprises the following steps:

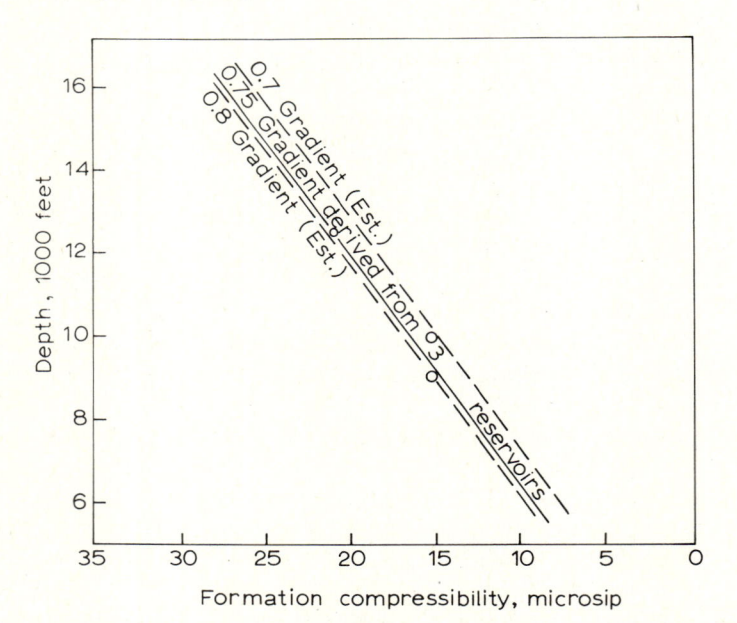

Fig.7.6. Formation compressibility versus depth in abnormally pressured reservoirs, e.g. 0.725-psi/ft [0.167 kg cm^{-2} m^{-1}] gradient in North Ossum field, Louisiana; 0.714-psi/ft [0.165 kg cm^{-2} m^{-1}] gradient in sand 34 W.M.I., southwestern Texas; and 0.811-psi/ft [0.187 kg cm^{-2} m^{-1}] gradient in sand 35 M.I. deep, southwestern Texas (after Hammerlindl, 1971). (Courtesy of the Society of Petroleum Engineers of A.I.M.E.)

RESERVOIR MECHANISMS 283

(1) Determine formation compressibility, c_f, from appropriate data (Fig.7.6), assuming no water influx.

(2) Compute gas compressibility data, c_{gi} and c_{g2}, when $P_2 \geqslant 0.5$ psi/ft of depth (see Craft and Hawkins, 1959, pp. 269—273):

$$c_{gi} = -\frac{\partial V}{V \partial P} = \frac{1}{P_i} - \frac{1}{Z_i} \cdot \frac{\Delta Z}{\Delta P} \tag{7.3}$$

where $\Delta Z/\Delta P$ = slope of gas deviation factor (Z) curve at Z_i, and ΔV = gas-volume change due to rock and water compressibility.

(3) Calculate effective compressibility, c_{eff}, at pressure P_i and P_2 as proposed by Craft and Hawkins (1959):

$$c_{effi} = (c_{gi} S_{gi} + c_{wi} S_{wi} + c_f)/S_{gi} \tag{7.4}$$

where c_w = compressibility of water, and S_g = gas saturation.

(4) Find the average value for (c_{eff}/c_g) which approximates:

$0.5[(c_{effi}/c_{gi}) + (c_{eff2}/c_{g2})]$

(5) Solve for real gas-in-place (GIP$_r$):

$$\text{GIP}_r = \text{GIP}_a/(c_{eff}/c_g) \tag{7.6}$$

where GIP$_a$ = apparent gas-in-place (GIP) estimate as obtained from the conventional P/Z plot using extrapolation of early-life reservoir performance data (see Fig.7.4).

(6) Finally, to predict the recoverable gas reserves, (a) determine by extrapolation on the P/Z plot the point corresponding to the normal pressure gradient (0.46 psi/ft in P/Z-value), (b) plot GIP$_r$ on abscissa at P/Z = 0, (c) connect this point by a straight line with the normal gradient point, and (d) determine gas reserves for any desired P/Z abandonment value.

Method B

P/Z correction basically includes compressibility in the expansion term of the material balance equation. The proposed procedure includes the following steps:

(1) Use the conventional P/Z plot to determine GIP$_a$.

(2) Compute the gas formation volume factor, B_{gi} and B_{g2} in scf/ft^3, from (T is formation temperature in °R):

$$B_g = 35.35 P/(ZT) \tag{7.7}$$

where pressures should be above the normal gradient.

(3) Determine formation compressibility, c_f, in the same fashion as shown in Method A.

(4) Solve for the ratio of conventional material balance equation to the one including both formation compressibility and water expansion terms, G_{pr}:

$$G_{pr} = \frac{B_{gi} - B_{g2}}{B_{gi} - B_{g2} + \dfrac{B_{g2} \cdot \Delta P(c_f + c_w S_{wi})}{1 - S_{wi}}} \qquad (7.8)$$

where ΔP = change in pressure, and S_w = water saturation.

(5) Determine real GIP as in step 5 and the recoverable gas reserves as shown in step 6 of Method A.

Field example

A practical field application of these two methods for the NS2B reservoir in the Ossum field, which was also used by Harville and Hawkins (1969) in their rock failure study, and using data given in Table 7.I, will be illustrated next. Note that both methods essentially give the same results.

Method A in NS2B reservoir. Formation compressibility, using values from Fig.7.3, gives:

$c_f = (VP_i - VP_2)/(VP_i \Delta P) = (583 \times 10^6 - 557.5 \times 10^6)/(583 \times 10^6 \times 2000)$

= 21.9 microsip

Gas compressibility:

$c_{gi} = (1/P_i) - \Delta Z/(Z_i \Delta P) = 1/8921 - 0.242/(1.472 \times 2000) = 30$ microsip

c_{g2} at 6921 psia = 48.7 microsip

Effective compressibility:

$c_{effi} = (c_{gi} S_{gi} + c_{wi} S_{wi} + c_f)/S_{gi}$

$= (30 \times 0.66 + 3 \times 0.34 + 21.9) \times 10^{-6}/0.66 = 64.8$ microsip

c_{eff2} at 6921 psia = $(48.7 \times 0.66 + 3 \times 0.34 + 21.9) \times 10^{-6}/0.66 = 83.5$ microsip

To be absolutely correct, a new S_w and S_g should be obtained at this pressure, but the accuracy of the S_{wi} and S_{gi} does not justify it.

Average compressibility:

$c_{effi}/c_{gi} = 64.8 \times 10^{-6}/(30 \times 10^{-6}) = 2.16$

$c_{eff2}/c_{g2} = 83.5 \times 10^{-6}/(48.7 \times 10^{-6}) = 1.71$

$(c_{eff}/c_g)_{ave} = (2.16 + 1.71)/2 = 1.935$

Real GIP:

$GIP_r = GIP_a/(c_{eff}/c_g)_{ave} = 220$ Bscf$/1.935 = 114$ Bscf

Method B in NS2B reservoir. Conventional P/Z plot gives early-life GIP:

GIP_a = 220 Bscf

Formation volume factor:

$B_{gi} = 35.35\, P_i/(Z_i T) = 35.35 \times 8921/(1.472 \times 708) = 302.5$ scf/ft³

B_{g2} at 6921 psia = 381.5 scf/ft³

Formation compressibility, using values from Fig.7.3, gives:

$c_f = (VP_i - VP_2)/(VP_i \Delta P) = (583 \times 10^6 - 557.5 \times 10^6)/(583 \times 10^6 \times 2000)$

= 21.9 microsip

Solve for G_{pr}:

$G_{pr} = (302.5 - 281.5)/\{302.5 - 218.5 + [(281.5 \times 2000)/(1 - 0.34)]$
    $\times (21.9 - 0.34 \times 3) \times 10^{-6}\} = 0.518$

Determine real GIP:

$GIP_r = GIP_a/G_{pr} = 220$ Bscf/0.518 = 114 Bscf

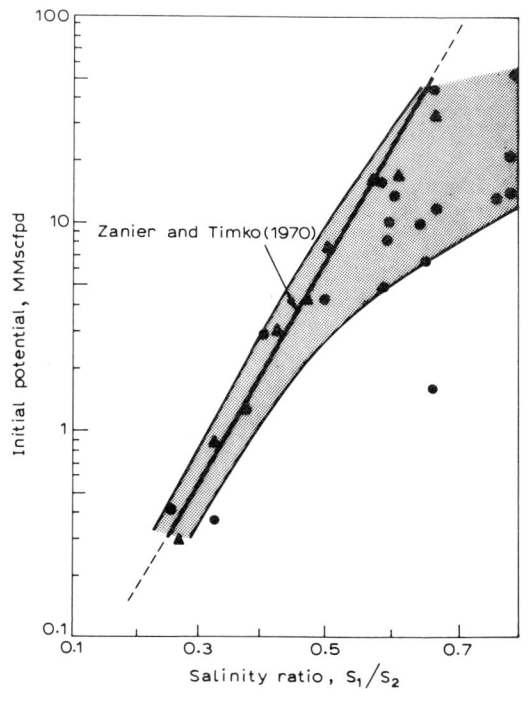

Fig.7.7. Initial well potential (in millions of standard cubic feet per day) versus formation water salinity in overpressured Morrow gas sand, Oklahoma, U.S.A. (after data by (▲) Zanier and Timko, 1970, and (●) Ferran, 1973). S_1 = salinity of overlying sand, S_2 = salinity in Morrow Sand.

Initial well potential versus formation water salinities

Zanier and Timko (1970) suggested the use of formation water salinities in predicting the initial performance of the potential Morrow gas sand in wells drilled in Blaine, Canadian, and Dewey counties, Oklahoma, U.S.A. Based upon these data and additional observations in a follow-up investigation by Ferran (1973), it appears that the increase in formation water salinity and the decrease in initial well potential can be correlated (Fig.7.7). Such data can also be used as an additional mapping technique to outline the areas of main interest in the Morrow Sand and similar geological environments.

Pressure depletion and abandonment pressures

Data by Mullins (1962) on several abandoned high-pressure gas reservoirs caused by excessive water cut are given in Table 7.II. As shown, abandonment pressures in these fields are close to hydrostatic or still even higher. Similar pressure data on three high-pressure Russian wells have been reported by Fertl (1971 — see Fig.7.8). The question, "Are these reservoirs adequately depleted?" has to be considered from both the reservoir performance and well completion points of view.

TABLE 7.II

Abandoned, high-pressure gas reservoirs, U.S. Gulf Coast area (after Mullins, 1962)

Reservoir	Depth (ft)	Cumulative gas production (MMscf)	Reservoir pressure (psia)	
			at abandonment	hydrostatic*
A	11,072	11,155	5,800	+650
B	11,550	427	4,900	−470
C	11,765	1,727	8,800	+3330
D	11,440	133	4,700	−620

*Positive sign indicates above, negative sign indicates below hydrostatic pressure.

In Chapter 8, we will illustrate drastic short-term pressure depletions of as much as several thousand psi during a few hours production test (see Fig.8.22). As illustrated in the same chapter, gas reservoirs in "superpressure" sand/shale environments are noncommercial since such intervals are frequently characterized by extremely high pressures and fast drawdown, indicative of low volume and very limited areal extent.

Furthermore, mechanical production problems often appear after short-term high rate flow periods. For example, sand and especially plastic shale "flow" through the perforations into the borehole plugging the tubing string. Despite repeated workovers, the results are always the same, and reservoirs

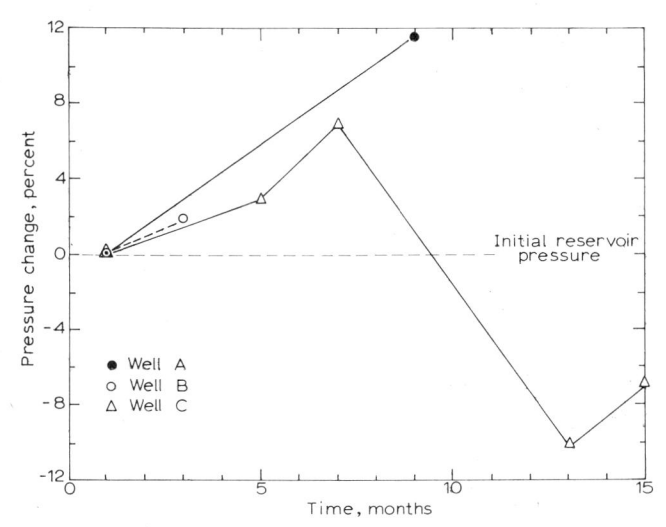

PRESSURE DEPLETION AND ABANDONMENT PRESSURES 287

Fig.7.8. Formation pressure versus production time in three overpressured wells in the U.S.S.R. (after Fertl, 1971). (Courtesy of the Society of Petroleum Engineers of A.I.M.E.)

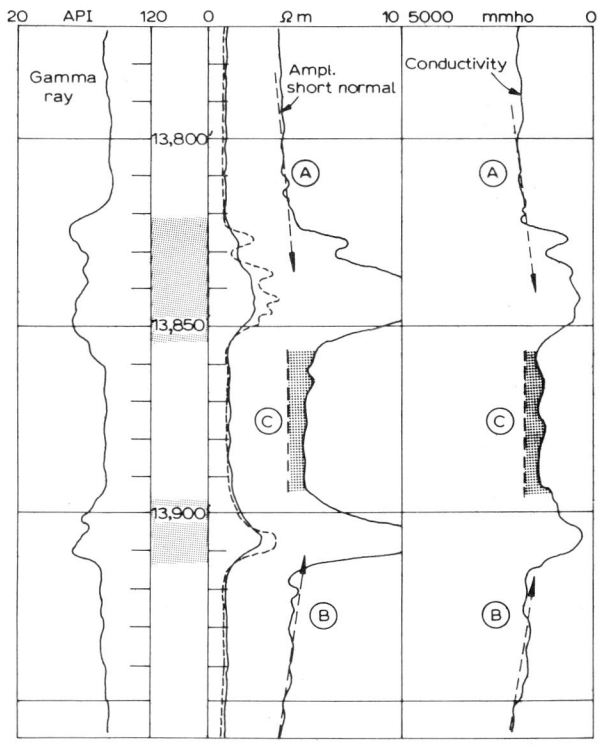

Fig.7.9. Induction-electric log in a high-pressure well in offshore Louisiana. Pressure gradients (A, B) approaching permeable sands, and the apparent lower-pressured shale (C) sandwiched between both sands. Second log track from the left gives the short normal and induction curves.

are frequently abandoned. Such observations are made along the U.S. Gulf Coast area both on- and offshore and in similar geologic environments elsewhere.

In the present chapter, the *mechanism* of *shale water influx* has been considered in detail from a reservoir engineering point of interest. It is certainly also of interest to investigate the possibility of detecting and/or monitoring such in-situ behavior with, say, conventional well-logging techniques (Fertl and Timko, 1970a,b, 1972).

The following field cases should be considered in the light of mathematical model studies by Bredehoeft and Hanshaw (1968) and Smith (1973) which suggested a varying pore pressure gradient in overpressured shale sections. These studies also suggest that the highest excess pressure is located near the center of the shale masses, whereas less excess pressure is found approaching permeable zones, such as sand and sandstones. Such a concept is supported by field observations in newly drilled wells; for example, in the U.S. Mid-Continent area (see Chapter 5, Figs.5.44 and 5.45), offshore Louisiana (Fig.7.9) and elsewhere.

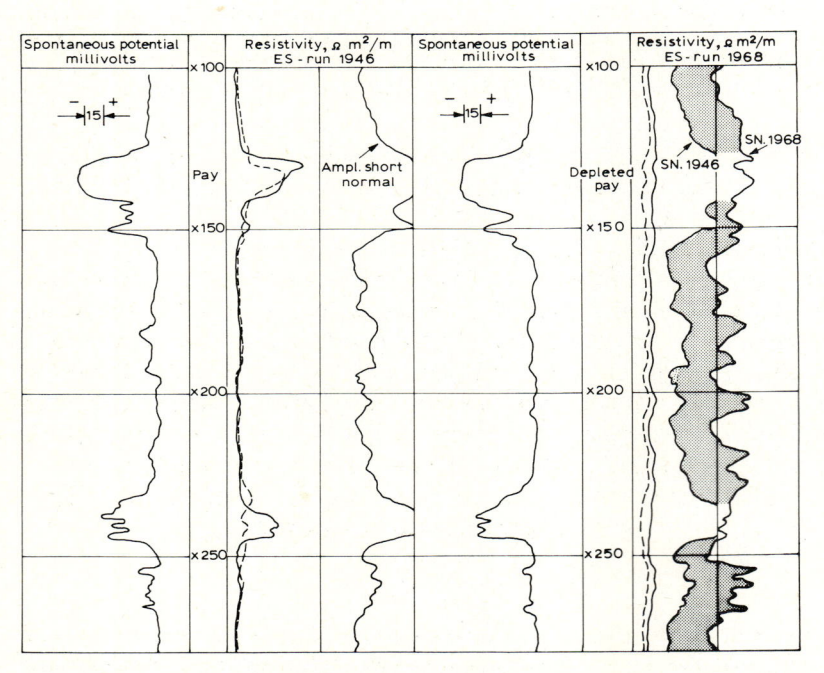

Fig.7.10. After over two decades of production, formation pressures have been reduced significantly in this Louisiana well, as indicated by increased shale resistivity (after Fertl and Timko, 1972). This observation by using electric logs was made possible when a sidetrack was drilled in 1968, only 50 ft [15.2 m] from the original well completed in 1946. (Courtesy of *World Oil.*)

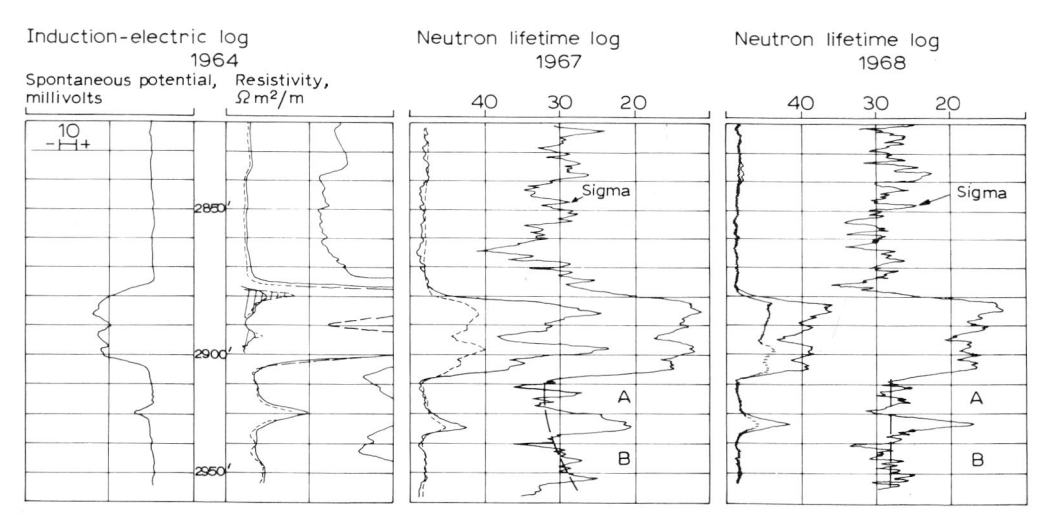

Fig.7.11. Pulsed neutron logs, run several years following the completion of a high-pressured Louisiana well, show effects of pressure depletion (after Fertl and Timko, 1970). The increase in sigma values in the pay is due to increased water saturations caused by production. The sigma decrease in shale A apparently was caused by increased compaction and decreased porosity (sigma is the macroscopic thermal neutron capture cross-section in 10^{-3}/cm). A = shale adjacent to permeable sand, B = shale distant to permeable sand.

The phenomenon of long-term pressure depletion over entire sections is shown in Fig.7.10 and has been also illustrated in Chapter 5 (Fig.5.29) previously.

On the other hand, Fig.7.11 shows *"shale water depletion"* restricted to the vicinity of the pay sand in an overpressured South Louisiana well. Two pulsed-neutron logs have been run a year apart to monitor hydrocarbon saturation changes in the pay which is being produced in several adjacent wells. Sigma-value changes in the sand are due to the increase in water saturation. Also note the changes in the adjacent shales. Zone A, the shale next to the pay, shows a marked sigma decrease due to increased compaction and porosity decrease (which suggests a shale water influx into the sand), whereas zone B, the shale some distance from the permeable sand, shows considerably less, if any, variations due to pressure drawdown.

References

Bass, D.M., 1972. Analysis of abnormally pressured gas reservoirs with partial water influx. *SPE 3850, 3rd Symp. on Abnormal Subsurface Pore Pressure*, Louisiana State Univ., Baton Rouge, La., May.

Bourgoyne, A.T., Hawkins, M.F., Lavaquial, F.P. and Wickenhauser, T.L., 1972. Shale water as a pressure support mechanism in superpressured reservoirs. *SPE 3851, 3rd Symp. on Abnormal Subsurface Pore Pressure*, Louisiana State Univ., Baton Rouge, La., May.

Bredehoeft, J.D. and Hanshaw, B.B., 1968. On the maintenance of anomalous fluid pressures, 1. Thick sedimentary sequences. *Geol. Soc. Am. Bull.*, 79: 1097—1106.

Bruns, J.R., Fetkovich, M.J. and Meitzen, V.C., 1965. The effect of water influx on P/Z-cumulative gas production curves. *J. Pet. Technol.*, 17: 287—291.

Craft, B.C. and Hawkins, M.F., 1959. *Applied Petroleum Reservoir Engineering*. Prentice-Hall, Englewood Cliffs, N.J., 437 pp.

Duggan, J.O., 1970. The Mobil-David (Anderson "L") field — an abnormally pressured gas reservoir in South Texas. *SPE 2938, 45th AIME Fall Meet.*, Houston, Texas, October; also *J. Pet. Technol.*, 24: 132—138, 1972.

Fatt, I., 1954. Compressibility of sandstones at low to moderate pressures. *Bull. Am. Assoc. Pet. Geol.*, 42: 1924—1957.

Ferran, L.H., 1973. *Evaluation of abnormally high and low pressured Morrow sands in northwestern Oklahoma using well logs and water sample data*. M.S. Thesis, Univ. of Tulsa, Tulsa, Okla., 110 pp.

Fertl, W.H., 1971. A look at abnormally pressured formations in the U.S.S.R. *SPE 3613, 46th AIME Fall Meet.*, New Orleans, La., October.

Fertl, W.H. and Timko, D.J., 1970a. Association of salinity variations and geopressures in soft and hard rocks. *Trans. Soc. Prof. Well Log Analysts*, Paper J, 24 pp.

Fertl, W.H. and Timko, D.J., 1970b. How abnormal pressure detection techniques are applied. *Oil Gas J.*, 68(2): 62—71.

Fertl, W.H. and Timko, D.J., 1972. How downhole temperatures, pressures affect drilling, 4. Pitfalls in overpressure prediction. *World Oil*, 175(4): 45—50.

Fowler, Jr., W.A., 1968. Pressures, hydrocarbon accumulation and salinities — Chocolate Bayou field, Brazoria County, Texas. *SPE 2226, 43rd AIME Fall Meet.*, Houston, Texas, September; also *J. Pet. Technol.*, 22: 411—423, 1970.

Hamby, T.W. and Smith, J.R., 1972. Contingency planning for drilling and producing high-pressure sour-gas wells. *J. Pet. Technol.*, 24: 347—356.

Hammerlindl, D.J., 1971. Predicting gas reserves in abnormally pressured reservoirs. *SPE 3479, 46th AIME Fall Meet.*, New Orleans, La., October.

Harville, D.W. and Hawkins, M.F., 1969. Rock compressibility and failure as reservoir mechanisms in geopressured gas reservoirs. *J. Pet. Technol.*, 21: 1528—1530.

Hurst, W., 1943. Water influx into a reservoir and its application to the equation of volumetric balance. *Trans. AIME*, 151: 57—73.

Hurst, W., 1969. On the subject of abnormally pressured gas reservoirs. *J. Pet. Technol.*, 21: 1509—1510.

Lavaquial, F.P., 1971. *Water influx into petroleum reservoirs from adjacent shales*. M.S. Thesis, Louisiana State Univ., Baton Rouge, La., 120 pp.

Mullins, J.D., 1962. Some problems of superhigh-pressure gas reservoirs in the Gulf Coast area. *J. Pet. Technol.*, 14: 935—938.

Muskat, M., 1937. *The Flow of Homogeneous Fluids Through Porous Media*. McGraw-Hill, New York, N.Y., 763 pp.

Pirson, S.J., 1958. *Oil Reservoir Engineering*. McGraw-Hill, New York, N.Y., 735 pp.

Schilthuis, R.J., 1936. Active oil and reservoir energy. *Trans. AIME*, 118: 33—53.

Smith, J.E., 1973. Shale compaction. *Soc. Pet. Eng. J.*, 13: 12—22.

Van Everdingen, A.F. and Hurst, W., 1949. The application of the Laplace transformation to flow problems in reservoirs. *Trans. AIME*, 186: 305—324.

Wallace, W.E., 1969. Water production from abnormally pressured gas reservoirs in South Louisiana. *J. Pet. Technol.*, 21: 969—983.

Wickenhauser, T.L., 1966. *Shale water as a pressure support mechanism in superpressure reservoirs*. M.S. Thesis, Louisiana State Univ., Baton Rouge, La., 115 pp.

Zanier, A.M. and Timko, D.J., 1970. Prediction of Morrow sand performance and geologic environment by well log salinity. *API Mid-Continent Spring Meet.*, Wichita, Kansas, Paper 851-44-5.

Chapter 8. APPRAISAL OF HYDROCARBON DISTRIBUTION AND RELATED ECONOMIC FACTORS IN NORMAL AND ABNORMAL FORMATION PRESSURE ENVIRONMENTS

Introduction

Wireline data in combination with the study of suitable source rocks, basic concepts of the origin of oil and gas, and geologic studies on delineating reservoir rocks (hydrocarbon traps) provide important information and guidelines for appraising the regional potential and probability of commercial hydrocarbon deposits. These concepts are also applicable and helpful in predicting the physical state of hydrocarbon systems, such as oil, condensate, or gas.

Numerous workers have investigated and/or reviewed basic concepts and considerations related to the origin of petroleum (Kartsev et al., 1954; Andreev et al., 1968; Dott, 1969; Hedberg, 1969; Vassoyevich et al., 1969; Cordell, 1972; Klemme, 1972; Pusey, 1973).

These and a multitude of similar studies have attempted to relate the occurrence of oil and gas to single parameters or their combination, such as source rock potential, geologic time and age, depth, pressure, temperature, sand/shale ratio, shale resistivity ratio, specific geochemical and hydrodynamic criteria, etc. In nature, however, a combination of several, superimposed factors will prevail, with individual parameters varying in importance.

Several of the more encouraging parameters, as discussed next, are readily available from wireline data, such as well logs.

Depth versus hydrocarbon accumulation

Effect of depth of reservoir rocks on hydrocarbon properties has been described for many areas since the early stages of the oil industry. For example, depth-dependent variations of petroleum properties were early suggested by Sorokin (1897) for Baku, U.S.S.R., described in Rumanian oil fields by Krejci-Graf (1934), and shown by Barton (1934) for the U.S. Gulf Coast area.

Today, worldwide average depths of the major hydrocarbon deposits are: 48% above 6000 ft [1829 m], 43% between 6000 and 9000 ft [1829 and 2744 m] and only 9% below 9000 ft [2744 m] (Klemme, 1972). However, one has to keep in mind that the decrease in total reserves with depth may

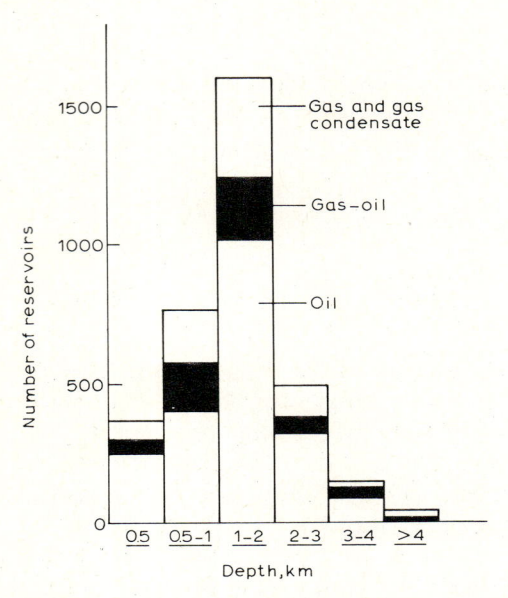

Fig.8.1. Depth-dependence of oil and gas reservoirs in the U.S.S.R. (after Vassoyevich et al., 1969).

be, to a certain extent, a reflection of the still less extensive exploration at great depth.

A similar depth-dependence is shown in Fig.8.1 for the oil and gas fields in the U.S.S.R. (Vassoyevich et al., 1969).

Generally, the regional distribution of hydrocarbons versus depth indicates an increase in gas reserves over oil reserves for both normal and abnormally high formation pressure environment. For example, Table 8.I lists record holders for deepest production in the United States. Note that since 1964 only gas wells have been record holders.

The increase in gas reserves with depth down to 14,000 ft [4268 m] is shown by Perry (1972) for southwest Louisiana (Fig.8.2). The decrease in reserves below that depth is a function of less exploration and, in general, probably smaller reservoirs in highly overpressured environments.

Depth-dependency of the oil and gas distribution in the Persian Gulf Basin, characteristic for many others, has been studied by Zvereva and Selitskii (1970) recently. About 80% of the oil reserves is concentrated on platform basin edges, and more than 98% of free gas reserves is associated with the leading depression (folded edge of the basin). Vertically, oil and gas accumulations in the basin are distributed in three zones:

(1) Oil-gas and oil pools in the interval 250—2500 m. More than 50% of oil reserves is concentrated there.

TABLE 8.I

Deep production statistics in the United States (after *World Oil*, 1973, February 15)

Year drilled	Well and location	Production	Producing depth (ft)*
1938	Fohs Oil Co.'s Buckley-Bourg 1, DeLarge field, South Louisiana	cond.	13,266
1943	Union Producing Co.-Fohs Oil Co.'s Fitzpatrick-Vizard 1, DeLarge field, South Louisiana	cond.	13,490
1944	Union Producing Co.-Fohs Oil Co.'s Buckley-Bourg 2, DeLarge field, South Louisiana	oil	13,503
1945	Shell Oil Co.'s Smith-State Unit 1-1, Weeks Island field, South Louisiana	oil	13,520
1946	Shell Oil Co.'s Smith-State Unit 1-2, Weeks Island field, South Louisiana	oil	13,778
1947	Shell Oil Co.'s Smith-State Unit 1-3, Weeks Island field, South Louisiana	oil	13,888
1947	The Texas Co.'s LaFourche Basin Levee Dist. 1, Queen Bess Island field, South Louisiana	cond.	13,904
1948	Pure Oil Co.'s Unit 1, West Poison Spider field, Wyoming	oil	14,307
1948	Denver Producing & Refg. Co.'s School Land 1-A, Cogar field, Caddo County, Oklahoma	gas	15,510
1949	Standard Oil Co. of California's Mushrush 5, Wasco field, Kern County, California	oil	15,530
1953	Shell Oil Co.'s Weeks-Gall Unit 1-1, Weeks Island field, South Louisiana	oil	17,122
1953	Shell Oil Co.'s Gonsoulin Minvielle-State 2, Weeks Island field, South Louisiana	oil	17,306
1953	Richfield Oil Corp.'s Cole Levee "A" 67-29, Coles Levee, North Field, California	oil	17,892
1956	Richardson & Bass et al.'s Humble-LL&E 1-L, Lake Washington field, South Louisiana	oil	21,465
1964	Forest Oil Co. and Robert J. Zonne et al.'s Charles J. Walker 1, Pecos County, West Texas	gas	21,793
1965	Union Oil Co. of Calif.'s No. 1 W.C. Tyrrell Unit 3, Pecos County, West Texas	gas	22,032
1967	Humble Oil & Refining Co.'s Henry Willbanks 11, Gomez field, Pecos County, West Texas	gas	22,610
1968	Gulf Oil Corp.'s "DD" Unit 1, Gomez field, Pecos County, West Texas	gas	22,752
1971	Texaco's E. Riggs Gas Unit 1	gas	23,040
1972	Union Oil Co. of California Bruner 1-33 NW SE NE, 33-11n-25w Beckham County	gas	24,548

*Bottom of perforated interval or bottom of open-hole completion interval.

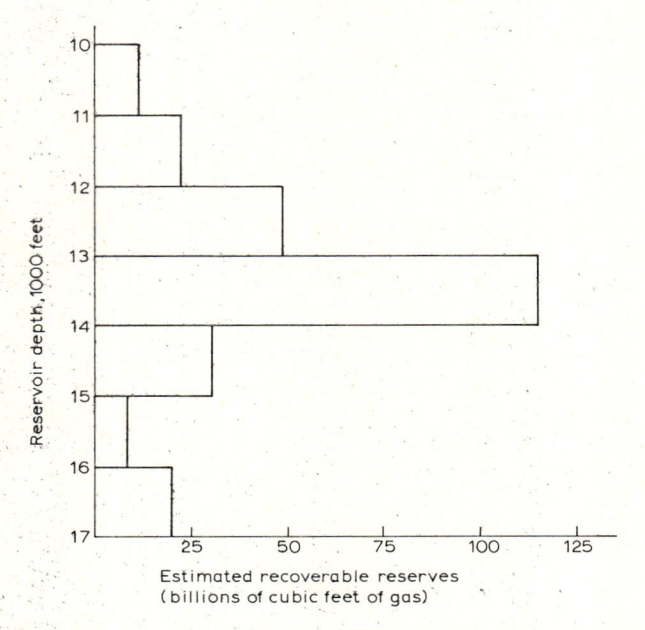

Fig.8.2. Estimated recoverable gas reserves versus depth in southwest Louisiana, U.S.A. (after Perry, 1972). (Courtesy of the Society of Petroleum Engineers of A.I.M.E.)

(2) Oil and gas-condensate pools at depths of 2500—4000 m, with substantial concentration of gas-condensate deposits.

(3) Probable zone of gas-condensate and gas deposits at depths exceeding 4000 m.

As yet, presence of the first two zones has been confirmed by exploratory drilling.

Formation temperature versus hydrocarbon distribution

The wide range of geothermal gradients (i.e. change of temperature with depth) from area to area over continental regions makes the selection of a universal "normal" gradient impractical (Moses, 1961; Landes, 1967).

In drilling for oil and gas, temperature gradients of 0.3°F/100 ft [0.55°C/100 m] to 6°F/100 ft [10.9°C/100 m] have been encountered, with gradients of 1.0°F/100 ft [1.8°C/100 m] to 1.7°F/100 ft [3.1°C/100 m] found most frequently. For example, superdeep wells (24,000 ft [7317 m] and deeper) in Beckham County, Oklahoma, have an average gradient of 1.5°F/100 ft [2.7°C/100 m]. A 28,500-ft [8689 m] well on the Ralph Lowe Estate, West Texas, exhibited an average 1.4°F/100 ft [2.55°C/100 m] temperature gradient. On the other hand, average gradients in the South China Sea area are as high as 2.5°F/100 ft to 3.5°F/100 ft [4.6°C/100 m to 6.4°C/100 m].

Reasonably accurate formation temperatures can be estimated from well logs. Thus, maximum recording thermometers should be used on *each* logging device. Often maximum recorded temperature is assumed to be the bottom-hole temperature (BHT). However, measured BHT is not necessarily absolute BHT unless the borehole has been static long enough so that borehole and adjacent formation temperatures stabilize. This stabilized temperature is rarely reached by the time well logs are run. Therefore, as a rule of thumb, log BHT will be less than actual formation temperature. Difference depends on several parameters and can range up to about 85°F [30°C].

To obtain the *true formation temperature*, an *extrapolation technique* has been proposed by Timko and Fertl (1972a) whenever multiple log runs are made at any particular depth. Besides maximum BHT data for each log run, it is necessary to record the time, after circulation stop, that the logging sonde was last on the bottom. Fig.8.3 illustrates this method, which is an adaption of Horner's method (1951) used in well testing, in an Indonesian offshore well, and in a Texan onshore borehole. There are almost identical temperature environments in both wells. However, the particular depth in the

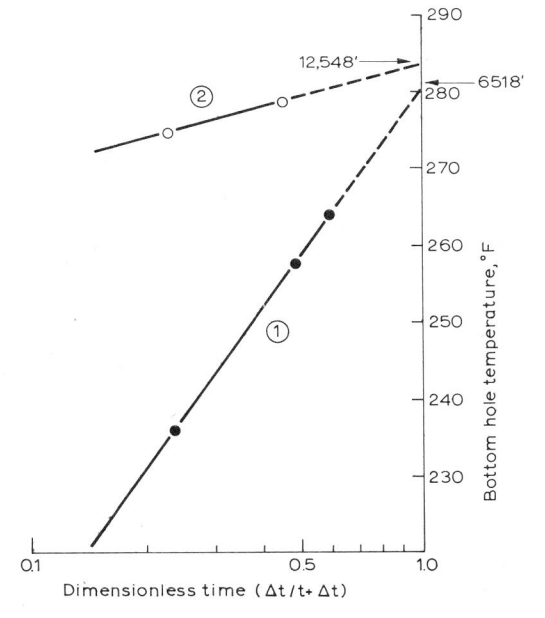

Fig.8.3. Extrapolation technique for true formation temperature. 1 = High-temperature well, South China Sea. Four logs were run to 6518 ft [1987 m]. First log, 4 hours after mud circulation stopped, recorded 218°F [103°C]. Log, 6 hours later, measured 264°F [128°C]. Straight-line extrapolation to infinite time (log 1.0) indicates a BHT of 281°F [138°C]. 2 = Deep Texan onshore well. Three logs were run to 12,548 ft [3826 m]. First-recorded temperature was 272°F [133°C], whereas actual stabilized BHT is 284°F [140°C]. t = circulating time, in hours; Δt = time after circulation stopped, in hours.

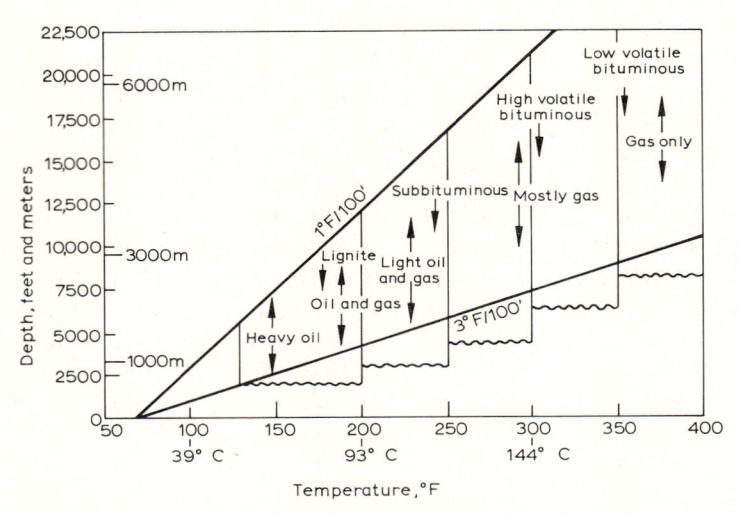

Fig.8.4. Depth/temperature relationships of coal and petroleum (after Klemme, 1972). (Courtesy of *Oil and Gas Journal*.)

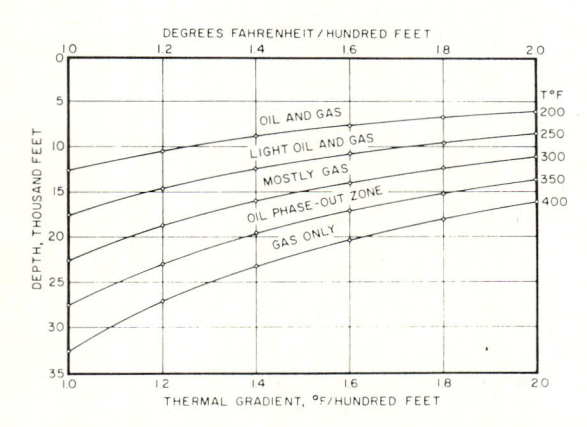

Fig.8.5. Physical state of hydrocarbons as a function of depth and geothermal gradient (after Landes, 1967). (Courtesy of the American Association of Petroleum Geologists.)

Indonesian well is 6518 ft [1987 m], whereas it is almost twice that value (12,548 ft [3826 m]) in the Texan well.

Terrestrial heat field distribution studies in potential hydrocarbon provinces around the world indicate oil fields to be generally confined to low-temperature zones, while gas fields occur in higher-temperature zones.

Fig.8.4 shows coal, oil, and gas combinations as related to temperature/depth relationships (Klemme, 1972). Similarly, Landes (1967) proposed a universal relationship of oil and gas distribution with depth and temperature.

TEMPERATURE VS. HYDROCARBON DISTRIBUTION 297

Landes' chart (Fig.8.5) agrees fairly well with oil and gas production in several basins around the world. However, we have reservations with Landes' statement that most temperature gradients become straight lines below 3000—4000 ft [915—1220 m] and that this gradient, once stabilized, can be extrapolated downward during drilling to forecast the content of prospective reservoirs.

Cautious application of the chart is recommended in regions of known or suspected overpressures, since recent experiences indicate that steeper than normal temperature gradients do occur (see also discussion of *Flow-line temperature*, Chapter 4). Fig.8.6 shows such a field case from offshore Louisiana, and Fig.8.7 illustrates the temperature gradient change in a deep Mid-Continent well.

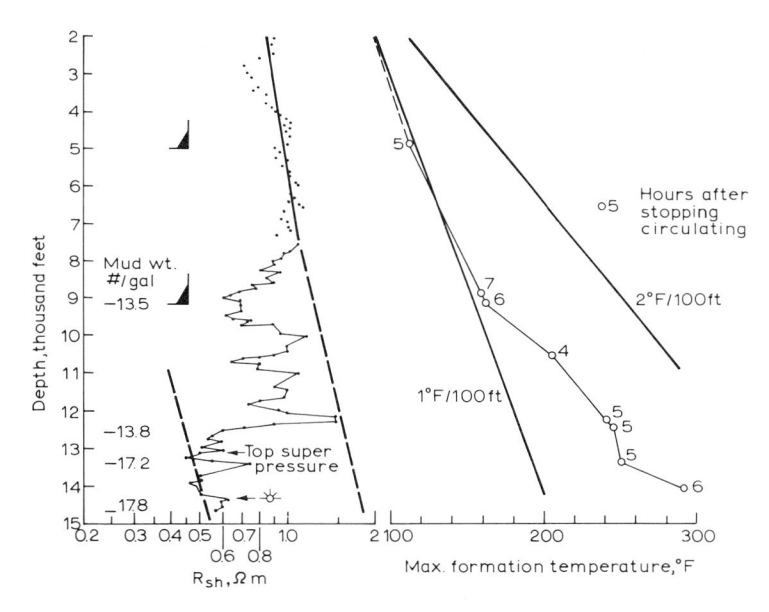

Fig.8.6. Conventional short normal shale plot indicates abnormal pressure below about 8000 ft [2439 m] and superpressures below 13,000 ft [3963 m], as characterized by a very high shale/sand ratio and >17 lb/gal [2.04 kg/dm^3] mud weight. Plot of log-derived formation temperature suggests an "elbow" region which corresponds closely to the high pressure (after Timko and Fertl, 1972b). (Courtesy of *World Oil*.)

Klemme (1972) recently discussed the influence of heat with respect to size of oil giants, formation and migration of hydrocarbons, and types of tectonic basins. He summarized his findings as follows:

"Present and past geothermal gradients in various hydrocarbon basins appear to have influence on the relative magnitude of hydrocarbon recovery. Considerable evidence suggests that high geothermal gradients in clastic

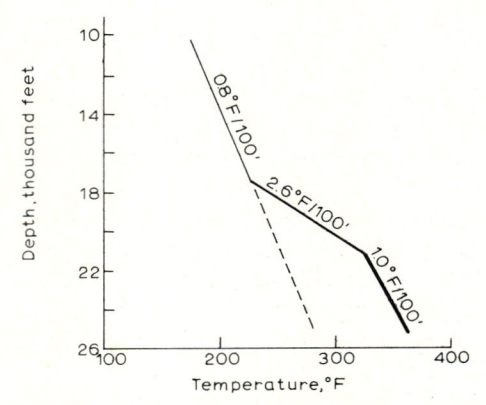

Fig.8.7. Temperature gradient change in a typical Mid-Continent well, U.S.A.

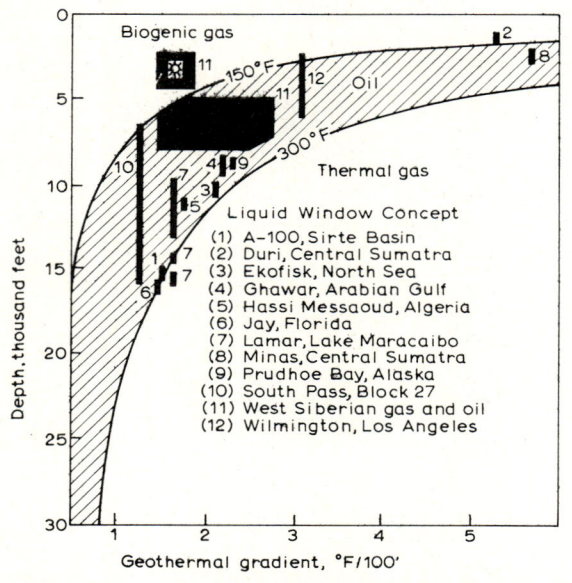

Liquid Window Concept
(1) A-100, Sirte Basin
(2) Duri, Central Sumatra
(3) Ekofisk, North Sea
(4) Ghawar, Arabian Gulf
(5) Hassi Messaoud, Algeria
(6) Jay, Florida
(7) Lamar, Lake Maracaibo
(8) Minas, Central Sumatra
(9) Prudhoe Bay, Alaska
(10) South Pass, Block 27
(11) West Siberian gas and oil
(12) Wilmington, Los Angeles

Fig.8.8. Giant fields of Mesozoic and younger age are found with geothermal gradients ranging from 1 to nearly 6°F/100 ft [1.8—10.9°C/100 m] (after Pusey, 1973). A "hot" gradient does not mean hot temperatures but indicates shallow production (e.g. Minas field). The vertical bar indicates the span of significant production in each field. Oil shows and completions are known at temperatures greater than 300°F [147°C], but >99% of the world's oil is found at temperatures cooler than 300°F [147°C], and most is found at temperatures of 250°F [121°C] or less. Many oil reservoirs are found at temperatures of less than 150°F [65.5°C], but invariably they are associated with either significant uplift after heating or, less commonly, long-distance migration updip. Huge gas reserves in the northern part of the West Siberian Basin are biogenic. Carbon isotope data clearly show that gases are highly deficient in ^{13}C and have formed at low temperatures. In the central and southern part of the basin, geothermal gradients are higher, and oil is found at temperatures of 130—250°F [54—121°C]. (Courtesy of *World Oil*.)

formations enhance processes of formation, migration, and entrapment of oil and gas.

The geothermal gradients in various basins, with exceptions, appear to be related to the magnitude of the earth's heat flow.

Variations in the earth's heat flow are considered compatible with the concept of sea-floor spreading and relate to the 'new global tectonics'.

A basin's depth of hydrocarbon occurrence and its temperature history are closely related. Several basin types associated with significantly high-heat flow zones are located along various continental-plate margins and areas of possible incipient sea-floor spreading or upwelling of basic material. These basins tend to yield more hydrocarbons per cubic mile of sediments than basins from low heat flow areas once all the proper geologic factors for hydrocarbon accumulation are present.

Temperature, modified by time, has been instrumental in the accumulation of many major (giant) accumulations of hydrocarbons."

Recently, Pusey (1973) proposed *a liquid window concept* based upon empirical data, with significant hydrocarbon generation starting at 150°F [65.5°C] and liquid hydrocarbon destruction beginning at temperatures exceeding 300°F [148.9°C]. A noteworthy exception is oil related to high temperatures (>300°F [>148.9°C]) as associated with carbonate rocks (Florida, U.S.A.; Lake Maracaibo, Venzuela) and/or overpressures. Pusey's liquid window concept is shown in Fig.8.8 for several oil and gas deposits from around the world.

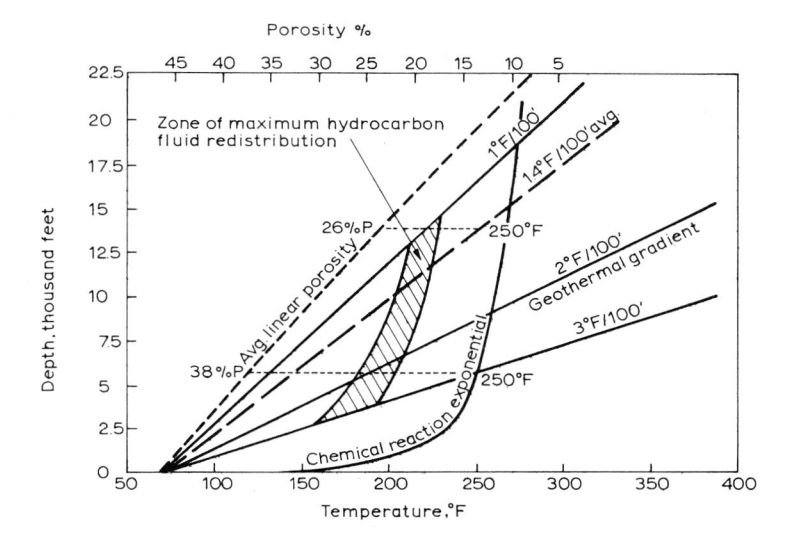

Fig.8.9. Hydrocarbon distribution model in clastic sediments based on data by Landes and Burst (after Klemme, 1972). Temperature increases the chemical reactions of hydrocarbon formation exponentially (the curve starts arbitrarily at 150°F). (Courtesy of *Oil and Gas Journal*.)

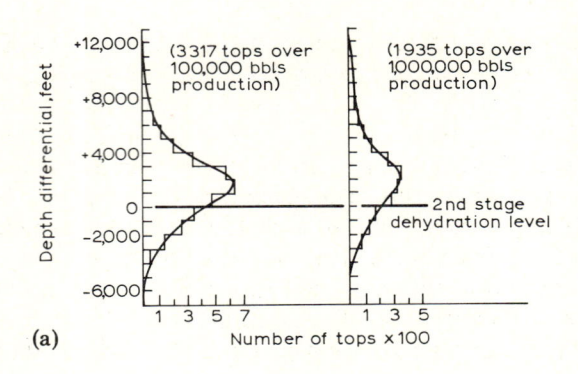

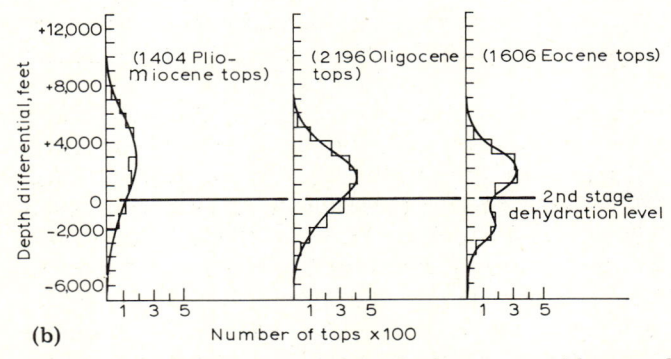

Fig.8.10. Correlation of top of producing zones with second-stage clay dehydration as a function of (a) field size and (b) geologic age, U.S. Gulf Coast area (after Burst, 1969). (Courtesy of the American Association of Petroleum Geologists.)

Finally, reference is made to Burst's (1969) shale compaction model (clay diagenesis, water escape model) (see Chapter 1, Fig.1.2) which was originally developed for the U.S. Gulf Coast area. The interval of maximum fluid distribution (i.e. second dehydration stage or, as termed by Russian scientists, zone of catagenesis) is limited to a temperature range of 150—266°F (65—130°C), and its depth is a function of the geothermal gradient prevailing (Fig.8.9). Statistical analyses of distance from top of producing zones to level of calculated second-stage clay dehydration in the U.S. Gulf Coast area (Burst, 1969) show consistent distribution, both as a function of field size and geologic age (Fig.8.10). Locating such clay diagenesis levels by geothermal gradient evaluation allows prediction of favorable hydrocarbon environments.

Formation pressure gradient versus hydrocarbon accumulations

In several areas of the world, study of formation pressure gradients sometimes indicates a correlation with the size and type of hydrocarbon deposit.

Such information, without doubt, is valuable in evolving operating techniques that will allow safe, economic development of gas and oil fields in many overpressured regions.

For example, the magnitude of overpressures encountered in the U.S. Gulf Coast area has become of special interest over the last two decades since modern operating techniques now enable the oil industry to test more efficiently the potential of these highly overpressured environments. Often such areas have been avoided previously because of associated drilling and completion problems and high risk and cost factors.

In this subject area, Cannon and Craze (1938) presented one of the first relationships of formation pressure versus depth for both Texas and Louisiana, with a maximum pressure gradient approaching 0.765 psi/ft [0.177 kg cm^{-2} m^{-1}]. In 1943, Denton reported a maximum gradient of 0.83 psi/ft [0.192 kg cm^{-2} m^{-1}] for the same region, a value which again was exceeded just a few years later (0.865 psi/ft [0.20 kg cm^{-2} m^{-1}], Cannon and Sullins, 1946).

Dickinson (1953) observed values up to 0.872 psi/ft [0.201 kg cm^{-2} m^{-1}] and suggested this value to approach the upper limit of overpressure gradients in shale/sand sequences; he suggested that it would be unlikely that 0.90 psi/ft [0.208 kg cm^{-2} m^{-1}] will be exceeded. Since then, however, even higher pressure gradients have been encountered in the U.S. Gulf Coast area.

In 1970, Fowler observed that the only criterion for commercial or non-commercial, barren sandstone reservoirs in the Chocolate Bayou field, Brazoria County, Texas, appears to be the variations in formation pressure gradient. Gradients less than 0.85 psi/ft [0.196 kg cm^{-2} m^{-1}] relate to commercial pay sands, whereas sands with gradients exceeding 0.85 psi/ft [0.196 kg cm^{-2} m^{-1}] are noncommercial.

Formation pressure—production correlations around the Lake Mongoulois dome in St. Martin Parish, Louisiana, have been studied by Classen (1968). Production has been found in normal, hydrostatic pressure environments (such as oil production on the east and north flanks) and associated with pressure gradients as high as 0.9 psi/ft [0.208 kg cm^{-2} m^{-1}] (dry-gas production).

In the U.S. Gulf Coast area, several commercial hydrocarbon accumulations were also encountered in overpressure environments with gradients as high as 0.9 psi/ft [0.208 kg cm^{-2} m^{-1}]. A typical field case is the East Cameron area, offshore Louisiana, where the presence of large reservoirs is well established.

Similar data may be found in the literature as early as 1949. McCaslin (1949) discussed two wells in Four Isle field, Terrebonne Parish, Louisiana, with pressure gradients of 0.955 psi/ft [0.221 kg cm^{-2} m^{-1}] and 0.984 psi/ft [0.227 kg cm^{-2} m^{-1}] while drilling to 14,564 ft [4440 m] and 13,230 ft [4034 m], respectively. Additional data from a high-pressure oil well in the Mississippi Delta area showed the well flowing 157 barrels per day of 38°

gravity crude with a 7900-psi [555.6 kg/cm^2] tubing pressure on a 3/32-inch choke. Production was from 12,994—13,050 ft [3962—3979 m] open-hole section; calculated original BHP was 12,635 psi [888.5 kg/cm^2] at 13,000 ft [3963 m].

Based on these data (and assuming they are correct), the pressure gradient in this well appears to be as high as 0.975 psi/ft [0.225 kg cm^{-2} m^{-1}]. Based on our experience in the Gulf Coast area, this value, which closely approaches the overburden pressure, appears to be one of the highest gradients ever encountered in a hydrocarbon zone in this region.

However, with such extremely high pressures, the presence of commercial gas and/or oil reservoirs in shale/sand sequences appears to be the exception rather than the rule. Frequently, such superpressured reservoirs yield water flows with oil and/or gas shows at tremendous flow capacity or have very high but short-lived hydrocarbon production with a rapid pressure drop. Such conditions occur in the U.S. Gulf Coast area; California, U.S.A.; Burma's Oligocene; the foothills of Andes and Himalayan mountain ranges; the Tertiary of Papua and New Guinea; the Maykop series (Miocene) in the U.S.S.R.; and elsewhere.

Besides the above data on the U.S. Gulf Coast area, additional data have been compiled by the Federal Power Commission (1966), Meyerhoff (1968), and Perry (1969). These data are summarized in Table 8.II, and the geometric and arithmetic means are plotted as a 0.05-psi/ft [0.0115 kg cm^{-2} m^{-1}] moving average for every 0.01 psi/ft [0.0023 kg cm^{-2} m^{-1}] increment in Fig.8.11 (Fowler, 1971, and personal communication, 1971).

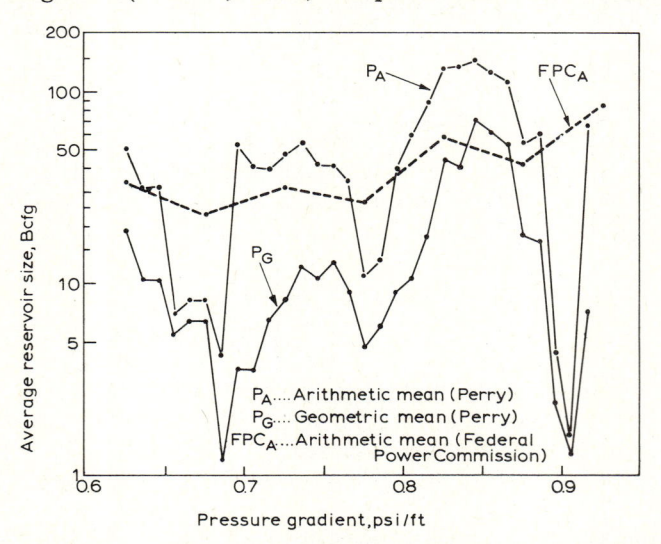

Fig.8.11. Arithmetic mean of hydrocarbon reservoir size versus pressure gradient (after W.A. Fowler, personal communication, 1971). Note that the size declines drastically at gradients exceeding 0.85 psi/ft [0.196 kg cm^{-2} m^{-1}]. Reservoir size is in billion cubic feet of gas (Bcfg).

TABLE 8.II

Correlation of average hydrocarbon reservoir size and formation pressure gradient in the U.S. Gulf Coast area

Number of reservoirs and their size*	Pressure gradient range (psi/ft)						
	0.60–0.65	0.65–0.70	0.70–0.75	0.75–0.80	0.80–0.85	0.85–0.90	>0.90
(1) Number of reservoirs	3	3	19	10	12	7	2
Geometric mean reservoir size (Bcfg)**	18.7	6.4	8.1	4.6	43.6	18.0	21.9
Arithmetic mean reservoir size (Bcfg)	49.8	8.1	47.0	10.8	131.1	53.4	97.2
(2) Number of reservoirs	2	1	2	3	5	1	1
Geometric mean reservoir size (Bcfg)	22.9	13.6	117.0	131.0	18.8	189	170
Arithmetic mean reservoir size (Bcfg)	34.4	13.6	177.1	228.2	155.9	189	170
(3) Number of reservoirs	49	60	78	55	79	57	23
Arithmetic mean reservoir size (Bcfg)	34	23	32	27	57	42	84

*Source: (1) Perry, 1969; (2) Meyerhoff, 1968; (3) Federal Power Commission, 1966.
**Billion cubic feet of gas.

According to Fowler these data "suggest that reservoir size decreases with increasing pressure gradient up to a pressure gradient of 0.65 or 0.70 psi/ft [0.15 or 0.162 kg cm^{-2} m^{-1}]. In the 0.70—0.85-psi/ft [0.162—0.196 kg cm^{-2} m^{-1}] range, reservoir size appears to increase with increasing gradients. At pressure gradients greater than 0.85 psi/ft [0.196 kg cm^{-2} m^{-1}] there is a suggestion from Perry's data that reservoir size decreases with increasing gradients. Meyerhoff's data in this range of pressure gradients are inadequate. The Federal Power Commission data show a continued increase; but there is some reason to believe that the Federal Power Commission data might conform better to Perry's data with careful editing and more detailed statistical work."

Formation temperature and pressure versus hydrocarbon accumulation

Formation temperature and pressure *together* are the important factors which significantly affect the physical state of hydrocarbons at depth. Both parameters can be determined from well logs.

It is interesting to recall the phase model of oil and gas at depth as proposed by Modelevskiy and Parnov (1967). Phase relationships between oil, oil-gas, gas-condensate-oil, as continuous series of the gaseous and liquid state, at *increasing formation pressures*, indicate the possibility of a still different state of the mixture, so far not encountered in nature: a single state at supercritical pressure and temperature (with all the gas dissolved in oil), termed *oil condensate* (Fig.8.12).

Data by Baybakov (1969) on the distribution of 220 hydrocarbon pools in a sedimentary sequence are schematically shown in Fig.8.13. With rising pres-

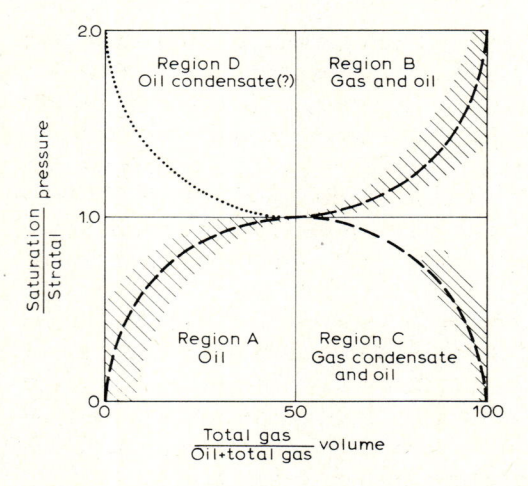

Fig.8.12. Proposed physical state of hydrocarbons at depth (after Modelevskiy and Parnov, 1967).

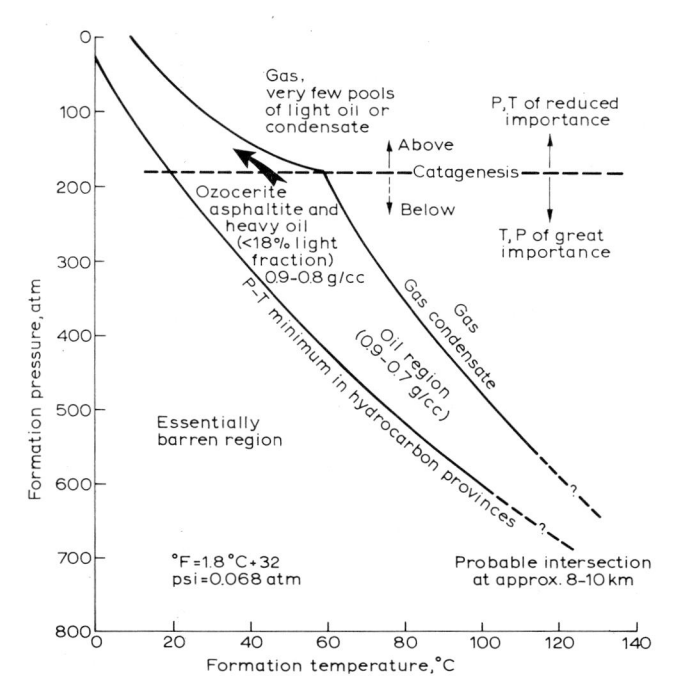

Fig.8.13. Relation of hydrocarbon state to formation temperature and pressure for 220 reservoirs in a sedimentary sequence indicates that temperature is the main factor governing hydrocarbon properties (after Baybakov, 1969). Note that oil reservoirs can exist at great depths with moderate pressures and low temperature gradients.

sure at constant temperature, the trend is from gas to gas condensate and light oil. This indicates the retarding effect of pressure on hydrocarbon destruction in a sedimentary sequence. However, without any doubt, temperature is the key factor for hydrocarbon transformation. Oil reservoirs can exist at great depth if the temperature gradient is low. Similarly, in the study of certain areas along the U.S. Gulf Coast, Timko and Fertl (1970) have found essentially normal pressure and temperature gradients to continue to great depths, and concluded that such deep horizons are still highly prospective for large oil accumulations.

A general misconception in the drilling industry is that high formation pressure gradients are associated with great depths. This is not necessarily so since essentially normal pore pressure gradients have been found at great depths in the U.S. Gulf Coast and elsewhere. Such deep horizons are still highly prospective for oil accumulations. Thus, depth itself is not the limiting factor for occurrence or absence of large, deep oil reserves.

For example, Fig.8.14 shows deep oil production in the Weeks Island area, Louisiana. Oil production in the well studied comes from 16,300 ft

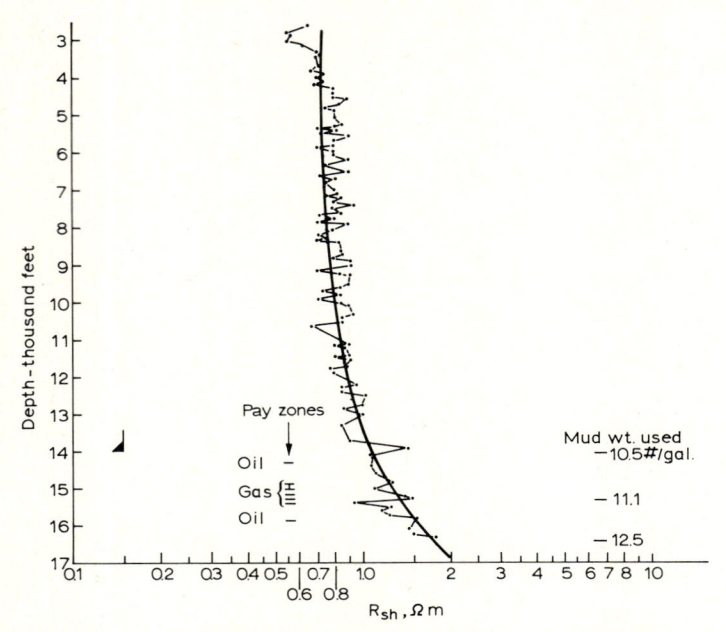

Fig.8.14. Deep oil production in an essentially normal-pressured environment with low geothermal gradient, Weeks Island area, Louisiana (after Timko and Fertl, 1970). (Courtesy of the Society of Petroleum Engineers of A.I.M.E.)

[4970 m], where a close to hydrostatic pressure environment and a low 1.0° F/ 100 ft [1.8° C/100 m] temperature gradient exist. Fig.8.15 is a plot for another well in the same field, about a mile away and structurally higher. Besides an average geothermal gradient of 2.0° F/100 ft [3.6° C/100 m], very high overpressures are encountered below 14,100 ft [4299 m]. As discussed in the shale ratio resistivity method later on, in this well the interval below 14,100 ft [4299 m] would not be prospective for oil, and below 14,600 ft [4451 m] it would not be prospective even for gas. This has been confirmed by well tests; yet, these two wells are traversing the same geologic horizons. However, their temperature and pressure regime is quite different.

Therefore, before condemning an entire area or complex of broken-up large structures, the relationship of formation pressure and temperature with structure must be studied and understood thoroughly. On the other hand, deep structures and deep sand development in shale/sand sequences are often regarded as highly prospective, unfortunately frequently without any considerations of thermodynamic conditions, such as temperatures and pressures.

In 1970, Timko and Fertl reported important pressure—temperature relationships in about 60 U.S. Gulf Coast wells. Fig.8.16 shows the temperature—depth relationship in these wells, with the average geothermal gradients

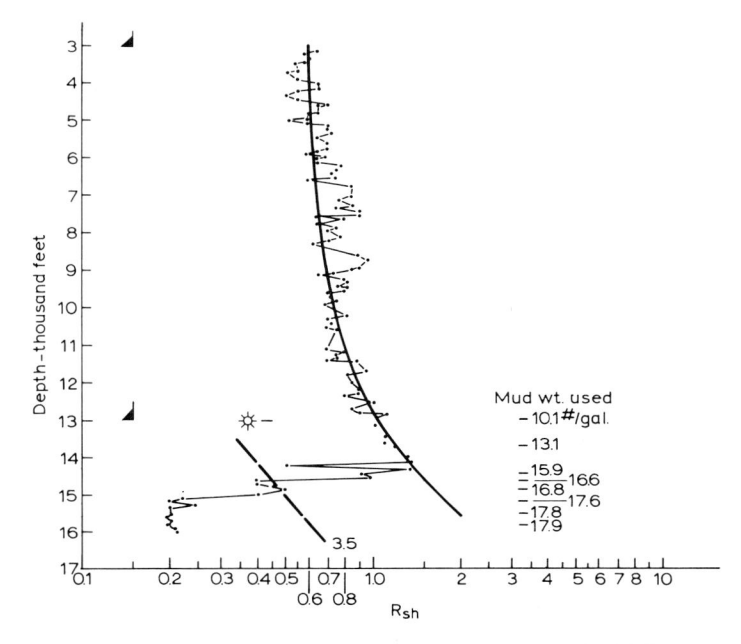

Fig.8.15. Nonprospective deep and overpressured, high-temperature intervals, Weeks Island area, Louisiana (after Timko and Fertl, 1970). Circle with radiating lines indicates the gas pay. (Courtesy of the Society of Petroleum Engineers of A.I.M.E.)

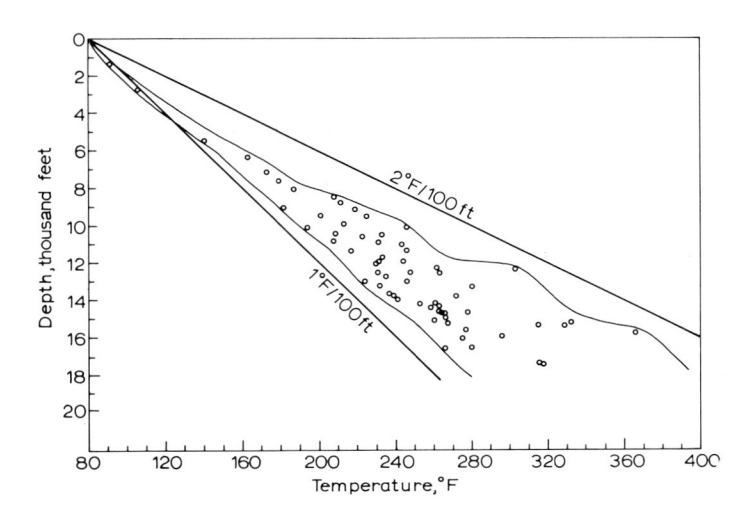

Fig.8.16. Borehole temperatures in 60 overpressured U.S. Gulf Coast wells (after Timko and Fertl, 1970). (Courtesy of the Society of Petroleum Engineers of A.I.M.E.)

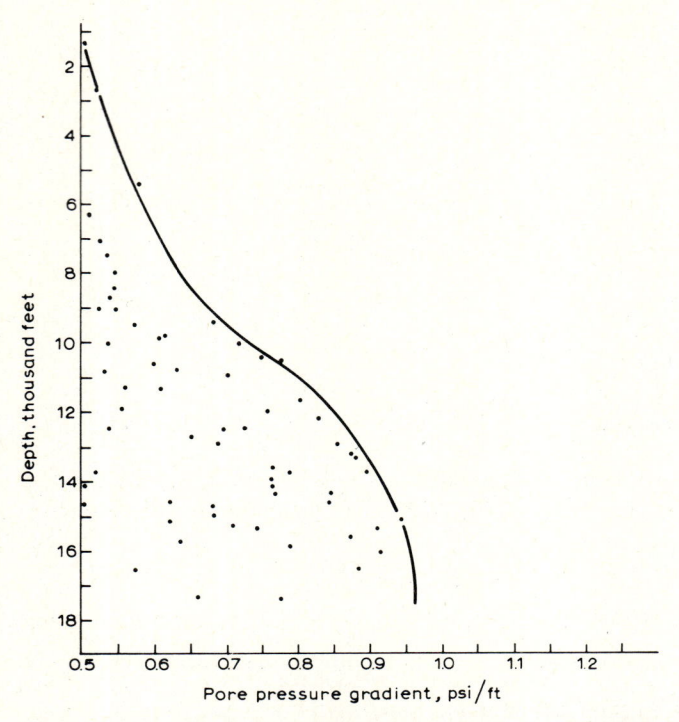

Fig.8.17. Pore pressure gradients versus depth in the same 60 wells as in Fig.8.16 (after Timko and Fertl, 1970). (Courtesy of the Society of Petroleum Engineers of A.I.M.E.)

varying from less than 1°F/100 ft [1.8°C/100 m] to about 1.8°F/100 ft [3.3°C/100 m]. Fig.8.17 illustrates the increase in abnormally high pore pressures with depth for the same wells. A sharp increase in the formation pressure is apparent at depths of about 8000 ft [2439 m] and deeper.

However, the most important findings are shown in the cross plot of formation pressure gradient and temperature (Fig.8.18). Note the bell-shaped hydrocarbon envelope, covering pressure gradients from 0.5 psi/ft [0.115 kg cm^{-2} m^{-1}] to as high as 0.94 psi/ft [0.217 kg cm^{-2} m^{-1}] and corresponding formation temperatures between 90 and 365°F (104 and 185°C). A temperature range of 215—290°F (102—143°C) coincides with the range of highest pressure gradients in hydrocarbon zones. This is particularly noteworthy as this temperature range lies mainly in the zone of second-stage clay dehydration as proposed by Burst (1969), i.e. in the zone of maximum fluid distribution. Fig.8.18 also shows that with further increase in temperature, the magnitude of overpressure gradients reverses its trend and declines in potential reservoir rocks, coinciding with a simultaneous increase in the volume of gas. At the same time, extremely high pressure gradients are encountered in high-temperature aquifers and zones with noncommercial oil and gas shows.

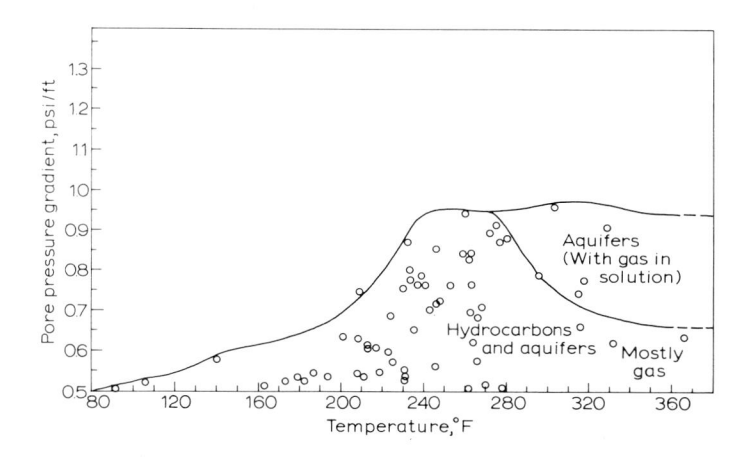

Fig.8.18. Pore pressure gradient versus formation temperature in 60 overpressured U.S. Gulf Coast wells (after Timko and Fertl, 1970). (Courtesy of the Society of Petroleum Engineers of A.I.M.E.)

Shale resistivity ratio method — a valuable concept for economic drilling decisions

Recently, Timko and Fertl (1970) have presented a generalized correlation between a "typical" U.S. Gulf Coast shale resistivity profile and the distribution of oil and gas fields in this area (Fig.8.19). Application of these findings can indicate whether it is possible for commercial production to exist below the depth to which the well has already been drilled and logged and whether it is economically attractive to continue drilling a borehole below a given depth in shale/sand sequences.

This statistically significant correlation, developed and tested over several years, is based upon analyses of hundreds of wells in all ranges of pressures, both commercially and noncommercially productive. The "typical" profile in Fig.8.19 is based entirely on the short normal curve (see Chapter 5) and does not apply to any other resistivity (or conductivity) measuring device. Similar correlations can be developed from acoustic velocity (sonic) log data if enough logs are present.

We have had more success in using the shale resistivity profile to determine favorable hydrocarbon environments rather than the actual pressure gradients, or required mud weight, as others have attempted previously.

One can use the *shale resistivity ratio* parameter, which is a function of the superimposed effects of formation pressure, temperature, and formation water salinity. Thus, thermodynamic *and* geochemical concepts are taken into account. Initially developed for the U.S. Gulf Coast area, additional experience has shown the model to hold true in California and several

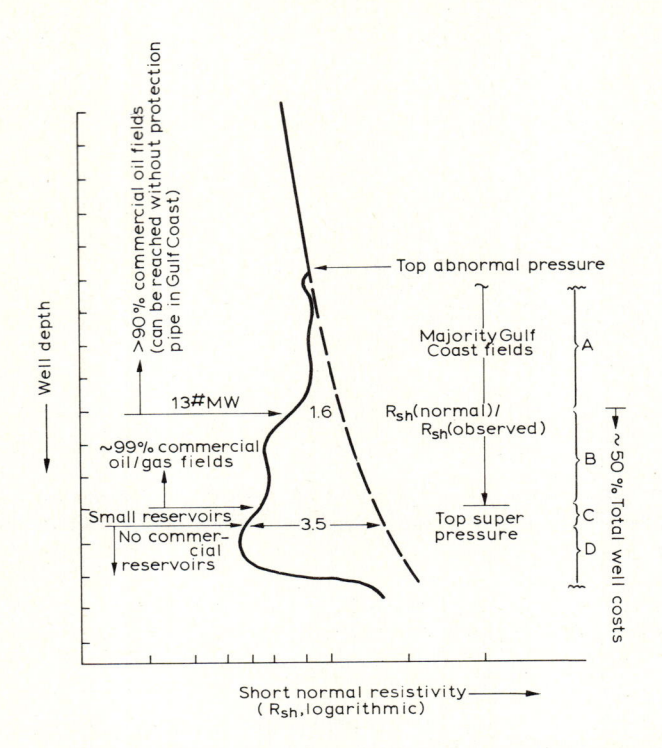

Fig.8.19. Typical Gulf Coast shale resistivity profile based on the short normal curve can be correlated to distribution of gas/oil reservoirs (after Timko and Fertl, 1970). Profile is based on hundreds of commercially and noncommercially productive wells analyzed over the past few years. Scales are omitted since specific values change depending on area. Significance of zones A, B, C, and D is discussed on p. 311. The numbers 1.6 and 3.5 represent the $R_{sh(normal)}/R_{sh(observed)}$ ratios. (Courtesy of the Society of Petroleum Engineers of A.I.M.E.)

other Tertiary basins with shale/sand sequences throughout the world. However, correlations similar to the U.S. Gulf Coast area have to be established first.

For example, Fig.8.20 shows the plot of shale resistivity ratios versus average pressure gradients as initially developed for the U.S. Gulf Coast area (Hottman and Johnson, 1965). In 1970, Timko and Fertl published a similar correlation for the local East Cameron area. For comparison, limited North Sea and South China Sea correlations and U.S. Gulf Coast correlations proposed by Eaton (1972), which take an overburden stress gradient limit into account, are also shown in Fig.8.20. These data clearly show the large variation in pressure gradients for a given shale resistivity ratio. Recognition of this fact is of greatest importance to both drilling operations (proper mud weight) and exploration, since application of the log-calculated gradient,

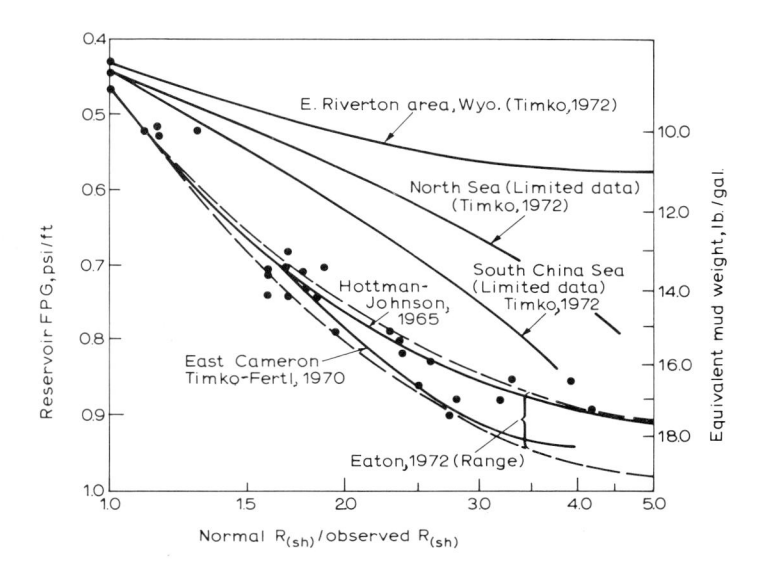

Fig.8.20. Use of shale resistivity ratios to detect favorable hydrocarbon environments has been more accurate than use of pressure gradients or equivalent mud weights (after Timko and Fertl, 1970). (Courtesy of the Society of Petroleum Engineers of A.I.M.E.)

unless modified for specific areas, can be misleading and costly. For this reason, the resistivity ratio method was developed and should be used rather than pressure gradients only if one attempts to forecast favorable downhole environments for commercial oil and gas distribution.

Completion guidelines

Based on additional experience, Timko and Fertl (1972c) have proposed the following completion guidelines:

Commercial oil fields

Most oil pay zones in the Gulf Coast area can be reached without a protection pipe. Statistical data show that 90% of the commercial oil fields are encountered in a normal, hydrostatic pressure environment and in abnormal pressures up to about 13 ppg [1.56 kg/dm^3], corresponding to a shale ratio of 1.6. Mud weight requirements exceeding 13 ppg [1.56 kg/dm^3] generally require a string of protection pipe to be set to avoid lost circulation or differentially stuck drill pipe.

Commercial gas and/or oil fields

About 99% of all commercial gas and oil reservoirs are found in Zones A and B. This observation is of great economic significance since the cost of successful completion in high-pressure formations may become exceedingly high.

As a matter of record, a proportionally large amount of money is usually spent in drilling the last few hundred feet, where the highest overpressures are usually encountered in a well. Knowing the local or regional relationship between shale resistivity ratio and distribution of hydrocarbons, a well-founded decision can be made as to whether a well should be deepened further.

Small hydrocarbon reservoirs

As may be observed from the "typical" Gulf Coast shale resistivity profile, a substantial number of commercial gas and/or oil reservoirs apparently occur in overpressured environments. However, with increasingly higher resistivity ratio, potential reservoirs become smaller in size. Pay may be of limited extent and produced as one or two well pools, and it generally falls in or near Zone C (see Fig.8.19).

Often, a well that falls in or near Zone C would be commercial by itself, but generally, because of the well's capability of producing at high rates, offsets are drilled which are noncommercial and ruin the economics of the overall venture. Wells falling in or near Zone C are candidates for detailed reservoir pressure studies to determine overall reservoir limits so decisions can be made as to further development.

Noncommercial hydrocarbon reservoirs

In the region of so-called *superpressures*, indicated as Zone D, we have not encountered commercial reservoirs. These intervals are frequently characterized by extremely high pressures and fast drawdown, indicative of low volume, and completion problems. Furthermore, most reservoir rocks exhibiting super-overpressures contain only rather fresh water with some gas in solution. Sidewall cores and well logs generally appear favorable for production since considerable gas is present in solution in these relatively fresh waters (as low as 3000 ppm NaCl).

One can simply state, therefore, that regardless of measured formation pressure gradients, the following conclusions can be drawn for sand/shale sequences:

Most commercial oil sands exhibit shale resistivity ratios less than 1.6 in adjacent shales and can generally be reached without a string of protection pipe.

Most commercial gas-sand reservoirs exhibit ratios in adjacent shales of about 3.0 and less. These could be the large, high pressure gradient reservoirs discussed by W.A. Fowler (personal communication, 1971) and Perry (1969). For instance, for the East Cameron area from Fig.8.20, a resistivity ratio of 2.8 is equivalent to a gradient of 0.9 psi/ft [0.208 kg cm^{-2} m^{-1}]. Even with this high pressure gradient, very large-size reservoirs have been encountered in this area.

Wells with ratios between 3.0 and 3.5 can be commercially gas productive and generally will produce as one or two well reservoirs. This situation could develop in large fields that are highly faulted, with faults limiting the individual reservoir size.

No commercial production is found when the shale resistivity ratio reaches and/or exceeds 3.5, no matter what the actual pressure gradient is. These wells are often highly productive initially and are characterized by an extremely fast pressure depletion. Further, it is very important to note that commercial production was not found in any well below the depth where the 3.5 ratio exists, even though the ratio drops below 3.5 and where pressure gradients trend back towards normal, provided no major unconformity is crossed.

Based on the above observations and experience, the conventional short normal logging curve can be used for the definition and distribution of the types of hydrocarbons in the U.S. Gulf Coast and other geologically similar areas. Knowledge of this statistical distribution can be used in economic decision-making, such as to stop or continue drilling in a favorable or unfavorable pressure environment.

However, once more it should be stressed that *these observations pertain to geologic basins with sand/shale sequences, and not to massive carbonate sections*, such as the Smackover Formation in Mississippi, West Texas, the Middle East, or elsewhere. This latter geologic environment should provide an additional area for challenging research.

Field case, offshore Texas

A typical shale resistivity profile of a dry hole drilled offshore of Aransas County is shown in Fig.8.21. The 3.5 shale ratio is found at a shallow depth of 6850 ft [2088 m], indicating the well has reached an unfavorable environment. As pointed out previously, no commercial production has been found within and below the 3.5 ratio cutoff point.

Statistical odds indicate the well should have been suspended at 6850 ft [2088 m] and drilling money saved invested in another well. Below 6850 ft [2088 m] drilling is expensive since mud weights exceeding 17 ppg [2.04 kg/dm^3] are needed.

This plot from offshore Texas indicates only a very narrow vertical interval of commercial gas and oil. Gas and oil are found almost invariably in sands within the shale pressure transition zone — in this case from about 4800 to 6850 ft [1463 to 2088 m] — and only the section from 4800 to 5800 ft [1463 to 1768 m] is attractive, since no sands are present in the upper Anahuac zone.

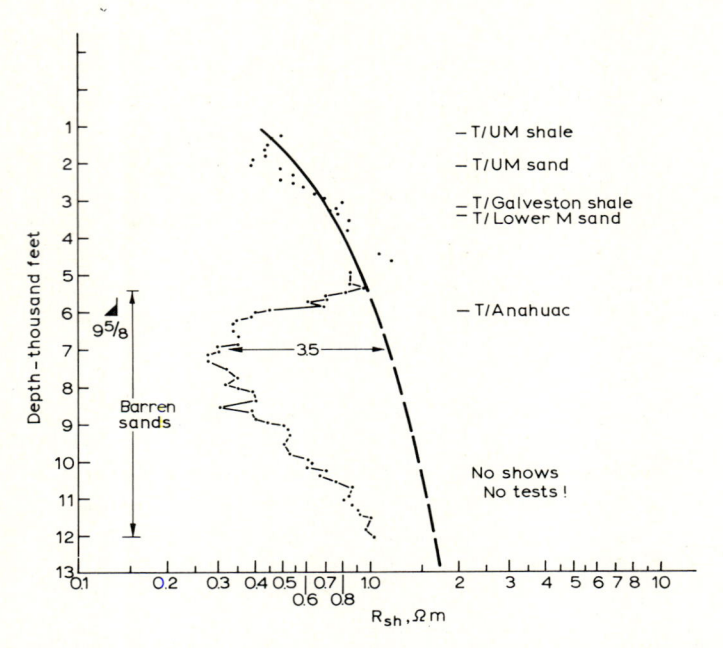

Fig.8.21. Shale resistivity profile in a Texan offshore well. Note the presence of 5000 ft [1524 m] of nonpetroliferous rocks (after Timko and Fertl, 1970). T indicates the top of the formation. (Courtesy of the Society of Petroleum Engineers of A.I.M.E.)

Field case, offshore Louisiana

A completion attempt was made at 14,373—14,377 ft [4382—4383.2 m] and 14,381—14,385 ft [4384.5—4385.7 m] in the well shown in Fig.8.22. These tested intervals are about 1300 ft [396 m] below the top of the 3.5 ratio cutoff point. The combined zones flowed gas, condensate, and water. Initial BHP was about 13,000 psi [914 kg/cm^2] declining to 6200 psi [436 kg/cm^2] in a matter of hours, essentially depleting during test. A fluid sample taken at the separator consisted of 1/3 oil, 1/3 water, and 1/3 solids. Solids below 2 microns in size were examined by X-ray diffraction analysis and showed 49% mixed-layered clays, 26% illite, 16% chlorite, and 9% kaolinite. The cost to test this interval in this unfavorable environment was near $200,000.

Field case, northern California

Fig.8.23 shows a rather shallow pressure transition zone (approximately at 4100 ft [1250 m]) with a slight pressure increase down to 4800 ft [1463 m]. A well-defined pressure cap is present at 4800—5400 ft [1463—1646 m].

SHALE RESISTIVITY RATIO METHOD   315

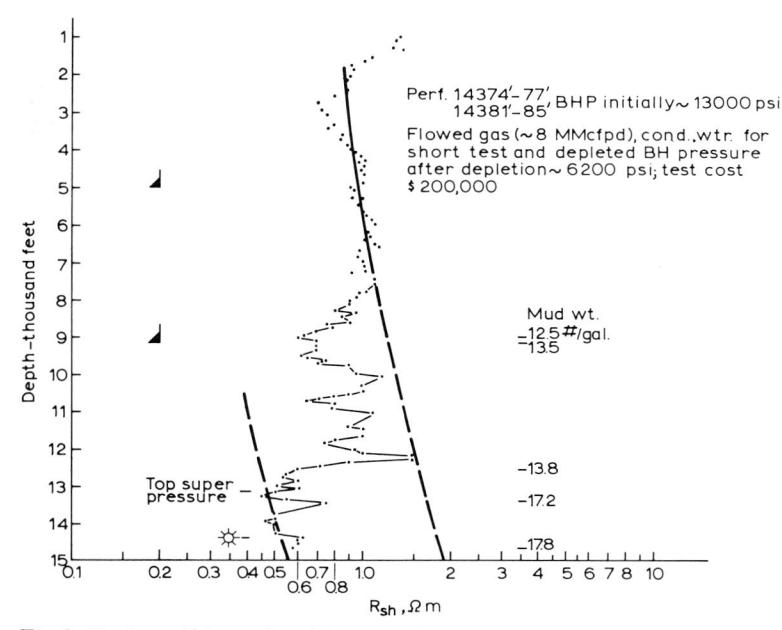

Fig.8.22. An offshore Louisiana well tested in an interval 1300 ft [396 m] below the top of the 3.5 ratio cutoff point (after Timko and Fertl, 1970). The well depleted on test. (Courtesy of the Society of Petroleum Engineers of A.I.M.E.)

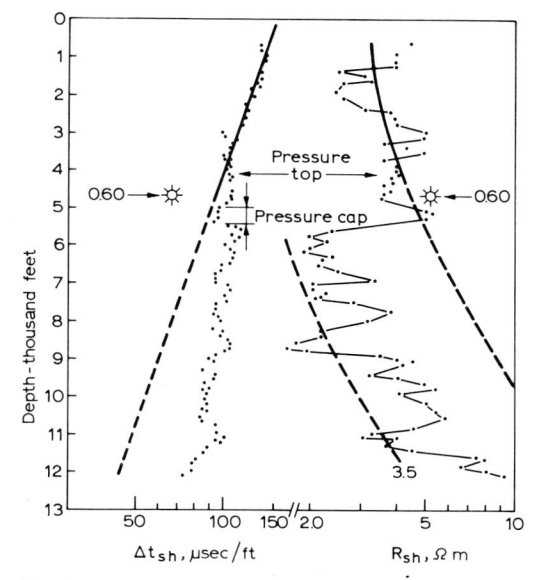

Fig.8.23. A northern California well which closely approached superpressures at 7000 ft [2134 m] and entered them at 8300 ft [2530 m] (after Timko and Fertl, 1972b). Although no commercial reservoirs have ever been found in the superpressure region, the well was drilled to a total depth of 12,300 ft [3750 m] and found to be dry. Circles with radiating lines indicate the gas pay; the numbers 0.6 and 3.5 represent the pressure gradient in psi/ft and the shale resistivity ratio, respectively. (Courtesy of *World Oil*.)

There is a rather abrupt increase in the pressure gradient immediately below the cap, with the location of the only potential pay zone again being in the transition zone. Superpressures were closely approached at 7000 ft [2134 m] and entered at 8300 ft [2530 m]. Despite this unfavorable hydrocarbon environment, the well was deepened to about 12,300 ft [3750 m], and no potential pays were encountered over the entire interval.

Use of mud log, digital well-log analysis, and shale resistivity ratio on a wildcat, U.S. Gulf Coast area

Fig.8.24 is a shale plot of a Vermilion Parish, Louisiana, wildcat well drilled recently to test deep prospective sands below 16,000 ft [4878 m]. Abnormal pressures were encountered below 11,000 ft [3354 m]. The cutoff shale resistivity ratio of 3.5 occurred at 14,000 ft [4268 m], indicating

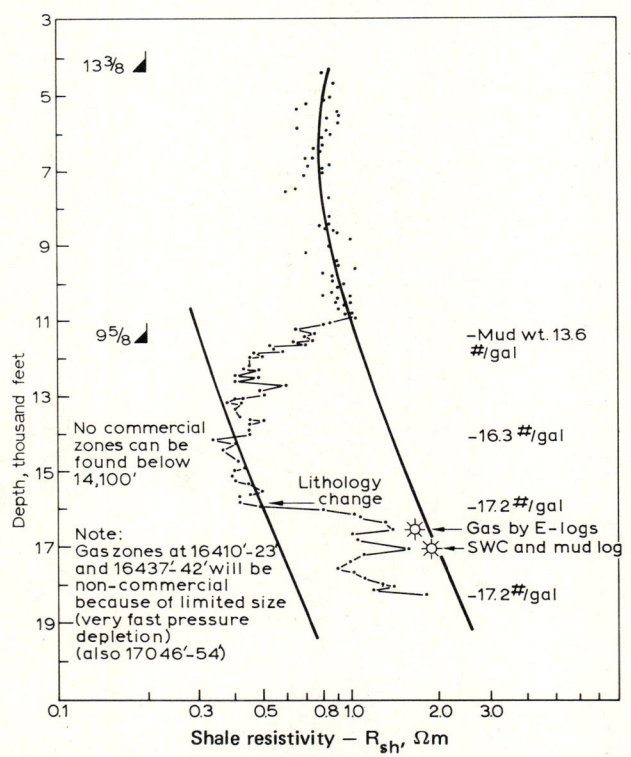

Fig.8.24. A Vermilion Parish, Louisiana, wildcat which could have been abandoned at 14,100 ft [4299 m] with significant monetary savings had credence been given to the 3.5 shale ratio cutoff point (after Timko and Fertl, 1972c). Conventional indicators caused the well to be drilled and cased much deeper at great expense. Although potential was considered good, based on conventional exploration techniques, one zone tested water and a second gas-condensate section depleted in three weeks. (Courtesy of *World Oil.*)

an unfavorable environment below this depth. Therefore, the well could have been abandoned at this point.

However, the well was continued to a depth below 18,000 ft [5488 m], encountering expensive delays because of extensive drilling problems. Shale resistivities increase below 16,000 ft [4878 m] in the absence of sands, indicating a possible pressure gradient decrease, but experience has shown that apparent gradient decrease almost always occurs in a formation temperature range of 260—300°F [127—149°C].

Rather than a pressure gradient decrease, a shale lithology change actually occurs. Because of high temperatures, the shales have become micaceous in nature, high in illites and chlorites which have been converted from less stable montmorillonite-type clays.

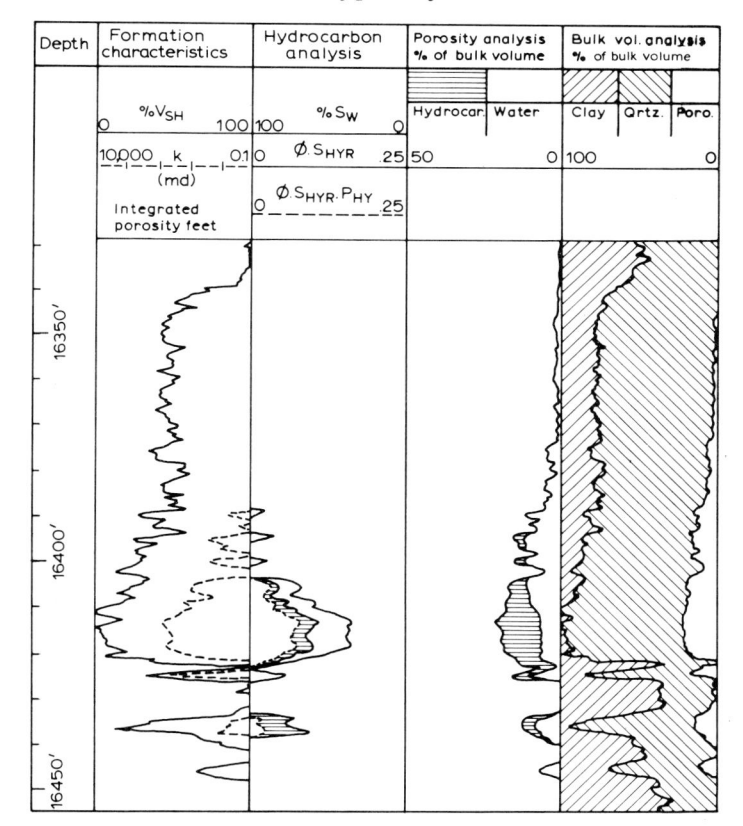

Fig.8.25. Computer printout of the log analysis for the upper section (16,410 ft [5003 m]) of the well in Fig.8.24 also indicates high potential for this interval and the section below 17,000 ft [5183 m] (after Timko and Fertl, 1972c). However, as pointed out in Fig.8.24, both zones were noncommercial. Once again, the 3.5 shale ratio cutoff point proved a more accurate indicator of potential than any other method. (Courtesy of *World Oil.*)

Extremely high pressures still exist below 16,000 ft [4878 m]. While drilling sands at 16,400 and 17,050 ft [5000 and 5198 m], commercial shows were picked up by the mud-logging unit. A sophisticated logging program was obtained through these sands and the graphical computer printout of the log analysis of the upper interval is shown in Fig.8.25.

As noted, the log analysis confirmed the mud log, not only in this upper zone but also in the lower zone below 17,000 ft [5182 m]. As shown in Fig.8.25, according to the log analysis of the sand at 16,402—16,424 ft [5000.6—5007.3 m], calculated porosities are 18—22%, and water saturations are 35%. Further, the analysis indicates the sand to be relatively shale-free with good permeability.

However, when one goes a step beyond conventional analysis and plots temperature and pressure gradient data on Fig.8.26 for the two sands of interest, data points fall in the noncommercial reservoir envelope. *Therefore, even though all conventional well data indicate productive sands, pressure and temperature data show the well to be of no commercial value*, confirming the shale resistivity ratio concept (Fig.8.24).

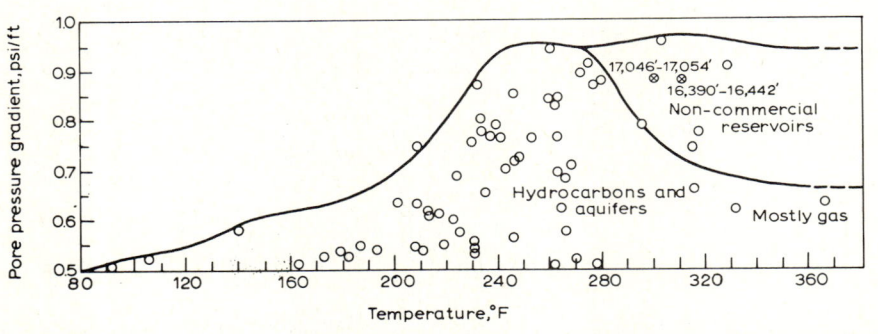

Fig.8.26. Relationship between temperature and pressure gradient for the well discussed in Figs.8.24 and 8.25 showing that the two reservoirs fall into the noncommercial envelope (after Timko and Fertl, 1972c). (Courtesy of *World Oil.*)

However, it is difficult to convince management not to run production casing on a rank wildcat that could prove up considerable acreage. Casing was run in this well with considerable difficulty and expense. The lower zone tested water with a small amount of gas and the pressure depleted on test.

However, the upper zone at 16,410 ft [5003 m] tested the rate of 4.4 million cubic feet of gas per day (MMcfd) (3-hour test) with 480 barrels of oil per day on a 10/64-inch choke and a surface tubing pressure of 8675 psi [610 kg/cm^2]. With such an encouraging test, one might start an offset and negotiate for a pipeline connection or, more cautiously, run a pressure buildup test to determine reservoir size.

A pressure buildup test was run on this well, indicating a limited reservoir. The well was placed on production at a rate of 2 MMcfd and the pressure depleted within three weeks.

Economics in overpressure environments

Experience in the U.S. Gulf Coast area indicates that, generally speaking, about 50% of all well costs have occurred when the borehole reached the intermediate protection-pipe setting depth. Fig.8.27 shows a well in Iberia Parish County, Louisiana.

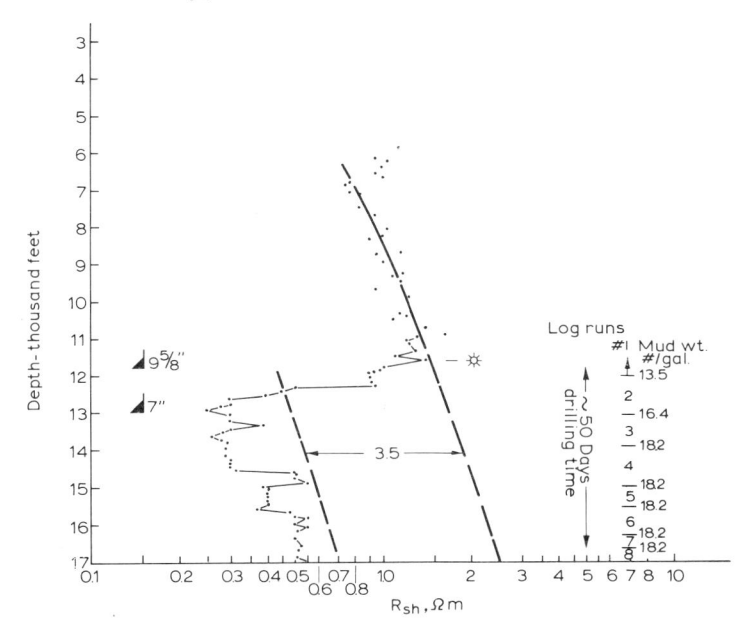

Fig.8.27. Additional needless expensive drilling was done and extra casing set in this Iberia Parish, Louisiana, wildcat since no production was likely below the 3.5 shale ratio cutoff point (after Timko and Fertl, 1970). (Courtesy of the Society of Petroleum Engineers of A.I.M.E.)

Production is from an essentially normal-pressured sand within the overpressured shale transition zone. This production also is well above the cutoff shale ratio of 3.5 and could have been reached with just a surface pipe set at 5000 ft [1524 m].

As discussed previously, little oil is found below the intermediate pipe setting depth, but the possibility of gas cannot be ruled out until the shale ratio of 3.5 is reached.

This well drilled about 5000 ft [1524 m] of superpressured formation below the shale cutoff, requiring another string of pipe (7-inch, at 13,000 ft

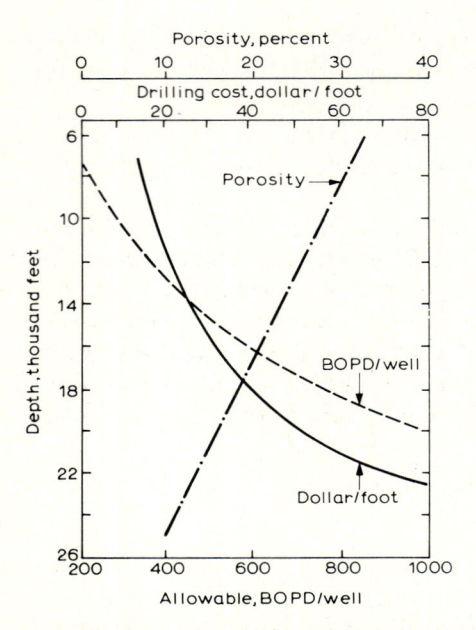

Fig.8.28. Trends in normal-pressure environments, U.S. Gulf Coast area (after Vidrine, 1971). BOPD = barrels of oil per day. (Courtesy of the American Association of Petroleum Geologists.)

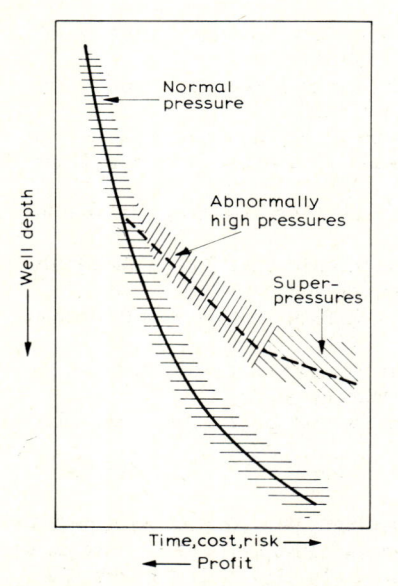

Fig.8.29. Generalized trends of deep-drilling costs and profit show drastic deviations where abnormal pressures are encountered (after Timko and Fertl, 1972b). More careful planning is required in such areas. (Courtesy of *World Oil*.)

[3963 m]) and high mud weights (up to 18.2 ppg [2.17 kg/dm^3]). Also, the time required to drill this lower, nonproductive part of the hole was about 50 days.

The bottom 5000 ft [1524 m] of the hole is in an unfavorable hydrocarbon environment, and the time and cost to drill it, in addition to the hazards, are unwarranted.

Recently, Vidrine (1971), while studying production potential of deep Miocene reservoir rocks in Louisiana, presented several generalized trends of three parameters with depth, representative of normal-pressure environments (Fig.8.28). Exploration in normal-pressure zones shows generally predictable trends for time, cost and risk. However, presence of abnormal pressures, especially superpressures, is a critical factor. Time, cost, and risks then increase drastically, greatly affecting profit (Timko and Fertl, 1972b). This is clearly illustrated in Fig.8.29.

References

Andreev, P.F., Bogomolov, A.I., Dobryanskiy, A.F. and Kartsev, A.A., 1968. Transformation of petroleum in nature. In: E.L. Ingerson (Editor), *International Sciences of Monographs in Earth Sciences*, Vol. 29. Pergamon Press, New York, N.Y., 290 pp.

Barton, D.C., 1934. Natural history of the Gulf Coast crude oils. In: W.E. Wrather and F.H. Lahee (Editors), *Problems of Petroleum Geology*. American Association of Petroleum Geologists, Tulsa, Okla., pp.109—155.

Baybakov, V.V., 1969. Effect of the thermodynamic conditions on the distribution of hydrocarbon pools. *Dokl. Akad. Nauk S.S.S.R.*, 187: 899—902.

Burst, J.F., 1969. Diagenesis of Gulf Coast clayey sediments and its possible relation to petroleum migration. *Bull. Am. Assoc. Pet. Geol.*, 53: 73—93.

Cannon, G.E. and Craze, R.C., 1938. Excessive pressures and pressure variations with depth of petroleum reservoirs in the Gulf Coast region of Texas and Louisiana. *Trans. AIME*, 127: 31—39.

Cannon, E. and Sullins, R.S., 1946. Problems encountered in drilling abnormal pressured formations. *Oil Gas J.*, 45(3): 120—124.

Classen, J.S., 1968. Formation pressure—production relation in Lake Mongoulois field. *SPE 2206, 43rd AIME Fall Meet.*, Houston, Texas, September.

Cordell, R.T., 1972. Depths of oil origin and primary migration: a review and critique. *Bull. Am. Assoc. Pet. Geol.*, 56: 2029—2067.

Denton, H.H., 1943. Abnormal salt water pressures on the Texas and Louisiana coast. *Field Lab.*, Southern Methodist Univ. Dallas, Texas, January, 45 pp.

Dickinson, G., 1953. Reservoir pressures in Gulf Coast Louisiana. *Bull. Am. Assoc. Pet. Geol.*, 37: 410—432.

Dott, R.H., 1969. Source book for petroleum geology, 1 and 2. *Am. Assoc. Pet. Geol., Mem.*, 5, 455 pp.

Eaton, B.A., 1972. A theory on the effect of overburden stress on geopressure prediction from well logs. *SPE 3719, 3rd Symp. on Abnormal Subsurface Pore Pressure*, Louisiana State Univ., Baton Rouge, La., May; also *J. Pet. Technol.*, 24: 929—934, 1972.

Federal Power Commission, 1966. *Reservoir Data Storage.* Exhibits H, South Louisiana Magnetic Tape File, Baton Rouge, La.

Fertl, W.H., 1971. A look at abnormally pressured formations in the U.S.S.R. *SPE 3613, 46th AIME Fall Meet.*, New Orleans, La., October.

Fowler, Jr., W.A., 1968. Pressures, hydrocarbon accumulation and salinities — Chocolate Bayou field, Brazoria County, Texas. *SPE 2226, 43rd AIME Fall Meet., Houston, Texas*, September; also *J. Pet. Technol.*, 22: 411—424, 1970.

Fowler, Jr., W.A., 1971. Discussion on the relationship between hydrocarbon accumulation and geopressure and its economic significance. *J. Pet. Technol.*, 23: 933—934.

Hedberg, H.D., 1969. Geologic aspects of origin of petroleum. *Bull. Am. Assoc. Pet. Geol.*, 53: 1755—1803.

Horner, D.R., 1951. Pressure build-up in wells. *Proc. 3rd World Pet. Congr.*, The Hague, 2: 503—522.

Hottman, C.E. and Johnson, R.K., 1965. Estimation of formation pressures from log-derived shale properties. *J. Pet. Technol.*, 17: 717—723.

Kartsev, A.A., Tabasaranskiy, Z.A., Subbota, M.I. and Mogilevskiy, G.A., 1954. *Geochemical Methods of Prospecting and Exploration for Petroleum and Natural Gas.* Univ. of California Press, Los Angeles, Calif., 347 pp. (1957, English translation from Russian).

Klemme, H.D., 1972. Heat influences size of oil giants, 1. *Oil Gas J.*, 69(29): 136, 141—144; Part 2, 70(30): 76—78.

Krejci-Graf, K., 1934. In: *Erdoel — Naturgeschichte eines Rohstoffes.* Springer Verlag, Berlin, 1955, 340 pp.

Landes, K.K., 1967. Eometamorphism and oil and gas in time and space. *Bull. Am. Assoc. Pet. Geol.*, 51: 828—841.

McCaslin, L.S., 1949. Tidewater's recordbreaking well. *Oil Gas J.*, 48(18): 58—59.

Meyerhoff, A.A. (Editor), 1968. Lafayette and New Orleans Geological Societies, geology of natural gas in South Louisiana. In: *Natural Gases of North America*, Vol. 1 — *Am. Assoc. Pet. Geol., Mem.*, 9, 225 pp.

Modelevskiy, M.S. and Parnov, Y.I., 1967. Physical state of oil and gas at depth. *Dokl. Akad. Nauk S.S.S.R.*, 175: 1372—1374.

Moses, P.L., 1961. Geothermal gradients now known in greater detail. *World Oil*, 152(6): 79—82.

Perry, D.R., 1969. *A correlation of reserves and drive mechanisms with reservoir pressure gradients in geopressured gas reservoirs in southwestern Louisiana.* M.S. Thesis, Univ. of southwestern Louisiana, Lafayette, La., 54 pp.

Perry, D.R., 1972. Statistical study of geopressured reservoirs in southwest Louisiana. *SPE 3888, 3rd Symp. on Abnormal Subsurface Pore Pressure*, Louisiana State Univ., Baton Rouge, La., May.

Pusey, W.C., 1973. How to evaluate potential gas and oil source rocks. *World Oil*, 176(5): 71—75.

Sorokin, 1897. In: A. Konchine, *Guide des excursions du VII Congrés Géologique Internationale*, p.24.

Timko, D.J. and Fertl, W.H., 1970. Hydrocarbon accumulation and geopressure relationship and prediction of well economics with log-calculated geopressures. *SPE 2990, 45th AIME Fall Meet.*, Houston, Texas, October; also *J. Pet. Technol.*, 23: 923—933, 1971.

Timko, D.J. and Fertl, W.H., 1972a. Implications of formation pressure and temperatures in the search and drilling for hydrocarbons. *4th Can. Well Logging Symp.*, Calgary, Alta., May, Paper E, 10 pp.

Timko, D.J. and Fertl, W.H., 1972b. How downhole temperatures, pressures affect drilling, 5. Predicting hydrocarbon environments with wireline data. *World Oil*, 175(5): 73—85.

REFERENCES

Timko, D.J. and Fertl, W.H., 1972c. How downhole temperatures, pressures affect drilling, 7. The shale resistivity ratio — a valuable tool for making economic drilling decisions. *World Oil*, 175(7): 59—63.
Vassoyevich, N.B., Korchagina, Y.I., Lopatin, N.V. and Chernyshev, V.V., 1969. Principal phase of oil formation. *Neft., Vestn. Mosk. Univ.*, 6: 3—27.
Vidrine, L.O., 1971. Production potential of deep Miocene rocks in southeastern Louisiana. *Bull. Am. Assoc. Pet. Geol.*, 55: 227—240.
Wallace, W.E., 1962. Water production from abnormally pressured gas reservoirs in South Louisiana. *Trans. Gulf Coast Assoc. Geol. Soc.*, 12: 187—193.
Zvereva, O.V. and Selitskiy, A.G., 1970. Some regularities in the spatial distribution of oil and gas in the oil/gas-bearing basin of the Persian Gulf. *Izv. Vyssh. Ucheb. Zaved., Geol. Razved.*, 10: 106—112.

Chapter 9. **GLOBAL OCCURRENCE AND EVALUATION OF ABNORMAL FORMATION PRESSURES**

Introduction

In the global search for hydrocarbon resources both on- and offshore, abnormal formation pressures have been encountered in all continents (Fig.9.1). Such abnormal formation pressures are defined as any departure from the normal hydrostatic pressure at any given depth. They may occur as shallow as only a few hundred feet below the surface or at depths far exceeding 20,000 ft [6098 m]. These abnormal pressures can be present in shale/sand sequences and/or massive evaporite-carbonate sections, and abnormally pressured formations are known to range in the geologic time scale from the Cenozoic era (Pleistocene age) to as old as the Paleozoic era (Cambrian age).

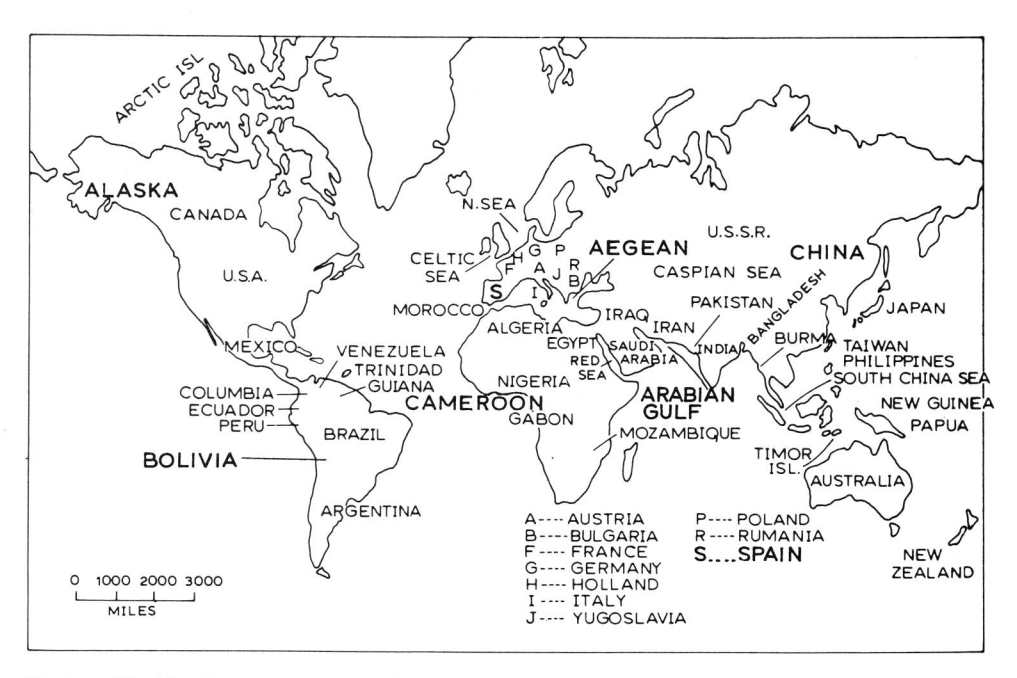

Fig.9.1. Worldwide occurrence of abnormal formation pressures (after Fertl, 1972). (Courtesy of the Society of Petroleum Engineers of A.I.M.E.)

Occurrence and magnitude of these abnormal-pressure environments have a profound impact on the oil industry. They greatly affect exploratory activities, drilling and well completion efforts, production operations, and associated reservoir engineering. As such, abnormal formation pressures are an important factor in managerial decision making.

In this chapter, information on the global occurrence of abnormal formation pressure environments has been compiled, and at the same time an attempt is made to show the potential of pressure evaluation methods using field examples specifically selected for different geologic areas.

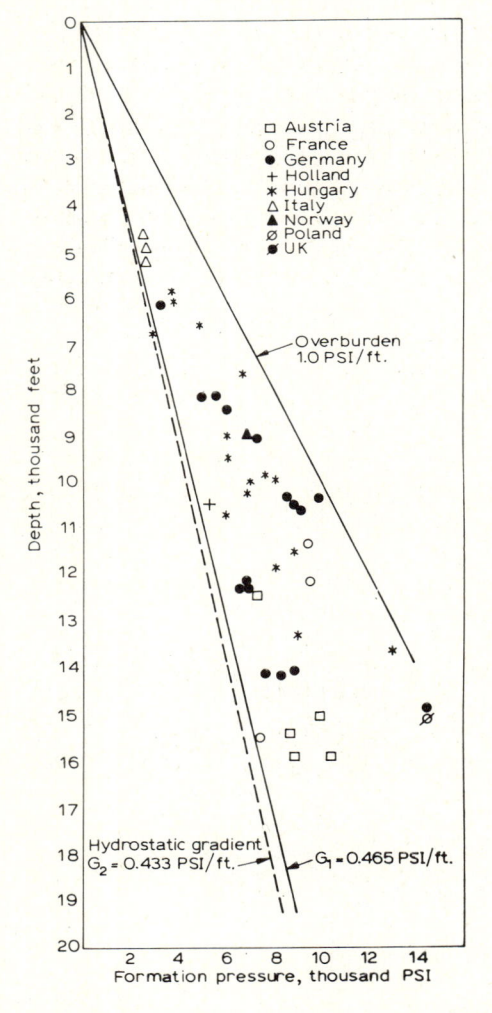

Fig.9.2. Selected abnormal formation pressure data, on- and offshore Europe (after Fertl, 1972). (Courtesy of the Society of Petroleum Engineers of A.I.M.E.)

Abnormal formation pressure environments in Europe

Drilling activities in both western and eastern Europe have encountered abnormal-pressure environments (Fig.9.2).

Onshore areas

Onshore areas with abnormal formation pressures include the Aquitaine Basin, *France*, with its deep sour-gas fields at Lacq, St. Marcet, Meillon, etc., located in a thick evaporite sequence along the north side of the Pyrenees Mountains. Besides reverse faulting and deep unconformities, abnormally high formation pressures have caused drilling problems. For example, at the deep Lacq well the initial reservoir pressure was 9420 psi [662 kg/cm^2] at 12,140 ft [3701 m] and at Meillon No. 1, 7330 psi [515 kg/cm^2] at 15,541 ft [4738 m]. Present-day technological challenges for ultra-deep drilling in this general area include (1) crooked-hole tendency in the "Flysch" section at the surface and below 13,000 ft [3963 m], (2) borehole instability in salt and red clays down to 12,000 ft [3659 m], (3) abnormally pressured shales covering the gas pay at 17,500 ft [5335 m], and (4) sour gas at high pressures from 18,800 to 21,000 ft [5732 to 6402 m]. For example, recent drilling on the Berenx and Came structures (Fig.9.3) has encountered BHP up to 15,000 psi [1055 kg/cm^2] at 20,000 ft [6098 m]. An additional problem is in casing design, drilling hydraulics, and other special procedures for handling H$_2$S.

Along the forelands of the *Alps* in several alpine basins, abnormal formation pressures, both high and subnormal, are found in formations of both Tertiary and Mesozoic age. It includes the Molasse Basin in *Germany*

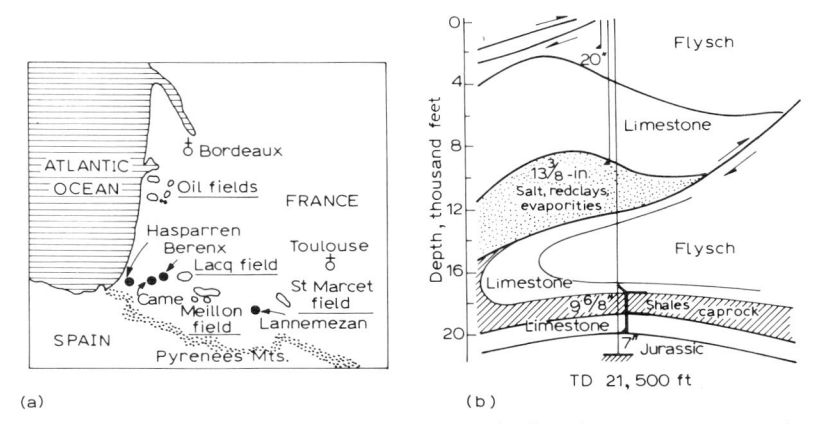

Fig.9.3. Geographic and geologic setting of ultra-deep, overpressured sour-gas field in France (after Guiraudet, 1972). (a) South Aquitaine sour-gas fields and prospects. (b) Berenx structure and No. 2 Berenx casing program.

(Bavaria) and *Austria* (Upper and Lower Austria), the prolific Vienna Basin in Austria, and the Po Basin in northern *Italy*. In general, conventional pressure evaluation methods are successfully applied in these areas as far as clastic sediments are concerned. However, complications have become apparent in the "Flysch" series and in the deep to ultra-deep massive gas-bearing carbonate sections which are often essentially devoid of pure shales. Furthermore, tectonic activities, presence of thick gas columns (i.e. fluid density contrast), variation in the piezometric fluid level (causing subnormal pressures in some Eocene and Jurassic formations in Bavaria), all still tend to complicate any simple, direct, and straightforward pressure evaluation method. However, Geier (1972) recently developed correlations, utilizing shale resistivities and depth as input parameters, which seem to approximate the expected formation pressure values in the Vienna Basin, Austria, and the Molasse Zone in Germany and Austria.

Another area of high drilling activities is the northwestern German Basin, which consists of several smaller troughs; the hydrocarbon-bearing formations of Tertiary and Lower Cretaceous age of the West Netherland Basin are separated from them by a swell region. In this area, which is all part of the major northwestern *European Permian Basin* (the latter also covers a large portion of the North Sea), abnormal pressures are found in formations of various age, including the Carboniferous, the gas-bearing Rotliegendes (Lower Permian), the Zechstein evaporite section (Upper Permian), the Buntsandstein (Lower Triassic), etc. These abnormal formation pressure environments are mainly due to the presence of salt domes or diapirs, other tectonic activities, and/or thick salt masses of large areal extent which frequently cover overpressured Permian shales and potential reservoir rocks. For example, the Groningen gas field in *Holland* — one of the world's largest — is overlain by the massive Zechstein evaporites which form the sealing cap of the slightly overpressured (approximately 0.5 psi/ft [0.115 kg cm^{-2} m^{-1}]) gas reservoir in the Permian Rotliegendes sandstone. In *northern Germany* (Fig.9.4) formation pressures are known to approach 0.8—0.9 psi/ft [0.185—0.207 kg cm^{-2} m^{-1}] and locally even higher. For example, pressures in the Main Dolomite (Hauptdolomit) are known to approach overburden load, which corresponds to a gradient of 1.0 psi/ft [0.231 kg cm^{-2} m^{-1}].

A quantitative formation pressure evaluation below the Zechstein (Upper Permian) appears to be possible if a combination of several parameters is used. These include rate of penetration, sonic travel time, lithology, and the extent of the massive overlying Zechstein series. Workable, empirical correlations have been developed for the Lower Triassic (Buntsandstein) formation. Limited success is obtained in the Permian Rotliegendes sandstone, whereas the Carboniferous as yet still presents major problems for any pressure prediction attempt. Furthermore, recent investigations of the massive Zechstein section (Upper Permian) using drilling, well-logging, and test data

EUROPE 329

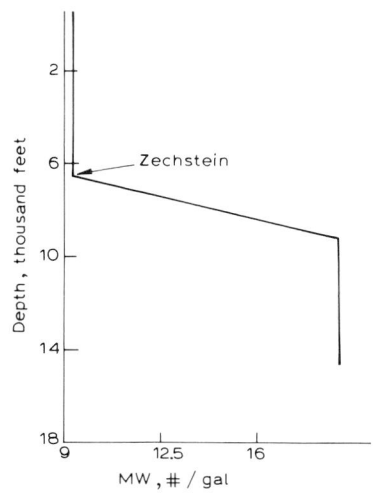

Fig.9.4. Typical formation pressure profile in North Germany (after Rehm, 1972). (Courtesy of the Society of Petroleum Engineers of A.I.M.E.)

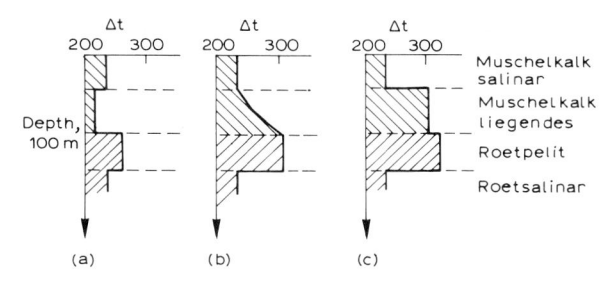

Fig.9.5. Typical acoustic travel-time variations (in μsec/m) in and below the Muschelkalk-salinar, North Germany (after Rizzi, 1973). (a) Normal pressure, (b) potential overpressure, (c) high overpressure.

have not resulted in any reliable quantitative pressure prediction methods. However, several empirical correlations have shown some promise in Upper and Middle Triassic carbonate formations such as Keuper and Muschelkalk. Typical Δt trends (μsec/m) in and below the Muschelkalksalinar are shown in Fig.9.5.

Similar to U.S. experiences in the Mid-Continent and West Texas areas, some of North Germany's drilling operations encounter repeated variations from normal to abnormally high pressure gradients, making properly engineered drilling operations essential.

The *Carpathian Mountains* extend from about 30 miles [48.3 km] east of Vienna, Austria, in an arc through Slovakia and northern Poland, finally crossing the Danube River to bend into the Balkan Mountains. Inside this arc are the Pannonian, Transylvanian, and several other small basins. *Hungary's* oil and gas fields are located inside this Carpathian mountain arc with the average well depth increasing steadily from 4113 ft [1254 m] in 1939 to 7616 ft [2322 m] in 1967. Recent exploratory efforts are focused on deeper targets (in 1968, the 15,000-ft [4573 m] mark was approached, and in 1972, a well was drilled to 19,680 ft [6000 m]). Much effort is spent on properly engineered drilling operations, with shale density measurements, penetration rate, *d*-exponent, pit level indicators, flowmeters, etc., along with well-logging techniques being utilized to maintain balanced drilling operations. Problems encountered in this area include unconsolidated Upper Pliocene sands and plastic shales to about 6000 ft [1829 m] and possibilities of lost circulation since Pliocene and Miocene formations were often deposited on fractured, cavernous Mesozoic carbonates (such as Triassic limestone). In general, formation pressure gradients of 0.86 psi/ft [0.199 kg cm^{-2} m^{-1}] and higher are always encountered along the base of Pliocene formations, which have been found to be as thick as 16,400 ft [5000 m] along the Mako Trench. This results in pressures exceeding 13,850 psi [974 kg cm^{-2} m^{-1}] at a depth of 16,000 ft [4878 m]. One additional handicap is the abnormally high formation temperature gradients. While showing wide variations over the entire basin, static formation temperatures (Fig.9.6) at depths of 13,100 ft [3994 m] to 16,400 ft [5000 m] are known to reach 430—480°F (221—243°C).

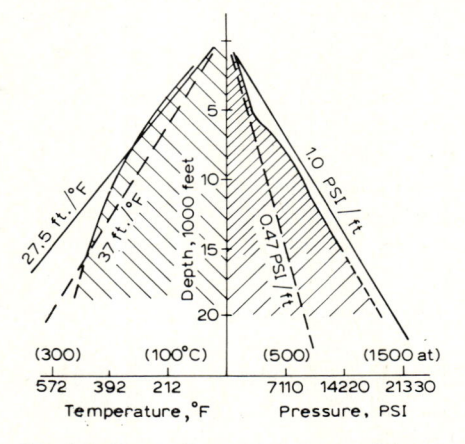

Fig.9.6. Formation pressures and temperatures in Hungarian oil and gas wells (modified after Alliquander, 1973). (Courtesy of *Oil and Gas Journal.*)

Along the forelands of the Carpathians, within the strongly folded and overthrusted foothill belt, abnormally high formation pressures have been encountered in *Rumania, Bulgaria,* and *Poland,* due to the presence and tectonic movement of salt and shale diapirs, interconnected hydrodynamic systems with higher-pressured waters at greater depth, etc. Besides sparse reports on major well control problems, such as blowouts and resulting well fires, no details are available on downhole pressure magnitudes and associated evaluation methods. Nevertheless, limited data on deep exploratory wells in Poland indicates pressures of about 14,500 psi [1020 kg cm^{-2} m^{-1}] and temperatures up to 400°F [204°C] at depths of about 15,000 ft [4573 m].

With the exception of the Kikinda blowout in 1971 and limited technical data on how to drill and achieve primary cementing under blowout conditions and loss of circulation, not much pertinent data have been found by the author with regard to drilling operations in *Yugoslavia*. However, it is our understanding that present drilling in the Dinaric Alps has penetrated fractures and caverns which are in direct hydraulic connection with the adjacent Adriatic Sea, thereby causing some well control problems.

Offshore Europe

Offshore drilling activities in Europe have also encountered abnormal formation pressures. One focus point is the *North Sea* area, which has become one of the most prolific oil and gas provinces. Exploration, at the present time, is carried out as far south as the projected seaward extension of the Groningen gas sands into offshore Holland and as far north as the 62° latitude in British and Norwegian waters.

In this region, abnormally high pressure environments are encountered *offshore of Holland, Germany*, the *United Kingdom, Ireland, Denmark* and *Norway*.

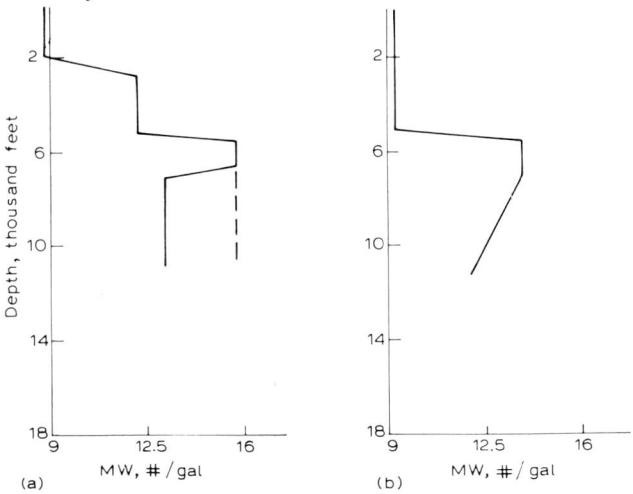

Fig.9.7. Typical formation pressure profiles as encountered in North Sea area (after Rehm, 1972). (a) Ekofisk area, (b) Zechstein evaporite section. (Courtesy of the Society of Petroleum Engineers of A.I.M.E.)

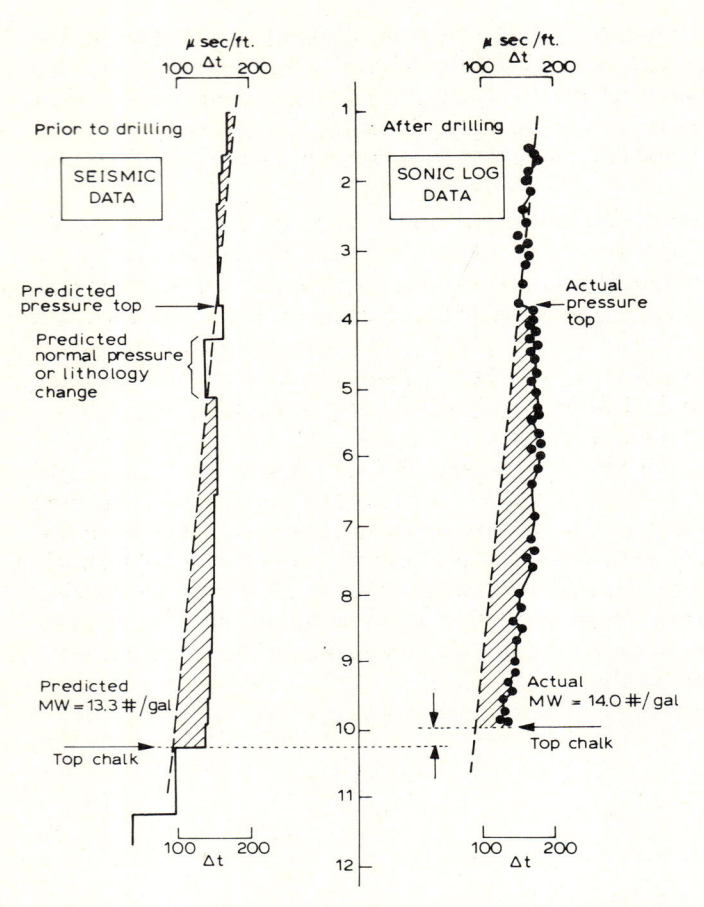

Fig. 9.8. Comparison of predicted and actual downhole pressure environment in the North Sea area (after Herring, 1973). (Courtesy of *Petroleum Engineer*.)

The magnitude of these pressures varies widely over the entire area in both Tertiary and Mesozoic formations. However, only limited pressure data have been released, including acknowledged overpressures and resulting well-control problems in gas-bearing Triassic sandstones, Upper Permian evaporites, etc. Furthermore, drilling offshore of Norway has encountered over-pressured Tertiary and Mesozoic oil and gas sands, but at the present time, the most prolific reservoir rocks are overpressured, chalky, often highly fractured limestones of Upper Cretaceous age in the oil- and gas-bearing structures offshore of Denmark. The same horizon is productive offshore of Norway with the more important pay zone in the overlying Danian chalk. For example, reported pressure data for the Ekofisk field give 7000 psi [492 kg cm^{-2} m^{-1}] at 10,000 ft [3048 m].

For localized United Kingdom and Norwegian waters, mud weights in the range of 13—15.5 lb/gal [1.56—1.86 kg/dm^3] have been necessary. Drilling may also become troublesome due to thick gumbo shales which may persist down to 7000 ft [2134 m]. Furthermore, shallow, highly pressured but low-volume gas pockets may often be encountered. Typical pressure profiles are shown in Fig.9.7.

Pressure prediction and evaluation methods used are quite similar to techniques applied in the U.S. As such, they are based on combinations of seismic, drilling, and logging parameters and are reliable. Fig.9.8 shows excellent agreement between predicted and actual pressures in a North Sea well.

Exploratory activities in the *Celtic Sea*, offshore Ireland, have encountered the presence of thick, folded and faulted sediments, promising hydrocarbon shows, and overpressured shales.

In the *Mediterranean*, abnormal pressures are found offshore of Spain and North Africa, and along the *Adriatic and Aegean Sea*. The Apennine Foredeep, located on the Adriatic side of the Italian peninsula, has been the scene of substantial hydrocarbon discoveries. Both on- and offshore abnormal-pressure environments have caused drilling problems, blowouts, and casing collapse. In recent years, application of modern pressure evaluation techniques has been highly successful. Fig. 9.9 illustrates this on an offshore well, located in the central part of the Adriatic Sea.

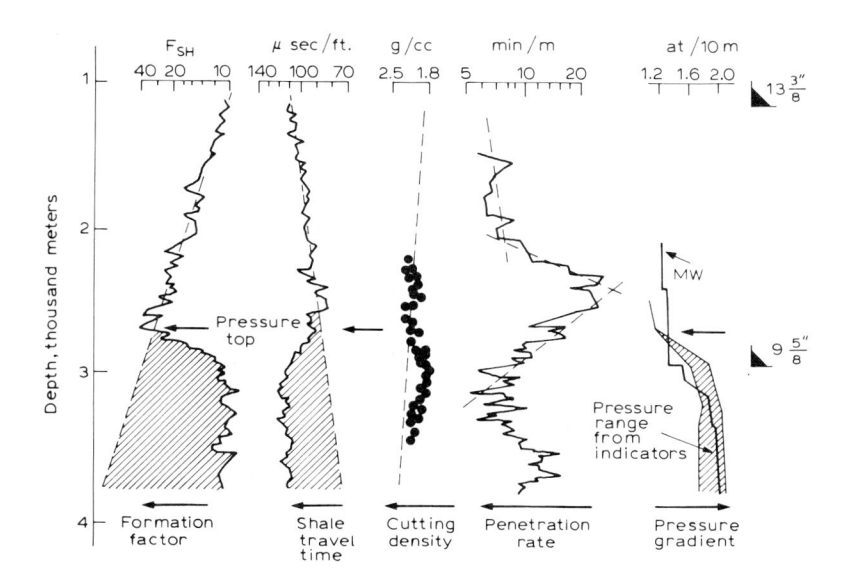

Fig.9.9. Overpressure environment in a central Adriatic well, Mediterranean Sea (after Rizzi, 1973).

Another Mediterranean area with abnormally high-pressure gas and oil shows has recently been encountered in *offshore Greece* in the *Aegean Sea*, such as in the South Kavala well No. 2.

Abnormal formation pressure environments in the Soviet Union

Recently, the author has examined data of over 100 Russian wells which penetrate Devonian, Carboniferous, Mesozoic, and Tertiary formations. In rocks of Devonian age, such as are found in the Saratov and Kuybyshev regions, pressures differ little from hydrostatic, occasionally exceeding it by not more than 15%. Carboniferous reservoir rocks behave similarly, with

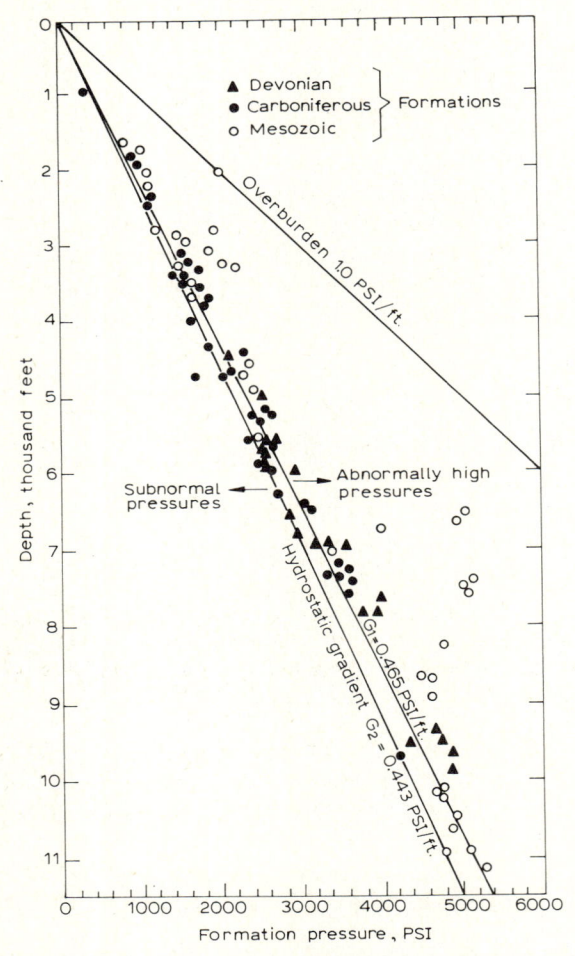

Fig.9.10. Abnormally pressured formations of Devonian, Carboniferous, and Mesozoic age in the U.S.S.R. (after Fertl, 1971).

SOVIET UNION 335

overpressures not exceeding hydrostatic by more than 20%. Thus, one concludes that in the Russian Platform, which is surrounded by the Ural and Caucasus mobile belt and contains many arches and depressions, Paleozoic formations of both Devonian and Carboniferous age, in general, exhibit hydrostatic pressures or only slightly above (Fig.9.10). Russian hydrocarbon-bearing formations of Mesozoic age (Fig.9.10) often show markedly different pressure conditions, whereas in the Dnieper-Donetz Basin the pressures approximate hydrostatic conditions; they exceed hydrostatic by 50—100% in the folded region of the Ciscaucasian area.

Pressure environments in the prolific oil and gas reservoirs of Tertiary age are grouped into 3 categories (Fig.9.11): (1) subnormal-pressure environ-

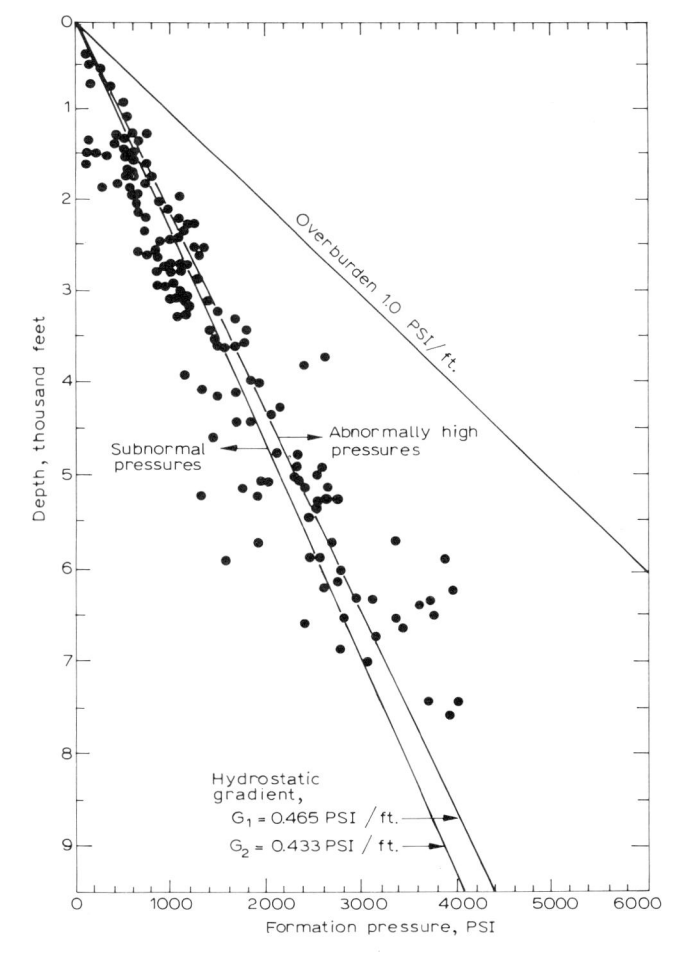

Fig.9.11. Abnormally pressured formations of Tertiary age in the U.S.S.R. (after Fertl, 1971). (Courtesy of the Society of Petroleum Engineers of A.I.M.E.)

ments occurring in Middle Miocene formations, e.g. of the Chokrak and Karagan stages; however, they occur less with increasing depth; (2) hydrostatic pressure environments, such as in the gas-bearing pay zones of Tortonian and Sarmatian age in the Ciscaucasian depression; and (3) overpressure environments, such as in several Maykopian beds. In Dagestan A.S.S.R., formation pressures closely approach overburden, and in the Checheno-Ingushian A.S.S.R., formation pressures locally exceed overburden. For example, at 5250 ft [1600 m] pressures were reported to range from 5870 to 7350 psi [413 to 517 kg/cm^2].

Soviet investigators have studied possible modes of overpressure origins to a large extent, and much literature has been published in the field of pressure prediction and evaluation (Anikiev, 1964; Zil'berman, 1972). Most of the methods, however, appear to be similar to the ones used in the United States.

Abnormal formation pressure environments on the North American continent

The hydrocarbon province of the *Arctic Islands* is centered around the elongated east—west Mesozoic Sverdrup Basin, with a complex of older Paleozoic arches and basins to the south and a younger Cenozoic shelf basin to the North. With most pressure data still proprietary, high-pressured gas blowouts have occurred on Melville and King Christian Islands, and pressure gradients of about 0.6 psi/ft [0.139 kg cm^{-2} m^{-1}] are known to be present on several islands.

Overpressures also occur in several *Canadian areas*. These include the Rainbow Lake area in western Canada, off the west coast, off the east coast, under the Nova Scotian Shelf, and in the Mackenzie Delta area both on- and offshore. Causes of these overpressures include rapid sedimentation, shale and salt diapirism, and the presence of major fault zones. A 17-lb/gal mud weight is about that required in the Delta area, whereas 18.5 lb/gal [2.22 kg/dm^3] mud seems to be able to handle anything encountered in the Canadian frontier areas so far. For example, in the initial Sable Island test, the first hole drilled into the Scotian Shelf, overpressures started at the base of the Naskapi Shale, calling for 15.5 lb/gal [1.86 kg/dm^3] mud. Another East Coast well encountered a gradient of 0.73 psi/ft [0.169 kg cm^{-2} m^{-1}] after penetrating a very short, abrupt transition zone from normal to overpressure. Fig.9.12 shows a typical pressure profile for this area.

Whereas most pressure data in rank wildcats drilled in these remote and new exploratory areas are proprietary, the pressure prediction and evaluation methods are well known and do not differ from those used in the United States.

Numerous publications have discussed occurrence and implications of abnormal formation pressures in the United States. Also, most of the predic-

NORTH AMERICA 337

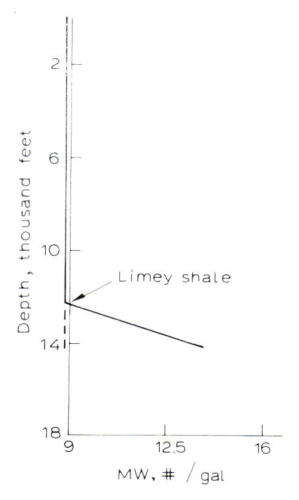

Fig.9.12. Typical mud weights used (formation pressure profile), offshore of eastern Canada (after Rehm, 1972). (Courtesy of the Society of Petroleum Engineers of A.I.M.E.)

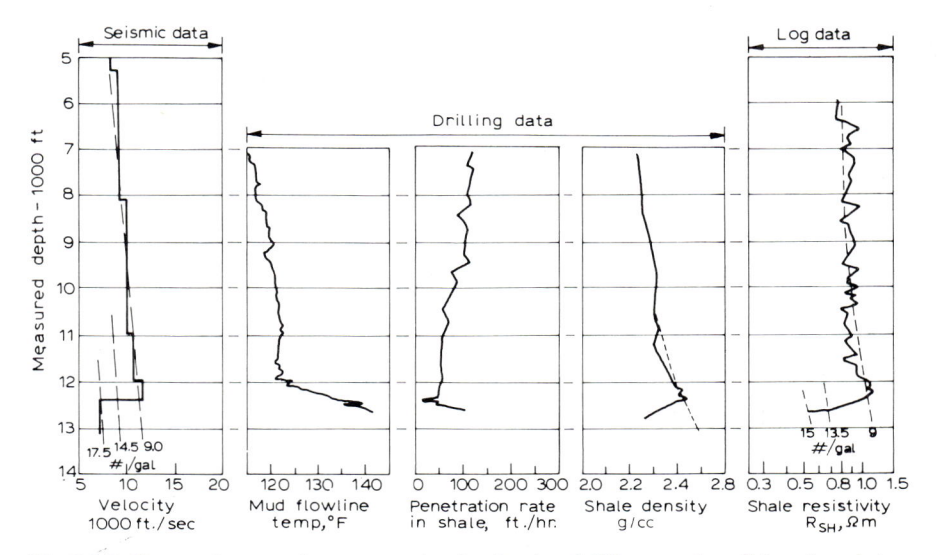

Fig.9.13. Comparison and agreement of seismic, drilling, and well-logging data used to predict and detect abnormal formation pressures, offshore Louisiana well (after Fertl and Timko, 1970). (Courtesy of *Oil and Gas Journal*.)

tion and evaluation methods used worldwide have been developed in the U.S. In the United States, abnormal pressures (both high and low) are found on- and offshore in shale/sand and carbonate sequences. This includes both shallow and ultra-deep exploration in Alabama, Alaska, Arkansas, California, Colorado, North and South Dakota, Florida, Louisiana, Mississippi, Montana, New

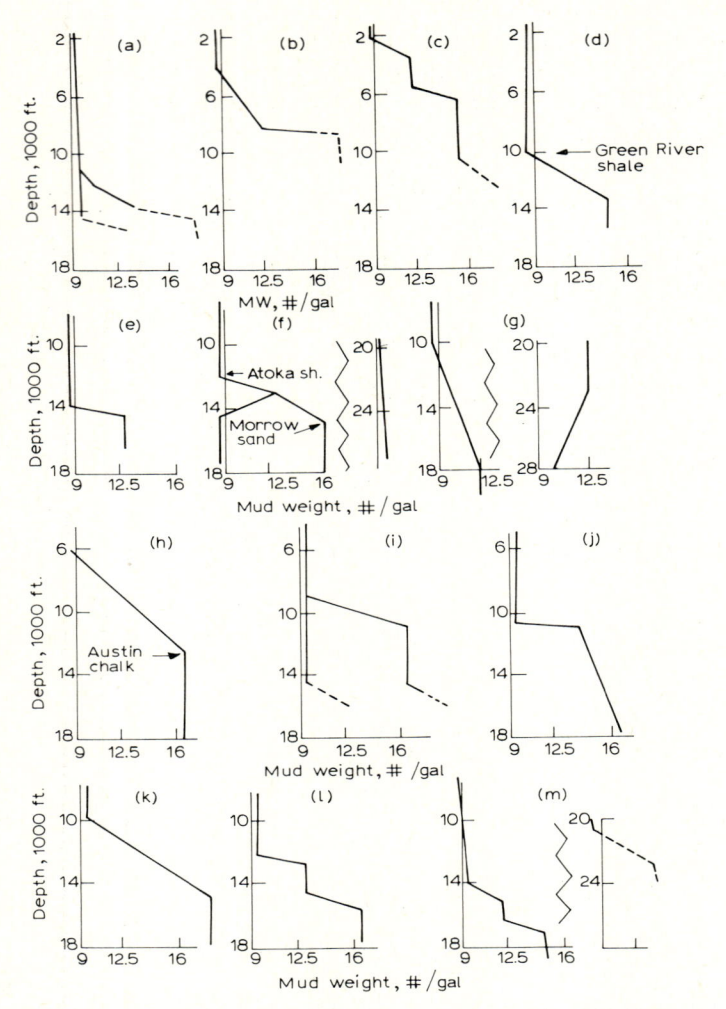

Fig.9.14. Typical formation pressure profiles in the United States (after Rehm, 1972). (a) Santa Barbara Channel, California; (b) Colusa County, California; (c) Humboldt County, California; (d) Uintah Basin, Utah; (e) Rio Blanco County, Colorado; (f) Anadarko Basin, Oklahoma; (g) Ralph Lowe Estate, West Texas; (h) Zapata County, South Texas; (i) offshore Upper Texas; (j) Vermilion Parish, Louisiana; (k) Terrebonne Parish, Louisiana; (l) Mississippi Delta, Louisiana; (m) Wayne County, Louisiana. Zig-zag lines in (f), (g), and (m) represent depth breaks. (Courtesy of the Society of Petroleum Engineers of A.I.M.E.)

Mexico, Oklahoma, Texas, Utah, Wyoming, West Virginia, and possibly elsewhere. Fig.9.13 shows various pressure prediction and detection data in an offshore Louisiana well. Typical pressure profiles in several U.S. areas are given in Fig.9.14.

SOUTH AMERICA 339

As expected, overpressures are also found in *Mexico*, south of the Texas border, and offshore. Mud weights up to 18 lb/gal [2.16 kg/dm^3] have become necessary for controlled drilling operations; formation temperatures approach 400°F [204°C].

Abnormal formation pressure environments in South America

Overpressures in this region frequently occur offshore, along the Andes Mountains, and elsewhere (Fig.9.15).

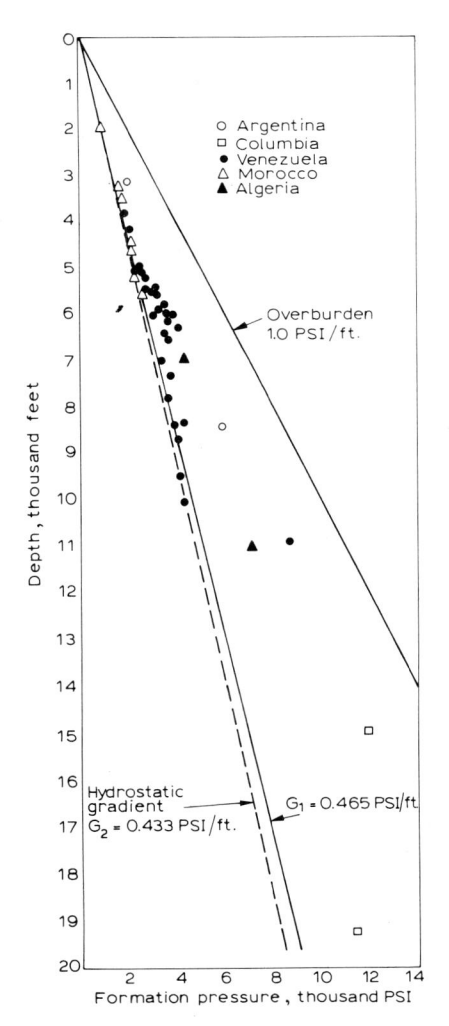

Fig.9.15. Selected abnormal formation pressures in South America and Africa (after Fertl, 1972). (Courtesy of the Society of Petroleum Engineers of A.I.M.E.)

In the *Trinidad-Tobago* area, overpressures are known in Tertiary shale/sand sequences (for example, 0.89 psi/ft [0.206 kg cm^{-2} m^{-1}] in the Forest Reserve oil field). Similar pressure gradients of 0.75—0.90 psi/ft [0.173—0.209 kg cm^{-2} m^{-1}] have been encountered in the foothills of the Andes in *Argentina*, *Bolivia*, and *Colombia*, both in shallow (above 2000 ft [610 m]) and very deep drilling. In *Venezuela*, overpressures were found while exploring for the Cretaceous potential in Lake Maracaibo, in the new offshore area to the north, and are encountered in East Venezuela. Mud requirements of 16—19.5 lb/gal [1.92—2.34 kg/dm^3] have become necessary. A particular problem is presented by the massive, heavy, black Cretaceous Colon Shale. Abnormal pressure data for the Anaco area fields are shown in Fig.9.16.

Overpressures also exist offshore; on the west coast in *Colombia*, *Ecuador*, *Peru*, and *Bolivia*, and along the east coast in *Venezuela*, *Guiana*, and *Brazil*. Similar to U.S. Gulf Coast experience extreme overpressures in shale/sand sequences relate to high-volume but rather limited hydrocarbon deposits which are characterized by rapid production declines. Fig. 9.17 shows data for a well drilled off the South American East Coast.

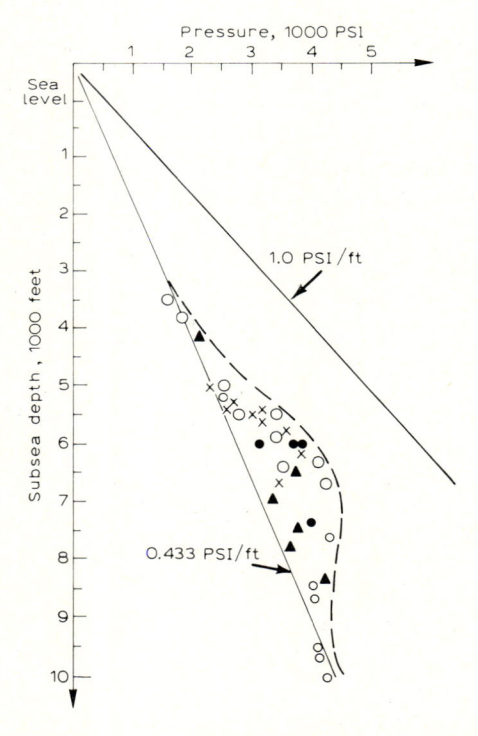

Fig.9.16. Abnormal formation pressures in Venezuela (modified after Funkhouser et al., 1948). ▲ = Santa Ana, ● = San Joaquin, ○ = Guario, and × = Santa Rosa oil fields (Anaco fields), Central Anzoategui region.

AFRICA 341

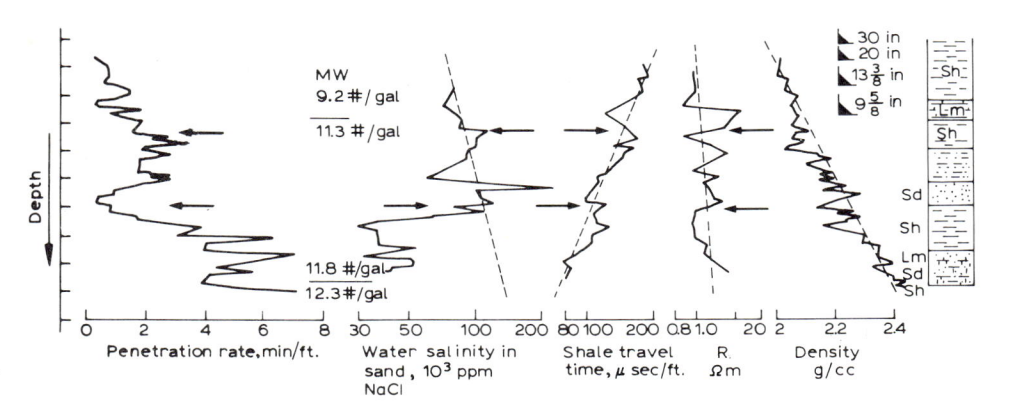

Fig.9.17. Overpressured formations in a South American well, offshore Guiana.

As elsewhere, a concerted effort has been undertaken over the last few years, both in personnel training and by using modern pressure evaluation concepts, to drill such overpressured formations properly.

Abnormal formation pressure environments in Africa

Only slight overpressures are found in tectonically broken up, elongated anticlinal folds covered by thick Miocene marls in onshore *Morocco*. Higher overpressures are present in several of the prolific fields in *Algeria*. For example, the 850-square mile [2210 km²] Hassi R'Mel anticline, covered by

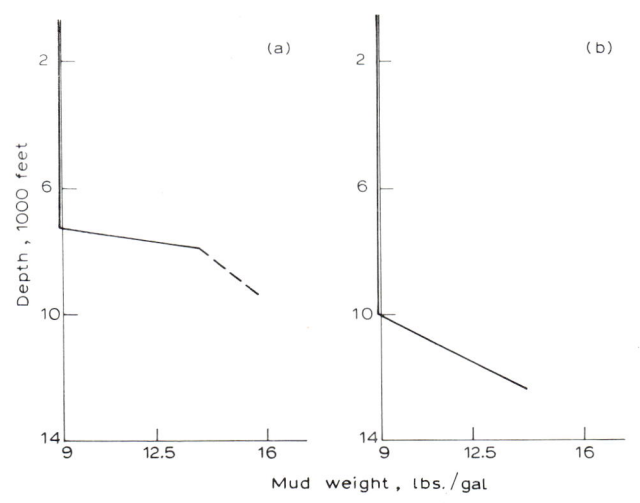

Fig.9.18. Typical formation pressure profiles in Africa (after Rehm, 1972). (a) Nigeria, (b) offshore East African coast. (Courtesy of the Society of Petroleum Engineers of A.I.M.E.)

Triassic salts, was found to be overpressured (approximately 0.62 psi/ft [0.143 kg cm^{-2} m^{-1}]). Also, drilling in the famous Hassi Messaoud field encountered gradients of 0.6—0.65 psi/ft [0.139—0.15 kg cm^{-2} m^{-1}]). Drilling operations in *Nigeria* and the Nile delta, *Egypt*, have found overpressure environments similar to offshore Louisiana. In addition, overpressures which range over wide areas of the *African West Coast* are localized on the *East Coast* (Mozambique, Madagascar); they are also found in the Red Sea area (offshore Ethiopia, onshore border region of Saudi Arabia, and Yemen) and can be encountered in both sand/shale sequences and massive carbonate sections (Figs.9.15 and 9.18).

Abnormal formation pressure environments in the Far East

Overpressured formations have been penetrated in Australia, Papua, and New Guinea, including the Papuan Basin, the Bowen Basin, the offshore

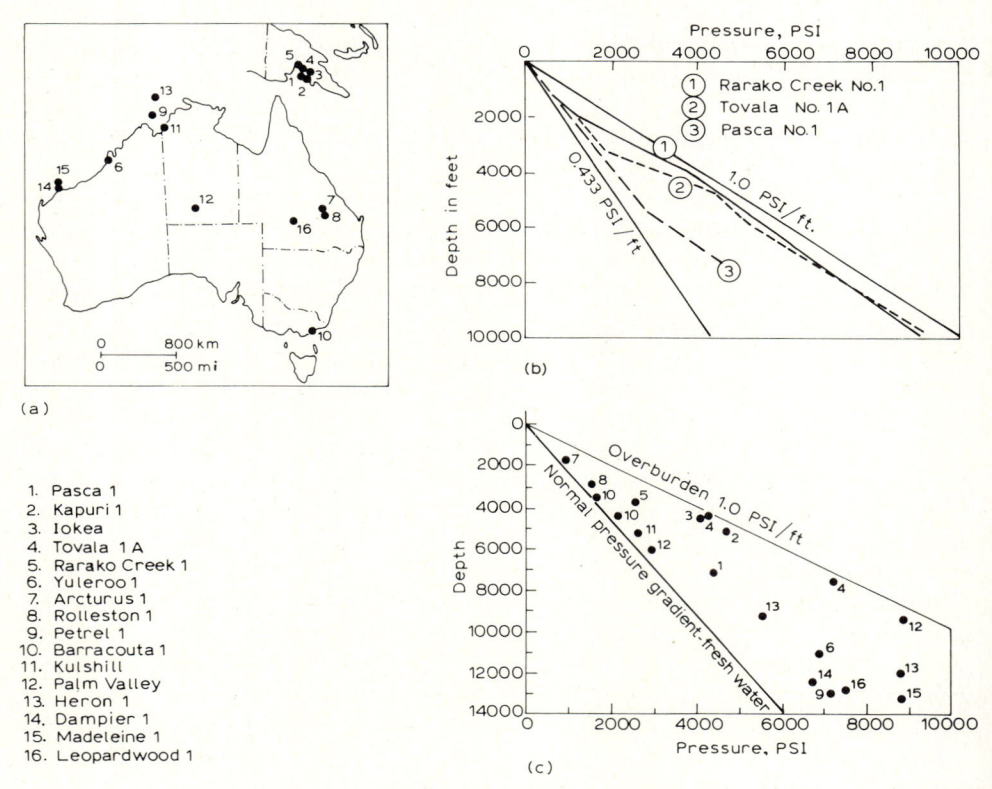

1. Pasca 1
2. Kapuri 1
3. Iokea
4. Tovala 1A
5. Rarako Creek 1
6. Yuleroo 1
7. Arcturus 1
8. Rolleston 1
9. Petrel 1
10. Barracouta 1
11. Kulshill
12. Palm Valley
13. Heron 1
14. Dampier 1
15. Madeleine 1
16. Leopardwood 1

Fig.9.19. Abnormal formation pressures in Australia and Papuan Basin (after Lepine and White, 1973). (a) Location and names of overpressured wells in the subject area. (b) Abnormal pressures recorded in Australian wells. (c) Pressure gradients of several wells drilled in the Papuan Basin.

portion of the Carnarvon Basin, the Bonaparte Gulf Basin, and the Perth Basin (Fig.9.19). Thus, occurrence of overpressures is widespread around *Australia*, both on- and offshore and in formations of different geological ages. In Queensland, overpressures have caused shallow blowouts (Arcturus No. 1, Rollestone No. 1) and pressure gradients of 0.6 psi/ft [0.139 kg cm^{-2} m^{-1}] are present at greater depth. On the Gippsland Shelf, gradients of 0.55 psi/ft [0.127 kg cm^{-2} m^{-1}] are present. In Western Australia, overpressures are common (Madeleine No. 1, Dampier No. 1, Yuleroo No. 1) with gradients of 0.62 psi/ft [0.143 kg cm^{-2} m^{-1}] below 11,000 ft [3354 m]. The same is true for the Northern Territory, both onshore (Palm Valley No. 1) and offshore (Petrel No. 1, Heron No. 1) where mud weight requirements sometimes exceed 14 lb/gal [1.67 kg/dm^3].

Severely faulted sediments of Tertiary and Mesozoic age are overpressured in *New Guinea* and *Papua*. Exploratory drilling in the delta area of the Mamberamo River in northern New Guinea encountered overpressures in excess of twice the hydrostatic. In general, overpressures are present in Pliocene/Miocene mudstones of the Aure Trough and Aptian Mudstone of Lower Cretaceous age. In Papua, the Rarako Creek No. 1 required 17 lb/gal [2.04 kg/dm^3] while penetrating a thrust fault and experiencing extremely heavy shale gas. Tovala No. 1 reported a 1.04-psi/ft [2.40 kg cm^{-2} m^{-1}] gradient at 4650 ft [1418 m]. Offshore Papua, in the Aure Trough, gradients of 0.91 psi/ft [2.10 kg cm^{-2} m^{-1}] were encountered. In massive gas-bearing reefal limestones, overpressures (0.62 psi/ft [0.143 kg cm^{-2} m^{-1}] in Pasca No. 1) are caused by the density contrast of the reservoir fluids. To overcome this problem, drilling operations have successfully utilized the "floating mud cap" technique, i.e. heavy mud floats in the annulus while drilling is achieved by pumping seawater down the drill pipe. This requires a delicate pressure balance, leads to high drilling fluid consumption, and results in loss of drill cuttings for geologic studies.

Geologic data on the island of *Timor*, 300 miles [483 km] off the northern coast of Australia, indicate presence of thick Tertiary and Mesozoic sediments, mud volcanoes, and hydrocarbon seeps. Limited drilling activities have encountered overpressures of about 0.65 psi/ft [0.15 kg cm^{-2} m^{-1}] at relatively shallow depth.

In *New Zealand*, mud volcanoes and a thick, rapidly deposited succession of Tertiary or Upper Cretaceous rocks have caused overpressure environments. Wells drilled near the north coast of Hawke Bay encountered pressure gradients of 0.8 psi/ft [0.185 kg cm^{-2} m^{-1}], particularly in the bentonitic mudstones of the Lower Tertiary. Slightly lower overpressures are found in the Upper Cretaceous formations. Also, overpressures and casing collapse are reported in the Northland Sedimentary Basin elsewhere.

Drilling experience in the *South China Sea* area indicates overpressures and abnormally high geothermal gradients (3°F/100 ft [5.5°C/100 m] and more) in several areas. Overpressures, widely varying in magnitude and requiring mud weights as high as 17.5 lb/gal [2.10 kg/dm³], occur on- and offshore of Sumatra, Java, Borneo, and Kalimantan. Gumbo shales, limey shales, volcanic ash beds, overpressured shallow gas pockets and deep overpressured environments, etc., present entirely different drilling challenges from area to area. Typical pressure profiles are shown in Fig.9.20.

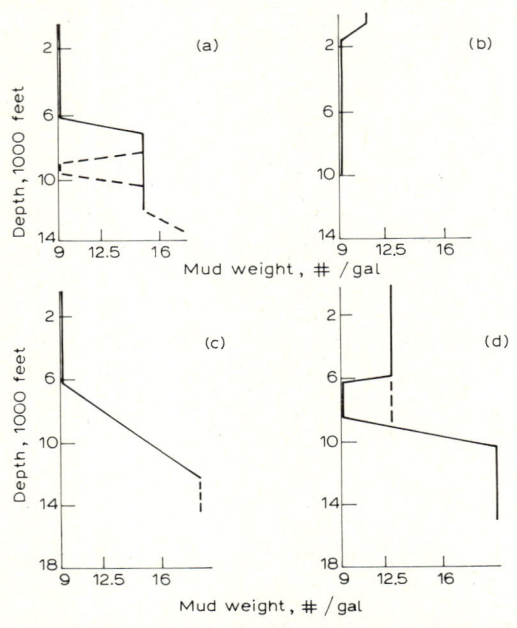

Fig.9.20. Typical formation pressure profiles in the South China Sea region (after Rehm, 1972). (a) Indonesia, Sumatra; (b) Indonesia, Java; (c) South China Sea; (d) Kalimantan marine lowlands. (Courtesy of the Society of Petroleum Engineers of A.I.M.E.)

Hydrocarbons have been found to be present in highly folded and faulted Miocene strata of southwestern Cebu, *Philippines*. Overpressures at very shallow depth (2100 ft [640 m]) result in gradients approaching 0.84 psi/ft [0.194 kg cm^{-2} m^{-1}].

On *Taiwan (Formosa)*, overpressures up to 0.8 psi/ft [0.185 kg cm^{-2} m^{-1}] are found in shale/sand sequences of Tertiary age, such as the Miocene Chuhuangkeng Formation, in both oil and gas fields (e.g. Chinshui, Chuhuangkeng, and Tiehchenshan-Tungshiao). In general, the highest pressure and temperature gradients occur in deep gas reservoirs (Table 9.I).

The Nagaoka Plain is located on Honshu, the main island of *Japan*. Main hydrocarbon reservoirs of Tertiary age are in volcanic and pyroclastic rocks, which exhibit a wide range of formation pressures (Table 9.II). Overpressures increase with the depth of these reservoir rocks and thickness of the overlying low-permeability mudstone beds which act as sealing caps. Gradients up to 0.8 psi/ft [0.185 kg cm^{-2} m^{-1}] are known.

TABLE 9.I

Formation pressures and gradients in gas and oil fields in Taiwan (after Chan, 1964)

Well No.	Zone	Depth mid-point (m)	Elevation a.s.l. (m)	Reservoir pressure (kg/cm^2)	Pressure gradient based on ground surface (kg cm^{-2} m^{-1})	Pressure gradient based on sea level (kg cm^{-2} m^{-1})
Chinshui gas field						
45	CS-13	2,354	115.68	231.8	0.0985	0.1036
54	CS-13	2,181	43.38	227.1	0.1041	0.1062
55	CS-13	2,283	63.72	232.3	0.1018	0.1047
56	CS-13	2,270	76.45	229.6	0.1011	0.1047
60	CS-13	2,222	112.37	229.9	0.1035	0.1090
60	CS-14Z	2,372	112.37	245.1	0.1033	0.1085
59	CS-14Z	2,313	66.56	232.5	0.1005	0.1035
59	CS-15	2,868	66.56	339.8	0.1188	0.1213
38	CS17A, 18	3,632	76.48	522.5	0.1439	0.1469
52	CS-18	3,393	112.37	505.8	0.1492	0.1541
58	CS-16, 17, 17A	3,121	91.29	476.1	0.1525	0.1571
58	CS-17A, 18	3,298	91.29	507.4	0.1537	0.1582
53	CS-30	4,416	37.83	669.9	0.1517	0.1530
Chuhuangkeng oil field						
108	CHK-2, 2A	1,847	237.33	237.0	0.1283	0.1472
108	CHK-3Z, 3	2,005	237.33	257.9	0.1286	0.1459
107	CHK-5Z, 5	2,026	186.87	331.7	0.1637	0.1804
106	CHK-6	2,911	215.80	405.6	0.1393	0.1505
106	CHK-7	3,062	215.80	420.1	0.1372	0.1476
106	CHK-9A	3,368	215.80	436.7	0.1297	0.1385
106	CHK-10	3,557	215.80	449.9	0.1265	0.1347
Tiehchenshan-Tungshiao gas field						
TCS-2	TT-1A	2,719	34.96	278.2	0.1023	0.1037
TCS-2	TT-3, 4A	3,357	34.96	466.0	0.1388	0.1403
TCS-2	TT-4B, 5A, 5B 5C	3,512	34.96	503.22	0.1433	0.1447
TCS-2	TT-6, 7	3,638	34.96	519.8	0.1429	0.1457
TCS-2	TT-9, 10, 10A	4,042	34.96	639.1	0.1581	0.1595
TCS-2	TT-12, 12A	4,214	34.96	698.5	0.1658	0.1671
TCS-2	TT-14, 15	4,390	34.96	704.2	0.1604	0.1617
TCS-2	TT-16, 17	4,535	34.96	709.1	0.1564	0.1576

TABLE 9.II

Bottom-hole pressures measured in several wells in the Nagaoka Plain, Japan (after Magara, 1968)

Name of field	Well No.	Formation	Depth (m)	Pressure (kg/cm^2)	Pressure gradient (kg cm^{-2} m^{-1})
Fujikawa		agglomerate	2,310.4	367.6	0.159
Kumoide		agglomerate	1,745.0	251.2	0.144
		sandstone	1,394.8	157.1	0.112
Sekihara	1	agglomerate	1,129.0	117.5	0.104
		sandstone	817.0	74.5	0.091
Katagai	1	agglomerate	1,052.5	99.8	0.105
Nishi-Nagaoka		agglomerate	1,172.2	117.7	0.10
		sandstone	1,144.5	106.1	0.093
Mitsuke		sandstone	1,201.3	125.5	0.104
		lava	1,744.9	264.3	0.151
Kurosaka		lava	1,816.0	256.6	0.142
Higashi-Sanjo		tuff	1,248.8	132.3	0.106

Abnormal formation pressure environments on the Indian subcontinent

In *Pakistan*, extreme overpressures are found in shallow Tertiary freshwater sediments in the foothills of the Himalayan Mountain ranges. At 5400 ft [1646 m], a pressure of 7000 psi [492 kg/cm^2] was reported. This 1.3-psi/ft [0.3 kg cm^{-2} m^{-1}] gradient in a shale/siltstone/sandstone sequence is related to tectonic stresses caused by the Himalayan folding and the extreme steepness of the potential pay zone. Similar overpressures are also found in Dhulian, Joya Mair, and other fields in Pakistan and across the border in northern India. Recent drilling offshore of Pakistan has also encountered overpressures.

In *India*, overpressures are found in zones of Eocene/Pliocene age, both in high shale/sand ratio sequences and tectonically greatly disturbed areas. In the folded belt of the Himalayan foothills, such as in Jammu, Punjab, and Assam, overpressures approach overburden in many wells. In the geosyncline region of the Bengal Basin, overpressures usually occur in the Debagram and Pandua formations below a massive shale section, requiring mud weights of up to 18.5 lb/gal [2.22 kg/dm^3].

In *Bangladesh*, frequent drilling problems are caused by overpressures. For example, thin sand streaks in massive shale sections at about 9000 ft [2744 m] exhibit pressure gradients of up to 0.99 psi/ft [0.229 kg cm^{-2} m^{-1}] in the Sylhet structure.

Overpressures in *Burma* are found in Tertiary rocks of the Irrawaddy and Chindwin River Basin and offshore in the Gulf of Martaban. Gradients in the Chauk field are 0.86 psi/ft [0.199 kg cm^{-2} m^{-1}] and in the central part of the Yenangyaung structure at 5000 ft [1524 m] it is as much as 1.0 psi/ft [0.231 kg cm^{-2} m^{-1}], whereas drilling below 9000 ft [2744 m] has required a 18.9 lb/gal [2.27 kg/dm^3] mud weight.

Abnormal formation pressure environments in the Middle East

In the Middle East, overpressures are caused by tectonics, shale and salt diapirs, massive overlying evaporite sequences, and fluid density contrast in

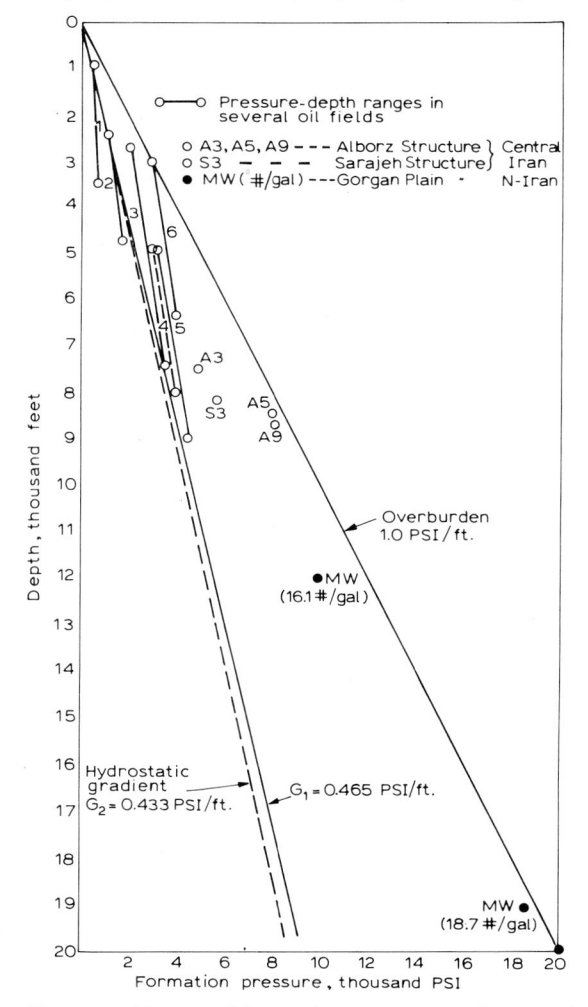

Fig.9.21. Abnormal formation pressure environments in Iran (after Fertl, 1972). (Courtesy of the Society of Petroleum Engineers of A.I.M.E.)

large structures. Overpressures occur in *Iran* in the northern Plains of Gorgan, in Central Iran, and elsewhere (Fig.9.21). Mud weights as high as 22 lb/gal [2.64 kg/dm^3] have been used, and pressure gradients ranging from 0.85 to 0.97 psi/ft [0.196 to 0.224 kg cm^{-2} m^{-1}] are known. Similar conditions are found in *Iraq*. In *Saudi Arabia*, overpressures plus abnormally high temperature gradients (2.5°F/100 ft [4.6°C/100 m]), also have been found on- and offshore along the Red Sea region.

Acknowledgements

The author is greatly indebted to numerous people in industrial, academic, and governmental positions around the world whose encouraging cooperation made this compilation of data possible. Additional data used in this review were selected from a vast amount of published data and the author's field experiences. However, the author has not attempted to reference all individual contributions with the exception of field cases used in this chapter. Needless to say, such a review will always show omissions due to proprietary and other reasons.

References

Alliquander, O., 1973. High pressures, temperatures plague deep drilling in Hungary. *Oil Gas J.*, 71(21): 97—100.

Anikiev, A.K., 1964. *Abnormally High Reservoir Pressures in Oil and Gas Fields.* Tr. VNIGRI (All-Union Petroleum Scientific Research Institute of Geol. Exploration), No. 233. Izd. Nedra, Leningrad, 167 pp.

Chan, Y.C., 1964. Preliminary study on the geothermal gradients and formation or reservoir pressures of oil and gas fields in northern Taiwan. *Pet. Geol. Taiwan*, 3: 127—139.

Fertl, W.H., 1971. A look at abnormally pressured formations in the U.S.S.R. *SPE 3613, 46th AIME Fall Meet.*, New Orleans, La., October.

Fertl, W.H., 1972. Worldwide occurrence of abnormal formation pressures, I. *SPE 3844, 3rd Symp. on Abnormal Subsurface Pore Pressure*, Louisiana State Univ., Baton Rouge, La., May.

Fertl, W.H. and Timko, D.J., 1970. How abnormal pressure detection techniques are applied. *Oil Gas J.*, 68(2): 62—71.

Funkhouser, H.J., Sass, L.C. and Hedberg, H.D., 1948. Santa Ana, San Joaquin, Guario, and Santa Rosa oil fields (Anaco fields), Central Anzoategui, Venezuela. *Bull. Am. Assoc. Pet. Geol.*, 32: 1851—1908.

Geier, O., 1972. Begruendungen fuer die An- oder Nichtanwendbarkeit der amerikanischen Methoden fuer Hochdruckzonen — Frueherkennung im adriatischen Raum, Wiener Becken und Sueddeutschland an hand einiger Beispiele des italienischen, oesterreichischen und sueddeutschen Raumes. Diplomarbeit, (Dissertation). Institut fuer Tiefbohrtechnik und Erdoelgewinnung, Montanistische Hochschule, Leoben, 95 pp.

Guiraudet, B., 1972. Drilling for ultra-deep sour gas in France. *Pet. Eng.*, 44: 41—45, September.

Herring, E.A., 1973. North Sea abnormal pressures determined from logs. *Pet. Eng.*, 45: 72—84, November.

REFERENCES

Lepine, F.H. and White, J.A.W., 1973. Drilling in overpressured formations in Australia and Papua. *Aust. Pet. Explor. Assoc. J.*, 13: 157—161.

Magara, K., 1968. Compaction and migration of fluids in Miocene mudstone, Nagaoka Plain, Japan. *Bull. Am. Assoc. Pet. Geol.*, 52: 2466—2501.

Mostofi, B. and Gansser, A., 1957. The story behind the 5 Alborz. *Oil Gas J.*, 55(3): 78—84.

Rehm, B., 1972. Worldwide occurrence of abnormal formation pressures, II. *SPE 3845, 3rd Symp. on Abnormal Subsurface Pore Pressure*, Louisiana State Univ., Baton Rouge, La., May.

Rizzi, P.W., 1973. Hochdruckzonen-Frueherkennung in Mitteleuropa. *Erdoel Erdgas Z.*, 89: 249—256.

Zil'berman, V.I., 1972. Indications of approach to horizons with abnormally high reservoir pressure during drilling of wells. *Neft. Khoz.*, 50: 12—14.

APPENDIX

Conversion factors between metric, A.P.I., and U.S. measures (after Schlumberger Company, 1970)

multiply	by	to find
acres	.4047	hectares
acres	43,560	sq feet
acres	4047	sq meters
acre-ft	7758	barrels
acre-ft	43,560	cu ft
acre-ft	3.259×10	gallons
atmospheres	76	cms of mercury
atmospheres	29.92	in. of mercury
atmospheres	33.93	ft of water
atmospheres	1.033	Kg/cm^2
atmospheres	14.70	psi
barrels (API)	1.289×10	acre-ft
barrels (API)	158987	cm^3
barrels (API)	5.615	cu ft
barrels (API)	42	gallons
barrels (API)	9702	cu. inches
barrels (API)	159	liters
barrels (API)	.1590	cu meters
bbl/day	5.615	ft^3/day
bbl/day	.02917	gal/min
bbl/day	6.625	liters/hr
bbl/day	.1590	m^3/day
bbl/day	.006625	m^3/hr
bbl/hr	.0936	ft^3/min
bbl/hr	.700	gal/min
bbl/hr	2.695	in^3/sec
bars	.9869	atmospheres
bars	1.020	Kg/cm^2
bars	14.50	psi
B.T.U.	778.57	ft-lbs
B.T.U.	.2520	Kg-calories
B.T.U.	.2930	watt-hrs
BTU/min	.02357	horsepower
BTU/min	.01758	kilowatts
BTU/min	12.97	ft-lbs per sec
centimeters	3.281×10^{-2}	feet
centimeters	.3937	inches
centimeters	.01	meters
centimeters	10	millimeters
cubic meters/hr	151.0	bbl/day
cubic meters/hr	847.8	ft^3/day
cubic meters/hr	10^3	liters/hr
cubic meters/hr	24	m^3/day
cubic meters/day	.2621	bbl/hr
cubic meters/day	6.2897	bbl/day

multiply	by	to find
cm of mercury	.01316	atmospheres
cm of mercury	.4461	ft of water
cm of mercury	.01360	Kg/cm^2
cm of mercury	.1934	psi
cm/sec	1.969	ft/min
cm/sec	.03281	ft/sec
cm/sec	.6	meters/min
cubic cm	3.531×10^{-5}	ft^3
cubic cm	6.102×10^{-2}	in^3
cubic cm	10^{-6}	m^3
cubic cm	2.642×10^{-4}	gallons
cubic cm	10^{-3}	liters
cubic cm	6.2897×10^{-6}	barrels
cubic feet	.1781	barrels
cubic feet	2.832×10^4	cm^3
cubic feet	7.481	gallons
cubic feet	1728	in^3
cubic feet	.02832	m^3
cubic feet	28.32	liters
cubic ft/min	10.686	bbl/hr
cubic ft/min	256.5	bbl/day
cubic ft/min	472	cm^3/sec
cubic ft/min	7.481	gal/min
cubic ft/min	.472	liters/sec
cubic inches	16.39	cm^3
cubic inches	5.787×10^{-4}	ft^3
cubic inches	1.639×10^{-5}	m^3
cubic inches	4.329×10^{-3}	gallons
cubic inches	1.639×10^{-2}	liters
cubic meters	6.2897	barrels
cubic meters	10^6	cm^3
cubic meters	264.2	gallons
cubic meters	6.102×10^4	in^3
cubic meters	35.31	ft^3
cubic meters	10^3	liters
cubic ft/day	1.18	liters/hr
cubic ft/day	1.18×10^{-3}	m^3/hr
cubic ft/day	.02832	m^3/day
cubic ft/day	.1781	bbl/day
cubic meters/day	1.471	ft^3/hr
cubic meters/day	35.31	ft^3/day
cubic meters/day	41.67	liters/hr
cubic meters/day	.04167	m^3/hr
days	1440	minutes

CONTINUED →

multiply	by	to find
days	86400	seconds
feet	30.48	cm
feet	12	in
feet	.3048	meters
feet of water	.02950	atmospheres
feet of water	.8826	in. of mercury
feet of water	.03048	Kg/cm^2
feet of water	62.43	lbs/ft^3
feet of water	.4335	psi
feet/min	.5080	cm/sec
feet/min	.01667	ft/sec
feet/min	.01829	Km/hr
feet/min	.3048	m/min
feet/sec	30.48	cm/sec
feet/sec	18.29	m/min
feet/hr	.008467	cm/sec
feet/hr	5.086 x 10^{-3}	m/min
feet/hr	.01667	ft/min
foot-pounds	1.285 x 10^{-3}	B.T.U.
foot-pounds	3.238 x 10^{-4}	Kg-calories
ft-lbs/min	3.030 x 10^{-5}	horsepower
ft-lbs/min	2.260 x 10^{-5}	kilowatts
ft-lbs/sec	1.818 x 10^{-3}	horsepower
ft-lbs/sec	1.356 x 10^{-3}	kilowatt
gallons	.02381	barrel
gallons	3785	cm^3
gallons	.1337	ft^3
gallons	231	in^3
gallons	3.785 x 10^{-3}	m^3
Kilograms	10^3	grams
Kilograms	2.205	lbs
Kilograms	1.102 x 10^{-3}	tons (short)
Kg-calories	3.986	B.T.U.
Kg-calories	3088	ft-lbs
Kg-calories	1.560 x 10^{-3}	h.p.-hrs.
Kg-calories	427	Kg-meters
Kg-calories	1.163 x 10^{-3}	Kilowatt-hrs
gallons	3.785	liters
gallon (Imp)	1.2009	gallon (U.S.)
gallons/min	1.429	bbl/hr
gallons/min	34.286	bbl/day
gallons/min	.1337	ft^3/min
gallons/min	192.5	ft^3/day
gallons/min	3.785	liters/min
gallons/min	90.84	liters/hr

multiply	by	to find
grain (av.)	.06480	grams
grains/gal	17.12	parts per million
grains/gal	142.9	lbs/million gallons
grains/gal	.01714	grams/liter
grams	15.432	grains
grams	10^{-3}	Kg
grams	.03215	ounces
grams	2.205 x 10^{-3}	lbs
grams/cm^3	62.43	lbs/ft^3
grams/cm^3	8.344	lbs/gallon
grams/cm^3	.03613	lbs/in^3
grams/liter	58.42	grains/gal
hectares	2.471	acres
hectares	1.076 x 10^5	ft^2
hectares	.010	Km^2
horsepower	42.40	BTU/min
horsepower	33000	ft-lbs/min
horsepower	550	ft-lbs/sec
horsepower	1.014	metric h.p.
horsepower	10.68	Kg-cal/min
horsepower	.7457	Kilowatts
horsepower	745.7	watts
horsepower-hr	2544	B.T.U.
horsepower-hr	641.1	Kg-cal
horsepower-hr	2.737 x 10^5	Kg-meters
horsepower-hr	.7455	Kilowatt-hrs
inches	2.540	cm
inches	8.333 x 10^{-2}	ft
in. of mercury	.03342	atmospheres
in. of mercury	1.133	ft of Water
in. of mercury	.03453	Kg/cm^2
in. of mercury	.4912	psi
in. of water	.002458	atmosheres
in. of water	.07349	in. of mercury
in. of water	.002538	Kg/cm^2
in. of water	.03609	psi
Kg-cal/min	.09358	horsepower
Kg-cal/min	.06977	Kilowatts
Kg/m^3	10^{-3}	gm/cm^3
Kg/cm^2	0.9678	atmospheres
Kg/cm^2	.9807	bars
Kg/cm^2	32.84	ft of water
Kg/cm^2	28.96	in of mercury
Kg/cm^2	14.22	psi
Kilowatts	56.88	B.T.U./min

CONVERSION FACTORS

multiply	by	to find
Kilowatts	4.427×10^4	ft-lbs/min
Kilowatts	737.8	ft-lbs/sec
Kilowatts	1.341	horsepower
Kilowatts	10^3	watts
Kilowatt-hours	3413	BTU
Kilowatt-hours	2.656×10^6	ft-lbs
Kilowatt-hours	1.341	h.p.-hours
Kilowatt-hours	860	Kg-calories
Kilowatt-hours	3.672×10^5	Kg-meters
liters	10^3	cm^3
liters	6.2897×10^{-3}	barrel
liters	.03531	ft^3
liters	.2642	gallons
liters	61.02	in^3
liters	10^{-3}	m^3
liters/hr	.1509	bbl/day
liters/hr	6.289×10^{-3}	bbl/hr
liters/hr	5.885×10^{-4}	ft^3/min
liters/hr	.8475	ft^3/day
liters/hr	10^{-3}	m^3/hr
liters/hr	.02400	m^3/day
mile	5280	ft
mile	1.609	Km
millimeters	.1	cm
millimeters	3.281×10^{-3}	feet
millimeters	.03937	in
minutes	6.944×10^{-4}	days
minutes	1.667×10^{-2}	hours
meters	3.281	feet
meters	39.37	inches
meters	10^3	mm
meters	6.214×10^{-4}	mile
meters/min	1.667	cm/sec
meters/min	3.281	ft/min
meters/min	196.9	ft/hr
meters/min	.05468	ft/sec
mile/hr	44.70	cm/sec
mile/hr	88	ft/min
mile/hr	26.82	m/min
parts/million	.05835	grains/gallon
parts/million	8.337	lbs/million gallons
pound	7000	grains
pound	453.6	grams
pound	.4536	Kg
pounds/ft^3	.1337	lbs/gallon

multiply	by	to find
pounds/ft^3	.01602	gm/cm^3
pounds/ft^3	16.02	Kg/m^3
pounds/ft^3	5.787×10^{-4}	lbs/in^3
pounds/in^2	.06805	atmospheres
pounds/in^2	2.309	feet of water
pounds/in^2	2.036	in. of mercury
pounds/in^2	51.70	mm of mercury
pounds/in^2	.07031	Kg/cm^2
pounds/in^2	144	lbs/ft^2
seconds	1.157×10^{-5}	days
seconds	2.778×10^{-4}	hrs
seconds	1.667×10^{-2}	min
sq cm	1.076×10^{-3}	ft^2
sq cm	.1550	in^2
sq cm	10^{-4}	m^2
sq cm	100	mm^2
square feet	2.296×10^{-5}	acres
square feet	929.0	cm^2
square feet	144	in^2
square feet	.09290	m^2
square inches	6.452	cm^2
square inches	6.944×10^{-3}	ft^2
square inches	645.2	mm^2
square meters	10.76	ft^2
square meters	2.471×10^{-4}	acres
square meters	1550	in^2
Temperature		
°Cent. + 273	1	°Kelvin (abs)
°Fahr. + 460	1	°Rankine (abs)
°Cent. + 17.8	1.8	°Fahrenheit
°Fahr. − 32	5/9	°Centigrade
Temperature Gradient		
°C/100 m	.5486	°F/100 ft
°F/100 ft.	1.823	°C/100m
tons (long)	1016	Kg
tons (long)	2240	lbs
tons (metric)	10^3	Kg
tons (metric)	2205	lbs
tons (short)	2000	lbs
viscosity, lb-sec/in^2	6.895×10^6	viscosity in cp
viscosity, lb-sec/ft^2	4.78×10^4	viscosity in cp
viscosity, centistokes	density	viscosity in cp
watts	.05688	BTU/min
watts	44.27	ft-lbs/min

CONTINUED →

multiply	by	to find
watts	.7378	ft-lbs/sec
watts	1.341×10^{-3}	horsepower
watts	.01433	Kg-cal/min
watts	10^{-3}	Kilowatts
watt-hours	3.413	BTU

multiply	by	to find
watt-hours	2656	ft-lbs
watt-hours	1.341×10^{-3}	h.p.-hrs
watt-hours	.860	Kg-calories
watt-hours	367.2	Kg-meters
watt-hours	10^{-3}	kilowatt-hour

WEIGHT AND PRESSURE TABLE

Oil, water, and mud specific gravity — weight and pressure table (after Schlumberger Company, 1970)

DE-GREES A.P.I.	SPE-CIFIC GRAVITY	WEIGHT (DENSITY)			FLUID HEAD		DE-GREES A.P.I.	SPE-CIFIC GRAVITY	WEIGHT (DENSITY)			FLUID HEAD	
		GALLON	CUBIC FOOT	BARREL	PRES-SURE PER FOOT	HEIGHT PER POUND			GALLON	CUBIC FOOT	BARREL	PRES-SURE PER FOOT	HEIGHT PER POUND
		Pounds			psi	Feet			Pounds			psi	Feet
60.	.739	6.16	46.1	259.	.320	3.13	10° A.P.I. OR PURE WATER →	1.00	8.34	62.4	350.	.433	2.31
59.	.743	6.20	46.4	260.	.322	3.11		1.01	8.4	62.8	353.	.436	2.29
58.	.747	6.23	46.6	262.	.324	3.09		1.03	8.6	64.3	361.	.447	2.24
57.	.751	6.26	46.8	263.	.325	3.08		1.06	8.8	65.8	370.	.457	2.19
56.	.755	6.30	47.1	265.	.327	3.06		1.08	9.0	67.3	378.	.468	2.14
55.	.759	6.33	47.4	266.	.329	3.04		1.10	9.2	68.8	386.	.478	2.09
54.	.763	6.36	47.6	267.	.330	3.03		1.13	9.4	70.3	395.	.488	2.05
53.	.767	6.40	47.9	269.	.332	3.01	SALT WATER →	1.15	9.6	71.8	403.	.490	2.00
52.	.771	6.43	48.1	270.	.334	2.99		1.154	9.625	72.0	404.	.500	2.00
51.	.775	6.46	48.3	271.	.336	2.98		1.18	9.8	73.3	412.	.509	1.96
50.	.780	6.51	48.7	273.	.338	2.96		1.20	10.0	74.8	420.	.519	1.93
49.	.784	6.54	48.9	275.	.340	2.94		1.22	10.2	76.3	428.	.530	1.89
48.	.788	6.57	49.2	276.	.341	2.93		1.25	10.4	77.8	437.	.540	1.85
47.	.793	6.61	49.5	278.	.343	2.92		1.27	10.6	79.3	445.	.551	1.81
46.	.797	6.65	49.8	279.	.345	2.90		1.29	10.8	80.8	454.	.561	1.78
45.	.802	6.69	50.0	281.	.348	2.87		1.32	11.0	82.3	462.	.571	1.75
44.	.806	6.72	50.3	282.	.349	2.87		1.34	11.2	83.8	470.	.582	1.72
43.	.811	6.76	50.6	284.	.351	2.85		1.37	11.4	85.3	479.	.592	1.69
42.	.816	6.81	50.9	286.	.354	2.82		1.39	11.6	86.8	487.	.603	1.66
41.	.820	6.84	51.2	287.	.355	2.82		1.41	11.8	88.3	496.	.613	1.63
40.	.825	6.88	51.5	289.	.357	2.80		1.44	12.0	89.8	504.	.623	1.61
39.	.830	6.92	51.8	291.	.359	2.79		1.46	12.2	91.3	512.	.634	1.58
38.	.835	6.96	52.1	292.	.362	2.76		1.49	12.4	92.8	521.	.644	1.55
37.	.840	7.01	52.4	294.	.364	2.75		1.51	12.6	94.3	529.	.655	1.53
36.	.845	7.05	52.7	296.	.366	2.73		1.53	12.8	95.8	538.	.665	1.50
35.	.850	7.09	53.0	298.	.368	2.72		1.56	13.0	97.2	546.	.675	1.48
34.	.855	7.13	53.3	299.	.370	2.70		1.58	13.2	98.7	554.	.686	1.46
33.	.860	7.17	53.6	301.	.372	2.69		1.61	13.4	100.	563.	.696	1.44
32.	.865	7.21	53.9	303.	.375	2.67		1.63	13.6	102.	571.	.706	1.42
31.	.871	7.26	54.3	305.	.377	2.65		1.65	13.8	103.	580.	.717	1.39
30.	.876	7.31	54.7	307.	.380	2.63		1.68	14.0	105.	588.	.727	1.38
29.	.882	7.36	55.1	309.	.382	2.62		1.70	14.2	106.	596.	.738	1.36
28.	.887	7.40	55.4	311.	.384	2.60		1.73	14.4	108.	605.	.748	1.34
27.	.893	7.45	55.7	313.	.387	2.58		1.75	14.6	109.	613.	.758	1.32
26.	.898	7.49	56.0	315.	.389	2.57		1.77	14.8	111.	622.	.769	1.30
25.	.904	7.54	56.4	317.	.392	2.55		1.80	15.0	112.	630.	.779	1.28
24.	.910	7.59	56.8	319.	.394	2.54		1.82	15.2	114.	638.	.790	1.27
23.	.916	7.64	57.2	321.	.397	2.52		1.85	15.4	115.	647.	.800	1.25
22.	.922	7.69	57.5	323.	.399	2.51		1.87	15.6	117.	655.	.810	1.23
21.	.928	7.74	57.9	325.	.402	2.49		1.89	15.8	118.	664.	.821	1.22
20.	.934	7.79	58.3	327.	.405	2.47	COMMON CEMENT SLURRY →	1.92	16.0	120.	672.	.831	1.20
19.	.940	7.84	58.7	329.	.407	2.46		1.94	16.2	121.	680.	.842	1.19
18.	.946	7.89	59.0	331.	.410	2.44		1.97	16.4	123.	689.	.852	1.17
17.	.953	7.95	59.5	334.	.413	2.42		1.99	16.6	124.	697.	.862	1.16
16.	.959	8.00	59.8	336.	.416	2.40		2.01	16.8	126.	706.	.873	1.15
15.	.966	8.06	60.3	339.	.419	2.39		2.04	17.0	127.	714.	.883	1.13
14.	.973	8.11	60.7	341.	.421	2.38		2.06	17.2	129.	722.	.894	1.12
13.	.979	8.16	61.0	343.	.424	2.36		2.09	17.4	130.	731.	.904	1.11
12.	.986	8.22	61.5	345.	.427	2.34		2.11	17.6	132.	739.	.914	1.09
11.	.993	8.28	61.9	348.	.430	2.33		2.13	17.8	133.	748.	.925	1.08

Temperature conversion formulas (after Langnes et al., 1972)

To obtain	Formula
°F (Fahrenheit)	(°C × 1.8) + 32
°C (Centigrade)*	(°F + 40)/1.8 − 40
°C	(°F − 32) × 0.5555
°K (Kelvin)	°C + 273.16
°R (Rankine)	°F + 459.688

*Or Celsius, which is preferred for international use.

References

Langnes, G.L., Robertson, J.O. and Chilingar, G.V., 1972. *Secondary Recovery and Carbonate Reservoirs.* Elsevier, Amsterdam, 304 pp.
Schlumberger Company, 1970. *Production Log Interpretation Manual.* 138 pp.

REFERENCES INDEX*

Adamson, L.G., 52, *96*
Alger, R.P., 184, 189, 198, 204, 205, *227*, *230*
Ali-Zade, A.A., 30, *48*
Allaud, L.A., 205, *227*
Allen, D.R., 5, *43*, *46*, 93, 95, *100*
Alliquander, O., 330, *348*
Althaus, V.E., 189, *227*
Amenson, H.L., 102, 114, *115*
Anderson, A.E., 189, *228*
Anderson, B.A., 140, *172*
Anderson, R.A., 250, 253, 254, 255, 256, *272*
Anderson, T.O., 141, *173*
Andreev, P.F., 291, *321*
Angelopulo, O.K., 140, *172*
Anikiev, K.A., 51, *96*, 336, *348*
Anonymous, 271, *272*
Athy, L.F., 193, *227*
Atwater, G.I., 31, *43*
Aud, B.W., 106, 107, 111, 112, 113, 114, *116*
Ausburn, B.E., 272, *274*

Back, W., 38, *43*
Bailey, E.H., 35, *43*
Bakhtin, V.V., *96*
Barker, C., 11, 42, *43*
Barrious, J.R., 264, *273*
Barrington, J., 36, *45*
Barton, D.C., 291, *321*
Bateman, S.J., 264, *273*
Bass, D.M., 281, *289*
Baugher, J.W., 16, 22, 23, 24, 30, 31, *45*
Baybakov, V.V., 304, 305, *321*
Bazer, D.A., 151, *173*
Bear, J., 63, 64, *96*
Beck, K.C., 36, *48*, 68, 73, *100*, 159, 165, *175*, 190, *230*
Becker, J., 267, *274*
Beitzel, J.E., 102, *115*
Bentor, Y.K., 73, *96*, 196, *227*

Bergan, R.A., 206, *230*
Berner, R.A., 57, 73, *96*, *100*
Berry, F.A.F., 17, 34, 35, *43*, 49, *96*
Berry, J.E., 198, *228*
Berry, L.N., 250, 251, 254, 272, *273*, *274*
Biot, M.A., 254, *272*
Birch, A.F., 144, *172*
Birchwood, K.M., 29, *48*
Bishop, W.F., 25, *43*
Bleakley, W.B., 42, *43*, 264, *273*
Bloch, F., 207, *227*
Blyth, C.R., 35, *43*
Boatman, W.A., 150, 151, 153, 154, *172*
Bogomolov, A.I., 291, *321*
Bogomolova, G.V., *96*
Bohn, J., 33, *47*
Bohor, B.F., 91, 94, *98*
Bolt, D.B., 124, *172*
Bolt, G.H., 50, *96*
Boone, D.E., 130, *172*
Borel, W.J., 159, *172*
Bottema, J.A., 40, *47*
Bourgoyne, A.T., 130, 132, 133, 134, *172*, 277, *289*
Boylston, L.E., 102, *115*
Bredehoeft, J.D., 21, 33, 35, 37, *43*, *45*, *47*, 59, 60, 61, *96*, 277, 288, *290*
Bredeson, D.H., 31, *45*
Breeze, A.F., 8, 9, 10, 34, *43*, 190, *227*
Bridgman, P.W., 144, *172*
Brighenti, G., 5, *44*
Brindley, G.W., 156, *173*
Brown, R.J.S., 208, *227*
Brown, W.O., 81, *100*
Bruce, C.H., 25, 26, 27, *44*
Bruist, E.H., 271, *273*
Bruns, J.R., 276, *290*
Bryant, W.R., *96*
Bullard, E.C., 144, *172*
Burak, L.A., *274*
Burley, J.D., *273*
Burst, J.F., 36, *44*, 209, *227*, 300, 308, *321*

---
*References to pages containing bibliographic details are italicized.

Bush, R.E., 145, 146, *175*

Cannon, G.E., 301, *321*
Caloi, P., 5, *44*
Caraway, W.H., 72, 81, *97*
Carpenter, C.B., 91, 94, *96*
Carver, R.E., 23, 25, *44*
Case, L.C., 196, *227*
Chamberlain, E., 41, *44*
Chan, Y.C., 345, *348*
Cheatham, J.B., 120, *173*, 267, *273*
Cheng, J.T., *96*
Chernyshev, V.V., 291, 292, *323*
Chilingar, G.V., 50, 52, 69, 70, 74, 92, 95, *96*, 99, *100*
Chilingarian, G.V., 49, 52, 58, 59, 60, 61, 65, 68, 70, 75, 78, 79, 93, 94, *96*, *100*, 231, *274*
Christ, C.K., 38, *44*, 169, *173*
Christman, S.A., 251, 252, 253, *273*
Clark, C.C., 41, *46*
Clark, H., 144, *172*
Classen, J.S., 23, *44*, 301, *321*
Clavier, C., 206, *227*
Clegg, J.D., 267, *273*
Clements, R.L., *273*
Cobb, J.B., 256, 257, *273*
Colburn, W.A., 35, *45*
Collins, A.G., 166, *172*
Combs, G.F., 128, 129, *172*
Cooper, H.H., 59, 65, 66, *96*
Cordell, R.T., 291, *321*
Costain, J.K., 102, 114, *116*
Couch, E.J., 264, *273*
Craft, B.C., 275, 282, 283, *290*
Craze, R.C., 301, *321*
Crittendon, R.D., 244, *273*

Darby, E.K., 102, 114, *115*
Darcy, H., 63, 64, 65, 66, *97*
Davis, J.M., 102, *115*
Davis, S.N., 5, *46*
Degens, E.T., 69, 70, *97*
Denton, H.H., 301, *321*
Desai, K.P., 189, *227*
De Sitter, L.U., 80, *97*
Dewan, J.T., 205, *227*
De Wiest, R.J.M., 64, *97*
De Witte, A.J., 189, *230*
Dickey, P.A., 22, 23, 24, *44*, 51, 69, 81, *97*, 193, *227*
Dickinson, G., 20, 21, 22, 23, 24, *44*, 52, 97, 301, *321*
Dietzman, W.D., 71, 72, 73, *97*
Dix, C.H., 102, 106, 114, *115*
Dobryanskii, A.F., 291, *321*
Doll, H.G., 184, 189, *227*
Dott, R.H., 291, *321*
Drouin, A.H., *273*
Duerbaum, H., 102, *115*
Duggan, J.O., 277, *290*
Dunoyer de Segonzac, G., 36, *44*
Dupin de Saint Cyr, P., 148, *172*
Dzhalilov, T.I., 38, *48*

Eaton, B.A., 243, 244, 245, 252, 258, 259, *273*, 310, 311, *321*
El-Hadidi, S., 135, *172*
Epperson, M.J., 267, *273*
Epstein, S., 70, *97*
Erdman, J.G., 52, *100*
Ershagi, I., 74, *96*
Evans, D.M., 12, *44*
Eucken, A., 144, *172*

Fatt, I., 91, 94, *97*, 281, *290*
Feather, J.N., 29, *45*
Feemster, W.E., 264, *274*
Ferran, L.H., 9, 10, *44*, 190, *227*, 285, 286, *290*
Fertl, W.H., 13, 14, 16, 21, 38, *44*, 79, 81, 82, 83, *97*, 137, 138, 145, 147, 154, 157, 158, 160, 161, 162, 163, 166, 167, 168, 170, 171, *172*, *173*, 186, 190, 192, 194, 195, 196, 198, 206, 207, 214, 216, 217, 222, 225, 226, *227*, *228*, 277, 286, 287, 288, 289, *290*, 295, 297, 305, 306, 307, 308, 309, 310, 311, 314, 315, 316, 317, 318, 319, 320, 321, *322*, 325, 326, 334, 335, 337, 339, 347, *348*
Fetkovich, M.J., 276, *290*
Flores, J.G., 7, *44*
Forgotson, J.M., 121, *173*
Forman, M.J., 31, *43*
Foster, J.B., 7, *44*, 190, 196, 198, *228*
Fowler, W.A., 23, *44*, 80, 82, *97*, 190, *228*, 277, *290*, 301, 302, 304, *322*
Friedman, M., 29, *45*
Friedman, T., 70, 71, 72, *97*
Froula, N.H., 41, *47*
Füchtbauer, H., 39, *44*
Funkhouser, H.J., 340, *348*

REFERENCES INDEX

Gabrysch, R.K., 6, *44*
Gaida, K.H., 68, 77, 78, *100*, 198, *230*
Galle, E.M., 120, *173*
Gansser, A., 29, *44*, *349*
Gardner, G.H.F., 198, *230*
Garnier, A.J., 120, *173*
Garrels, A., 38, *44*
Garrels, R.M., 169, *173*
Gates, G.L., 72, 81, *97*
Geier, O., 328, *348*
Germann, F.E., 165, *173*
Gibbs, M.A., 246, 263, *273*
Gibson, R.E., 60, 61, *97*
Gill, J.A., 158, *173*
Gilluly, J., 5, *44*
Glasstone, S., 33, 34, *44*
Gnirk, P.F., 120, *173*
Goddard, R.D., 141, *173*
Goeken, R.J., 268, *273*
Goins, W.C., 142, 143, 148, 149, *173*, *174*
Goldsmith, R.C., 138, *173*
Gondouin, M., 189, *228*
Goodman, M.A., 42, *45*, 264, *273*
Goolsby, D.A., 19, *46*
Gorkun, V.N., 29, *45*
Graf, D.L., 70, 71, 72, *97*
Grant, U.S., 5, *44*
Gray, K.E., 94, *99*
Green, C.H., 102, *116*
Gregory, A.R., 94, *99*, 198, *230*
Gresens, R.L., 39, *45*
Gretener, P.E., 29, 32, 41, *45*
Griffin, D.G., 151, *173*
Gstalder, S., 135, *173*, *174*
Guest, R.J., 141, *173*
Guiraudet, B., 327, *348*
Gullikson, D.M., 72, 81, *97*
Guseva, A.N., 190, *229*
Guyod, H., 145, *173*

Hagen, D.C., 102, *116*
Hager, R.V., 29, 30, *45*
Halbouty, M.T., 31, *45*
Ham, H.H., 183, 184, 203, *228*
Hamby, T.W., 268, 269, 270, *290*
Hammerlindl, D.J., 281, 282, *290*
Handin, J., 29, 30, *45*
Hang, P.T., 156, *173*
Hanger, D.M., 257, 258, 267, *273*
Hanshaw, B.B., 21, 35, 37, 38, 43, *45*, 48, 59, 60, 61, *96*, 277, 288, *290*
Hanway, J., 43, *45*

Hardin, F.R., 25, 31, *45*
Hardin, G.C., 25, 31, *45*
Harkins, K.L., 16, 22, 23, 24, 30, 31, *45*
Harville, D.W., 278, 279, 280, 284, *290*
Hawkins, M.E., 71, 72, 73, *97*, 275, 277, 278, 279, 280, 282, 283, 284, *289*
Healy, J.H., 33, *47*
Hedberg, H.D., 49, 51, *98*, 193, *228*, 291, *322*, 340, *348*
Hedberg, W.H., 68, 82, *98*, 193, 198, *228*
Herbert, W.E., 135, *173*
Herget, G., 49, *98*
Herring, E.A., 332, *348*
Hicks, W.G., 110, 114, *116*, 198, *228*
Hilchie, D.W., *228*
Hill, G.A., 35, *45*
Hill, H.J., 189, *228*
Hopkinson, E.C., 206, *230*
Horner, D.R., 295, *322*
Hoshino, K., *98*
Hospers, J., 27, *45*
Hottman, C.E., 68, 81, *98*, 106, *116*, 178, 179, 181, 182, 199, 201, 226, *228*, 310, 311, *322*
Hower, J., 36, *47*, 209, *229*
Hoyle, W.R., 204, *227*
Hubbert, M.K., 21, 27, *45*, *47*, 49, 64, 83, 86, *98*, 196, *228*, 231, 238, 239, *273*
Hughes, D.J., 31, *45*
Hunt, J.M., 70, *97*
Hurst, W., 276, 277, *290*

Ibrahim, M.A., 53, 54, 55, *98*
Inami, K., *98*
Ingram, D.S., 250, 253, 254, 255, 256, *272*
Ioffe, V.I., 140, *173*
Irmay, S., 63, *96*
Issenmann, O., 141, *173*
Iwamura, S., *98*

Jacob, C.E., 64, *98*
Jageler, A., 209, *228*
Jean, L.A., 267, *273*
Johnson, H.A., 31, *45*
Johnson, R.K., 68, 81, *98*, 106, *116*, 178, 179, 181, 182, 199, 201, 226, *228*, 310, *322*
Jones, C.W., 71, 72, 73, *97*
Jones, B.R., 209, 210, 211, *228*
Jones, P.H., 16, 21, 22, 23, 34, 35, *45*,

144, 145, 165, 166, *173*, 190, 191, *228*
Jorden, J.R., 121, 122, 123, *173*
Judd, W.R., 5, *45*
Jürgenson, L., 83, 85, 86, 88, 89, *98*

Kalinko, M.K., 30, *45*, *98*
Karpova, G.V., *98*
Kartsev, A.A., 165, *174*, 291, 311, *321*, *322*
Kastryulin, N.S., 30, *48*
Katz, D.L., 52, 53, 54, 55, *98*
Kaufman, H., 106, *116*
Kazintsev, E.E., 75, 76, *98*
Keller, H.H., 264, *273*
Kelly, J., 240, 241, 242, 243, 250, *273*
Kennedy, J.L., 119, 130, *174*
Kerr, P.F., 36, *45*
Kirk, W.L., 263, 267, *273*
Kissin, I.G., 165, *174*
Kjellstrand, J.A., 140, *175*
Klaveness, A., 115, *116*
Klemme, H.D., 291, 296, 297, 299, *322*
Kloeckner, A., 267, *274*
Kluyshin, G.P., 264, *274*
Knight, J.W., 35, *45*
Knight, L., 50, 92, *96*
Knowles, C.R., 41, *47*, 265, *274*
Knutson, C.F., 91, 94, *98*
Kohlhaas, C.A., 94, *98*
Koide, H., *98*
Konyukhov, A.I., 36, *47*
Korchagina, Y.I., 291, 292, *323*
Korringa, J., 208, *228*
Korunova, V.V., 75, 76, 77, *98*
Kozelskiy, I.T., *98*
Krasintseva, V.V., 75, 76, 77, *98*
Krumbol'dt, T.S., 190, 193, 198, *230*
Krynine, D.P., 5, *45*
Kryukov, P.A., 73, 74, 75, *99*
Kugler, H.G., 28, 29, *46*
Kuksov, A.K., *274*
Kunz, K.S., 220, *228*
Kurt, E.T., 72, 73, *100*
Kuznetsov, S.I., 165, *174*
Kvet, R., 140, *174*

Landes, K.K., 294, 296, *322*
Lang, W.B., 165, *174*
Langseth, M.G., 144, *174*
Laubscher, H.P., 32, 38, *46*
Lavaquial, F.P., 277, *289*, *290*

Lehner, F., 271, *273*
Leonardon, E.G., 189, *229*
Lepine, F.H., 342, *348*
Leve, G.W., 19, *46*
Levin, F.K., 114, *116*
Levorsen, A.I., 6, *46*, 196, *228*
Lewis, C.R., 145, *174*
Lewis, R.L., 159, *172*
Lindsey, H.E., 264, 268, *273*, *274*
Lindtrop, N.T., 190, *229*
Lo, K.Y., 88, *99*
Long, G., 68, *99*
Looney, J.D., 102, *115*
Lopatin, N.V., 291, 292, *323*
Loren, J.D., 208, *229*
Louden, L.R., 14, 38, *46*, 106, 107, 111, 112, 113, 114, *116*, 217, *229*
Low, P.F., 34, *48*
Lowry, W.D., 39, *46*
Lucas, D.R., 272, *274*
Lucon, C., 141, *173*
Lutz, J., 135, *174*

Mabutt, J.A., 19, *47*
MacGregor, J.R., 185, *229*
MacKay, J.R., 40, *46*
MacPherson, L.A., 250, 251, 254, *273*
Magara, K., 21, *46*, 79, *99*, 346, *349*
Mahony, B.J., 264, *273*
Manheim, F.T., 68, 72, 74, 76, *99*
Marenko, Y.I., 38, *46*
Maricelli, J.J., 80, 81, *100*, 190, *230*
Marsden, S.S., 5, *46*
Martin, G.B., 22, 23, *46*
Martirosova, A.O., *96*
Matthews, W.R., 106, 107, 111, 112, 113, 114, *116*, 122, 147, 150, *174*, 186, 187, 188, 200, 223, *229*, 240, 241, 242, 243, 250, *273*
Matveev, A.K., *98*
Maurer, W.C., 120, 121, *174*
Maxey, G.B., 35, *43*
Maxwell, A.E., 144, *172*
May, J.E., 115, *116*
Mayell, M.J., 266, *273*
Mayuga, M.N., 5, *46*
McCaslin, L.S., 301, *322*
McClendon, R., 106, 107, 111, 112, 113, 114, *116*, 127, 128, *174*
McCulloh, T.H., 114, *116*, 209, *229*
McEver, J.W., 267, *273*
McGinnis, L.D., 41, *46*

REFERENCES INDEX 361

McLamore, R.T., 271, *273*
McNeal, R.P., 35, *46*
Mead, J.L., 242, *273*
Meade, R.H., 50, 92, *99*
Meents, W.F., 70, 71, 72, *97*
Meinzer, O.E., 19, 42, *46*
Meitzen, V.C., 276, *290*
Mellor, M., 41, *46*
Meyerhoff, A.A., 302, 303, 304, *322*
Meyers, J.D., 23, *46*
Miller, F.G., 52, 94, 95, *98, 99*
Miller, R.T., *273*
Mitchell, F.R., 272, *274*
Mitchell, J.K., 50, *99*
Mitsui, S., *98*
Modelevskiy, M.S., 304, *322*
Mogilevski, G.A., 291, *321*
Mondshine, T.C., 157, 158, *174*
Moore, C.A., 11, *46*
Moore, E.J., 189, *227, 229*
Moore, P.L., 259, *274*
Moran, J.H., 220, *228*, 272, *274*
Morgan, J.P., 28, *46*
Moriarty, D.G., 263, *274*
Morris, D.A., 52, *100*
Moses, P.L., 294, *322*
Mostofi, B., *349*
Mottley, J.R., 267, *274*
Mounce, D.W., 189, *229*
Muckleroy, J.A., 135, *174*
Muehlberger, W.R., 30, *46*
Mueller, F., 40, *46*
Muller, S.W., 40, *46*
Mullins, J.D., 286, *290*
Murray, G.E., 23, 26, 28, *47*
Musgrave, A.W., 102, 110, 114, *116*
Muskat, M., 275, *290*
Myers, R.L., 80, 81, *99*
Myers, G.M., 130, *172*, 190, 192, *229*

Nagel, W.A., 205, *230*
Nakano, Y., 41, *47*
Nakao, K., 41, *46*
Neglia, S., 68, *99*
Nettleton, L.L., 209, *228*
Nevins, M.J., 156, *174*
Nukiyama, S., 144, *174*

O'Brien, T.B., 143, *174*
Ocamb, R.D., 25, *47*
Oomkens, E., 32, *47*
Oshry, H.I., 206, *230*

O'Sullivan, R.B., 32, *48*
Ott, H., 267, *274*
Overton, H.L., 79, 80, *99*, 141, 159, *174*, 190, *229*

Paine, W.R., 22, 23, 24, *44*, 69, 81, *97*
Pandey, G.N., 52, *99*
Paramonov, V.S., 140, *172*
Parnov, E.I., 190, *229*
Parnov, Y.I., 304, *322*
Pearson, C., 72, 73, *100*
Pearson, C.A., 72, 73, *97*
Pechorin, O.M., 140, *174*
Peck, R.P., 55, *100*
Pehoushek, F., 272, *274*
Penhollow, J.O., 105, *116*
Pennebaker, E.S., 102, 103, 106, 108, 109, 113, 114, *116*
Perkins, T.K., 41, *47*, 265, *274*
Perrier, R., *99*
Perry, E., 36, *47*, 209, *229*
Perry, D.R., 292, 294, 303, *322*
Peterson, H., 267, *274*
Philip, J.R., 66, 67, *99*
Phillipp, W., 39, *47*
Picard, M.D., 165, *174*
Pickett, G.R., 198, *229*
Pilkington, P.E., 198, 228
Pirson, S.J., 35, 42, *47*, 178, 197, *229*, 275, *290*
Pixler, B.O., 137, *174*
Plaskova, A.G., 190, *229*
Podio, A.L., 94, *99*
Pol'ster, L.A., 190, *229*
Ponomarev, V.A., *274*
Posokhov, E.V., 72, 73, *99*
Postma, G.W., 114, *116*
Postnikov, V.G., 38, *46*
Powers, M.C., 36, *47*, 209, *229*
Purcell, E.M., 206, *229*
Pusey, W.C., 291, 298, 299, *322*

Quiblier, J., *99*
Quichaud, C., 135, 174

Raghavan, R., 52, 64, *99*
Raleigh, C.B., 33, *47*
Rall, C.G., 71, 72, *99*
Ramsay, J.G., 87, 88, *99*
Ratcliffe, E.W., 144, *174*
Raymer, L.L., 204, *227*
Raynal, J., 135, *173, 174*

Raynaud, M., 135, *174*
Record, L., *273*
Reed, W.E., 70, *97*
Rehm, B., 246, *274*, 329, 331, 337, 338, 341, 344, *349*
Rehm, W.A., 106, 107, 111, 112, 113, 114, *116*, 127, 128, 143, *174*
Reid, C.A., 233, 242, *274*
Reuter, J.H., 70, *97*
Revelle, R., 144, *172*
Reynolds, E.B., 104, 105, 110, 111, 112, 115, *116*, 217, 219, *229*
Richard, J.J., 29, *47*
Ridd, M.F., 28, 29, *47*
Rieke, H.H., 49, 52, 58, 59, 60, 61, 65, 68, 70, 74, 75, 78, 79, 93, 94, *96*, *99*, *100*, 231, *274*
Rizer, J.A., 130, *172*
Rizzi, P.W., 333, *349*
Roberts, J.L., 27, *47*
Robertson, E.C., 57, 67, *100*
Robertson, J.O., 74, *99*
Robinson, J.D., 208, *229*, 272, *274*
Rochon, J.A., 41, *47*, 265, *274*
Rochon, R.W., 137, 154, *174*
Rodionov, V.A., 140, *175*
Rose, S.C., 145, *174*
Rubey, W.W., 21, 27, *45*, *47*, 49, 83, 86, *98*, 196, *228*
Rubino, E., 68, *99*
Runge, R.J., 272, *274*
Russel, W.L., 7, 47
Rust, W.M., 189, *229*

Sahay, B., 194, *229*
Samedov, N.K., 38, *48*
Samedov, S.S., 190, *229*
Samuels, S.G., 50, *100*
Sarmiento, R., 106, *116*
Sass, L.C., 340, *348*
Sawabini, C.T., 74, 78, 93, 94, 95, *96*, *100*
Sayles, F.L., 68, 72, *99*
Schilthuis, R.J., 276, *290*
Schlumberger, C., 189, *229*
Schmidt, G.W., 68, 83, *100*, 190, 193, 198, *229*
Scorer, J.D.T., 95, *100*
Scott, J.B., 198, *228*
Seevers, D.O., 208, 209, *229*
Selitskii, A.G., 292, *323*
Senturia, S.D., 208, *229*

Seward, J.M., 71, 72, 73, *97*
Shalavin, A.M., 264, *274*
Shelton, J.W., 29, *47*
Sherman, H., 205, *230*
Shimp, N.F., 70, 71, 72, *97*
Shiram, C.R., 69, 81, *97*
Shirley, O.J., 121, 122, 123, *173*
Shishkina, O.V., 77, *100*
Shishov, V.A., *274*
Shriram, C.R., 22, 23, 24, *44*
Shugart, T.R., 114, *116*
Siever, R., 73, *100*
Simard, G.L., 189, *228*
Simkhaev, V.Z., 190, *229*
Siryk, I.M., 29, *45*
Skelly, M.G., 140, *175*
Skempton, A.W., 50, *100*
Slack, H.A., 102, 114, *116*
Slatyer, R.O., 19, *47*
Smith, J.E., 21, *47*, 52, 62, 63, *100*, 277, 288, *290*
Smith, J.R., 268, 269, 270, *290*
Smith, N.J., 209, *230*
Smith, T.K., 247, 248, 250, 251, 253, 254, *274*
Snavely, P.D., 35, *43*
Somerton, W.H., 120, *175*
Sorby, H.C., 49, *100*
Sorokin, 291, *322*
Speel, L., 266, 267, *274*
Spencer, G.B., 91, 94, *96*
Stein, F.C., 266, *273*
Subbota, M.I., 291, *322*
Sukhanov, V.B., *274*
Sukharev, G.M., 190, 193, 198, *230*
Sullins, R.S., 301, *321*
Suman, G.O., 271, *273*
Suter, H.H., 28, *47*
Szasz, S.E., 189, *229*

Tabasaranskii, Z.A., 291, *322*
Tanguy, D.R., 184, 198, *230*
Taylor, D.B., 247, 248, 250, 251, 253, 254, *274*
Tek, M.R., 52, *99*
Teodorovich, G.I., 36, *47*
Terzaghi, K., 54, 55, *100*, 254, *274*
Thomeer, J.H.M., 40, *47*
Thompson, T.L., 49, *100*
Thorsen, C.E., 25, *48*
Thurber, C.H., 101, *116*
Timko, D.J., 14, 21, 38, *44*, 79, 80, 81,

REFERENCES INDEX

82, 83, *97*, 145, 146, 147, 154, 157, 158, 160, 161, 162, 163, 166, 167, 168, 170, 171, *172*, *173*, *174*, 186, 190, 192, 194, 195, 196, 198, 206, 207, 214, 216, 217, 219, 222, 225, 226, *227*, *228*, *229*, *230*, 277, 285, 286, 288, 289, *290*, 295, 297, 305, 306, 307, 308, 309, 310, 311, 314, 315, 316, 317, 318, 319, 320, 321, *322*, 337, *348*
Timm, B.C., 80, 81, *100*, 190, *230*
Timur, A., 208, *230*
Tittle, C.W., 205, *230*
Tittman, J., 204, 205, *230*
Tixier, M.P., 184, 189, 198, 204, *227*, *228*
Trusheim, F., 30, *48*
Tsimel'zon, I.O., 114, *116*

Van der Knaap, W., 5, *48*, 91, 92, *100*
Vander Stoep, D.M., 114, *116*
Van der Vlis, A.C., 5, *48*, 91, 92, *100*
Van Everdingen, A.F., 276, *290*
Van Everdingen, R.O., 35, *48*
Van Hook, W.A., 264, *274*
Van Lingen, N.H., 120, *173*
Van Moort, J.C., 36, *48*
Van Olphen, H., 162, *175*
Van Siclen, D.C., 190, 192, *229*
Vassoyevich, N.B., 291, 298, *323*
Vidrine, L.O., 320, 321, *323*
Viskovskiy, Y.A., 190, *229*
Vogiatzis, J.P., 272, *274*
Von Engelhardt, W., 68, 77, 78, *100*, 193, 198, *230*
Voroshilov, E.A., 38, *48*
Vries, D.A., 144, *175*

Wahl, J.S., 204, *230*
Wallace, W.E., 81, *100*, 185, 186, 187, 213, *230*, 277, *290*, *323*
Walton, M.S., 32, *48*
Wardlaw, H.W.R., 129, *175*
Warren, J.E., 189, *230*
Watkins, J.W., 72, 73, *100*
Weaver, C.E., 36, *48*, 68, *100*, 159, 165, *175*, 190, *230*
Weintritt, D.J., 156, 158, *173*, *174*

Werner, K., 267, *274*
West, E.R., 268, *274*
West, S.S., 106, *116*
Whalen, H.E., 7, *44*, 190, 196, 198, *228*
Wheeler, R., 263, *274*
White, D.E., 35, *43*, 70, *100*
White, J.A.W., 342, *349*
White, W.A., 35, *43*
Whitney, B.F., 189, *229*
Wichmann, P.A., 206, *230*
Wickenhauser, T.L., 277, *289*, *290*
Williams, D.G., 81, *100*
Williamson, A.S., 271, *273*
Willis, D.G., 231, 238, 239, *273*
Wilson, C.C., 29, *48*
Wilson, G.J., 145, 146, *175*
Winslow, A.G., 6, *48*
Woeber, A.F., 105, *116*
Wood, D.B., 42, *45*, 264, *273*
Wood, J.J., 81, *100*
Wood, L.A., 6, *48*
Woods, E.W., 217, *229*
Woods, H.B., 120, *173*
Worthington, A.E., 272, *274*
Wright, J., 71, 72, 73, *99*, *100*
Wyllie, M.R.J., 178, 189, 198, *230*

Yakubov, A.A., 30, *48*
Yancy, J.D., 102, 114, 116
Yoshizawa, Y., 144, *174*
Youmans, A.H., 206, *230*
Young, A., 34, *48*
Young, F.S., 129, 135, *173*, *175*

Zamora, M., 126, *175*
Zanier, A.M., 195, 196, 217, 219, *229*, *230*, 250, 253, 254, 255, 256, *272*, 285, 286, *290*
Zartman, R.E., 165, *175*
Zaslavsky, D., 63, *96*
Zen, E.A., 35, *48*
Zhuchkova, A.A., 74, 75, *99*
Zierfuss, H., 144, 145, *175*
Zil'berman, V.I., 336, *349*
ZoBell, C.E., 165, *175*
Zoeller, W.A., 130, *175*
Zvereva, O.V., 292, *323*

SUBJECT INDEX

Abnormal formation pressure
   detection and evaluation, 101—230
   —, geophysical methods, 101—116
   —, drilling parameters, 117—176
   —, well logging techniques, 177—230
   effects on
   —, drilling and completion, 231—274
   —, economics, 319—321
   —, hydrocarbon distribution, 291—319
   —, reservoir engineering, 275—290
   occurrence (worldwide), 325—348
   origin, 5—43
Abnormally high formation pressure
   origin(s), 14—43
   biochemical, 42
   depositional environment, 21—23
   diagenesis, 36—38
   —, clayey, sediments, 36, 37
   —, secondary precipitation of cementing
      materials, 38—40
   —, sulfates, 37, 38
   —, volcanic ash, 38
   earthquakes, 32, 33
   massive areal rock salt deposits, 40
   osmotic phenomena, 33—36
   —, osmotic flow, 34
   —, theoretical pressures across clay
      membrane, 35
   paleopressures, 23
   permafrost, 40—42
   —, pingo, 40, 41
   —, talik, 40
   piezometric fluid level, 19
   rate of sedimentation, 21—23
   —, U.S. Gulf Coast, 22
   prepressuring of reservoir rock, 21
   salt diapirism, 30—32
   —, piercement salt dome, 30
   —, Weeks Island, U.S.A., 31
   sandstone dikes, 32, 33
   shale diapirism, 28—30
   —, mud lumps, 28
   —, mud volcanoes, 28—30
   —, —, criteria, 29, 30
   —, —, schematic, 28
   structure of reservoir, 19—21
   tectonic activities, 23—28
   —, contemporaneous faults, 24—26
   —, fault seals, 14, 23—28
   thermodynamic, 42
Acoustic logging, see Geophysical well
   logging
Adriatic Sea, 331, 333
Aegean Sea, 334
Africa, 27, 36, 40, 339, 341, 342
Alabama, U.S.A., 6, 337
Alaska, U.S.A., 265, 298, 337
   North Slope, 265
   Prudhoe Bay, 265, 298
Alberta, Canada, 33, 265
Alborz oil field, Iran, 347
Algeria, 298, 325, 341, 342
   Hassi Messaoud oil field, 298, 342
   Hassi R'Mel oil field, 341
Alpine basin, Europe, 327
Alps, Europe, 327, 331
   Dinaric, 331
Amarillo—Texas Panhandle oil field,
   U.S.A., 6
Amazon River, 28
Anaco oil field, Venezuela, 340
Anadarko Basin, U.S.A., 11, 12, 106, 112,
   113, 261—263, 338
Anahuac Formation, U.S.A., 22, 26, 313
Anderson "L" Formation, U.S.A., 277
Andes, South America, 302, 339, 340
Anomalia Formation, U.S.A., 23
A-100 oil well, Lybia, 298
Appalachian region, U.S.A., 7
Appenine Foredeep, Italy, 333
Apsheron Peninsula, U.S.S.R., 29
Aptian Mudstone, Far East, 343
Aquitanian Basin, France, 135, 327
Aransas County, Texas, U.S.A., 313
Arbuckle Formation, U.S.A., 72, 107
Arctic, 40, 264, 265, 325, 336
   King Christian Island, 336
   Melville Island, 336

Sverdrup Basin, 336
Arcturus No.1, Australia, 342, 343
Areal rock-salt deposits, 40
Argentina, 325, 340
Arkansas, U.S.A., 5, 25, 71, 337
Artesian Basin, U.S.A., 19
Assam, India, 346
Assumption, Louisiana, U.S.A., 201
Atlantic Ocean, 77
Atoka Formation, U.S.A., 9, 106, 107, 110, 233
Aure Trough, New Guinea, 343
Australia, 19, 325, 342, 343, 346
    Bonaparte Gulf Basin, 343
    Bowen Basin, 342
    Carnarvon Basin, 343
    Gippsland Shelf, 343
    Great Artesian Basin, 19
    Northern Territory, 343
    Perth Basin, 343
    Queensland, 19, 343
Austria, 325, 328, 329
    Vienna Basin, 328, 329
Azerbaidzhan S.S.R., U.S.S.R., 29, 30

Baku, U.S.S.R., 291
Balkan Mountains, 329
Bangladesh, 325, 346
    Sylhet structure, 346
Barnett Formation, U.S.A., 233
Barracouta No.1 oil well, Australia, 342
Bartlesville—Dewey oil field, U.S.A., 71
Bartlesville Sand, U.S.A., 72, 73
Basin(s), 5, 12, 16, 18, 19, 35, 36, 70, 74, 135, 154, 233, 242, 243, 246, 247, 257, 260—263, 292, 298, 310, 327, 328, 330, 338, 342, 343, 346, 347
    Alpine, Europe, 327
    Anadarko, U.S.A., 11, 12, 106, 112, 113, 261—263, 338
    Aquitanian, France, 135, 327
    Artesian, U.S.A., 19
    Bengal, India, 346
    Bonaparte Gulf, Australia, 343
    Bowen, Australia, 342
    Carnarvon, Australia, 343
    Chindwin River, Burma, 347
    Delaware, U.S.A., 233, 242, 243, 246, 257, 260—263
    Denver—Julesburg, U.S.A., 12
    Dnieper-Donetz, U.S.S.R., 335
    European Permian, 328
    —, German, Germany, 328
    —, West Netherland, Europe, 328
    Great Artesian, Australia, 19
    Great Basin, U.S.A., 18
    Green River, U.S.A., 246, 247
    Gulf of Mexico, U.S.A., 16
    Illinois, U.S.A., 35, 72
    Irrawaddy, Burma, 347
    Michigan, U.S.A., 70
    Molasse, Germany, 327
    North Dakota, U.S.A., 19
    Northland Sedimentary, New Zealand, 343
    Pannonian, Europe, 330
    Papuan, Papua, 36, 342
    Paradox, U.S.A., 36
    Permian, U.S.A., 35, 154
    Persian Gulf, 292
    Perth, Australia, 343
    Piceance, U.S.A., 262
    Po, Italy, 5, 328
    Powder River, U.S.A., 246, 247
    San Juan, U.S.A., 35
    Santa Cruz, U.S.A., 74
    Sirte, Lybia, 298
    Sverdrup, Arctic, 336
    Tertiary, 106, 126, 310
    Transylvanian, Europe, 330
    Uinta, U.S.A., 262, 338
    Vienna, Austria, 328
    Western Sedimentary, Canada, 35
    West Siberian, U.S.S.R., 298
    Wind River, U.S.A., 246, 247
Bavaria, Germany, 328
Bayou Long, U.S.A., 165
Baytown, Texas, U.S.A., 6
Beckham County, U.S.A., 263, 293, 294
Bengal Basin, India, 346
Berenx oil field, France, 327
Bertha Rogers No.1 oil well, U.S.A., 265
Black Sea, U.S.S.R., 75, 76
Blaine County, U.S.A., 8, 286
Block(s)
    Salinas, U.S.A., 18
    San Emigdio—Sierran, U.S.A., 18
    Sierran—Klamath, U.S.A., 18
Bolivia, 340
Bolivar Coast, Venezuela, 5
Bonaparte Gulf Basin, Australia, 343
Bone Spring Formation, U.S.A., 246, 257
Borneo, South China Sea, 344
Bowen Basin, Australia, 342

SUBJECT INDEX 367

Brazil, 325, 340
Brazoria County, U.S.A., 82, 277, 301
Brown Dolomite, U.S.A., 106, 107
Brushy Canyon Formation, U.S.A., 246
Buckner Formation, U.S.A., 38
Bulgaria, 325, 330
Burma, 302, 325, 347
    Chaulk oil field, 347
    Chindwin River Basin, 347
    Gulf of Martaban, 347
    Irrawaddy Basin, 347
    Yenangyaung oil field, 347

Caddo County, U.S.A., 293
Calcasieu Parish, U.S.A., 68, 83
California, U.S.A., 3, 5, 17, 18, 33, 72–74, 92, 252, 253, 293, 302, 309, 314, 315, 337, 338
    Cape Mendocino, 18
    Central Valley, 17
    Colusa County, 338
    Great Valley, 17, 18
    Humbold County, 338
    Kern County, 293
    Long Beach Harbor, 5
    Los Angeles, 5
    Sacramento Valley, 17
    San Joaquin Valley, 17, 35
    Santa Barbara Channel, 3, 252, 253, 338
    Santa Clara Valley, 92
    Tehoma County, 315
    Ventura County, 72
Calveras Fault, U.S.A., 19
Cambrian, see Geologic time scale
Came oil field, France, 327
Cameroun, Africa, 36
Canada, 5, 33, 35, 265, 325, 336, 337
    Alberta, 5, 33, 265
    Mackenzie Delta, 265, 336
    Naskapi Shale, 336
    Nova Scotian Shelf, 336
    Rainbow Lake area, 336
    Sable Island, 336
    Viking Formation, 5, 33
Canadian County, U.S.A., 286
Cape Mendocino, U.S.A., 18
Capitan Reef, U.S.A., 246, 257
Caprian Formation, U.S.A., 246
Carboniferous, see Geologic time scale
Carnarvon Basin, Australia, 343
Carpathian Mountains, Europe, 329, 330

Casing and tubing requirements, 266–268
Casing point selection, 233–235, 257–259
Casing programs, 256–262
    Anadarko Basin, U.S.A., 260–262
    Delaware Basin, U.S.A., 257, 260, 261
    Rocky Mountain area, U.S.A., 260–262
    South Texas well, U.S.A., 258, 261
Caspian Sea, U.S.S.R., 140
Caucasus Mountains, U.S.S.R., 30, 335
Cebu Island, 344
Cementing operations, 262–264
Cenozoic, see Geologic time scale
Central America, 140
Central Anzoategui region, Venezuela, 340
Central Asia, 30, 140
Central Valley, U.S.A., 17
Chambers, Texas, U.S.A., 201
Chaulk oil field, Burma, 347
Checheno-Ingushian A.S.S.R., U.S.S.R., 336
Chemistry of interstitial fluids, 67–73
Cherry Canyon Formation, U.S.A., 107
Chindwin River Basin, Burma, 347
Chinshui oil field, Taiwan, 344, 345
Chocolate Bayou oil field, U.S.A., 82, 277, 301
Chokrak Formation, U.S.S.R., 12, 336
Chuhuangkeng Formation, Taiwan, 344
Chuhuangkeng oil field, Taiwan, 344, 345
Cimarron Anhydrite, U.S.A., 107
Cimarron County, U.S.A., 7
Ciscaucasia S.S.R., U.S.S.R., 36, 335, 336
Clovelly salt dome, U.S.A., 32
Cody Formation, U.S.A., 246
Cogar oil field, U.S.A., 293
Coles Levee oil field, U.S.A., 293
Colon Shale, Venezuela, 340
Colorado, U.S.A., 5, 12, 33, 36, 260–262, 337, 338
    Denver, 12
    Plateau, 5
    Rangely, 33
    Rio Blanco County, 338
Columbia, 325, 340
Colusa County, U.S.A., 338
Compaction of argillaceous sediments, 50–100
    chemistry of interstitial fluids, 67–73
    compaction models, 52–55
    experimental data, 50, 73–80
    –, fluids squeezed out, 50, 73–80
    –, –, kaolinite, 50, 74

—, —, marine mud, 73, 76, 77
—, —, Maykop clay, U.S.S.R., 75, 76
—, —, montmorillonite, 50, 78
—, interstitial solutions related to seawater, 69—73
fluid flow, 63—66
—, Darcy's equation, 63, 64, 66
mathematical analysis, 55—63
oilfield water composition, 71—73
overburden potential, 66—67
salinity distribution, 80—83
shale model(s), 53—55, 59, 60
—, Katz and Ibrahim, 53—55
—, Bredehoeft and Hanshaw, 59, 60
Completion and drilling concepts, see Drilling and completion concepts
Completion and production considerations, 264—268
    casing and tubing requirements, 266—268
    hydrates, 265
    packer fluid requirements, 266
    permafrost-freezeback pressure, 264, 265
    precipitation of salt, sulfur, etc., 265, 266
Compressibility, 89—95
    experimental data, 92—94
    in reservoir engineering, 95
    mathematical equations, 89—91
    terminology, 89—91
Conductivity curve, see Geophysical well logging
Contemporaneous fault systems, 24—27
Contingency planning, 268—270
    evacuation conditions under H_2S, 269
    evacuation plan, 270
Cote Blanche salt dome, U.S.A., 32
Cox No. 1 oil well, U.S.A., 268, 272
Cretaceous, see Geologic time scale
Cuttings, 150—172
    shape, 158
    size, 158
    slurry resistivity, 158—161
    specific ion distribution, 170—172
    volume, 158
    water color, 164
Czechoslovakia, 140
    Moravia, 140

$\delta^{18}O$, 70
Dagestan A.S.S.R., U.S.S.R., 336

Dakota Formation, U.S.A., 246
Dampier No. 1 oil well, Australia, 342, 343
Danian Chalk Formation, North Sea, 332
Danube River, Europe, 28
Darcy's equation, 63—65
Debagram Formation, India, 346
Decompressional expansion, 9, 10
DeLarge oil field, U.S.A., 293
Delaware Basin, U.S.A., 233, 242, 243, 246, 247, 257, 260—263
Delaware Sand, U.S.A., 246, 257
Denmark, 331, 332
Denver, U.S.A., 12
Denver—Julesburg Basin, U.S.A., 12
Denver Producing and Refg. Co.'s School Land 1-A oil well, U.S.A., 293
Depositional environment, 21—23
Des Moines Formation, U.S.A., 107
Devonian, see Geologic time scale
Dewey County, U.S.A., 8, 286
d-Exponent, 122—125
Dhulian oil field, Pakistan, 346
Diagenesis, 36—38
    clayey sediments, 36, 37
    —, subsurface water distribution, 36
    secondary precipitation of cementation materials, 38—40
    sulfates, 37, 38
    volcanic ash, 38
Divnoe area, U.S.S.R., 76
Dnieper-Donetz Basin, U.S.S.R., 335
Dog Creek Shale, U.S.A., 107
Downhole gravity data, 209, 210
Drag, 136, 137
Drilling and completion concepts, 231—274
    casing programs, 256—262
    cementing operations, 262—264
    completion and production considerations, 264—268
    contingency planning, 268—270
    fracture pressure gradient prediction, 238—256
    pore pressure versus fracture pressure, 231—238
    relief-well planning, 270—272
Drilling data, 117—175
    drilling mud parameters, 137—150
    —, chloride ion, 147, 148
    —, flow-line mud weight, 141
    —, flow-line temperature, 144—147
    —, —, field data, 145, 146

SUBJECT INDEX 369

—, —, heat conductivity values, 145
—, hole fill-up, 148, 149
—, —, drilling pipe displacement data, 149
—, mud flow rate, 149, 150
—, mud-gas cutting, 137—140
—, novel concepts, 147
—, pit level, 147, 148
—, pressure kicks, 141—144
—, —, gas rise with expansion, 142
—, —, gas rise without expansion, 143
—, resistivity, 147
—, total pit volume, 147, 148
drilling parameters, 120—137
—, d-exponent, 122—125
—, drag, 136, 137
—, drilling porosity and formation pressure logs, 130—136
—, —, Bourgoyne's model, 130—134
—, —, SNAP log, 135, 136
—, drilling (penetration) rate, 120—122
—, drilling rate equations, 128—130
—, logging-while-drilling concepts, 136
—, modified d-exponent, 125—128
—, torque, 136
shale cuttings parameters, 150—172
—, bicarbonate—overpressure relationship, 165—168
—, —, field observations, 166, 167
—, —, Gulf Coast formation brines, U.S.A., 167
—, filtration rate of shale slurry, 161—164
—, litho-function plotting, 164, 165
—, moisture index, 164
—, redox and pH potentials, 168—170
—, —, definitions, 169
—, —, field example, U.S.A., 169, 170
—, shale bulk density, 150—156
—, —, density gradient column, 150
—, —, field case studies, 151, 154—157
—, —, mercury pump technique, 150
—, —, mud balance technique, 150—153
—, shale factor, 156—158
—, —, Mondshine's shale classification, 157
—, shale slurry resistivity, 158—161
—, —, onshore Texas field test, 160
—, —, offshore Louisiana field test, 160, 161
—, —, test procedure, 159, 160
—, shale water color, 164

—, shape, 158
—, size, 158
—, specific ion distribution, 170—172
—, volume, 158
Drilling mud parameters, see Drilling data
Drilling parameters, see Drilling data
Drilling rate equations, 128—130
Drill pipe displacement data, 149
Drill pipe sticking, 235—237
Duchesne River Formation, U.S.A., 262
Duri oil field, Sumatra, 298

Earthquakes, 32, 33
East Baton Rouge, U.S.A., 201
East Cameron area, U.S.A., 301, 310—312
East Riverton area, U.S.A., 311
Economic considerations, 309—321
    abnormal pressure environments, 319—321
    Iberia Parish field case, U.S.A., 319
    normal pressure environments, 320
    shale resistivity ratio, 309—319
Ecuador, 325, 340
Egypt, 325, 342
Ekofisk oil field, North Sea, 298, 331, 332
El Dorado oil field, U.S.A., 71
Electrical prospecting, 115
Electrical surveys, see Geophysical well logging
Elk City Formation, U.S.A., 107
Ellenburger Formation, U.S.A., 246, 257
Eocene, see Geologic time scale
Ethiopia, 342
Europe, 28, 326—334, 336
    Alpine Basin, 327
    Alps, 327, 331
    Balkan Mountains, 329
    Carpathian Mountains, 329, 330
    Danube, 28
    Flysch Formation, 327, 328
    Keuper Formation, 329
    Main Dolomite (Hauptdolomit), 328
    Permian shales, 40
    Rotliegendes Formation, 328
    Sarmatian Formation, 336
    Tortonian Formation, 336
    Zechstein Formation, 14, 328, 331
European Permian Basin, Europe, 328
Evacuation, 268—270
    conditions, 269
    plan, 270

Far East, 216, 342—346
 Aptian Mudstone, 343
Farrar No.1 oil well, U.S.A., 106, 107, 111—113
Fault(s), 14—16, 18, 19, 23—28
 Calveras, U.S.A., 19
 Hayward, U.S.A., 19
 Hluk, Moravia, Czechoslovakia, 140
 San Andreas, U.S.A., 18, 19
 seals, 14—16
Federal Power Commission, U.S.A., 302—304
Filtration rate of shale slurry, 161—164
Florida, U.S.A., 19, 298, 299, 337
Flow line, 141, 144—147, 149, 150
 mud weight, 141
 rate, 149, 150
 temperature, 144—147
Fluid flow, 63—66
Flysch Formation, Europe, 327, 328
Fohs Oil Co.'s Buckley-Bourg 1 oil well, U.S.A., 293
Fold(s), 18
 Himalayan, Asia, 346
 Kettleman, U.S.A., 18
Forest Oil Co. and Robert J. Zanne et al.'s Charles J. Walker 1 oil well, U.S.A., 293
Forest Reserve oil field, Trinidad, 340
Formation(s), *also* Reservoir rocks, 5, 8, 9, 11, 12, 14, 17, 18, 22, 23, 26, 29, 34, 38—40, 72, 73, 75, 76, 106, 107, 110, 188, 196, 218, 223, 224, 233, 246, 257, 261, 262, 267, 275, 278—282, 284, 285, 302, 313, 327—329, 331, 332, 336, 340, 343, 344, 346
 Anahuac, U.S.A., 22, 26, 313
 Anderson "L", U.S.A., 277
 Anomalina, U.S.A., 23
 Aptian Mudstone, Far East, 343
 Arbuckle, U.S.A., 72, 107
 Atoka, U.S.A., 9, 106, 107, 110, 233
 Barnett, U.S.A., 233
 Bartlesville Sand, U.S.A., 72, 73
 Bone Spring, U.S.A., 233, 246
 Brown Dolomite, U.S.A., 106, 107
 Buckner, U.S.A., 38
 Buntsandstein, Germany, 328
 Brushy Canyon, U.S.A., 246
 Capitan Reef, U.S.A., 246
 Caprian, U.S.A., 246
 Cherry Canyon, U.S.A., 246
 Chester, U.S.A., 107

Chokrak, U.S.S.R., 5, 336
Chuhuangkeng, Taiwan, 344
Cimarron Anhydrite, U.S.A., 107
Cody, U.S.A., 246
Colon Shale, Venezuela, 340
Danian Chalk, North Sea, 332
Dakota, U.S.A., 246
Debagram, India, 346
Delaware Sand, U.S.A., 246
Des Moines, U.S.A., 107
Dog Creek Shale, U.S.A., 107
Duchesne River, U.S.A., 262
Elk City, U.S.A., 107
Ellenburger, U.S.A., 233, 246
Flysch, Europe, 327, 328
Fort Union, U.S.A., 246
Franciscan, U.S.A., 17, 18
Frio, U.S.A., 22, 26, 188
Fronteer, U.S.A., 246
Fusselman, U.S.A., 233
Galveston Shale, U.S.A., 314
Great Valley, U.S.A., 17, 18
Green River, U.S.A., 262
Grubb, U.S.A., 94
Haynesville Salt, U.S.A., 110
Hinnant, U.S.A., 258
Hunton, U.S.A., 12, 107, 267
Karagan, U.S.S.R., 5, 336
Keuper, Europe, 329
Lansing—Kansas City, U.S.A., 94
Lower M-Sand, U.S.A., 000
Madison, U.S.A., 246
Main Dolomite (Hauptdolomit), Europe, 328
Maykop, U.S.S.R., 75, 76, 302, 336
Meramec, U.S.A., 107
Mesa Verde, U.S.A., 246
Migura, U.S.A., 218
Minnelusa, U.S.A., 246
Mississippian Lime, U.S.A., 11, 12, 233
Morrow, U.S.A., 8—11, 34, 106, 107, 196, 233, 267, 285, 286
Muddy, U.S.A., 246
Muschelkalk Liegendes, Germany, 329
Muschelkalk Salinar, Germany, 329
Nacatoch, U.S.A., 72, 73
Naskapi Shale, Canada, 336
Norphlet, U.S.A., 110
NS2B, U.S.A., 275, 278—280, 284
Osage, U.S.A., 107
Osinskii, U.S.S.R., 39
Pandua, India, 346

SUBJECT INDEX 371

Permian, Germany, 40
Permian Shales, Europe, 40
Phosphoria, U.S.A., 246
Pico, U.S.A., 72, 73
Pleistocene, Italy, 188
Pliocene, Italy, 188
Queen City Sand, U.S.A., 258
Repetto, U.S.A., 94
Roetpelit, Germany, 329
Roetsalinar, Germany, 329
Rotliegendes, Europe, 328
Rustler, U.S.A., 246
Sarmatian, Europe, 336
Sespe, U.S.A., 72, 73, 281
Shannon, U.S.A., 246
Simpson, U.S.A., 107, 233, 246
Smackover, U.S.A., 267, 313
Springer, U.S.A., 29, 106, 107, 223, 224, 233, 267
Ste. Genevieve, U.S.A., 72
Strawn, U.S.A., 233
Sussex, U.S.A., 246
Sylvan, U.S.A., 107
Tansil, U.S.A., 246
Tensleep, U.S.A., 246
34 W.M.I., U.S.A., 282
35 M.I. deep, U.S.A., 282
Tonkawa Sand, U.S.A., 107
Tortonian, Europe, 336
Uinta, U.S.A., 262
Vicksburg, U.S.A., 22, 23
Viking, Canada, 5, 33
Viola, U.S.A., 107
Waddell, U.S.A., 257
Wasatch, U.S.A., 262
Wellington Anhydrite, U.S.A., 107
Wilcox, U.S.A., 72, 73, 188, 257, 261, 262, 267
Wolfcamp, U.S.A., 107, 246
Woodbine, U.S.A., 94
Woodford, U.S.A., 107, 233
Yates, U.S.A., 246
Zechstein, Europe, 14, 328, 331
Formation factor method, 7, 196—198
Formation pressure, 3—5
Formation pressure gradient versus hydrocarbon distribution, 300—304
Formation temperature, 294—300
coal versus petroleum, 296
determination, 295
physical state of hydrocarbons, 296, 298

—, model(s), 296—300
—, —, Burst, 300
—, —, Klemme, 296, 297, 299
—, —, Landes, 296
—, —, Pussey, 298
Formosa, see Taiwan
Fort Union Formation, U.S.A., 246
Four Isle oil field, U.S.A., 301
Fracture pressure gradient prediction, 238—256
comparison of methods, 250
method(s), 238—256
—, Anderson, Ingram, and Zanier, 253—256
—, —, fracture pressure gradient log, 256
—, —, U.S. Gulf Coast field data, 254, 255
—, Christman, 251—253
—, —, Santa Barbara Channel, 252, 253
—, Eaton, 243—247
—, —, gradient in, Delaware Basin, U.S.A., 246
—, —, —, U.S. Gulf Coast area, 244, 245
—, —, —, Wyoming, U.S.A., 246
—, empirical data correlations, 240
—, Hubbert and Willis, 238—240
—, MacPherson and Berry, 250, 251
—, Matthews and Kelly, 240—243
—, —, Delaware Basin, U.S.A., 242
—, —, U.S. Gulf Coast area, 242
—, —, variable matrix stress coefficient, 241
—, Taylor and Smith, 247—250
France, 135, 267, 325, 327
Aquitanian Basin, 135, 327
Berenx oil field, 327
Came oil field, 327
Lacq oil field, 327
Pyrenees, 267, 327
St. Marcet oil field, 327
Franciscan Formation, U.S.A., 17
Frio Formation, U.S.A., 22, 26, 188
Fronteer Formation, U.S.A., 246
Fujikawa oil field, Japan, 346

Galveston, U.S.A., 5, 201
Galveston Shale, U.S.A., 314
Geologic time scale, 2, 5, 7, 9, 11, 12, 14, 18, 19, 21, 22, 25, 26, 29, 31, 33, 36, 40, 51, 52, 70—73, 80, 106, 107, 110, 127, 178, 181, 182, 186—188, 195, 199—201, 221, 246, 257, 261, 262,

298, 302, 310, 325, 327—330, 332, 334—336, 340, 342—344, 346, 347
Cenozoic era, 2, 5, 12, 14, 18, 19, 21, 22, 26, 29, 31, 52, 70—73, 80, 127, 178, 181, 182, 186—188, 195, 199—201, 221, 261, 262, 302, 310, 325, 327, 328, 330, 332, 334—336, 340, 342, 344, 346, 347
—, Quaternary, 14, 21, 22, 186—188, 325
—, —, Pleistocene, 14, 21, 22, 186—188, 325
—, Tertiary, 2, 12, 22, 29, 31, 52, 70, 80, 106, 127, 178, 184, 188, 195, 199—201, 221, 261, 262, 302, 310, 327, 328, 330, 332, 334—336, 340, 343, 344, 346, 347
—, —, Eocene, 72, 80, 195, 328, 346
—, —, Miocene, 5, 12, 22, 26, 31, 52, 80, 127, 178, 181, 182, 195, 199—201, 221, 261, 262, 302, 330, 336, 340, 343, 344
—, —, Oligocene, 22, 72, 73, 178, 181, 182, 195, 199, 200, 201, 221, 302
—, —, Pliocene, 22, 52, 71—73, 80, 188, 195, 330, 343, 346
Mesozoic era, 5, 22, 25, 29, 33, 36, 71—73, 195, 298, 327—330, 332, 334—336, 342, 343
—, Cretaceous, 5, 22, 29, 33, 36, 71—73, 195, 328, 332, 343
—, Jurassic, 25, 36, 195, 328
—, Triassic, 195, 329, 332, 342
Paleozoic era, 7, 9, 11, 12, 14, 29, 40, 51, 70—73, 107, 110, 195, 246, 257, 325, 328, 334, 335
—, Cambrian, 14, 70—72, 107, 195, 325
—, Carboniferous, 328, 334, 335
—, Devonian, 11, 195, 246, 257, 334, 335
—, Mississippian, 9, 11, 12, 29, 72, 107, 110, 195, 246, 257
—, Ordovician, 71, 72, 107, 195
—, Pennsylvanian, 7, 11, 12, 29, 71—73, 107, 110, 195, 246, 257
—, Permian, 7, 11, 12, 40, 107, 195, 328
—, Silurian, 107, 195, 246, 257
Precambrian era, 12, 195
Geophysical methods, 101—116
electrical prospecting, 115
gravity data, 114, 115
seismic data, 102—114

—, field case studies, 104—114
—, formation tops, 105—107
—, interval velocity, 101
—, mathematical concepts, 103, 104
—, pitfalls, 114
—, pressure predictions, 108—113
velocity ranges, 101
Geophysical well logging, 7, 198—204
acoustic (sonic) surveys, 198—204
—, basic principles, 198, 199
—, possible limitations and pitfalls, 203, 204
—, quantitative pressure evaluation, 199—203
—, —, equivalent depth method, 202
—, —, Ham's method, 203
—, —, Hottman and Johnson method, 199—202
density measurements, 204
—, basic principle, 204
—, quantitative pressure evaluation, 204
downhole gravity data, 209, 210
electrical surveys, 7, 178—198
—, conductivity, 184—189
—, —, basic considerations, 184, 185
—, —, possible pitfall, 189
—, —, quantitative pressure evaluation, 185—189
—, shale formation factor, 7, 196—198
—, —, basic concepts, 196
—, —, possible limitations, 198
—, —, quantitative pressure evaluation, 197, 198
—, short normal, 178—184
—, —, basic principles, 178—180
—, —, possible pitfalls, 184
—, —, quantitative pressure evaluation, 180
—, —, —, equivalent depth method, 181—183
—, —, —, Ham's method, 183, 184
—, —, —, Hottman and Johnson method, 180, 181
—, spontaneous potential (SP), 7, 189—196
—, —, basic principle, 189, 190
—, —, salinity principle, 190—196
—, —, —, correlation with geologic age, 195
—, —, —, field cases, 192, 194, 195
—, —, —, formation water composition, U.S. Gulf Coast, 191

SUBJECT INDEX 373

limitations and pitfalls, 212—225
—, considerations in plotting shale parameters, 220—225
—, possible pitfalls, 212—220
nuclear magnetic resonance technique, 207—209
neutron measurements, 204, 205
pulsed neutron measurements, 205—207
—, basic principles, 205, 206
—, quantitative pressure evaluation, 206, 207
Geopressure, see Abnormal formation pressure
Germany, 14, 265, 267, 323, 327—329, 331
   Bavaria, 328
   Muschelkalk Liegendes Formation, 329
   Muschelkalk Salinar Formation, 329
   Roetpelit Formation, 329
   Roetsalinar Formation, 329
Ghawar oil field, Arabian Gulf, 298
Gippsland Shelf, Australia, 343
Gomez oil field, U.S.A., 293
Grand Isle, U.S.A., 80
Gravity data, 114, 115, 209, 210
Great Artesian Basin, Australia, 19
Great Basin, U.S.A., 18
Great Salt Lake, U.S.A., 73
Great Valley, U.S.A., 17, 18
Great Valley Formation, U.S.A., 17
Greece, 333
Green No. 1 oil well, U.S.A., 263
Green River Basin, U.S.A., 246, 247
Green River Shale, U.S.A., 262
Groningen gas field, Holland, 328, 331
Grubb Formation, U.S.A., 94
Guario oil field, Venezuela, 340
Guiana, 325, 340, 341
Gulf Coast, U.S.A., 2, 5, 6, 14, 16, 19, 21, 22, 24, 31, 36, 38, 52, 72, 79, 80, 105, 108, 109, 112, 124, 125, 132, 135, 137, 140, 153, 165—169, 179, 181—183, 186, 187, 190, 191, 193, 194, 197, 199—203, 210, 217, 221—223, 239—245, 250, 254—256, 260, 261, 275, 276, 286, 288, 291, 300, 301, 302, 305—307, 309—313, 316, 319, 320, 340
Gulf of Martaban, Burma, 347
Gulf of Mexico, U.S.A., 16, 22, 31, 34, 35, 52, 72, 165, 191

Gulf Oil Corp.'s "DD" Unit 1 oil well, U.S.A., 293

Hall-Gurney oil field, U.S.A., 71
Ham's method, 203
Hassi Messaoud oil field, Algeria, 298, 342
Hassi R'Mel oil field, Algeria, 341
Hawkey Bay, New Zealand, 343
Haynesville Salt, U.S.A., 110
Hayward Fault, U.S.A., 19
Heron No. 1 oil well, Australia, 342, 343
Higashi-Sanjo oil field, Japan, 346
Himalayan folding, Asia, 346
Himalayan Mountains, Asia, 194, 302, 346
Hinnant Formation, U.S.A., 258
Hluk Fault, Czechoslovakia, 140
Hole fill-up, 148, 149
Holland, 325, 328, 331
   Groningen gas field, 328, 331
Honshu Island, Japan, 344
Hottman and Johnson method, 199—202
Humble Oil and Refining Co.'s Henry Willbanks 11 oil well, U.S.A., 293
Humbold County, U.S.A., 338
Hungary, 325, 330
   Mako Trench, 330
Hunton Formation, U.S.A., 12, 107, 267
Hydrates, 265
Hydrocarbon distribution, 291—323
   depth, 291—294
   —, gas reserves, southwest Louisiana, U.S.A., 294
   —, oil and gas reservoirs, U.S.S.R., 292
   —, production statistics, U.S.A., 293
   formation pressure gradient, 300—304
   —, reservoir size in U.S. Gulf Coast, 302, 303
   formation temperature, 294—300
   —, coal versus petroleum, 296
   —, determination, 295
   —, physical state of hydrocarbons, 296, 298
   —, —, model(s), 296—300
   —, —, —, Burst, 300
   —, —, —, Klemme, 296, 297, 299
   —, —, —, Landes, 296
   —, —, —, Pussey, 298
   shale resistivity ratio method, 309—319
   —, completion guidelines, 311—313
   —, —, commercial gas and/or oil fields, 311, 312
   —, —, commercial oil fields, 311

—, —, non-commercial reservoirs, 312
—, field cases, 313—319
—, —, California, 314, 315
—, —, offshore Louisiana, 314, 315
—, —, offshore Texas, 313
—, —, U.S. Gulf Coast area, 316—319
—, other Tertiary Basins, 311
—, U.S. Gulf Coast resistivity profile, 310
temperature and pressure, 304—309
—, field cases, 306—309
—, —, U.S. Gulf Coast area, 307—309
—, —, Weeks Island, U.S.A., 306, 307
—, model(s), 304, 305
—, —, Baybakov, 305
—, —, Modelevskiy and Parnov, 304
Hydrostatic pressure, 1, 2
average gradients, 2

Iberia Parish, U.S.A., 20, 178, 182, 193, 194, 319
Illinois Basin, U.S.A., 35, 72
India, 194, 325, 346
  Assam region, 346
  Bengal Basin, 346
  Debagram Formation, 346
  Jammu region, 346
  Pandua Formation, 346
  Punjab region, 346
Indian subcontinent, 325, 346, 347
  Bangladesh, 325, 346
  Burma, 325, 347
  India, 325, 346
Indonesia, 127, 325
Interstitial fluids related to seawater, 69—73
Iokea oil well, Papua, 342
Iran, 5, 15, 21, 325, 347
  Alborz oil field, 347
  Plains of Gorgan, 347, 348
  Sarajeh oil field, 347
Iraq, 325, 348
Ireland, 333
Irrawaddy Basin, Burma, 347
Israel, 196
Italy, 5, 188, 189, 325, 328, 333
  Appenine Foredeep, 333
  Pleistocene Formations, 188
  Pliocene Formations, 188
  Po Delta, 5, 328

Jammu, India, 346

Japan, 325, 344, 346
  Fujikawa oil field, 346
  Higashi-Sanjo oil field, 346
  Honshu Island, 344
  Katagri oil field, 346
  Kumoide oil field, 346
  Kurosaka oil field, 346
  Mitsuke oil field, 346
  Nagaoka Plain, 344, 346
  Nishi-Nagaoka oil field, 346
Java, South China Sea, 38, 140, 344
Jay oil field, U.S.A., 298
Jefferson, Texas, U.S.A., 182, 201
Jefferson Parish, Louisiana, U.S.A., 178, 180, 182
Joya Mair oil field, Pakistan, 346
Jura Mountains, Switzerland, 38
Jurassic, see Geologic time scale

Kalimantan, South China Sea, 125, 344
Kansas, U.S.A., 71, 196
Kapuri No. 1 oil well, Papua, 342
Karachukhar oil field, U.S.S.R., 38
Karagan Formation, U.S.S.R., 12, 336
Katagai oil field, Japan, 346
Katy, Texas, U.S.A., 6
Kern County, U.S.A., 293
Kettleman Fold, U.S.A., 18
Keuper Formation, Europe, 329
Keyes oil field, U.S.A., 7
Kicks, 139, 141—144
Kikinda blow out, Yugoslavia, 331
King Christian Island, Arctic, 336
Kuibyshev region, U.S.S.R., 334
Kulshill oil well, Australia, 342
Kumoide oil field, Japan, 346
Kurosaka oil field, Japan, 346

Lacq oil field, France, 327
Lafayette, Parish, U.S.A., 182, 275, 278—280
Lafourche, U.S.A., 201
Lake Maracaibo, Venezuela, 5, 298, 299
Lake Mongoulois oil field, U.S.A., 301
Lake Washington oil field, U.S.A., 32, 293
Lamar oil field, Venezuela, 298
Lansing—Kansas City Formation, U.S.A., 94
Leopardwood No. 1 oil well, Australia, 342
Litho-function plotting, 164, 165
Logging-while-drilling, 136
Loktaban, U.S.S.R., 30

SUBJECT INDEX 375

Long Beach Harbor, U.S.A., 5
Los Angeles, U.S.A., 5, 298
Lost circulation, 237, 238
Lost Hills oil field, U.S.A., 17
Louisiana, U.S.A., 3, 20—22, 25, 31, 72,
   73, 80, 81, 83, 104, 112, 126, 127, 131,
   154, 157, 159, 160, 165, 178, 180,
   182—188, 191—195, 197, 199, 201,
   206, 213, 217, 224, 247, 250, 251,
   254, 257, 259—261, 268, 275, 277—
   280, 282, 287—289, 292—294, 297,
   301, 305—307, 310—312, 314—316,
   319, 321, 336—338, 342
   Assumption, 201
   Bayou Long, 165
   Calcasieu, 68, 83, 201
   Cameron, 182, 201
   East Baton Rouge, 201
   East Cameron area, 301, 310—312
   Grand Isle, 80
   Iberia Parish, 20, 178, 182, 193, 194,
     319
   Jefferson Davis, 182
   Jefferson Parish, 178, 180
   Lafayette Parish, 182, 275, 278—280
   Lafourche, 201
   Mississippi Delta, 338
   Plaquemines, 201
   Ship Shoal, 80
   South Pass, 159, 165
   St. Martin Parish, 182, 201, 301
   St. Mary Parish, 182, 201
   Terrebonne Parish, 80, 182, 201, 301,
     338
   Vermilion Parish, 201, 213, 316, 338
   Wayne County, 338
Lower M-Sand, U.S.A., 000
Lybia, 298
   Sirte Basin, 298

Mackenzie Delta, Canada, 265, 336
Madagaskar, 342
Madeleine No. 1 oil well, Australia, 342, 343
Madison Formation, U.S.A., 246
Main Dolomite (Hauptdolomit), Europe, 328
Mako Trench, Hungary, 330
Mamberamo River, Far East, 343
Markovo oil field, U.S.S.R., 39
Morocco, 325
Matagorda County, U.S.A., 192, 193

Maykop Formation, U.S.S.R., 74, 76, 302, 336
Mediterranean Sea, 333
Meillon oil field, France, 327
Melville Island, Arctic, 336
Meramec Formation, U.S.A., 107
Mesa Verde Formation, U.S.A., 246
Mesozoic, see Geologic time scale
Mexico, 325, 339
Michigan Basin, U.S.A., 70
Mid-Continent, U.S.A., 2, 106, 182, 190,
   194, 217, 236, 237, 262, 288, 297,
   298, 329
Middle East, 6, 14, 19, 21, 40, 83, 292,
   313, 347, 348
   Persian Gulf Basin, 292
Minas oil field, Sumatra, 298
Minnelusa Formation, U.S.A., 246
Miocene, see Geologic time scale
Mississippi, U.S.A., 71, 110, 247, 263,
   265, 267, 268, 271, 272, 301, 313,
   337, 338
   Cox No. 1 blow-out, 268, 271, 272
   Delta, 301, 338
   Pine Woods, 268
   River, 28, 247
Mississippian, see Geologic time scale
   Lime, U.S.A., 11, 12, 233
Mitsuke oil field, Japan, 346
Model(s),
   Anderson, Ingram, and Zanier, 253—256
   Baybakov, 305
   Bourgoyne, 130—134
   Bredehoeft and Hanshaw, 59, 60
   Christman, 251—253
   Eaton, 243—247
   Hammerlindl, 281—285
   Hubbert and Willis, 238—240
   Katz and Ibrahim, 53—55
   Klemme, 296, 297, 299
   Landes, 296
   MacPherson and Kelly, 250, 251
   Modelevskiy and Parnov, 304
   Pussey, 298
   Taylor and Smith, 247—250
   Terzaghi and Peck, 53
Modified d-exponent, 125—128
Moisture, 66, 67, 164
   index, 164
   ratio, 66, 67
Mondshine's shale classification, 157

Montana, U.S.A., 337
Morocco, 325, 341
Moravia, Czechoslovakia, 140
Morrow Formation, U.S.A., 8—11, 34, 107, 196, 267, 285, 286, 335
Mountain(s)
  Alps, Europe, 327, 331
  —, Dinaric, Europe, 331
  Andes, South America, 302, 338, 340
  Appenine, Italy, 333
  Balkan, Europe, 329
  Carpathian, Europe, 329, 330
  Caucasus, U.S.S.R., 335
  Himalayan, Asia, 302, 346
  Jura, Switzerland, 36
  Pyrenees, Europe, 267, 327
  Rocky Mountains, U.S.A., 2, 12, 182, 260—262
  Wichita, U.S.A., 6
  Ural, U.S.S.R., 335
Mountain View oil field, U.S.A., 33
Mozambique, 342
Muddy Formation, U.S.A., 246
Mud-gas cutting, 137—140
Mud volcano, 28—30, 51, 140
Muschelkalk Liegendes, Germany, 329
Muschelkalk Salinar, Germany, 329

Nacatoch Formation, U.S.A., 72, 73
Nagoako Plain, Japan, 344, 346
Napoleonville salt dome, U.S.A., 32
Naskapi Shale, Canada, 336
Nevada, U.S.A., 18
New Guinea, 36, 302, 325, 342, 343
  Aure Trough, 343
New Mexico, U.S.A., 35, 36, 337
New Zealand, 140, 325, 343
  Hawke Bay, 343
  Northern Sedimentary Basin, 343
Niger River, Africa, 27, 28
Nigeria, 27, 325, 342
Nile River, Africa, 28, 342
Nishi-Nagaoka oil field, Japan, 346
Norphlet Sand, U.S.A., 110
North American continent, 8, 12, 336—339
North Dakota, U.S.A., 19
  Basin, 19
Northern Territory, Australia, 343
Northland Sedimentary Basin, New Zealand, 343
North Sea area, 3, 14, 19, 108, 112, 124, 130, 146, 179, 202, 298, 310, 311, 325, 328, 331, 332
North Slope, Alaska, U.S.A., 265
Norway, 331—333
Nova Scotian Shelf, Canada, 336
NS2B Formation, U.S.A., 275, 278—280, 284
Nuclear magnetic resonance technique, see Geophysical well logging
Nueces County, U.S.A., 22, 277

Occurrence of abnormal pressures, 325—349
  Africa, 341, 342
  Europe, 326—334
  —, offshore, 331—334
  —, onshore, 327—331
  Far East, 342—346
  Indian subcontinent, 346, 347
  Middle East, 347, 348
  North American continent, 8, 12, 336—339
  South America, 339—341
  Soviet Union, 13, 334—336
Oil field(s), 6, 7, 17, 22, 33, 38, 39, 71, 73, 82, 275, 277—280, 282, 284, 293, 298, 301, 305, 306, 327, 340—342, 344—347
  Alborz, Iran, 347
  Amarillo—Texas Panhandle, U.S.A., 6
  Anaco, Venezuela, 340
  Anderson "L", U.S.A., 277
  Bartlesville—Dewey, U.S.A., 71
  Berenx, France, 327
  Came, France, 327
  Chaulk, Burma, 347
  Chinshui, Taiwan, 344, 345
  Chocolate Bayou, U.S.A., 82, 277, 301
  Chuhuangkeng, Taiwan, 344, 345
  Cogar, U.S.A., 293
  Coles Levee, U.S.A., 293
  DeLarge, U.S.A., 293
  Dhulian, Pakistan, 346
  Duri, Sumatra, 298
  Ekofisk, North Sea, 298
  El Dorado, U.S.A., 71
  Forest Reserve, Trinidad, 340
  Four Isle, U.S.A., 301
  Fujikawa, Japan, 346
  Ghawar, Arabian Gulf, 298
  Gomez, U.S.A., 293
  Groningen, Holland, 328, 331

SUBJECT INDEX 377

Guario, Venezuela, 340
Hall-Gurney, U.S.A., 71
Hassi Messaoud, Algeria, 298, 342
Hassi R'Mel, Algeria, 341
Higashi-Sanjo, Japan, 346
Jay, U.S.A., 298
Joya Mair, Pakistan, 346
Karachukhar, U.S.S.R., 38
Katagai, Japan, 346
Keyes, U.S.A., 7
Kumoide, Japan, 346
Kurosaka, Japan, 346
Lacq, France, 327
Lake Mongoulois, U.S.A., 301
Lake Washington, U.S.A., 293
Lamar, Venezuela, 298
Lost Hills, U.S.A., 17
Markovo, U.S.S.R., 39
Meillon, France, 327
Minas, Sumatra, 298
Mitsuke, Japan, 346
Mountain View, U.S.A., 33
Nishi-Nagaoka, Japan, 346
Ossum, U.S.A., 275, 278—280, 282, 284
Petronilla, U.S.A., 22
Pistol Ridge, U.S.A., 298
Prudhoe Bay, U.S.A., 298
Queen Bess Island, U.S.A., 293
Rangely, U.S.A., 33
San Joaquin, Venezuela, 340
Santa Ana, Venezuela, 340
Santa Rosa, Venezuela, 340
Sarajeh, Iran, 347
Sekihara, Japan, 346
Soso, U.S.A., 71
South Pass Block 27, U.S.A., 298
St. Louis, U.S.A., 71
St. Marcet, France, 327
Surakhan, U.S.S.R., 38
Sylhet, Pakistan, 346
Tiehchenshan-Tungshiao, Taiwan, 344, 345
Tinsley, U.S.A., 71
Wasco, U.S.A., 293
Weeks Island, U.S.A., 293, 305, 306
Wesson, U.S.A., 71
West Poison Spider, U.S.A., 293
West Siberia, U.S.S.R., 298
Wilmington, U.S.A., 73, 298
Yenangyaung, Burma, 347
Oil well(s), 10, 12, 106, 107, 111—113,
263, 265, 268, 271, 272, 293, 298, 333, 336, 342, 343, 346
A-100, Lybia, 298
Arcturus No. 1, Australia, 342, 343
Barracouta No. 1, Australia, 342
Bertha Rogers No. 1, U.S.A., 265
Cheyenne-Arapaho No. 1, U.S.A., 10
Cox No. 1, U.S.A., 268, 271, 272
Dampier No. 1, Australia, 342, 343
Denver Producing & Refg. Co.'s School Land 1-A, U.S.A., 293
Farrar No. 1, U.S.A., 106, 107, 111—113
Ferguson No. 1, U.S.A., 10
Fohs Oil Co.'s Buckley-Bourg 1, U.S.A., 293
Forest Oil Co. and Robert J. Zonne et al.'s Charles J. Walker 1, U.S.A., 293
Green No. 1, U.S.A., 263
Gulf Oil Corp.'s "DD" Unit 1, U.S.A., 293
Heron No. 1, Australia, 342, 343
Hill No. 1, U.S.A., 10
Humble Oil & Refining Co.'s Henry Willbanks 11, U.S.A., 293
Iokea, Papua, 342
Kapuri No. 1, Papua, 342
Kulshill, Australia, 342
Leopardwood No. 1, Australia, 342
Madeleine No. 1, Australia, 342, 343
McNeely-Fahl No. 1, U.S.A., 10
Palm Valley No. 1, Australia, 342, 343
Pasca No. 1, Papua, 342, 343
Paul No. 1, U.S.A., 10
Petrel No. 1, Australia, 342, 343
Pure Oil Co.'s Unit 1, U.S.A., 293
Rarako Creek No. 1, Papua, 342, 343
Richardson & Bass et al.'s Humble-LL & E1-L, U.S.A., 293
Richfield Oil Corp.'s Cole Levee "A" 67-29, U.S.A., 293
Rocky Mountain Arsenal, U.S.A., 12
Rollestone No. 1, Australia, 342, 343
Sable Island, Canada, 336
Shell Oil Co.'s Gonsoulin Minvielle-State 2, U.S.A., 293
—, Smith-State Unit 1-1, U.S.A., 293
—, —, 1-2, U.S.A., 293
—, —, 1-3, U.S.A., 293
—, Weeks-Gall Unit 1-1, U.S.A., 293
South Kavala No. 2, Greece, 333
Standard Oil of California's Mushrush 5,

U.S.A., 293
Texaco's E. Riggs Gas Unit 1, U.S.A., 293
Texas Co.'s LaFourche Basin Levee District 1, U.S.A., 293
Tovala No. 1, Papua, 342, 343
Union Oil Co. of California Brunner 1, U.S.A., 293
—, No. 1 W.C. Tyrrell Unit 3, U.S.A., 293
Union Producing Co.-Fohs Oil Co.'s Buckley Bourg 2, U.S.A., 293
—, Fitzpatrick-Vizard 1, U.S.A., 293
Yuleroo No. 1, Australia, 342, 343
Oklahoma U.S.A., 5—12, 29, 34, 71—73, 106, 110, 112—114, 126, 194, 196, 223, 224, 233, 236, 260—263, 265, 267, 285, 286, 293, 294, 338
  Beckham County, 263, 293, 294
  Blaine County, 8, 286
  Caddo County, 293
  Canadian County, 286
  Cimarron County, 7
  Dewey County, 8, 286
Oligocene, see Geologic time scale
Ordovician, see Geologic time scale
Osage Formation, U.S.A., 107
Osinskii Formation, U.S.A., 39
Osmotic phenomena, 33—36
Ossum oil field, U.S.A., 275, 278—280, 282, 284
Overburden load, 2, 3

Pacific Ocean, 77
Packer fluids, 266
Pakistan, 15, 325, 346
  Dhulian oil field, 346
  Joya Mair oil field, 346
Paleozoic, see Geologic time scale
Paleopressure, 23
Palm Valley No. 1 oil well, Australia, 342, 343
Pandua Formation, India, 346
Panhandle, U.S.A., 5, 6, 156, 260—262
Papua, 15, 36, 302, 325, 342, 343
  Papuan Basin, 36, 342
Paradox Basin, U.S.A., 36
Pasadena, U.S.A., 6
Pasca No. 1 oil well, Papua, 342
Pecos County, U.S.A., 260, 293
Penetration rate, 120—122

Pennsylvanian, see Geologic time scale
Peripherical water influx, 281
Permafrost, 40—42
  pingo, 40, 41
  talik, 40
Permafrost-freezeback pressure, 41, 42
Permian, see Geologic time scale
Permian Basin, U.S.A., 35, 154
Persian Gulf Basin, Middle East, 292
Perth Basin, Australia, 343
Peru, 325, 340
Petrel No. 1 oil well, Australia, 342, 343
Petronilla oil field, U.S.A., 22
Philippines, 344
  Cebu Island, 344
Phosphoria Formation, U.S.A., 246
Piceance Basin, U.S.A., 262
Pico Formation, U.S.A., 72, 73
Piezometric fluid level, 19
Pingo, Arctic, 40, 41
Pistol Ridge oil field, U.S.A., 71
Pit level, 147, 148
Plains of Gorgan, Iran, 347, 348
Plaquemines, U.S.A., 201
Pleistocene, see Geologic time scale
Pliocene, see Geologic time scale
Po Basin, Italy, 5, 328
Poland, 325, 329, 330, 331
Pore pressure, see Formation pressure
Powder River Basin, U.S.A., 246, 247
Precambrian, see Geologic time scale
Pre-Caucasus, U.S.S.R., 75, 76
Pressure concepts, 1—5
  formation (pore), 3—5
  hydrostatic, 1, 2
  overburden, 2, 3
Pressure depletion and abandonment pressure, 286—289
Pressure seals, 14—16
Pressure-temperature-density diagram for water, 11
Prudhoe Bay, U.S.A., 265, 298
Pulsed neutron measurements, see Geophysical well logging
Punjab, India, 346
Pure Oil Co.'s Unit 1 oil well, U.S.A., 293
Pyrenees Mountains, Europe, 267, 327

Quaternary, see Geologic time scale
Queen Bess Island oil field, U.S.A., 293
Queen City Sand, U.S.A., 258

SUBJECT INDEX

Queensland, Australia, 19, 343

Rainbow Lake area, Canada, 336
Ralph Lowe Estate, U.S.A., 294, 338
Rangely, U.S.A., 33
  oil field, 33
Rate of sedimentation, 21—23
Rarako Creek No. 1 oil well, Papua, 342, 343
Redox and pH potentials, 168—170
Red Sea, 342, 348
Reeves County, U.S.A., 260
Relief-well planning, 270—272
Remanent magnetization logging, 272
Repetto Formation, U.S.A., 94
Repressuring of reservoirs, 21
Reservoir rocks, see Formation(s)
Reservoir engineering, 275—290
  initial well potential versus formation water salinities, 285, 286
  pressure depletion and abandonment pressure, 286—289
  —, offshore Louisiana, U.S.A., 296
  —, U.S. Gulf Coast area data, 287—289
  —, U.S.S.R. wells, 287
  reservoir mechanism(s), 276—285
  —, Hammerlindl's model, 281—285
  —, peripheral water influx, 281
  —, rock compressibility and rock failure, 278—281
  —, shale water influx, 276—278
  —, —, effect on P/Z-plots, 276, 277
Reservoir mechanisms, see Reservoir engineering
Richardson & Bass et al.'s Humble-LL & E1-L oil well, U.S.A., 293
Richfield Oil Corp.'s Cole Levee "A" 67-29 oil well, U.S.A., 293
Rio Blanco County, U.S.A., 338
Rio Grande Embayment, U.S.A., 26
River(s), 27, 28, 34, 247, 342, 343
  Amazon, South America, 28
  Danube, Europe, 28
  Mamberamo, Far East, 343
  Mississippi, U.S.A., 28, 247
  Niger, Africa, 27, 28
  Nile, Africa, 28, 342
Rocky Mountains, U.S.A., 2, 12, 182, 260—262
Rocky Mountain Arsenal oil well, U.S.A., 12
Roetpelit Formation, Germany, 329
Roetslinar, Germany, 329

Rollestone No. 1, Australia, 342, 343
Rotliegendes Formation, Europe, 328
Rumania, 291, 325, 330
Russia, see U.S.S.R.
Rustler Formation, U.S.A., 246, 257

Sable Island, Canada, 336
  oil well, 336
Sacramento Valley, U.S.A., 17
Salinas block, U.S.A., 18
Salinity principle, 190—196
Salt diapirism, 30—32
Salt deposition, 40
Salt domes, 31, 32
  Clovelly, U.S.A., 32
  Cote Blanche, U.S.A., 32
  Napoleonville, U.S.A., 32
  Weeks Island, U.S.A., 31, 32
  White Castle, U.S.A., 32
San Andres Fault, U.S.A., 18, 19
Sandstone dikes, 32
San Emigdio—Sierran block, U.S.A., 18
San Joaquin oil field, Venezuela, 340
San Joaquin Valley, U.S.A., 17, 33
San Juan Basin, U.S.A., 35
Santa Ana oil field, Venezuela, 340
Santa Barbara Channel, U.S.A., 3, 252, 253, 338
Santa Clara Valley, U.S.A., 92
Santa Cruz Basin, U.S.A., 74
Santa Rosa oil field, Venezuela, 340
Sarajeh oil field, Iran, 347
Saratov region, U.S.S.R., 334
Sarmatian Formation, Europe, 336
Saudi Arabia, 298, 325, 342, 348
  Ghawar oil field, 298
Seismic, 101—116
  field case studies, 104—114
  formation tops, 105—107
  interval velocity, 101
  mathematical concepts, 103, 104
  pitfalls, 114
  pressure predictions, 108—113
Sekihara oil field, Japan, 346
Sespe Formation, U.S.A., 72, 73, 281
Shale bulk density measurements, 150—156
Shale cuttings parameters, see Drilling data
Shale diapirism, 28—31, 51
  mud lumps, 28
  mud volcanoes, 28—31, 51
Shale factor, 156—158

Shale resistivity ratio method, 309—319
Shale water influx, 276—278
Shannon Formation, U.S.A., 246
Shell Oil Co.'s Gonsoulin Minvielle-State 2
  oil well, U.S.A., 293
  Smith-State Unit 1-1 oil well, U.S.A.,
    293
  —, 1-2 oil well, U.S.A., 293
  —, 1-3 oil well, U.S.A., 293
  Weeks-Gall Unit 1-1 oil well, U.S.A.,
    293
Ship Shoal, U.S.A., 80
Short normal curve, see Geophysical well
  logging
Siberia, U.S.S.R., 298
Sierran—Klamath block, U.S.A., 18
Silurian, see Geologic time scale
Simpson Formation, U.S.A., 107, 246, 257
Sirte Basin, Lybia, 298
Slovakia, 329
Smackover Formation, U.S.A., 267, 313
SNAP log, 135, 136
Sonic surveys, see Geophysical well logging
Soso oil field, U.S.A., 71
South America, 140, 155, 339—341
South China Sea area, 19, 38, 108, 125,
  140, 146, 179, 194, 202, 294, 295,
  310, 311, 343
South Dakota, U.S.A., 337
South Kavala No. 2 oil well, Greece, 333
South Pass, U.S.A., 159, 165
South Pass Block 27 oil field, U.S.A., 298
Spontaneous potential (SP) curve, see
  Geophysical well logging
Springer Formation, U.S.A., 29, 106, 107,
  267
Standard Oil of California's Mushrush 5
  oil well, U.S.A., 293
Ste. Genevieve Formation, U.S.A., 72
St. Louis oil field, U.S.A., 71
St. Marcet oil field, France, 327
St. Martin Parish, U.S.A., 182, 301
Strawn Formation, U.S.A., 233
Stress, 4, 53, 83—89
  deviatoric, 87—89
  hydrostatic, 83—86
  mathematical concept, 83—89
  spheric, 86—87
  state, 4, 83
  terminology, 83
  total stress, 84—86
  —, field, 84—86

—, tensor, 86—89
volumetric, 86, 89
Structure of reservoir, 19—21
Subsidence, 5, 6, 18
Subnormal formation pressures, 5—13
  Anadarko Basin, U.S.A., 8, 11, 12
  decompressional expansion, 9, 10
  pressure-temperature-density diagram
    for water, 11
  Tertiary formations, U.S.S.R., 13
Subpressures, see Subnormal pressures,
  5—13
Sumatra, South China Sea, 298, 344
  Duri oil field, 298
  Minas oil field, 298
Surakhan oil field, U.S.S.R., 38
Surpressures, see Abnormally high pressures
Sussex Formation, U.S.A., 246
Sverdrup Basin, Arctic, 336
Switzerland, 36
  Jura Mountains, 36
Sylhet oil field, Pakistan, 346
Sylvan Formation, U.S.A., 107

Taiwan, 325, 344, 345
  Chinshui oil field, 345
  Chuhangkeng Formation, 345
  Chuhangkeng oil field, 345
  Tiehchenshan-Tungshiao oil field, 345
Talik, Arctic, 40
Tansil Formation, U.S.A., 246
Tectonic activities, 23—28
Tehoma County, U.S.A., 315
Tensleep Formation, U.S.A., 246
Terrebonne Parish, U.S.A., 80, 182, 301
Tertiary, see Geologic time scale
Tertiary basins, 106, 126, 184, 190, 310
Terzaghi and Peck compaction model, 55
Texaco's E. Riggs Gas Unit 1, U.S.A., 293
Texas Co.'s LaFourche Levee District 1
  oil well, U.S.A., 293
Texas, U.S.A., 3, 5, 6, 21, 22, 25—27, 72,
  73, 82, 83, 105, 126, 146, 147, 156,
  160—162, 182, 188, 189, 191—193,
  199, 201, 215, 218, 233, 240—245,
  247, 250, 254, 256, 257, 260—263, 267,
  268, 277, 282, 293—295, 301, 313,
  314, 329, 338, 339
  Aransas County, 160, 313
  Baytown, 6
  Brazoria County, 82, 277, 301
  Chambers, 201

SUBJECT INDEX 381

DeWitt County, 218
Galveston, 5, 201
Jefferson, 182, 201
Katy, 6
Matagorda County, 192, 193
Nueces County, 22, 277
Panhandle, 5, 6, 156
Pasadena, 6
Pecos County, 260, 293
Ralph Lowe Estate, 294, 338
Reeves County, 260
Rio Grande Embayment, 26
Ward County, 260
Winkler County, 260
Zapata County, 338
34 W.M.I. Formation, U.S.A., 282
35 M.I.-deep Formation, U.S.A., 282
Tiehchenshan-Tungshiao oil field, Taiwan, 344, 345
Timor, Far East, 325, 343
Tinsley oil field, U.S.A., 71
Tobago, 340
Tonkawa Sand, U.S.A., 107
Torque, 136
Tortonian Formation, Europe, 336
Total pit volume, 147, 148
Tovala No.1 oil well, Papua, 342, 343
Triassic, see Geologic time scale
Trinidad, 325, 340
    Forest Reserve oil field, 340

Uinta Basin, U.S.A., 5, 33, 262, 338
    Formation, U.S.A., 262
ULSEL logging, 272
Union Oil Co. of California Brunner 1 oil well, U.S.A., 293
    No.1 W.C. Tyrrell Unit 3 oil well, U.S.A., 293
Union Producing Co.-Fohs Oil Co.'s Fitzpatrick-Vizard 1 oil well, U.S.A., 293
    Buckley-Bourg 2 oil well, U.S.A., 293
United Kingdom, 331, 333
United States, see individual states of America
Ural Mountains, U.S.S.R., 335
U.S. Gulf Coast, see Gulf Coast, U.S.A.
U.S.S.R., 5, 12—15, 29, 30, 36, 38—40, 42, 75, 76, 140, 286, 287, 291, 292, 325, 334—336
    Apsheron Peninsula, 29
    Azerbaidzhan S.S.R., 29, 30

Baku, 291
Black Sea, 75, 76
Caspian Sea, 42, 140, 325
Caucasus, 30, 335
Checheno-Ingushian A.S.S.R., 336
Chokrak Formation, 5, 12
Ciscaucasia S.S.R., 36, 335, 336
Dagestan A.S.S.R., 336
Divnoe area, 76
Dnieper-Donetz, 335
Karachakhura oil field, 38
Karagan Formation, 5, 12
Kuibyshev region, 334
Loktaban, 30
Markovo oil field, 39
Maykop Formation, 74, 76, 302
Osinskii Formation, 39
Pre-Caucasus, 75, 76
Russian Platform, 335
Saratov region, 334
Siberia, 298
Surakhan oil field, 38
Ural, 335
Utah, U.S.A., 5, 18, 36, 73, 260—262, 338
    Great Salt Lake, 73

Velocity ranges in sediments, 101
Venezuela, 5, 153, 155, 298, 299, 325, 340
    Anaco oil fields, 340
    Bolivar Coast, 5
    Central Anzoategui region, 340
    Colon Shale, 340
    Guario oil field, 340
    Lake Maracaibo, 5, 298, 299, 340
    Lamar oil field, 298
    San Joaquin oil field, 340
    Santa Ana oil field, 340
    Santa Rosa oil field, 340
Ventura County, U.S.A., 72
Vermilion Parish, U.S.A., 201, 213, 316
Vicksburg Formation, U.S.A., 22, 23
Vienna Basin, Austria, 328, 329
Viking Formation, Canada, 5, 33
Viola Formation, U.S.A., 107
Void ratio versus pressures, 50

Waddell Formation, U.S.A., 257
Ward County, U.S.A., 260
Wasatch Formation, U.S.A., 262
Wasco oil field, U.S.A., 293

Wayne County, U.S.A., 338
Weeks Island oil field, U.S.A., 20, 293, 305—307
   salt dome, U.S.A., 20, 31, 32
Wellington Anhydrite, U.S.A., 107
Well logging, see Geophysical well logging
Wesson oil field, U.S.A., 71
West Netherland Permian Basin, Europe, 328
West Poison Spider oil field, U.S.A., 293
Western Sedimentary Basin, Canada, 35
West Siberian oil fields, U.S.S.R., 298
West Virginia, U.S.A., 338
White Castle salt dome, U.S.A., 32
Wichita Mountains, U.S.A., 6
Wilcox Formation, U.S.A., 72, 73, 188, 257, 261, 262, 267
Wilmington oil field, U.S.A., 73
Wind River Basin, U.S.A., 246, 247
Winkler County, U.S.A., 260

Wolfcamp Formation, U.S.A., 107, 246, 257
Woodbine Formation, U.S.A., 94
Woodford Shale, U.S.A., 107
Wyoming, U.S.A., 246, 247, 262, 293, 311, 338
   East Riverton area, 311

Yates Formation, U.S.A., 246, 257
Yemen, 342
Yenangyaung oil field, Burma, 347
Yugoslavia, 325, 331
   Dinaric Alps, 331
   Kikinda blow-out, 331
Yuleroo No. 1 oil well, Australia, 342, 343

Zapata County, U.S.A., 338
Zechstein Formation, Germany, 14, 265, 267